RADIO COMMUNICATIONS CONCEPTS: ANALOG

RADIO COMMUNICATIONS CONCEPTS: ANALOG

RALPH S. CARSON
University of Missouri–Rolla

WILEY
A Wiley-Interscience Publication
JOHN WILEY & SONS
New York / Chichester / Brisbane / Toronto / Singapore

Library of Congress Cataloging in Publication Data:

Carson, Ralph S.
Radio communications concepts : analog / by Ralph S. Carson.
p. cm.
"A Wiley-Interscience publication."
Bibliography: p.
Includes index.
1. Radio. 2. Electronics. I. Title.
TK6550.C336 1990
621.384—dc20 89-34210
CIP

ISBN 0-471-62169-2

Printed in the United States of America

10 9 8 7 6 5 4 3 2 1

CONTENTS

PREFACE

The nature of undergraduate radio communications courses changed during the years following the Second World War. Previously, subject matter pertinent to radio communications was typically included in undergraduate electronics courses. There, the principal topics on which emphasis was placed were the analysis, design, and operation of the electronics circuits. Radio communications theory was usually developed only enough to provide students with a measure of insight into what functions the electronics circuits were supposed to accomplish.

In the quest to upgrade the quality, performance, and training of undergraduates in the field of communications, efforts were made to simplify the sophisticated mathematical treatments of the Fourier transform, convolution, probability theory, statistical analysis, and signal processing so that undergraduates could begin to learn something about how those powerful methods related to communications systems. The era of the communications specialist was initiated.

The primary objective of the emerging communications specialists in undergraduate education was then, and still is, to prepare students for additional study, systems analysis, and research at the graduate level. The importance of their great achievements is attested to by their many intricate high-level theoretical developments (little of which can be claimed by a designer with only an undergraduate degree) implemented by high-technology electronics. Doing justice to a carefully selected collection of the above-listed topics with reasonable mathematical rigor leaves no time for other important system- and circuit-oriented details. Besides, the communications theorists generally profess no particular interest or background in those technical aspects of radio communications that form the basis for most of this book.

Also, apparently intimidated by the abstract nature of the material advanced by the communications specialists, the electronics circuits specialists renounced their formerly assumed responsibility of teaching some theory and applications of radio communications systems. Their reasonable assumption was that if certain topics could be of any significant importance to radio communications systems, then the communications specialists would deal with them in no uncertain fashion. Unfortunately, however, such useful treatments have not been forthcoming during the intervening decades.

Consequently, since others have not come to grips with the topics that are deemed to be, upon judicious reflection, educationally necessary and valid, we herewith resolutely launch our own campaign to highlight and to share some of the more practical aspects of the fundamental concepts that undergird analog radio communications systems. Much of the material to be presented has been developed, used, known, forgotten, relearned, misused, redeveloped, ignored, and, at times, correctly applied, over and over throughout the history of radio communications. Unfortunately, the material has never seemed to find its proper niche in the undergraduate curriculum, even though professional designers affirm its indispensable character. My hope is that teachers, their students, and practicing designers of communications circuits and systems will benefit considerably from the information contained herein and thereby gain a better understanding of some worthwhile matters being omitted from present courses.

Chapter 1 begins with a brief review of some definitions and terminology. This is followed by an in-depth treatment of the linear mixing of sinusoidal signals and waves. Among the concepts discussed are: periodicity, waveforms, beating, symmetry, phase and group velocities, and standing waves. Then, laboratory considerations about how to combine signal-generator outputs in resistive mixers and hybrid combiners are described. Next, a brief introduction to circuits using time-varying resistance and time-varying capacitance are presented. This is followed by a discussion of the relationship between propagation time, time delay or advance, and phase shift in distributed and lumped systems. Then, amplitude distortion, phase distortion, and group delay are defined and illustrated. The chapter ends with a discussion of group delay in amplitude-modulated signals.

Chapter 2 introduces the systems classifications of fixed vs. time-varying, instantaneous vs. dynamic, and linear vs. nonlinear. Then, incremental linearity of a nonlinear system is discussed. Power-law and exponential-law nonlinearities are described, and the use of Taylor's series to represent nonlinearity is presented, followed by the distinction between static and dynamic system characteristics. Next, details of the method of least squares are explained to aid in the calculations of the coefficients in the power-series representation for a nonlinear static characteristic. This is followed by calculations of the nonlinearity products when the input is a single sinusoidal signal. A simple method of expanding cos x and sin x is shown and demonstrated, and harmonic distortion is defined. The coefficients of the trigonometric Fourier series for a typical odd-function type of electronic-device nonlinearity are determined and the compression effect is shown. Next, the nonlinearity products produced by multiple input signals are determined. These are then used to explain compression, desensitization, and blocking—effects so often present to an undesirable degree in electronic systems.

Chapter 3 discusses the topic of random noise. After an introduction to ports and transfer functions, noise spectra are described. Some of the properties of thermal noise, shot noise, and other noise sources are developed. Noise calculations are made with energy-storage elements as well as resistive elements present. The concept of noise bandwidth is described and it is shown how to use a table of integrals to determine noise bandwidth when the transfer function is known. The chapter ends with some simplified noise calculations for a transistor amplifier.

Chapter 4 is composed of two parts. Part 1 is a brief introduction to gamma and Bessel functions, and tables of numerical values for both are given. Readers are led through the development of recursion formulas and Bessel functions of several different types. This extra amount of detail not only gives the readers more confidence in their ability to manipulate the various forms of Bessel functions, it also provides background for the expansions of terms like $\cos(\beta \sin \omega t)$, $\sin(\beta \sin \omega t)$, and $e^{x \cos \omega t}$. The expansions of the sinusoidal-like terms are needed in the study of angle modulation. The expansion of the exponential term is needed to help explain compression in electronic circuits that maintain their bias currents more-or-less constant by means of negative feedback. Part 2 pertains to Butterworth, Chebyshev, and linear phase-response filters. Rolloff, attenuation, pole locations, and shape factors are developed for lowpass filters. Then, transformations to highpass, bandpass, and bandreject filters are presented.

Processes associated with amplitude modulation and demodulation are discussed in Chapter 5. Topics include double-sideband suppressed carrier, double-sideband with carrier, and single-sideband. Circuits known as balanced modulators, product detectors, mixers, and singly balanced and doubly balanced switching-mode detectors are developed step by step. Spectra, power relationships, and modulation efficiency are examined. The diode envelope detector is analyzed in detail to demonstrate some of its more obscure deficiencies. Then, the single-sideband processes of sideband elimination through filtering and phasing are described, and the analysis of a wideband 90° phase-shifting network is given. The chapter ends with demodulation and testing of single-sideband systems.

Chapter 6 is on angle-modulation processes. Phase modulation and modulators are described first, followed by frequency modulation and modulators. The spectra of PM and FM signals modulated by two sinusoidal modulating waveforms are presented in detail. Programs written in BASIC are included to calculate the required Bessel functions and spectral amplitudes. Various rules for determining the approximate bandwidth are described and compared. Power relationships and ratios are developed. Analysis of PM and FM signals modulated with nonsinusoidal waveforms is shown. Then, phase and frequency detectors are described. The chapter ends with an introduction to the phase-locked loop (PLL) and to its applications as a frequency detector and frequency synthesizer.

Chapter 7 is a collection of noise-related topics that pertain primarily to the use of noise factor, noise figure, noise measure, and noise temperature of an operating system. Topics include available power gain, signal-to-noise ratios, operating noise factor, noise factors of cascaded amplifiers, lossless and lossy networks, noise measurements, electronic noise sources, and equivalent input-noise temperature.

After a brief history of radio communications, early radio receivers and solid-state electronics, Chapter 8 introduces the superheterodyne principle, oscillator injection, and reciprocal mixing. The superheterodyne characteristics of sensitivity, selectivity, and fidelity are described. The influence of adjacent-channel selectivity and image rejection on the choice of the intermediate frequency is examined. Then, the double-conversion process is described. An automatic gain control system is presented and its steady-state operation is analyzed. Diode attenuators are examined in detail. This is followed by explanations of hang AGC, diplexers, bandpass tuning, variable bandwidth, and the Wadley loop.

A second treatment of the superheterodyne receiver appears in Chapter 9 where spurious responses are examined in detail. After defining some mixer terminology, the phenomena of coincidence and congruence are described. Harmonic numbers for four types of coincidence are found and spurious response charts are explained. The effects of IF and RF selectivities, individually and combined, are developed. Spurious response charts for congruence and details of a new method of trace construction are given. The chapter ends with a discussion of cross-modulation.

Chapter 10 is devoted to other deficiencies of the superheterodyne principle that are not always recognized as being present. The production of spurious responses is examined in detail. After a brief review of intermodulation (IM) products, dB, and dBm, the concepts of the second- and third-order intercept points are presented. They are used to determine the minimum discernible input signal strength and the dynamic range with the aid of graphical constructions. Relationships between power, IM distortion, intercept points, and correlation coefficients are developed so that cascaded intercept points can be calculated by two different methods.

Each chapter contains worked examples and ends with a list of problems. There is adequate material for solid courses that are two semesters or three quarters in length. The courses could either precede, substitute for, or follow the customary required or elective communications offerings for undergraduates with a leaning toward analog electronics.

Appendix A contains derivations pertaining to beat periods and true periods in the linear mixing of two sinusoidal signals.

Appendix B demonstrates Gauss's method of successive elimination in the solution of simultaneous linear equations.

Appendix C is a short table of integrals that is useful in evaluating noise bandwidth.

Appendix D develops the theory behind elliptical and circular operating paths that are often encountered instead of DC and AC load lines.

Appendix E contains the listings of two BASIC computer programs. The first program evaluates Bessel functions $J_n(x)$ for a specified value of x. The second program determines the spectral amplitudes in angle modulation when the modulating signal is composed of two sinusoidal components.

Answers for selected problems are given in Appendix F. A bibliography of some useful references follows Appendix F.

Although a few circuits are described and analyzed, the main thrust of the book is not primarily the analysis and design of radio circuitry. Instead, emphasis is placed on those aspects of communications systems design that allow technical specifications to be established for the radio circuitry. The main mathematical thrust of *Radio Communications Concepts*: *Analog* is by trigonometric methods from the beginning, and students are not intimidated by this because its procedural details are built on familiar mathematics. This allows students to be introduced to the subtle and new (to them) communications concepts at the lowest possible level of difficulty consistent with their maturity, and it also allows sound theory to be developed in such a form that it can be put to immediate practical use in secure fashion. The receiver is the weak link in a radio communications system and therefore the general design of the analog receiving system is given priority.

To gain the most from this book, the reader should have a rudimentary understanding of basic electronic devices and their circuits, the ability to follow calculations of the coefficients for a trigonometric Fourier series in one variable, and the capacity to recognize a differential equation but not necessarily to solve it. These requirements mean that a student in electrical engineering will probably be at least a second-semester junior. Graduate students could benefit from this material if they are interested in the applications aspects of analog systems. In industry, engineers and managers unacquainted with the design of analog receiving systems will find much valuable and useful information, because no similar textbook has been available for them to have studied earlier.

Most of this material has been used in both an elective course and in the customary communications course from time to time at the University of Missouri–Rolla with gratifying results. Students were generally enthusiastic about it. I thank all those who offered encouragement and who made corrections to and suggested improvements in earlier versions of these chapters.

RALPH S. CARSON

Rolla, Missouri
September 1989

RADIO COMMUNICATIONS CONCEPTS: ANALOG

1

LINEAR MIXING

1-1 REVIEW OF DEFINITIONS AND TERMINOLOGY

Waveform If the variable x is a function of time t, represented by $x(t)$, then t is the independent variable in the region or rectangular coordinate system having t as the abscissa, known as the *time domain.* A graphical representation of $x(t)$ in the time domain is called the *waveform,* the waveshape, or simply the shape of the function. When $x(t)$ describes the variation of some physical quantity such as voltage, current, force, displacement, etc., it is usually referred to as a *signal,* sometimes as a *wave,* and often as a *tone* (especially when it is sinusoidal and whether or not it is audible). Signals can be electrical or nonelectrical in nature, and an infinite variety of waveforms is possible. Fortunately, the name of the signal often indicates its waveform, as in sinusoid, square wave, sawtooth, positive pulse, and triangular wave. Note that the word *wave* is frequently used in connection with any variations of a physical quantity that recur regularly.

Signals A waveform that is defined for any value of time is known as a continuous-time signal, or, more often, as an *analog* signal. For example, when temperature is sensed continuously by a thermocouple, or sound pressure by a microphone, or mechanical displacement by a strain gage, the output electrical equivalents of the original physical quantities are analog signals. Information about the original physical quantities is then contained in the amplitude, polarity, frequency, or phase of the output signals. The devices that do the sensing are called *transducers,* because they transform energy from one physical form into another. Some analog signals such as pulses, square waves, and sawtooth waves undergo abrupt changes in amplitude and therefore are discontinuous functions of the continuous-time variable. This book is a compendium of key topics that undergrid the processing of analog signals in radio communications.

A discrete-time signal is one that is defined only at certain discrete values of time and is undefined, zero, or of no consequence whatever its amplitude, everywhere else. Such a signal can be produced by sampling an analog signal by means of an ideal switch that is closed at the discrete-time instants (usually equally spaced) and is open at all other times. Discrete-time signals and their important constituents known as digital signals will not be considered in this book.

Amplitude Amplitude is expressed in many ways. Every waveform must have instantaneous amplitudes that show the detailed variations of the signal from instant to instant in the strength, intensity, value, size, extent, or range of the quantity that the waveform represents. Peak amplitude indicates only the maximum excursion of the instantaneous amplitudes in either the positive or negative direction. Amplitude is sometimes expressed as the average value of the waveform over a specified time interval. The effective or root-mean-square (rms) amplitude is found by taking the square root of the average value over the specified time interval of the squared original waveform. Peak-to-peak amplitude is the algebraic difference between the positive and negative peak amplitudes. The ratio of peak amplitude to rms amplitude is known as the crest factor. The ratio of rms amplitude to the average value for a specified time interval is sometimes called the form factor. Depending on the situation, the various amplitudes may change with time and with the value, type, and number of components, and how they are interconnected. Zero amplitude may have a special and nontrivial meaning, such as in a balanced bridge circuit.

Period A waveform $x(t)$ is said to be periodic with a temporal period T_0 at a certain fixed location if, for every time t in the range $-\infty < t < \infty$,

$$x(t + kT_0) = x(t) \tag{1-1}$$

where the index $k = 1, 2, 3, \ldots$, and T_0 is the smallest, positive, finite, time interval for which (1-1) is true. Thus, T_0 is the fundamental (or primitive) period, that is, a period no proper fraction of which is itself a period. According to this definition for periodicity, $x(t)$ must be a steady-state waveform extending over the entire time domain and it has no beginning and no ending. A constant $x(t)$ is not regarded as being periodic; it is a case of degenerate periodicity.

The repetitive nature of periodic waveforms is reinforced visually by using the index k integers other than unity. Any regularly recurrent sequence of events that repeats at the same constant interval can be identified as having a period, but that period may not be equal to the fundamental period. Thus, a sequence extending over two fundamental periods of a periodic waveform is periodic with a period $2T_0$, and a sequence extending over k fundamental periods of a periodic waveform is periodic with a period kT_0. Unless otherwise noted, the word *period* will be used to denote the fundamental period.

The definition of periodicity is a convenient mathematical abstraction but it is not something that can be demonstrated exactly in the laboratory. Practically, there are limits to how far back in time a waveform could have been generated and observed, and how far into the future the generated waveform will exist without any disturbance that ruins the periodicity. But this obstacle is eliminated when the system under test has transients, caused by making connections to a signal generator, that decay to insignificance and disappear in a small time interval. Then, connecting the operating signal generator, taking measurements a few minutes or so later, and disconnecting the signal generator a few minutes after that, means that the measurements were made for a waveform that could not be distinguished from one that existed over the entire time domain. What happened to the waveform this morning or a week ago, or what may happen to it this afternoon or next week is of no concern and has nothing to do with the measurements now or in the near future.

Sometimes a waveform displays a regularly recurring pattern only during a certain restricted time interval and it may be zero, undefined, or behave differently in other intervals. During that certain interval, however, it can be said that the waveform has *restricted periodicity* within that specified interval even though, taken as a whole, the waveform does not satisfy the definition of a periodic function throughout the entire time domain. Interpretations and deductions based on such restricted periodicity must be clearly stated to avoid confusion and possible misunderstanding about the meaning of the restricted periodicity of the waveform. For example, the Federal Communications Commission color television standards provide for subcarrier-synchronizing information consisting of "bursts" of at least eight periods of the subcarrier frequency at a predetermined phase transmitted during the "back porch" interval of a blanking pulse following each horizontal synchronizing pulse. Here the subcarrier sinusoids of at least eight periods have restricted periodicity within the "back porch" interval only. Since the color "bursts" occur following each horizontal synchronizing pulse, the "bursts" themselves appear to be periodic provided the sync generator is warmed up and transients caused by making the connections have disappeared. Waveforms that are periodic for $t > 0$ and zero for $t < 0$ are said to be semiperiodic.

Cycle A *cycle* is the portion of a waveform that occurs in an interval corresponding to a period for which the waveform is said to be periodic. It represents a complete sequence of events that regularly recur in a repeatable pattern. Therefore, a period is the number of seconds required to complete one cycle of the waveform.

Frequency *Frequency* is the name given to the number of times per second a periodic event, that is, a cycle, occurs. There are two ways to express frequency that depend on whether it is more convenient in a particular situation to use the number of cycles per second or the number of radians per second.

(a) Frequency can be given by the number of cycles per second of a waveform when it is said to be periodic, and this is sometimes called *cyclic frequency*. Designated by the lower case letter f (or sometimes by the lower-case Greek letter nu, ν) frequency is measured in cycles per second, or hertz, abbreviated Hz. Cycles per second is the reciprocal of seconds per cycle, so frequency and period are reciprocals. Thus, $f_0 = 1/T_0$ and $T_0 = 1/f_0$.

(b) *Radian frequency,* also called angular velocity in mechanics, is the number of radians per second of the waveform when it is said to be periodic, and is designated by the lower-case Greek letter omega, ω. Its unit of measurement is the radian per second, and $\omega_0 = 2\pi f_0 = 2\pi/T_0$ because there are 2π radians in a complete cycle of a sinusoid. Also, $\omega(t) = d\theta(t)/dt$. If the angle $\theta(t) = \omega_0 t$, that is, if it is a linear function of time, then $\omega(t) = d(\omega_0 t)/dt = \omega_0$, so that a fixed frequency results. But if $\theta(t)$ is some nonlinear function of time such as $\tan^{-1} \omega_0 t$ or a random phase jitter, then $\omega(t)$ is no longer constant but varies from instant to instant, so that an instantaneous frequency can be obtained from $\omega(t) = d\theta(t)/dt$.

If the function $x(t)$ also depends explicitly on frequency, as in $x(t) = \cos \omega_0 t$, then f or ω becomes the independent variable in the region or rectangular coordinate system having f or ω as the abscissa, known as the *frequency domain,* and the amplitude (peak or rms) is plotted at the single frequency f_0 or ω_0 for which that amplitude applies. The resulting figure is an *amplitude spectrum* of the function $x(t)$.

It is helpful to mention a few common and well-known frequencies to put the concept of frequency in better perspective. The audible sound frequency range extends from about 16 Hz to 16 kHz. In the United States the power-system frequency is 60 Hz, the standard AM radio broadcast band covers 535 kHz to 1605 kHz, and the standard FM radio broadcast band is from 88 MHz to 108 MHz. A frequency–wavelength comparison chart is shown on the inside front cover.

Phase Phase is a particular stage in a series of periodic changes or movements. Phase refers to the progress of any cyclic phenomenon or periodic waveform relative to an arbitrary starting point or origin. The starting point is usually taken to be at the last previous passage of the waveform through zero from the negative to the positive direction. Each point on a cycle of a periodic waveform is a phase of that waveform. If $x(t)$ is a periodic function, then $x(t - t_1)$ is also a periodic function. Each phase of $x(t)$ occurs at a later time in $x(t - t_1)$, the difference in time occurrence of each phase being t_1. This time is the time-phase difference between $x(t)$ and $x(t - t_1)$. If t_1 is a positive value, then $x(t - t_1)$ is said to lag $x(t)$ by the time phase t_1 and $x(t)$ is said to lead $x(t-t_1)$ by the time phase t_1. If t_1 is a negative value, then $x(t-t_1)$ is said to lead $x(t)$ by the time phase t_1 and $x(t)$ is said to lag $x(t-t_1)$ by the time phase t_1.

The time-phase difference between $x(t)$ and $x(t-t_1)$ for a particular value t_1 can also be expressed as the fractional part t_1/T_0 of the period T_0 through which t_1 has advanced relative to an arbitrary origin. This fractional part is sometimes expressed in degrees of phase difference as $360t_1/T_0$ and sometimes in radians of phase difference given by $2\pi t_1/T_0$. The signals $x(t)$ and $x(t-t_1)$ may be two sinusoids having the same frequency and the phase-angle difference between them may be a function of time and of frequency, in which case it can be plotted in the time domain and in the frequency domain. When the amount of phase-angle difference is plotted versus frequency in the frequency domain, a *phase spectrum* is obtained.

Sinusoids The unit circle $e^{j\omega_0 t} = \cos \omega_0 t \pm j \sin \omega_0 t$ in the complex plane and its decomposition into the circular functions $\cos \omega_0 t$ and $\sin \omega_0 t$ have occupied a dominant position in electrical engineering beginning with Hertz's use of the exponential function to solve Maxwell's equations for dipole radiation. These are the functions that are meant when the term *sinusoid* is used.

There are several reasons for the dominance of sinusoids. The natural response of an underdamped second-order system is a damped sinusoid, but if there are no losses (i.e., for zero damping) the response is a pure sinusoid. In nature, the motion of a pendulum, bouncing of a ball, vibration of a violin string, and the ripples produced when a liquid surface is disturbed are examples of (damped) sinusoidal motions. When sinusoids having the same frequency are added, the sum is also a sinusoid with the same frequency. By means of the trigonometric Fourier series, non-sinusoidal waveforms can be represented by the sum of sinusoids that have the proper frequencies, amplitudes, and phases. The time derivatives and integrals of sinusoids remain sinusoidal. Finally, sinusoidal signals are relatively easy to generate and to control.

Spectrum The word *spectrum* refers to the entire range or extent of anything, such as the complete musical scale; the gamut of emotions, dialects, or writings of a group of people; the ordered arrangement of visible colored bands, together with the invisible components formed by passing "white" light through a prism; the content of a curriculum; and the radio frequencies from about 30 kHz through roughly 300 GHz. In electrical engineering, *spectrum* most often refers to the different frequencies (including zero) of the various sinusoidal components whose *sum* represents the original signal $x(t)$.

When a given signal $x(t)$ is decomposed into the *sum* of sinusoidal components by determining the amplitude and phase quantities at the frequencies involved, then f or ω becomes the independent variable (i.e., the abscissa) in the region or rectangular coordinate system known as the *frequency domain.* When the amplitude quantities are plotted as ordinate, the resulting figure is called the *amplitude spectrum* of $x(t)$. When the phase quantities are plotted as ordinatc, the resulting figure is called the *phase spectrum* of $x(t)$.

There are only a few ways to determine the spectrum of a given signal $x(t)$. The simplest procedure to use depends on the mathematical form available, if any, for $x(t)$. If $x_1(t) = V_1 \cos 2\pi f_1 t + V_2 \cos 2\pi f_2 t$, such as is obtained from the linear mixing of two sinusoids, it is already decomposed into the sum of sinusoidal components whose amplitudes, phases, and frequencies are specified. Using trigonometric identities to cast that sum into another mathematical form can *not* make any change whatsoever in the spectrum of the given $x_1(t)$. If $x_2(t) = 2V_0(\cos 2\pi f_1 t)(\cos 2\pi f_2 t)$, the spectrum is readily found by using the trigonometric identity $2 \cos A \cos B = \cos (a + B) + \cos (A - B)$. Thus, $x_2(t) = V_0 \cos 2\pi(f_2 + f_1)t + V_0 \cos 2\pi(f_2 - f_1)t$, from which the spectrum is obtained. If $x(t)$ is expressed as a periodic nonsinusoidal waveform such as a square wave, sawtooth, or pulse train, appeal is made to the procedures of the trigonometric Fourier series to find the spectrum, because that series is the sum of sinusoidal components that represents $x(t)$. When a mathematical form for $x(t)$ is not available (or even if it is), the amplitude spectrum can be measured in the laboratory with a wave analyzer or a spectrum analyzer. The phase spectrum of some signals can be measured with a Fourier analyzer.

Phasor A signal $v(t) = A \cos (\omega_0 t + \theta)$ can be represented as

$$
\begin{aligned}
v(t) &= \text{Re}\, Ae^{j(\omega_0 t + \theta)} \\
&= \text{Re}\, Ae^{j\theta} e^{j\omega_0 t}
\end{aligned}
$$

where Re stands for "the real part of." The quantity $Ae^{j\theta}$ contains the amplitude and angle information of $v(t)$, whose frequency is $\omega_0 = 2\pi f_0$ and is called a *phasor*. Thus the phasor

$$V(f_0) = Ae^{j\theta}$$

is a complex number in exponential form and when expressed in polar form becomes

$$V(f_0) = A\underline{/\theta}$$

and in rectangular form as

$$V(f_0) = A \cos \theta + jA \sin \theta$$

The ratio of the phasors $V(f_0)$ and $I(f_0)$ that represent the voltage and current at frequency $\omega_0 = 2\pi f_0$ yields the impedance at that frequency:

$$Z(f_0) = \frac{V(f_0)}{I(f_0)} = R + jX$$

R and X are the real and imaginary series components of the impedance and are called resistance and reactance, respectively. Impedance, resistance, and reactance are measured in ohms (Ω). The admittance is

$$Y(f_0) = \frac{I(f_0)}{V(f_0)} = G + jB$$

G and B are the real and imaginary parallel components of the admittance and are called conductance and susceptance, respectively, and are measured in siemens† (S).

Positive resistance or positive conductance absorbs power, whereas negative resistance or negative conductance delivers power. The signs of reactance and susceptance may be either positive or negative. Inductive reactance and capacitive susceptance are both positive, while capacitive reactance and inductive susceptance are negative.

1-2 PERIOD CALCULATIONS

To determine whether the sum of two sinusoids is periodic with period T_0 it is necessary to find the smallest value of T_0 for which the two sinusoids again have the same instantaneous amplitudes and slopes as they had at an arbitrary starting point. Consider the waveform

$$x(t) = 2.1 \cos (2\pi \times 1000t - 20°) - 3.4 \sin (2\pi \times 1200t + 45°)$$

At a starting point such as $t = 0$ the cosine term is at $2.1 \cos 20°$ and is increasing in value, and the sine term is at $-3.4 \sin 45°$ and is decreasing in value. The cosine term has a period $T_1 = 1/1000$ s and the sine term has a period $T_2 = 1/1200$ s. If $x(t)$ is periodic with a period T_0 the cosine term must complete a whole number of periods nT_1 of the lower frequency in the same time T_0 that the sine term completes a whole number of periods mT_2 of the higher frequency, such that the cosine term is again at $2.1 \cos 20°$ and increasing in value and the sine term is again at $-3.4 \sin 45°$ and decreasing in value in the same manner as at $t = 0$. For consistency we will associate the letter m with the higher frequency or the lower period component. Thus, if

† Previously called the mho (℧).

$x(t)$ is periodic, then

$$T_0 = nT_1 = mT_2 \tag{1-2}$$

In order for the period T_0 in (1-2) to be the fundamental period, n and m must be the *smallest* integers that satisfy (1-2). Then, substituting, $n/1000 = m/1200$, from which is obtained

$$1200n = 1000m$$

An equality is obtained if $n = 1000$ and $m = 1200$ indicating that the sum $x(t)$ is periodic simply because integers have been found for m and n. To obtain the *smallest* integers, divide the numbers 1000 and 1200 by their highest common factor 200. Therefore

$$m = 6 \quad \text{and} \quad n = 5$$

The two numbers 5 and 6 do not have a common factor, therefore 5 and 6 are relatively prime and are the smallest integers that satisfy (1-2). Thus the period of $x(t)$ is

$$T_0 = \frac{5}{1000} = \frac{6}{1200} = 5 \text{ ms}$$

The fundamental frequency is $f_0 = 1/T_0 = 1/(5 \text{ ms}) = 200$ Hz. Each of the original frequencies is a harmonic of this fundamental frequency. Thus, $f_1 = 1000 \text{ Hz} = 5(200)$, so that f_1 is the fifth harmonic of the fundamental frequency. Note that the harmonic numbers are the same integer values as m and n. The two original frequencies have a common measure, 200 Hz in this example. Hence, the original sinusoids are said to be *commensurable* because their frequencies are measured by the same fundamental frequency. Further, whenever the ratio of the two original frequencies can be expressed as the ratio of two relatively prime integers, those integers are the harmonic numbers m and n.

Observe that in finding the period of $x(t)$ the amplitudes, signs, and phase angles of the original sinusoids were not explicitly involved. This will always be true if the amplitudes, signs, and phase angles do not change with time. Further, the sinusoids could be sines or cosines in any combination. However, for the addition of sinusoids that each have a constant amplitude, frequency (or period), and phase, to yield a periodic waveform, their frequencies (or periods) must be rational numbers (i.e., expressible as a ratio of integers) or rational multiples of the same irrational (e.g., of $\sqrt{3}$) or transcendental number (i.e., π or the naperian e). If the frequencies (or periods) otherwise contain rational, irrational, or transcendental numbers in any combination, the waveform that results from their addition never exactly repeats itself and the resultant period is infinite.

Example 1-1

If $x(t) = 1.2\cos(21.3 \times 10^3\pi t + 20°) - 8.1\sin(16.14 \times 10^3\pi t - 32°)$ is periodic, find the period, the fundamental frequency, and the harmonic numbers.

Solution The cosine term has a period $T_1 = 2/21.3$ ms and the sine term has a period $T_2 = 2/16.14$ ms. Therefore, $T_0 = m(2/21.3)$ ms $= n(2/16.14)$ ms. Divide by 2 ms and rearrange to obtain $16.14m = 21.3n$. Divide each side by 3. Thus, $5.38m = 7.1n$. Since n and m must be integers, eliminate the decimals by multiplying by 100. Therefore, $538m = 710n$ involves integers only. Divide this by 2 to make the resulting numbers relatively prime. Thus, $269m = 355n$, so that $m = 355$ and $n = 269$ are the smallest integers that can be found. Factorizations of integers up to 409 are given in Table 1-1. Hence $x(t)$ is periodic, and the period is $T_0 = 355 \times 2/21.3$ ms $= 269 \times 2/16.14$ ms $= 33.33$ ms. The fundamental frequency is $f_0 = 1/T_0 = 30$ Hz. The frequency of the cosine term is $f_1 = 21.3/2$ kHz $= 10.65$ kHz $= 355 \times 30$, so f_1 is the 355th harmonic of 30 Hz. Similarly, f_2 is the 269th harmonic of 30 Hz. The two original sinusoids are commensurable.

There are many ways to find divisors of numbers. For example, any number is divisible by 2 if its rightmost digit is divisible by 2. A number is divisible by 4 when the number formed by its last two digits is divisible by 4. The general pattern can be extended to higher powers of 2: a number is divisible by 8 when the number formed by its last three digits is divisible by 8, and so forth. A number is divisible by 5 when it ends in 0 or 5; divisible by 25 if the number formed by its last two digits is divisible by 25; divisible by 125 if the number formed by its last three digits is divisible by 125, and so forth. A number is divisible by 3 when the sum of its digits is divisible by 3. An even number that is divisible by 3 is also divisible by 6. A number is divisible by 9 if the sum of its digits is divisible by 9. To see whether a number is divisible by 7, detach its last digit, double it, and subtract that result from the number that remains after detaching the last digit. If the answer is divisible by 7, then the original number is divisible by 7. To see whether a number is divisible by 11, add every other digit in the number starting at the leftmost digit, then add the remaining digits. When the difference between the two sums is divisible by 11, the original number is divisible by 11.

Euclid's algorithm provides a direct and easy method for finding the greatest common divisor of two numbers. First, divide the larger number by the smaller number and note the remainder. If the remainder is not 0, divide that remainder into the divisor and again note what the next remainder is. If that remainder is not 0, divide it into the second divisor and note the next remainder. So long as the remainder is not 0, continue

TABLE 1-1 Factorizations of Integers up to 409

N	0	1	2	3	4	5	6	7	8	9	N
0	—	1	2	3	2^2	5	$2\cdot3$	7	2^3	3^2	0
1	$2\cdot5$	11	$2^2\cdot3$	13	$2\cdot7$	$3\cdot5$	2^4	17	$2\cdot3^2$	19	1
2	$2^2\cdot5$	$3\cdot7$	$2\cdot11$	23	$2^3\cdot3$	5^2	$2\cdot13$	3^3	$2^2\cdot7$	29	2
3	$2\cdot3\cdot5$	31	2^5	$3\cdot11$	$2\cdot17$	$5\cdot7$	$2^2\cdot3^2$	37	$2\cdot19$	$3\cdot13$	3
4	$2^3\cdot5$	41	$2\cdot3\cdot7$	43	$2^2\cdot11$	$3^2\cdot5$	$2\cdot23$	47	$2^4\cdot3$	7^2	4
5	$2\cdot5^2$	$3\cdot17$	$2^2\cdot13$	53	$2\cdot3^3$	$5\cdot11$	$2^3\cdot7$	$3\cdot19$	$2\cdot29$	59	5
6	$2^2\cdot3\cdot5$	61	$2\cdot31$	$3^2\cdot7$	2^6	$5\cdot13$	$2\cdot3\cdot11$	67	$2^2\cdot17$	$3\cdot23$	6
7	$2\cdot5\cdot7$	71	$2^3\cdot3^2$	73	$2\cdot37$	$3\cdot5^2$	$2^2\cdot19$	$7\cdot11$	$2\cdot3\cdot13$	79	7
8	$2^4\cdot5$	3^4	$2\cdot41$	83	$2^2\cdot3\cdot7$	$5\cdot17$	$2\cdot43$	$3\cdot29$	$2^3\cdot11$	89	8
9	$2\cdot3^2\cdot5$	$7\cdot13$	$2^2\cdot23$	$3\cdot31$	$2\cdot47$	$5\cdot19$	$2^5\cdot3$	97	$2\cdot7^2$	$3^2\cdot11$	9
10	$2^2\cdot5^2$	101	$2\cdot3\cdot17$	103	$2^3\cdot13$	$3\cdot5\cdot7$	$2\cdot53$	107	$2^2\cdot3^3$	109	10
11	$2\cdot5\cdot11$	$3\cdot37$	$2^4\cdot7$	113	$2\cdot3\cdot19$	$5\cdot23$	$2^2\cdot29$	$3^2\cdot13$	$2\cdot59$	$7\cdot17$	11
12	$2^3\cdot3\cdot5$	11^2	$2\cdot61$	$3\cdot41$	$2^2\cdot31$	5^3	$2\cdot3^2\cdot7$	127	2^7	$3\cdot43$	12
13	$2\cdot5\cdot13$	131	$2^2\cdot3\cdot11$	$7\cdot19$	$2\cdot67$	$3^3\cdot5$	$2^3\cdot17$	137	$2\cdot3\cdot23$	139	13
14	$2^2\cdot5\cdot7$	$3\cdot47$	$2\cdot71$	$11\cdot13$	$2^4\cdot3^2$	$5\cdot29$	$2\cdot73$	$3\cdot7^2$	$2^2\cdot37$	149	14
15	$2\cdot3\cdot5^2$	151	$2^3\cdot19$	$3^2\cdot17$	$2\cdot7\cdot11$	$5\cdot31$	$2^2\cdot3\cdot13$	157	$2\cdot79$	$3\cdot53$	15
16	$2^5\cdot5$	$7\cdot23$	$2\cdot3^4$	163	$2^2\cdot41$	$3\cdot5\cdot11$	$2\cdot83$	167	$2^3\cdot3\cdot7$	13^2	16
17	$2\cdot5\cdot17$	$3^2\cdot19$	$2^2\cdot43$	173	$2\cdot3\cdot29$	$5^2\cdot7$	$2^4\cdot11$	$3\cdot59$	$2\cdot89$	179	17
18	$2^2\cdot3^2\cdot5$	181	$2\cdot7\cdot13$	$3\cdot61$	$2^3\cdot23$	$5\cdot37$	$2\cdot3\cdot31$	$11\cdot17$	$2^2\cdot47$	$3^3\cdot7$	18

19	$2 \cdot 5 \cdot 19$	191	$2^6 \cdot 3$	193	$2 \cdot 97$	$3 \cdot 5 \cdot 13$	$2^2 \cdot 7^2$	197	$2 \cdot 3^2 \cdot 11$	199	19
20	$2^3 \cdot 5^2$	$3 \cdot 67$	$2 \cdot 101$	$7 \cdot 29$	$2^2 \cdot 3 \cdot 17$	$5 \cdot 41$	$2 \cdot 103$	$3^2 \cdot 23$	$2^4 \cdot 13$	$11 \cdot 19$	20
21	$2 \cdot 3 \cdot 5 \cdot 7$	211	$2^2 \cdot 53$	$3 \cdot 71$	$2 \cdot 107$	$5 \cdot 43$	$2^3 \cdot 3^3$	$7 \cdot 31$	$2 \cdot 109$	$3 \cdot 73$	21
22	$2^5 \cdot 5 \cdot 11$	$13 \cdot 17$	$2 \cdot 3 \cdot 37$	223	$2^5 \cdot 7$	$3^2 \cdot 5^2$	$2 \cdot 113$	227	$2^2 \cdot 3 \cdot 19$	229	22
23	$2 \cdot 5 \cdot 23$	$3 \cdot 7 \cdot 11$	$2^3 \cdot 29$	233	$2 \cdot 3^2 \cdot 13$	$5 \cdot 47$	$2^2 \cdot 59$	$3 \cdot 79$	$2 \cdot 7 \cdot 17$	239	23
24	$2^4 \cdot 3 \cdot 5$	241	$2 \cdot 11^2$	3^5	$2^2 \cdot 61$	$5 \cdot 7^2$	$2 \cdot 3 \cdot 41$	$13 \cdot 19$	$2^3 \cdot 31$	$3 \cdot 83$	24
25	$2 \cdot 5^3$	251	$2^2 \cdot 3^2 \cdot 7$	$11 \cdot 23$	$2 \cdot 127$	$3 \cdot 5 \cdot 17$	2^8	257	$2 \cdot 3 \cdot 43$	$7 \cdot 37$	25
26	$2^2 \cdot 5 \cdot 13$	$3^2 \cdot 29$	$2 \cdot 131$	263	$2^3 \cdot 3 \cdot 11$	$5 \cdot 53$	$2 \cdot 7 \cdot 19$	$3 \cdot 89$	$2^2 \cdot 67$	269	26
27	$2 \cdot 3^3 \cdot 5$	271	$2^4 \cdot 17$	$3 \cdot 7 \cdot 13$	$2 \cdot 137$	$5^2 \cdot 11$	$2^2 \cdot 3 \cdot 23$	277	$2 \cdot 139$	$3^2 \cdot 31$	27
28	$2^3 \cdot 5 \cdot 7$	281	$2 \cdot 3 \cdot 47$	283	$2^2 \cdot 71$	$3 \cdot 5 \cdot 19$	$2 \cdot 11 \cdot 13$	$7 \cdot 41$	$2^5 \cdot 3^2$	17^2	28
29	$2 \cdot 5 \cdot 29$	$3 \cdot 97$	$2^2 \cdot 73$	293	$2 \cdot 3 \cdot 7^2$	$5 \cdot 59$	$2^3 \cdot 37$	$3^3 \cdot 11$	$2 \cdot 149$	$13 \cdot 23$	29
30	$2^2 \cdot 3 \cdot 5^2$	$7 \cdot 43$	$2 \cdot 151$	$3 \cdot 101$	$2^4 \cdot 19$	$5 \cdot 61$	$2 \cdot 3^2 \cdot 17$	307	$2^2 \cdot 7 \cdot 11$	$3 \cdot 103$	30
31	$2 \cdot 5 \cdot 31$	311	$2^3 \cdot 3 \cdot 13$	313	$2 \cdot 157$	$3^2 \cdot 5 \cdot 7$	$2^2 \cdot 79$	317	$2 \cdot 3 \cdot 53$	$11 \cdot 29$	31
32	$2^6 \cdot 5$	$3 \cdot 107$	$2 \cdot 7 \cdot 23$	$17 \cdot 59$	$2^2 \cdot 3^4$	$5^2 \cdot 13$	$2 \cdot 163$	$3 \cdot 109$	$2^3 \cdot 41$	$7 \cdot 47$	32
33	$2 \cdot 3 \cdot 5 \cdot 11$	331	$2^2 \cdot 83$	$3^2 \cdot 37$	$2 \cdot 167$	$5 \cdot 67$	$2^4 \cdot 3 \cdot 7$	337	$2 \cdot 13^2$	$3 \cdot 113$	33
34	$2^2 \cdot 5 \cdot 17$	$11 \cdot 31$	$2 \cdot 3^2 \cdot 19$	7^3	$2^3 \cdot 43$	$3 \cdot 5 \cdot 23$	$2 \cdot 173$	347	$2^2 \cdot 3 \cdot 29$	349	34
35	$2 \cdot 5^2 \cdot 7$	$3^3 \cdot 13$	$2^5 \cdot 11$	353	$2 \cdot 3 \cdot 59$	$5 \cdot 71$	$2^2 \cdot 89$	$3 \cdot 7 \cdot 17$	$2 \cdot 179$	359	35
36	$2^3 \cdot 3^2 \cdot 5$	19^2	$2 \cdot 181$	$3 \cdot 11^2$	$2^2 \cdot 7 \cdot 13$	$5 \cdot 73$	$2 \cdot 3 \cdot 61$	367	$2^4 \cdot 23$	$3^2 \cdot 41$	36
37	$2 \cdot 5 \cdot 37$	$7 \cdot 53$	$2^2 \cdot 3 \cdot 31$	373	$2 \cdot 11 \cdot 17$	$3 \cdot 5^3$	$2^3 \cdot 47$	$13 \cdot 29$	$2 \cdot 3^3 \cdot 7$	379	37
38	$2^2 \cdot 5 \cdot 19$	$3 \cdot 127$	$2 \cdot 191$	383	$2^7 \cdot 3$	$5 \cdot 7 \cdot 11$	$2 \cdot 193$	$3^2 \cdot 43$	$2^2 \cdot 97$	389	38
39	$2 \cdot 3 \cdot 5 \cdot 13$	$17 \cdot 23$	$2^3 \cdot 7^2$	$3 \cdot 131$	$2 \cdot 197$	$5 \cdot 79$	$2^2 \cdot 3^2 \cdot 11$	397	$2 \cdot 199$	$3 \cdot 7 \cdot 19$	39
40	$2^4 \cdot 5^2$	401	$2 \cdot 3 \cdot 67$	$13 \cdot 31$	$2^2 \cdot 101$	$3^4 \cdot 5$	$2 \cdot 7 \cdot 29$	$11 \cdot 37$	$2^3 \cdot 3 \cdot 17$	409	40

dividing the remainders into the divisors until a remainder is 0. The last divisor that produced a remainder of 0 is the greatest common divisor of the original two numbers.

Example 1-2

If $x(t) = 2\cos 2\pi \times 3.0t - \sin 2\pi \times 7.5t + 1.5\cos 2\pi \times 9.3t$ is periodic, find the period, the fundamental frequency, and the harmonic numbers.

Solution

$$T_0 = pT_1 = nT_2 = mT_3$$

$$\frac{p}{3.0} = \frac{n}{7.5} = \frac{m}{9.3}$$

Three equations are obtained given by

$$75p = 30n, \qquad 93p = 30m, \qquad 93n = 75m$$

from which are obtained separately $p = 30$, $n = 75$, and $p = 30$, $m = 93$, and $n = 75$, $m = 93$. Therefore $x(t)$ is periodic because integers for p, n, and m have been found. Their largest common factor is 3 and so the smallest integers that are relatively prime are $p = 10$, $n = 25$, and $m = 31$. Hence the fundamental period is $T_0 = 10/3 = 25/7.5 = 31/9.3 = 3.33$ s and the fundamental frequency is $f_0 = 1/3.33 = 0.3$ Hz. The harmonic numbers are $p = 10$, $n = 25$, and $m = 31$.

1-3 LINEAR MIXING

The process of combining two or more different electrical signals is called *mixing*. The circuit or device in which mixing takes place is called a *mixer*. Mixing can occur in both linear and nonlinear mixers, leading to, respectively, linear mixing and nonlinear mixing. The principle of superposition applies in a linear mixer and the superposition of two or more signals *is* linear mixing. Linear mixing is equivalent to algebraic addition.

Examples of linear mixing abound. Linear mixing occurs when several signals are induced in a receiving antenna and coupled into the input of a receiver. Linear mixing occurs when the outputs of several microphones in a studio are combined in resistive faders and attenuators. Linear mixing occurs when several small-signal voltages are applied to the various inputs of an op-amp summer.

The results of linear mixing are significantly different from the results of nonlinear mixing. Nonlinear mixing cannot be fully appreciated unless linear mixing is thoroughly understood.

1-3.1 Linear Mixing of Two Sinusoids

Consider the superposition of the two sinusoids given by

$$v(t) = V_1 \cos \omega_1 t + V_2 \cos \omega_2 t \tag{1-3}$$

where $\omega_1 = 2\pi f_1$ and $\omega_2 = 2\pi f_2$. If the two frequencies are far apart, with f_2 greater than about $3f_1$, the sum waveform is simply the higher-frequency signal riding on the lower-frequency signal. Typical resultant waveforms are shown in Fig. 1-1 for several conditions as indicated. Lines drawn to connect the points of maximum instantaneous excursion and to connect the points of minimum instantaneous excursion yield the *envelopes,* so named because those lines envelop or enclose the waveform completely. Since the sum is given by (1-3), when $\cos \omega_2 t = +1$, then $v(t) = V_1 \cos \omega_1 t + V_2$ is the equation of the line for the maximum instantaneous excursion; and when $\cos \omega_2 t = -1$, then $v(t) = V_1 \cos \omega_1 t - V_2$ is the equation of the line for the minimum instantaneous excursion, when $f_2 > f_1$. No new frequencies have

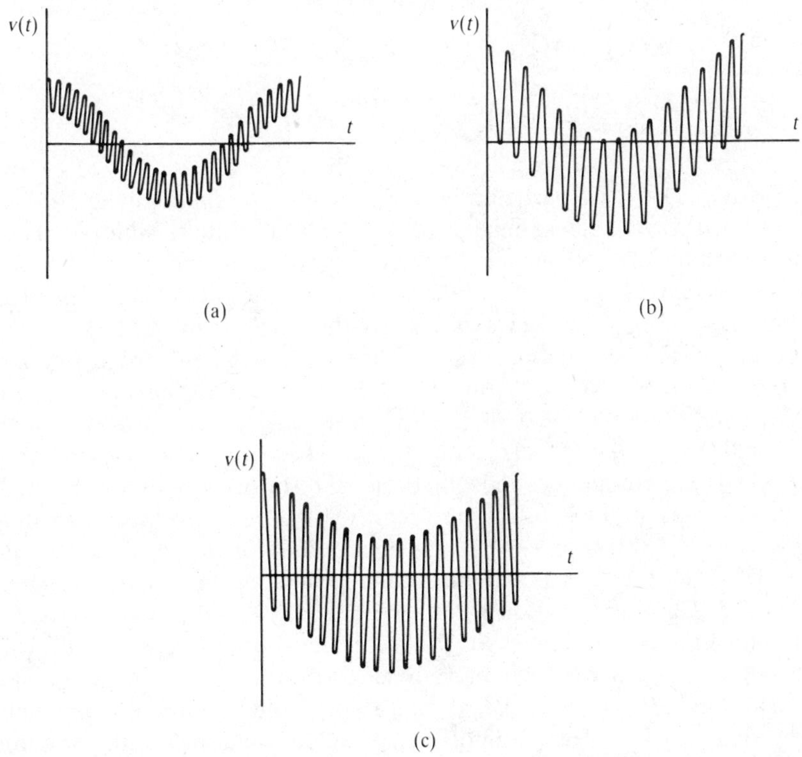

Fig. 1-1 Waveforms for $v(t) = V_1 \cos \omega_1 t + V_2 \cos \omega_2 t$ when $f_2 \gg f_1$. (a) $V_1 > V_2$; (b) $V_1 = V_2$; (c) $V_2 > V_1$.

been created by this linear mixing process: the spectrum still contains only the same two original frequencies.

The resultant waveform becomes more interesting when f_2 and f_1 are more nearly the same. Suppose $f_2 > f_1$. Let the average frequency be $f_a = (f_2 + f_1)/2$ and let $\Delta f = (f_2 - f_1)/2$. Then $f_1 = f_a - \Delta f$ or $\omega_1 = \omega_a - \Delta\omega$, and $f_2 = f_a + \Delta f$ or $\omega_2 = \omega_a + \Delta\omega$, where $\Delta\omega = 2\pi\,\Delta f$. Thus (1-3) becomes

$$v(t) = V_1 \cos(\omega_a - \Delta\omega)t + V_2 \cos(\omega_a + \Delta\omega)t \tag{1-4}$$

Using familiar trigonometric identities for the cosines of the sum and difference angles in (1-4) and combining terms in that result yields

$$v(t) = [(V_1 + V_2)\cos \Delta\omega t]\cos \omega_a t + [(V_1 - V_2)\sin \Delta\omega t]\sin \omega_a t \tag{1-5}$$

Two cases can be identified because of the nature of (1-5): Case 1 for which $V_1 = V_2$, and Case 2 for which $V_1 \neq V_2$.

Case 1

When $V_1 = V_2 = V$, the second term in (1-5) disappears so that

$$\begin{aligned} v(t) &= 2V \cos \Delta\omega t \cos \omega_a t \\ &= 2V \cos\left[\frac{2\pi(f_2 - f_1)t}{2}\right] \cos\left[\frac{2\pi(f_2 + f_1)t}{2}\right] \end{aligned} \tag{1-6}$$

Equation (1-6) is particularly convenient for finding the resultant waveform for the linear mixing of two equal-amplitude sinusoids having nearly the same frequencies because $v(t)/2V$ is the *product* of two sinusoids having widely different frequencies instead of the sum of two sinusoids whose frequencies are nearly the same as in (1-4). This makes the graphical construction easier. The sinusoid having frequency $(f_2 - f_1)/2$ is slowly varying compared to the sinusoid having frequency $(f_2 + f_1)/2$, and the ratio of the two frequencies is $(f_2 + f_1)/(f_2 - f_1)$. If $f_2 = 800$ Hz and $f_1 = 600$ Hz, then $(f_2 + f_1)/(f_2 - f_1) = 7$. During one period of the lower-frequency sinusoid there are seven periods of the higher frequency as in Fig. 1-2. Lines representing the envelopes are drawn given by $\pm|\cos \frac{1}{2}[2\pi(f_2 - f_1)t]|$ and then points for the product are found at the numerous zero and maximum points of the higher frequency sinusoid. Careful attention should be devoted to the regions where the envelope passes through zero. When the resultant waveform is drawn, it is seen to resemble an interference pattern. There are regions of constructive interference called *loops* where the amplitude is maximum, and regions of destructive interference called *nodes* where the amplitude is zero. Each of the regularly recurring amplitude pulsations is called a *beat* and the process of linearly combining two sinusoids relatively close

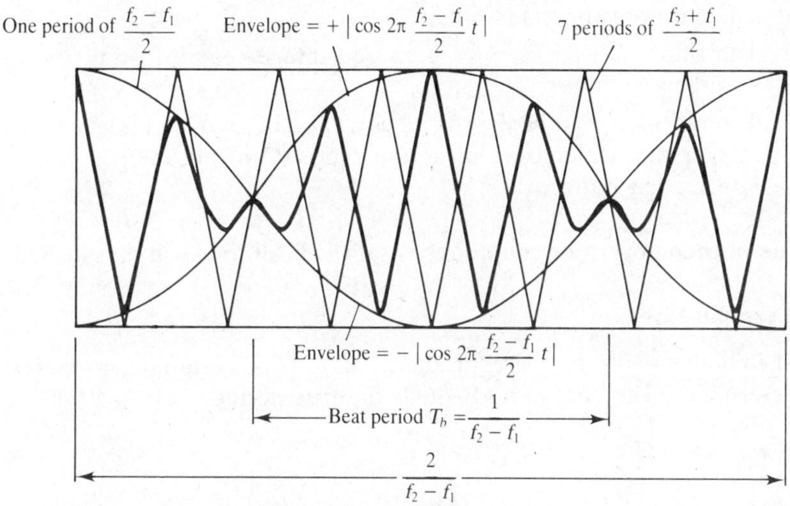

Fig. 1-2 Example of linear mixing showing that the beat period equals the true period when $f_2/f_1 = 4/3$.

in frequency is called *beating*. The beat period is found to be

$$T_b = \frac{1}{f_2 - f_1} \tag{1-7}$$

The rate at which beats occur is called the *beat frequency* and it is the reciprocal of the beat period. Thus

$$f_b = \frac{1}{T_b} = f_2 - f_1 \tag{1-8}$$

There are no beats when $f_2 = f_1$ and the condition is known as *zero beat*.

In Fig. 1-2, if the waveform for a beat period is considered to be moved one beat period later or earlier, exactly the same waveform is obtained as is supposed to be there. This means that the beat period T_b is the same as the true period T_0 obtained from the linear mixing of the original sinusoids. Thus, $T_0 = m/800 = n/600$, from which is obtained $m = 4$ and $n = 3$, so that $T_0 = 1/200$ s. From (1-7), $T_b = 1/200$ s, showing that indeed $T_b = T_0$. Beating is said to be *pure* when the beat period is the same as the true period. It is shown in Appendix A that the beat period is the same as the true period if:

1. The beat frequency $(f_2 - f_1)$ is the highest common divisor of both f_2 and f_1 and those ratios yield the harmonic numbers.

2. The harmonic numbers differ by 1.
3. The ratio $(f_2+f_1)/(f_2-f_1)$ is an odd integer equal to the sum of the harmonic numbers.
4. The ratio f_2/f_1 can be expressed as a ratio of an even integer to an odd integer that differ by 1 or as the ratio of an odd integer to an even integer that differ by 1.

If one of the above four conditions is satisfied, all four will be satisfied.

Example 1-3

Sketch the beats that are produced if $f_2 = 900$ Hz and $f_1 = 700$ Hz and determine if the beat period equals the true period.

Solution

$$\frac{(f_2+f_1)}{(f_2-f_1)} = \frac{1600}{200} = 8$$

This ratio is an even integer, so the beat period is not equal to the true period. The resultant waveform showing the beats is given in Fig. 1-3, where it is seen that the beat period is not equal to the true period. The beat period is $T_b = 1/200$ s. Since $T_0 = m/900 = n/700$, from which are obtained $m = 9$ and $n = 7$, the true period is $T_0 = 1/100$ s. A sequence of $m - n = 9 - 7 = 2$ beat periods is required to obtain a true period (see Appendix B).

The graphical construction becomes more tedious when the ratio $(f_2+f_1)/(f_2-f_1)$ is not a simple integer, but the procedure is exactly the same. Thus, for $f_2 = 13$ Hz and $f_1 = 10$ Hz, the ratio is $23/3 = 7\frac{2}{3}$ and this means that $7\frac{2}{3}$ periods of one sinusoid must be drawn during one period of the other. Careful attention to constructional details is necessary for accurate drawing of the resultant waveform.

The advantage of being able to draw the resultant beating waveform is lost when the ratio $(f_2+f_1)/(f_2-f_1)$ is such a large number that individual

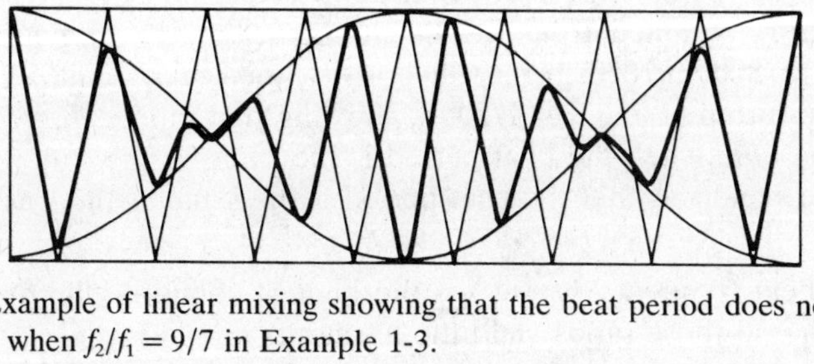

Fig. 1-3 Example of linear mixing showing that the beat period does not equal the true period when $f_2/f_1 = 9/7$ in Example 1-3.

periods of the higher frequency $(f_2+f_1)/2$ cannot be resolved during an entire period of the lower frequency $(f_2-f_1)/2$. In this situation, if the resultant waveform in the regions of destructive interference is desired, it will be convenient to subdivide the time scale so that only a portion of a period of the higher-frequency signal is shown during which time the variation of the lower-frequency signal is essentially linear. Additional details about the resultant waveform are given in Appendix A. It is also shown there that: (a) when pure beating occurs with f_2/f_1 given by the ratio of an even integer to an odd integer that differ by 1, a node of the envelope always occurs $\frac{1}{4}$ period earlier than a complete number of periods of the frequency $(f_2+f_1)/2$, equal to half the even integer; when pure beating occurs with (b) f_2/f_1 given by the ratio of an odd integer to an even integer that differ by 1, a node of the envelope always occurs $\frac{1}{4}$ period later than a complete number of periods of the frequency $(f_2+f_1)/2$ equal to half the even integer.

It must be emphasized that in this linear mixing only the two original frequencies f_2 and f_1 are present and measurable. The frequencies $(f_2+f_1)/2$, $(f_2-f_1)/2$, and (f_2-f_1), as well as the lines drawn to show the envelope, are not actual physical entities and are not measurable.

The superposition of two equal-amplitude sinusoidal signals having slightly different frequencies at the input of a system is a good method of testing that system for nonlinearity. If the system is linear, the envelopes of the output waveform are sinusoidal as shown in Figs. 1-2 and 1-3. Any observable distortion in the envelopes indicates that some kind of nonlinearity is present. This testing method is called the *two-tone* test, regardless of whether the input frequencies are audible sounds or not.

Case 2

When $V_1 \neq V_2$, (1-5) becomes

$$v(t) = A(t)\cos\omega_a t + B(t)\sin\omega_a t \tag{1-9}$$

where

$$A(t) = (V_1 + V_2)\cos\Delta\omega t \tag{1-9a}$$

and

$$B(t) = (V_1 - V_2)\sin\Delta\omega t \tag{1-9b}$$

Thus (1-9) can be expressed as

$$v(t) = E(t)\cos[\omega_a t - \theta(t)] \tag{1-10}$$

where

$$E(t) = \sqrt{A^2(t) + B^2(t)} \tag{1-10a}$$

and

$$\theta(t) = \tan^{-1}\frac{B(t)}{A(t)} \tag{1-10b}$$

Substituting (1-9a) and (1-9b) into (1-10a) and (1-10b) yields

$$E(t) = \sqrt{V_1^2 + V_2^2 + 2V_1V_2 \cos 2\,\Delta\omega t} \tag{1-11a}$$

for the envelope and

$$\theta(t) = \tan^{-1}\left(\frac{V_1 - V_2}{V_1 + V_2}\tan \Delta\omega t\right) \tag{1-11b}$$

for the angle. When $V_1 > V_2$, let $V_2 = kV_1$, and when $V_2 > V_1$, let $V_1 = kV_2$, so that $0 < k < 1$ in both situations. Then (1-11a) and (1-11b) become

$$E(t) = V_q\sqrt{1 + 2k \cos 2\,\Delta\omega t + k^2} \tag{1-12a}$$

for the envelope and

$$\theta(t) = \tan^{-1}\left[(-1)^{q+1}\frac{1-k}{1+k}\tan \Delta\omega t\right] \tag{1-12b}$$

for the angle, where the index $q = 1$ when $V_1 > V_2$ and the index $q = 2$ when $V_2 > V_1$.

The envelope $E(t)$ varies at the rate $2\,\Delta f = f_2 - f_1$, which is the beat frequency. When $\cos 2\,\Delta\omega t = +1$, $E(t) = V_{\max} = \pm V_q(1 + k)$; and when $\cos 2\,\Delta\omega t = -1$, $E(t) = V_{\min} = \pm V_q(1 - k)$. The envelope never goes to zero unless $k = 1$, that is, when $V_1 = V_2$ as was considered in Case 1 above. The normalized envelope can be obtained from (1-12a) as

$$\frac{E(t)}{V_q} = \pm\sqrt{1 + k^2 + 2k \cos 2\,\Delta\omega t} \tag{1-13}$$

and is plotted in Fig. 1-4 for several values of the parameter k. It is observed that the envelope shape is generally narrow in the vicinity of $V_{\min}$ and broad in the vicinity of $V_{\max}$. If $k \ll 1$, (1-13) becomes

$$\frac{E(t)}{V_q} \cong \sqrt{1 + 2k \cos 2\,\Delta\omega t}$$

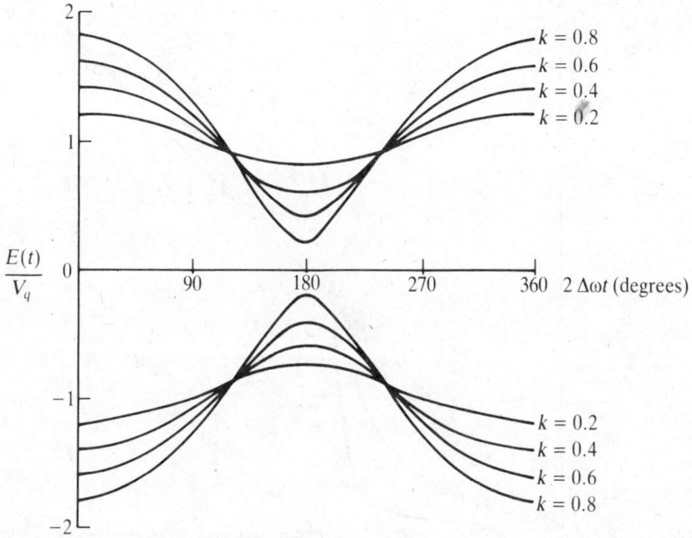

Fig. 1-4 Normalized envelope shapes in linear mixing of unequal-amplitude sinusoids for several values of the parameter k.

But $\sqrt{1+x} \cong 1 + x/2$ for small x, therefore

$$\frac{E(t)}{V_q} \cong 1 + k \cos 2\,\Delta\omega t \tag{1-14}$$

which shows that the envelope becomes sinusoidal but with small amplitude when $k \ll 1$.

The phase angle given by (1-12b) is plotted in Fig. 1-5 for several values of the parameter k when $q = 1$ and $V_1 > V_2$. If $V_2 > V_1$ and $q = 2$, the sign of $\theta(t)$ changes to the negative of the values plotted in Fig. 1-5. It is seen that an additional angle of 180° occurs per interval of twice the beat period.

The nonlinear variation of $\theta(t)$ over a period of the frequency $(f_2 - f_1)/2$ leads to a variation in the instantaneous frequency ω_i of the cosine in (1-10). Thus

$$\omega_i = \omega_a - \frac{d\theta(t)}{dt} \qquad \text{or} \qquad f_i = f_a - \frac{1}{2\pi}\frac{d\theta(t)}{dt}$$

Taking the time derivative of (1-12b) with $q = 1$ and simplifying yields

$$f_i = f_a - F(t)\,\Delta f \tag{1-15}$$

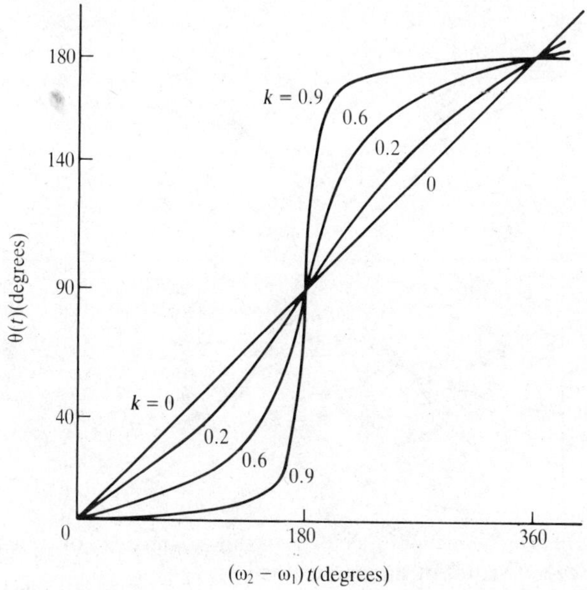

Fig. 1-5 Variation of $\theta(t)$ over a period of the beat frequency.

where

$$F(t) = \frac{\left(\dfrac{1-k^2}{2k}\right)}{\left(\dfrac{1+k^2}{2k}\right) + \cos 2\,\Delta\omega t} \tag{1-16}$$

It is the variation of the time function $F(t)$ that accounts for the variation of the instantaneous frequency f_i. The value of $F(t)$ is always positive because $0 < k < 1$. The function $F(t)$ is plotted in Fig. 1-6 for several values of the parameter k. The instantaneous frequency alternately increases and decreases during each beat period. The average value of $F(t)$ over a beat period is found to equal 1. Therefore the average frequency is, for index $q = 1$ and $V_1 > V_2$,

$$(f_i)_{\text{ave}} = f_a - [F(t)]_{\text{ave}}\,\Delta f = f_a - \Delta f = f_1$$

which is the frequency of the higher-amplitude voltage. If the index $q = 2$ and $V_2 > V_1$,

$$(f_i)_{\text{ave}} = f_a + [F(t)]_{\text{ave}}\,\Delta f = f_a + \Delta f = f_2$$

which, again, is the frequency of the higher-amplitude voltage.

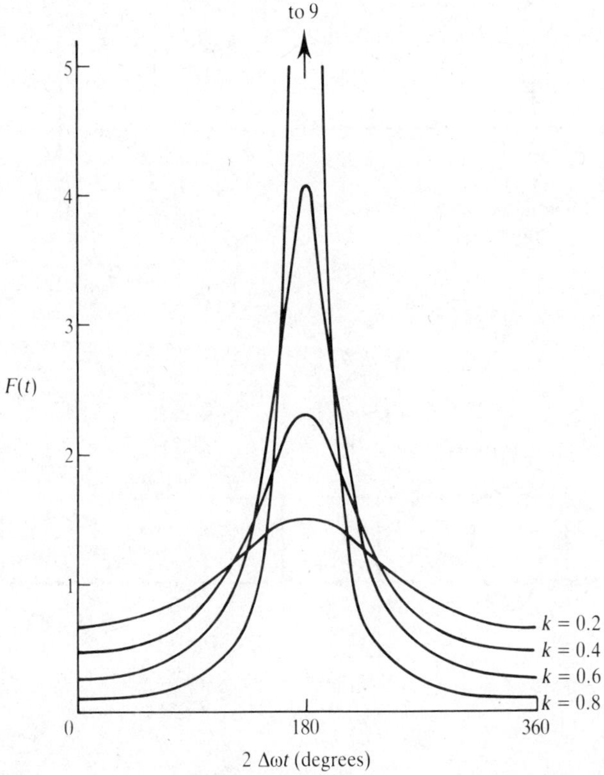

Fig. 1-6 The function $F(t)$ of Eq. (1-16) that determines instantaneous frequency $f_i = f_a - F(t)\,\Delta f$.

1-3.2 Linear Mixing of Multiple Sinusoids

The spectrum of n equal-amplitude sinusoids whose frequencies differ by Δf is shown in Fig. 1.7. The case of an even number of components is represented in the spectrum of Fig. 1-7a and that of an odd number of components is represented in Fig. 1-7b. In both cases, let f_a be the *average* frequency of the n equally spaced sinusoids so that the different frequencies are symmetrically located above and below f_a. For the even number of sinusoids shown in Fig. 1-7a, the sum of all the components is

$$\begin{aligned} x(t) = A\Big\{&\cos\left(\omega_a + \frac{\Delta\omega}{2}\right)t + \cos\left(\omega_a + 3\frac{\Delta\omega}{2}\right)t + \cdots \\ &+ \cos\left[\omega_a + (n-1)\frac{\Delta\omega}{2}\right]t \\ &+ \cos\left(\omega_a - \frac{\Delta\omega}{2}\right)t + \cos\left(\omega_a - 3\frac{\Delta\omega}{2}\right)t + \cdots \\ &+ \cos\left[\omega_a - (n-1)\frac{\Delta\omega}{2}\right]t\Big\} \end{aligned} \tag{1-17}$$

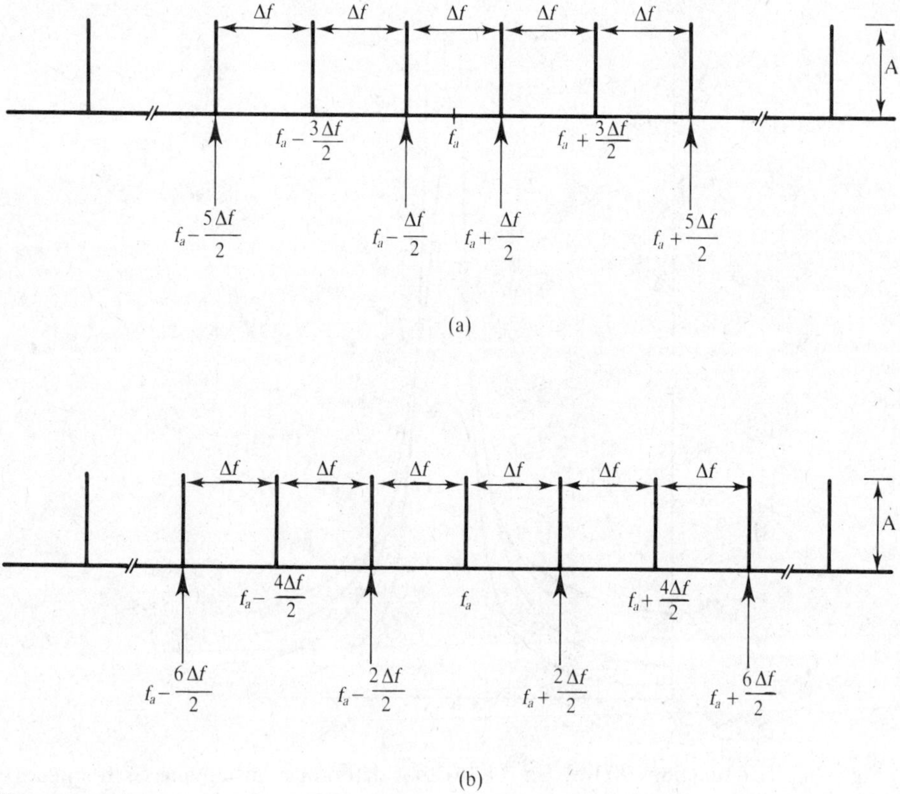

Fig. 1-7 Line spectra of n equal-amplitude equally spaced sinusoids: (a) n even; (b) n odd.

where $\Delta\omega = 2\pi\,\Delta f$. Using the trigonometric identity $\cos(a+b) + \cos(a-b) = 2\cos a \cos b$ repeatedly in (1-17) results in

$$x(t) = 2A\left[\cos\frac{\Delta\omega}{2}t + \cos\frac{3\,\Delta\omega}{2}t + \cdots + \cos\frac{(n-1)\,\Delta\omega}{2}t\right]\cos\omega_a t \quad \text{(1-18)}$$

in which all coefficients of $\Delta\omega/2$ are odd integers. In Table 1-2 it is found that a cosine polynomial having only odd integer multiples of an angle θ can be represented in compact form by

$$\frac{\sin(k+1)\theta}{2\sin\theta} = \cos\theta + \cos 3\theta + \cos 5\theta + \cdots + \cos k\theta \quad \text{(1-19)}$$

TABLE 1-2 Summations of Some Trigonometric Series

$\dfrac{\sin (k+1)\theta}{2 \sin \theta} = \cos \theta + \cos 3\theta + \cos 5\theta + \cdots + \cos k\theta,$	k odd
$\dfrac{\sin [(2k+1)(\theta/2)]}{2 \sin \theta} = \frac{1}{2} + \cos 2\theta + \cos 3\theta + \cdots + \cos k\theta,$	all k

Therefore, using (1-19) with $k = n - 1$ and $\theta = \Delta\omega/2\,t$ in (1-18) yields

$$x(t) = A\left[\frac{\sin \dfrac{n\,\Delta\omega}{2} t}{\sin \dfrac{\Delta\omega}{2} t}\right] \cos \omega_a t, \qquad \text{for } n \text{ even} \tag{1-20}$$

For the odd number of sinusoids shown in Fig. 1-7b, the sum of all components is

$$\begin{aligned} x(t) = A\Big\{&\cos \omega_a t + \cos (\omega_a + \Delta\omega)t + \cos (\omega_a + 2\,\Delta\omega)t + \cdots \\ &+ \cos\left[\omega_a + \frac{(n-1)}{2}\Delta\omega\right]t \\ &+ \cos (\omega_a - \Delta\omega)t + \cos (\omega_a - 2\,\Delta\omega)t + \cdots \\ &+ \cos\left[\omega_a - \frac{(n-1)}{2}\Delta\omega\right]t\Big\} \end{aligned} \tag{1-21}$$

where n is odd and $\Delta\omega = 2\pi\,\Delta f$. Using the trigonometric identity $\cos (a + b) + \cos (a - b) = 2 \cos a \cos b$ repeatedly in (1-21) yields

$$\begin{aligned} x(t) = 2A\Big[&\tfrac{1}{2} + \cos \Delta\omega t + \cos 2\,\Delta\omega t \\ &+ \cos 3\,\Delta\omega t + \cdots + \cos \frac{(n-1)}{2}\Delta\omega t\Big] \cos \omega_a t \end{aligned} \tag{1-22}$$

where n is odd. It is found in Table 1-2 that

$$\frac{\sin (2k+1)(\theta/2)}{2 \sin (\theta/2)} = \tfrac{1}{2} + \cos \theta + \cos 2\theta + \cos 3\theta + \cdots + \cos k\theta \tag{1-23}$$

Therefore, using (1-23) with $k = (n-1)/2$ and $\theta = \Delta\omega t$ in (1-22) results in

$$x(t) = A\left[\frac{\sin\frac{n\,\Delta\omega}{2}t}{\sin\frac{\Delta\omega}{2}t}\right]\cos\omega_a t, \qquad \text{for } n \text{ odd} \tag{1-24}$$

It is seen that (1-24) is exactly the same as (1-20). Therefore, for any number n, even or odd, of equal-amplitude sinusoids whose frequencies differ by Δf, that are to be added so that linear mixing results,

$$x(t) = A\left[\frac{\sin\frac{n\,\Delta\omega}{2}t}{\sin\frac{\Delta\omega}{2}t}\right]\cos\omega_a t, \qquad \text{for all } n \tag{1-25}$$

where f_a is the average frequency and $\Delta\omega = 2\pi\,\Delta f$.

The envelope $E(t)$ is given by the slowly varying coefficient of $\cos\omega_a t$ in (1-25). Thus,

$$E(t) = \pm A\left|\frac{\sin\frac{n\,\Delta\omega}{2}t}{\sin\frac{\Delta\omega}{2}t}\right| \tag{1-26}$$

This is quickly sketched by drawing n periods of the numerator sinusoid during the same time as one period of the denominator and then dividing instant by instant. Applying l'Hospital's rule at the instants where both numerator and denominator are simultaneously zero yields $\pm nA$ for the envelope amplitudes there. The envelopes for six sinusoids and for seven sinusoids superimposed are given in Fig. 1-8, where beating is seen. The beat period is the reciprocal of the frequency difference between the sinusoids.

1-3.3 Symmetry in Linear Mixing

Details about the symmetry of periodic waveforms produced by the linear mixing of sinusoids can be readily deduced. Consider

$$x(t) = V_1\cos 2\pi f_1 t + V_2\cos 2\pi f_2 t \tag{1-27}$$

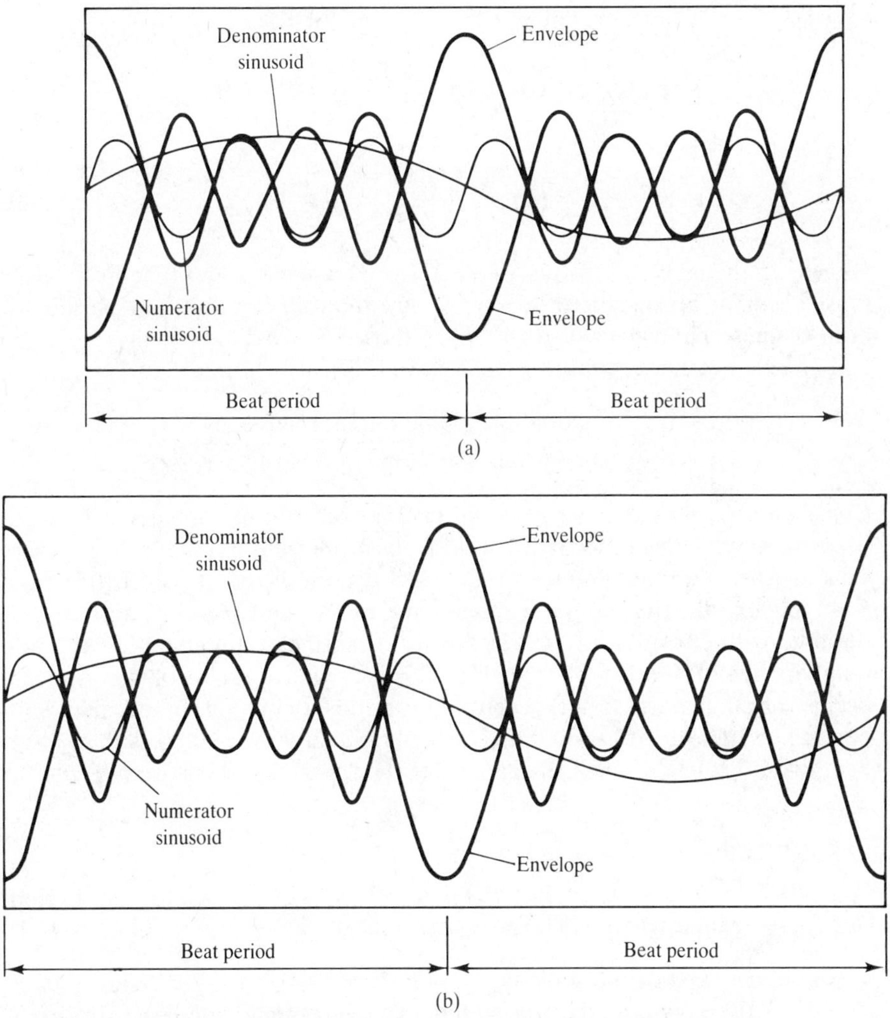

Fig. 1-8 Envelopes for (a) six sinusoids, (b) seven sinusoids.

Assume that $f_2 > f_1$ and that $x(t)$ is periodic with period T_0. It is clear that both cosines have their positive peaks simultaneously at $t = 0$ and periodically every T_0 seconds thereafter, so that $x(t)_{max} = V_1 + V_2$. We wish to determine whether it is also possible for the negative peaks of the cosines in (1-27) to occur simultaneously at some instant $t = t_-$ and periodically thereafter so that $x(t)_{min} = -x(t)_{max} = -(V_1 + V_2)$. If so, then the positive and negative extents of $x(t)$ have the same magnitudes. For this to occur at $t = t_-$ means that $\cos 2\pi f_1 t_- = -1$ and $\cos 2\pi f_2 t_- = -1$. Therefore, $2\pi f_1 t_- = (2p + 1)\pi$ and $2\pi f_2 t_- = (2q + 1)\pi$, because the angles must be odd multiples of π for the cosines to be equal to -1 each. Thus,

$$t_- = \frac{2p+1}{2f_1} = (2p-1)\left(\frac{T_1}{2}\right), \qquad p = 0, 1, 2, \ldots \tag{1-28}$$

and

$$t_- = \frac{2q+1}{2f_2} = (2q+1)\left(\frac{T_2}{2}\right), \qquad q = 0, 1, 2, \ldots \tag{1-29}$$

Hence, if the negative peaks of (1-27) are to occur simultaneously, they must occur at an instant t_- given by odd multiples of the half periods of both cosines. The values of p and q are those required for equality of (1-28) and (1-29). Further, equating (1-28) and (1-29), it is found that

$$\frac{f_2}{f_1} = \frac{2q+1}{2p+1} = \frac{\text{an odd harmonic number like } 3, 5, 7, \ldots}{\text{a smaller odd harmonic number like } 1, 3, 5, \ldots} \tag{1-30}$$

Consequently, the negative peaks of (1-27) occur simultaneously if f_2/f_1 can be expressed as the ratio of two relatively prime *odd* numbers that are the same as the harmonic numbers required for periodicity. If $f_1 = 1000$ Hz and $f_2 = 1400$ Hz, the harmonic numbers are $m = 7$ and $n = 5$. Thus, $x(t)$ is periodic with period $T_0 = 0.005$ s. Since the harmonic numbers are the odd numbers 7 and 5 and $f_2/f_1 = 1400/1000 = 7/5$, the negative peaks in (1-27) occur simultaneously after seven half-periods of 1400 Hz and five half-periods of 1000 Hz at $t_- = 0.0025$ s and periodically every 0.005 s thereafter. See Fig. 1-3 for $f_2/f_1 = 9/7$. In Fig. 1-2, $x(t)_{\min} \neq x(t)_{\max}$ because $f_2/f_1 = 4/3$.

Example 1-4

$x(t) = 3\cos 2\pi 30t + 2\cos 2\pi 90t + \cos 2\pi 270t$. (a) Calculate the period; (b) Determine whether $x(t)_{\min} = -x(t)_{\max}$.

Solution (a) For periodicity, $T_0 = k/30 = n/90 = m/270$. Thus, $90k = 30n$, $270k = 30m$, and $270n = 90m$. The harmonic numbers are $k = 1$, $n = 3$, and $m = 9$, and the period is $T_0 = 1/30$ s. (b) At $t = 0$ each cosine is equal to 1. Thus, $x(t)_{\max} = 3 + 2 + 1 = 6V$, and it occurs every 1/30 s. If $\cos 2\pi 30t_- = -1$, $\cos 2\pi 90t_- = -1$, and $\cos 2\pi 270t_- = -1$, then $2\pi 30t_- = (2p+1)\pi$, $2\pi 90t_- = (2q+1)\pi$, and $2\pi 270t_- = (2r+1)\pi$. Hence,

$$t_- = \frac{2p+1}{2(30)} = \frac{2q+1}{2(90)} = \frac{2r+1}{2(270)}$$

Setting up frequency ratios results in

$$\frac{90}{30} = \frac{3}{1} = \frac{2q+1}{2p+1}, \qquad \frac{270}{30} = \frac{9}{1} = \frac{2r+1}{2p+1}, \qquad \frac{270}{90} = \frac{9}{3} = \frac{2r+1}{2q+1}$$

Note that here the ratio (9/3) cannot be reduced to (3/1). From these frequency ratios it is seen that $2p+1=1$, $2q+1=3$, and $2r+1=9$, the same as the harmonic numbers, so that $t_- = 1/60$ s and it occurs at 1 half-period of 30 Hz, 3 half-periods of 90 Hz, and 9 half-periods of 270 Hz. The individual negative peaks all occur at $t_- = 1/60$ s and periodically every 1/30 s thereafter, so that $x(t)_{\text{min}} = -6$.

Symmetry situation is more complicated when sines are involved. Consider

$$y(t) = V_1 \sin 2\pi f_1 t + V_2 \sin 2\pi f_2 t \tag{1-31}$$

Assume that $f_2 > f_1$ and that $y(t)$ is periodic. If the positive peaks of the sines in (1-31) occur simultaneously at $t = t_+$ and periodically thereafter, then $y(t)_{\text{max}} = V_1 + V_2$, $\sin 2\pi f_1 t_+ = +1$, and $\sin 2\pi f_2 t_+ = +1$. Therefore,

$$t_+ = \frac{4k+1}{4f_1} = (4k+1)\frac{T_1}{4}, \qquad k = 0, 1, 2, \ldots$$

$$t_+ = \frac{4\ell+1}{4f_2} = (4\ell+1)\frac{T_2}{4}, \qquad \ell = 0, 1, 2, \ldots$$

(i.e., t_+ occurs at odd multiples of the quarter periods of both sines).

$$\frac{f_2}{f_1} = \frac{4\ell+1}{4k+1} = \frac{\text{an odd harmonic number like } 5, 9, 13, \ldots}{\text{a smaller odd harmonic number like } 1, 5, 9, 13, \ldots} \tag{1-32}$$

If the negative peaks in (1-31) occur simultaneously at $t = t_-$ and periodically thereafter, then

$$t_- = \frac{4m+3}{4f_1} = (4m+3)\frac{T_1}{4}, \qquad m = 0, 1, 2, \ldots$$

$$t_- = \frac{4n+3}{4f_2} = (4n+3)\frac{T_2}{4}, \qquad n = 0, 1, 2, \ldots$$

(i.e., t_- occurs at different odd multiples of the quarter-periods of both sines from those obtained above).

$$\frac{f_2}{f_1} = \frac{4n+3}{4m+3} = \frac{\text{an odd harmonic number like } 7, 11, 15, \ldots}{\text{a smaller odd harmonic number like } 3, 7, 11, \ldots} \tag{1-33}$$

In order for $y(t)_{\text{max}} = V_1 + V_2$ and $y(t)_{\text{min}} = -y(t)_{\text{max}} = -(V_1 + V_2)$ to occur at their respective instants, both (1-32) and (1-33) must be satisfied. If $f_2/f_1 = 5/1$ in (1-32), then $f_2/f_1 = 15/3$ in (1-33). If $f_2/f_1 = 7/3$ in (1-33), then $f_2/f_1 = 21/9$ in (1-32). The harmonic numbers differ by integer multiples of 4.

1-4 WAVES

In the general sense intended here, a wave is characterized by instantaneous amplitudes that depend on both the instant of time and on the spatial location. For example, an undamped sinusoidal wave advancing in the $+z$ direction in a uniform medium is given by

$$x_1(t, z) = A \cos(\omega t - \beta z) \tag{1-34}$$

where β is a spatial phase constant. This wave is plotted in Fig. 1-9 as functions of z for several different instants of time. It is noticed that the wave at any instant is a sinusoidal function of the spatial angle βz and that the wave repeats itself spatially every 2π radians, for which the distance is called the *wavelength* λ in the medium. Thus

$$\beta\lambda = 2\pi$$

from which is obtained

$$\beta = \frac{2\pi}{\lambda} \tag{1-35}$$

The phase constant β represents 2π radians of spatial phase change per wavelength.

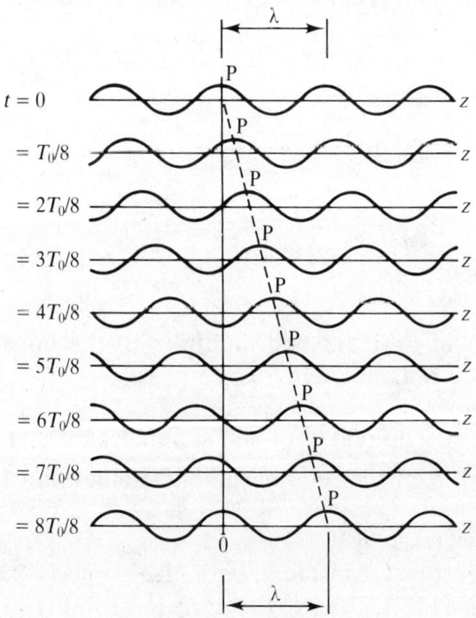

Fig. 1-9 The wave $A\cos(\omega t - \beta z)$ at several instants separated by $T_0/4$. A point of constant phase such as point P advances at the phase velocity.

The wave expressed by (1-34) exhibits a spatial periodicity with spatial period λ at a certain time as well as a temporal periodicity with period T_0 at a certain distance.

Select a point P on the wave for $t = 0$ as in Fig. 1-9. Then, as t increases, z must also increase in order for point P to remain at the same phase point on the various waves at the different instants shown. It is seen that point P moves in the positive z direction as t increases and this is interpreted to be a wave called a *traveling wave.*

1-4.1 Phase Velocity

Since point P remains at a constant phase,

$$\omega t - \beta z = \text{a constant}$$

Differentiating with respect to t yields

$$\frac{dz}{dt} = \frac{\omega}{\beta} \tag{1-36}$$

which represents the velocity at which a point of constant phase such as point P moves. This velocity is called the *phase velocity* v_p. Thus,

$$v_p = \frac{\omega}{\beta} \tag{1-37}$$

The phase velocity of radio waves and of light in free space (i.e., a vacuum) is usually designated by the lower-case letter c and its value is approximately 3×10^8 meters per second. The phase velocity in some other medium will depend on the relative dielectric constant and the relative permeability of the medium.

Substituting (1-35) into (1-37) results in

$$v_p = f\lambda \tag{1-38}$$

Thus, in free space, wavelength in meters is related to frequency in megahertz by

$$\lambda = \frac{300}{f(\text{MHz})}\ \text{m} \tag{1-39a}$$

and the wavelength in centimeters is related to frequency in gigahertz by

$$\lambda = \frac{30}{f(\text{GHz})}\ \text{cm} \tag{1-39b}$$

Further, observe that point P in Fig. 1-9 moves a distance equal to a wavelength as t increases by one period T_0. Therefore, since velocity equals distance traveled divided by the time interval,

$$v_p = \frac{\lambda}{T_0} = f\lambda$$

which is the same as obtained in (1-38).

Showing that the sinusoidal wave represented by

$$x_2(t, z) = B \cos(\omega t + \beta z)$$

can be interpreted as a wave motion in the $-z$ direction is left to the problems (see Prob. 1-9).

It must be emphasized that in this situation no actual physical entity is traveling at the phase velocity. The wave $x(t, z)$ may represent a voltage, an electric field, a current, or a magnetic field. Under the steady-state conditions that have been assumed, the voltage, the current, or the field is not moving spatially but is simply changing instantaneous spatial values in such a way that a point of constant phase has a velocity v_p in a direction called the *direction of propagation.*

1-5 LINEAR MIXING OF WAVES

Consider the superposition of two sinusoidal voltage waves traveling in the $+z$ direction given by

$$v(t, z) = V_1 \cos(\omega_1 t - \beta_1 z) + V_2 \cos(\omega_2 t - \beta_2 z) \tag{1-40}$$

Assume that $f_2 > f_1$. Let $\Delta f = (f_2 - f_1)/2$ and also let the average frequency be $f_a = (f_2 + f_1)/2$. Then

$$f_1 = f_a - \Delta f \qquad \text{or} \qquad \omega_1 = \omega_a - \Delta\omega$$

where $\Delta\omega = 2\pi\, \Delta f$, and

$$f_2 = f_a + \Delta f \qquad \text{or} \qquad \omega_2 = \omega_a + \Delta\omega$$

The wave at frequency f_2 has a phase constant β_2 and the wave at frequency f_1 has a phase constant β_1. Let $\Delta\beta = (\beta_2 - \beta_1)/2$ and let the average phase constant be $\beta_a = (\beta_2 + \beta_1)/2$. Then

$$\beta_1 = \beta_a - \Delta\beta$$

and

$$\beta_2 = \beta_a + \Delta\beta$$

Therefore (1-40) becomes

$$v(t, z) = V_1 \cos [(\omega_a - \Delta\omega)t - (\beta_a - \Delta\beta)z] \\ + V_2 \cos [(\omega_a + \Delta\omega)t - (\beta_a + \Delta\beta)z]$$

which can be rearranged as

$$v(t, z) = V_1 \cos [(\omega_a t - \beta_a z) - (\Delta\omega t - \Delta\beta z)] \\ + V_2 \cos [(\omega_a t - \beta_a z) + (\Delta\omega t - \Delta\beta z)]$$

Using trigonometric identities for the cosines of sum and difference angles and combining terms in that result yields

$$v(t, z) = [(V_2 + V_1) \cos (\Delta\omega t - \Delta\beta z)] \cos (\omega_a t - \beta_a z) \\ + [(V_1 - V_2) \sin (\Delta\omega t - \Delta\beta z)] \sin (\omega_a t - \beta_a z) \qquad (1\text{-}41)$$

Two cases can be identified because of the nature of (1-41): Case I for which $V_1 = V_2$, and Case II for which $V_1 \neq V_2$.

Case I

When $V_1 = V_2 = V$, the second term in (1-41) disappears so that

$$v(t, z) = 2V \cos (\Delta\omega t - \Delta\beta z) \cos (\omega_a t - \beta_a z) \qquad (1\text{-}42)$$

The two cosine factors in (1-42) represent the product of two traveling waves and indicate that a traveling form of *spatial* beating occurs. The high-frequency wave whose frequency is ω_a travels at the phase velocity $v_p = \omega_a/\beta_a = f_a\lambda_a$. The low-frequency wave is the envelope of the beats and it has a fixed shape but travels as time increases.

1-5.1 Group Velocity

A point of constant phase *on the envelope* requires that

$$\Delta\omega t - \Delta\beta z = \text{a constant}$$

Taking the time derivative results in an envelope velocity

$$\frac{dz}{dt} = \frac{\Delta\omega}{\Delta\beta}$$

If $\Delta\beta$ is negative, the envelope travels in the $-z$ direction.

Group velocity is defined to be the limit of the envelope velocity as $\Delta\omega$ becomes small. Thus, for a particular frequency, the group velocity is

$$v_g = \lim_{\Delta\omega \to 0} \frac{\Delta\omega}{\Delta\beta} = \frac{d\omega}{d\beta} \tag{1-43}$$

The concept of a group velocity applies to all frequencies within a narrow bandwidth that is small compared to the average frequency, because the envelope encloses a grouping (i.e., an interference pattern) that is formed by the combination of the individual waves.

Group velocity differs from phase velocity whenever phase velocity depends on the wavelength. Since $\omega = 2\pi f = 2\pi v_p/\lambda$ and $\lambda = 2\pi/\beta$, then

$$v_g = \frac{d\omega}{d\beta} = \frac{d\omega}{d\lambda}\frac{d\lambda}{d\beta} = \frac{2\pi}{\lambda^2}\left(\lambda \frac{dv_p}{d\lambda} - v_p\right)\left(\frac{-2\pi}{\beta^2}\right)$$

which simplifies to

$$v_g = v_p - \lambda \frac{dv_p}{d\lambda} \tag{1-44}$$

A dispersive medium is one in which the phase velocity depends on the wavelength. Normally dispersive media have $dv_p/d\lambda > 0$, so that $v_g < v_p$. Anomalously dispersive media have $dv_p/d\lambda < 0$, so that $v_g > v_p$. Nondispersive media have $dv_p/d\lambda = 0$, so that $v_g = v_p$. The concept of group velocity is lost when the individual components are so widely spread in frequency that $\Delta\omega/\Delta\beta$ is not even approximately equal to $d\omega/d\beta$ at the average frequency.

Case II

When $V_1 \neq V_2$, (1-41) becomes

$$v(t, z) = A(t, z)\cos(\omega_a t - \beta_a z) + B(t, z)\sin(\omega_a t - \beta_a z) \tag{1-45}$$

where

$$A(t, z) = (V_2 + V_1)\cos \Delta\omega t - \Delta\beta z) \tag{1-45a}$$

and

$$B(t, z) = (V_1 - V_2)\sin(\Delta\omega t - \Delta\beta z) \tag{1-45b}$$

Thus (1-45) can be expressed as

$$v(t, z) = E(t, z)\cos[\omega_a t - \beta_a z - \theta(t, z)] \tag{1-46}$$

where

$$E(t, z) = \sqrt{A^2(t, z) + B^2(t, z)} \tag{1-46a}$$

and

$$\theta(t, z) = \tan^{-1} \frac{B(t, z)}{A(t, z)} \tag{1-46b}$$

Substituting (1-45a) and (1-45b) into (1-46a) and (1-46b) yields

$$E(t, z) = \sqrt{V_1^2 + V_2^2 + 2V_1V_2 \cos 2(\Delta\omega t - \Delta\beta z)} \tag{1-47a}$$

for the envelope, and

$$\theta(t, z) = \tan^{-1}\left[\frac{V_1 - V_2}{V_1 + V_2} \tan(\Delta\omega t - \Delta\beta z)\right] \tag{1-47b}$$

for the angle.

A point of constant phase on the envelope still moves at the group velocity if $\Delta\omega$ is small. At a fixed location such as $z = 0$, the envelope amplitude $E(t, 0)$ and the angle $\theta(t, 0)$ vary with time in the same manner as discussed in Case 2, Section 1-3.1. At a certain instant such as $t = 0$, the envelope amplitude $E(0, z)$ and the angle $\theta(0, z)$ vary spatially in a similar manner to that in Case 2 time variations. The envelope is narrow at locations in the vicinity of V_{min} and broad at locations in the vicinity of V_{max}. The details will not be pursued further here because of their similarity to those of Case 2 of Section 1-3.1.

1-5.2 Standing Waves

Consider the linear mixing of two identical-frequency, undamped sinusoidal voltage waves traveling in opposite directions. Thus

$$v(t, z) = V_1 \cos(\omega t - \beta z) + V_2 \cos(\omega t + \beta z) \tag{1-48}$$

Using familiar trigonometric identities for the cosines of the sum and difference angles, (1-48) becomes

$$v(t, z) = [(V_1 + V_2) \cos \beta z] \cos \omega t + [(V_1 - V_2) \sin \beta z] \sin \omega t \tag{1-49}$$

Again, two cases can be identified because of the nature of (1-49): Case A for which $V_1 = V_2 = V$, and Case B for which $V_1 \neq V_2$.

Case A

When $V_1 = V_2 = V$, the second term in (1-49) disappears so that

$$v(t, z) = 2V \cos \beta z \cos \omega t \tag{1-50}$$

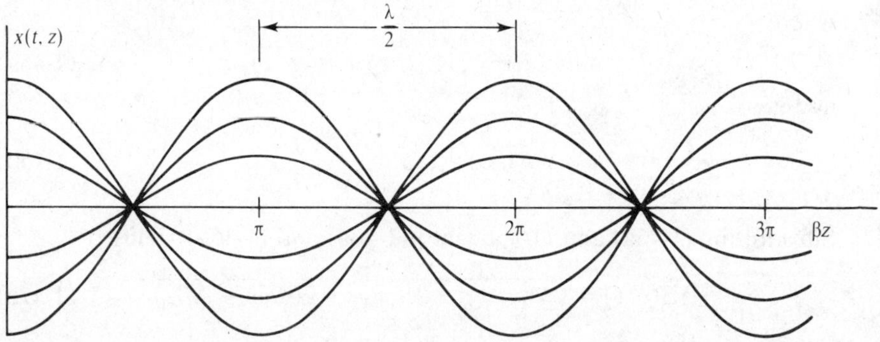

Fig. 1-10 Standing waves at various instants produced by the linear mixing of two oppositely directed equal-amplitude traveling waves (not shown).

This is plotted in Fig. 1-10 for several instants of time. It is seen that when βz equals odd-integer multiples of $\pi/2$ radians, the value of $v(t, z)$ is always zero (a node) for any instant of time. At all other locations, such as where βz equals even-integer multiples of $\pi/2$ radians for which $v(t, z)$ attains its maximum (an antinode) of $\pm 2V\,|\cos \omega t|$, the value of $v(t, z)$ is sinusoidal in its time variations. The location of a constant space-phase point does not change from one instant to another. When such spatial stationarity occurs, the result is designated as a *standing wave.*

Case B

When $V_1 \neq V_2$, (1-43) becomes

$$v(t, z) = A(z)\cos \omega t + B(z)\sin \omega t = E(z)\cos[\omega t - \theta(z)]$$

where

$$A(z) = (V_1 + V_2)\cos \beta z$$

$$B(z) = (V_1 - V_2)\sin \beta z$$

$$E(z) = \sqrt{A^2(z) + B^2(z)} = \sqrt{V_1^2 + V_2^2 + 2V_1V_2\cos 2\beta z}$$

$$\theta(z) = \tan^{-1}\frac{B(z)}{A(z)} = \tan^{-1}\left[\frac{V_1 - V_2}{V_1 + V_2}\tan \beta z\right]$$

The spatial variations of the envelope $E(z)$ and of the angle $\theta(z)$ can be examined in a similar manner to the temporal variations detailed in Case 2 and will not be repeated here. A standing wave is created in which the locations of two adjacent points for $V_{max} = V_1 + V_2$ and the locations of two adjacent points for $V_{min} = V_1 - V_2$ are separated by $\lambda/2$. The envelope is narrower at locations in the vicinity of V_{min} than in the vicinity of V_{max}. This is a useful fact to keep in mind when taking measurements using a slotted-line apparatus to probe the voltage standing wave. The ratio $(V_1 + V_2)/(V_1 - V_2) = V_{max}/V_{min}$ is called the *voltage standing wave ratio, VSWR,* and it is an important operating characteristic of transmission lines.

1-6 LABORATORY CONSIDERATIONS

Quite often it is necessary to apply the sum of two or more sinusoids to the input terminals of a system to test how that system responds to the input signals. Usually the signal generators that are employed for the test do not have their outputs balanced to ground. Most signal generators have unbalanced outputs suitable for using with 50-Ω coaxial cable interconnections, the outside shield being grounded. This grounding of one side of each output prevents connecting the outputs of two or more signal generators directly in series. Isolating a signal generator from ground by means of an isolation transformer in the input power line is not an appropriate solution for high-frequency operation.

The appropriate solution for linear mixing does not tamper with the normal signal-generator grounds in any manner whatsoever, but employs the principle of superposition. When two or more independent signal sources are applied to separate input ports of a linear time-invariant system, the output at another port is the sum (i.e., the superposition) of the individual outputs produced by each signal source acting alone when the outputs of the other signal sources are set equal to zero. Superposition, therefore, is actually linear mixing in its most basic form.

1-6.1 A Resistive Mixer

The resistive 3-port network shown in Fig. 1-11 may be used as a linear mixer under certain conditions to be discussed in what follows. Each port is taken to be between a network terminal and ground. The two signal generators are represented by their Thévenin equivalent circuits composed of E_a and R_a and E_b and R_b. Each generator is to be terminated in a resistance equal to its own output resistance R_a or R_b. The network is terminated in a resistance R_a equal to the output resistance of one of the signal generators and the output resistance of the network is to be equal to the terminating resistance. Thus, given R_a and R_b, the problem is to find R_1 and R_2 to satisfy the specified conditions and to investigate the effects of the signal sources on the output voltage v_o and on each other.

The term *available power* is often used to designate the maximum average power that a signal source or network can deliver to its matched load resistance. The conditions specified above allow each signal generator

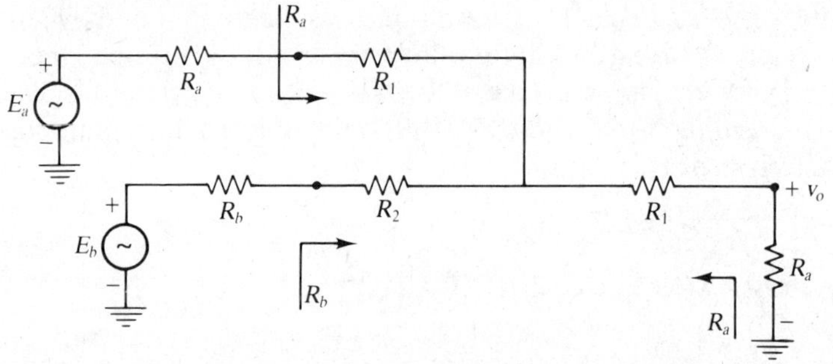

Fig. 1-11 A resistive mixer.

to deliver its own available power to the network, and the network in turn delivers its available power to the load resistance.

Because of symmetry, the input resistance at port a and the output resistance at the output port are the same and equal to

$$R_1 = R_1 + \frac{(R_b + R_2)(R_a + R_1)}{R_b + R_2 + R_a + R_1} \tag{1-51}$$

The input resistance of port b is

$$R_b = R_2 + \frac{R_1 + R_a}{2} \tag{1-52}$$

Solve (1-51) and (1-52) simultaneously to obtain

$$R_1 = \frac{R_a^2}{4R_b - R_a} \tag{1-53}$$

and

$$R_2 = R_b \frac{4R_b - 3R_a}{4R_b - R_a} \tag{1-54}$$

The denominator of (1-53) must be positive (i.e., $R_b > \frac{1}{4}R_a$) for R_1 to be positive, and the numerator of (1-54) must also be positive for R_2 to be positive. Hence, for both R_1 and R_2 to be positive,

$$R_b > \tfrac{3}{4}R_a \tag{1-55}$$

is a condition that must be satisfied for the network to meet the original specifications.

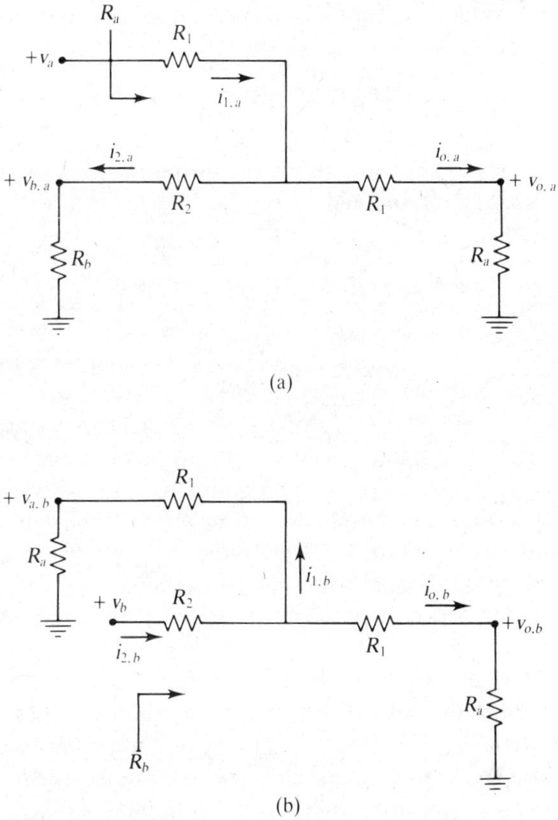

Fig. 1-12 Equivalent circuits of the resistive mixer: (a) signal source at port *a* only; (b) signal source at port *b* only.

The output voltage produced by the signal source at port *a* only is, from Fig. 1-12a,

$$v_{o,a} = v_a\left(1 - \frac{R_a}{2R_b}\right) \tag{1-56}$$

and the voltage produced across the other signal generator by v_a is

$$v_{b,a} = \frac{v_a}{2} \tag{1-57}$$

The output voltage produced by the signal source at port *b* only is, from Fig. 1-12b,

$$v_{o,b} = v_b \frac{R_a}{2R_b} \tag{1-58}$$

and the voltage produced across the other signal generator by v_b is

$$v_{a,b} = v_b \frac{R_a}{2R_b} \tag{1-59}$$

Therefore, using superposition, the total output voltage when both signal sources are applied simultaneously is

$$v_o = v_{o,a} + v_{o,b} = v_a\left(1 - \frac{R_a}{2R_b}\right) + v_b \frac{R_a}{2R_b} \tag{1-60}$$

and linear mixing has been accomplished. (Remember that $R_b > \frac{3}{4}R_a$.) When $R_a = R_b$ the output voltage is $v_o = (v_a + v_b)/2$.

Notice that each signal generator causes some of its output voltage to appear across the other signal generator's output resistance. If a significant portion of the voltage $v_{a,b}$ or $v_{b,a}$ can find its way back into the frequency-determining and amplifying circuitry of the signal generator, it can thereby undesirably influence the frequency and phase stability of the signal generated by that signal source, and perhaps also instigate distressing nonlinear effects. This problem can be attacked two ways.

1. Increase the external isolation† between the signal generators by choosing a well-shielded linear mixer that can do the job. For example, from (1-57), the voltage $v_{b,a}$ is only 6 dB less than v_a, and this amount may not be enough to protect generator *b*.
2. Select signal generators that have large internal isolation between their frequency-determining-and-amplifying circuitry and the output circuitry. The net effect of such "buffering" is to render the output resistance of the electronic circuitry in the milliohm range or less. Then the desired output resistance is obtained by inserting a precision resistor of the appropriate value in series with the connection to the output terminal. For example, the internal voltage division between 5 mΩ output resistance of the electronic circuitry and a 50-Ω series resistor that establishes the output resistance of the signal generator produces 20 log (50.005/0.005) = 80 dB of internal isolation.

In all cases, good internal or external isolation can be achieved only by paying careful attention to minimizing other electrical and magnetic coupling paths by resorting to such things as shielding, bypassing, filtering, component placement and orientation, lead dress, grounding, shortest-distance wiring, and prayer.

† The word *buffer* is often used in situations where it is desirable to reduce the possibility of interaction between different components. For example, a buffer amplifier is used between a crystal oscillator and the first power amplifier in a transmitter.

Example 1-5

(a) Design the resistive mixer of Fig. 1-11 to combine the outputs of signal generators having $R_a = 50\ \Omega$ and $R_b = 300\ \Omega$ in a load resistor $R_L = R_a = 50\ \Omega$.
(b) Calculate the input resistance of each port.
(c) Calculate the isolation in dB between the generators.
(d) Calculate the dB loss from each generator terminal voltage to the output load voltage.

Solution (a) The condition $R_b > \frac{3}{4}R_a$ is satisfied. Therefore, from (1-53),

$$R_1 = \frac{50^2}{4(300) - 50} = 2.17\ \Omega$$

and from (1-54),

$$R_2 = 300\left[\frac{4(300) - 3(50)}{4(500) - 50}\right] = 273.91\ \Omega$$

(b) The input resistance of port a is

$$R_{\text{in},a} = 2.17\left[\frac{(52.17)(573.91)}{52.17 + 573.91}\right] = 50\ \Omega$$

which matches the 50-Ω generator as expected. The input resistance of port b is

$$R_{\text{in},b} = 273.91 + \frac{52.17}{2} = 300\ \Omega$$

which matches the 300-Ω generator as expected. The output resistance of the mixer is

$$R_{\text{out}} = 2.17 + \frac{(52.17)(573.91)}{52.17 + 573.91} = 50\ \Omega$$

which matches the load resistor as expected.
(c) From (1-57), $v_{b,a} = v_a/2$, therefore

$$20 \log \frac{v_a}{v_{b,a}} = 20 \log 2 = 6\ \text{dB}$$

Thus, this resistive network attenuates the output voltage of generator *a* by 6 dB at the terminals of generator *b*. From (1-59), $v_{a,b} = v_b R_b/(2R_a) = v_b 300/100 = 3v_b$. Therefore

$$20 \log \frac{v_b}{v_{a,b}} = 20 \log 3 = 9.54 \text{ dB}$$

Thus, this resistive network attenuates the output of generator *b* by 9.54 dB at the terminals of generator *a*. Whether 6 dB and 9.54 dB are adequate depends on the internal design of the generators, and that property is not specified in this example.
(d) From (1-60),

$$v_o = v_a\left(1 - \frac{50}{600}\right) + v_b \frac{50}{600} = 0.917v_a + 0.0833v_b$$

The resistive network attenuates v_a by 20 log (1/0.917) = 0.756 dB and v_b by 20 log (1/0.835) = 21.584 dB at the output terminals. If v_b is made 0.917/0.0833 = 11.01 times larger (or 21.584 − 0.756 = 20.83 dB) in amplitude than v_a, then both terms that comprise v_o will have the same amplitude.

1-6.2 The Hybrid Combiner

The possibility of undesirable interaction between the frequency-determining and amplifying circuitry of the signal generators is probably the main objection to using the resistive mixer of Fig. 1-11, whereas its suitability for combining the outputs of signal generators having different output resistances is an advantage. The isolation between signal generators having identical output resistances can be increased considerably by the bridge-type *hybrid combiner* (so named because the bridge circuit uses a mixture of resistors and transformer windings) shown in Fig. 1-13. The signal generators connected to the input ports *a* and *b* each have output

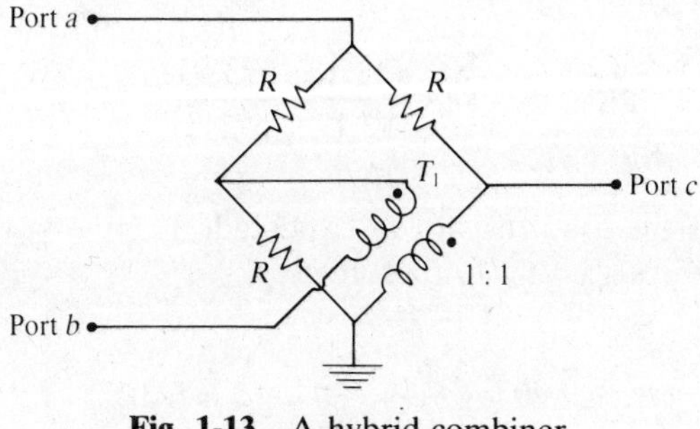

Fig. 1-13 A hybrid combiner.

resistance R. If one signal generator has a smaller output resistance than the other, add sufficient external series resistance to the smaller value to make the sum equal to the larger value, which is taken to be equal to R. The load resistance at port c and the bridge resistors are each equal to R.

Transformer T_1 can be made with about 10 bifilar turns of a small-size enamel wire on a high-permeability ($\mu_r \geq 125$) ferrite toroid core about 3/8″ OD (a bifilar construction is composed of two insulated wires side-by-side wound simultaneously as one). The bifilar construction assures that the dots are at the same end of the bifilar winding. As usually defined, when current enters the dotted terminal of one winding, the voltage induced in the other winding is positive at its dotted end. The turns ratio of T_1 is 1:1.

Assuming an ideal transformer and no stray capacitances that couple signals here and there, the hybrid combiner prevents any of the output of either signal generator from appearing across the other. Ideally, infinite isolation between the signal generators is possible. Actually, with practical devices, isolation on the order of 30 dB is obtained.

An analysis of the operation when only source voltage V_a is applied to port a uses the circuit of Fig. 1-14. Loop currents have been chosen so that there are equal currents in the 1:1 transformer windings as required. The load winding has the voltage $(I_2 + I_3)R$ across it, and so that voltage also appears across the other winding. Applying KVL around the three loops yields

$$\begin{aligned} 3RI_1 - RI_2 + RI_3 &= V_a \\ -RI_1 + 3RI_2 + RI_3 &= 0 \\ RI_1 + RI_2 + 3RI_3 &= V_a \end{aligned} \tag{1-61}$$

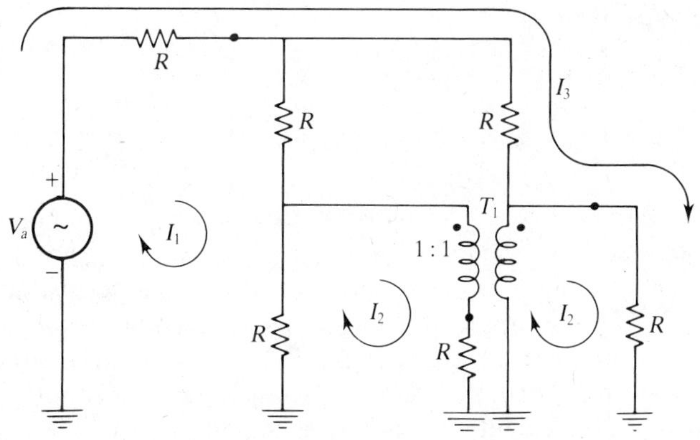

Fig. 1-14 Possible loop currents with a signal generator at port a and the other ports terminated in R.

Solve (1-61) for the currents to obtain

$$I_1 = \frac{V_a}{4R}, \qquad I_2 = 0, \qquad I_3 = \frac{V_a}{4R}$$

The fact that $I_2 = 0$ means that the source voltage V_a does not cause any voltage to appear across the terminals of the other signal generator. This is a direct consequence of the fact that the resistors that comprise the bridge circuit also cause the bridge to be balanced. The isolation of port b from port a is complete.

The resistance seen by the voltage source V_a is

$$\frac{V_a}{I_1 + I_3} = 2R$$

Half of this is composed of the output resistance R of the signal generator and therefore the other half represents the equivalent load resistance R of the bridge circuit. Hence, the signal generator at port a delivers its available power to the hybrid combiner. The current to port c is

$$I_2 + I_3 = 0 + I_3 = \frac{V_a}{4R}$$

The total input current is

$$I_1 + I_3 = \frac{V_a}{2R}$$

The attenuation ratio θ is

$$\theta = \frac{I_1 + I_3}{I_2 + I_3} = \frac{V_a/2R}{V_a/4R} = 2$$

Consequently there is $20 \log 2 = 6$ dB of attenuation from port a to the output port e.

When a similar analysis is performed with the signal generator at port b (see Problem 1-17) it is found that port a is completely isolated from the source voltage V_b applied at port b, and that there is an attenuation of 6 dB from port b to port c. Further, when a signal generator is connected to port c (see Problem 1-18) the output resistance is found to equal R and there is 6 dB of attenuation between the signal at port c and either of the other ports when properly terminated in resistor R.

1-7 LINEAR TIME-VARYING COMPONENTS

In the circuits that have just been described, it has been assumed that the components are linear and time-invariant. It is worthwhile, however, to investigate some effects that occur when resistance, inductance, and capacitance change in value from instant to instant according to some controlling signal. Fortunately, Ohm's and Kirchhoff's laws and the usual passive sign conventions concerning voltage polarities and current directions also apply to time-varying circuits. Ohm's law for a time-varying resistance $R(t)$ is

$$v_R(t) = R(t)\, i_R(t) \tag{1-62}$$

The voltage $v_L(t)$ across a time-varying inductance $L(t)$ is given by the time rate of change of flux linkage $\phi(t) = L(t)\, i_L(t)$. Thus,

$$v_L(t) = \frac{d\phi(t)}{dt} = \frac{d}{dt}[L(t)\, i_L(t)] \tag{1-63}$$

$$= L(t)\frac{di_L(t)}{dt} + \frac{dL(t)}{dt} i_L(t) \tag{1-63a}$$

Also,

$$i_L(t) = \frac{1}{L(t)} \int v_L(t)\, dt \tag{1-63b}$$

The current $i_C(t)$ through a time-varying capacitance $C(t)$ is given by the time rate of change of the electric charge $q(t) = C(t)\, v_C(t)$. Thus

$$i_C(t) = \frac{dq(t)}{dt} = \frac{d}{dt}[C(t)\, v_C(t)] \tag{1-64}$$

$$= C(t)\frac{dv_C(t)}{dt} + \frac{dC(t)}{dt} v_C(t) \tag{1-64a}$$

Also,

$$v_C(t) = \frac{1}{C(t)} \int i_C(t)\, dt \tag{1-64b}$$

The similarities in the properties between time-varying inductance and time-varying capacitance are obvious.

When a circuit is constructed from some combination of linear time-invariant and linear time-varying components, the output signal and its time derivatives will be related to the input signal and its time derivatives such that the overall circuit is linear (albeit time-varying) and superposition (i.e., linear mixing) of various input signals is applicable. But the time-varying nature of some of the components introduces some significant differences from the time-invariant situations that need to be examined in detail.

1-7.1 Time-varying Resistance

Assume that a resistance $R(t)$ is composed of the sum of constant and time-varying components, that is,

$$R(t) = R + r(t) \tag{1-65}$$

Further, let the value of the time-varying component $r(t)$ be directly proportional to a controlling signal $m(t)$, so that

$$r(t) = k_r m(t) \tag{1-66}$$

where k_r represents a resistance deviation constant in ohms per unit amplitude of $m(t)$. The numerical value of the parameter k_r depends on the specific mechanism (herein unspecified) whereby the resistance is made to be time-varying. Then, if $m(t) = A \cos(\omega_1 t + \theta)$,

$$r(t) = k_r A \cos(\omega_1 t + \theta) = \Delta R \cos(\omega_1 t + \theta)$$

in which $\Delta R = k_r A$ represents the maximum deviation in the resistance from its constant value R. Hence,

$$R(t) = R + \Delta R \cos(\omega_1 t + \theta) \tag{1-67}$$

where $\Delta R \leq R$ because negative resistance is not necessarily produced by this process. Now apply a current source $i(t) = I \cos \omega_2 t$ to the time-varying resistance as shown in Fig. 1-15. Since Ohm's law applies to the time-varying case also,

$$v(t) = R(t)\, i(t) = [R + \Delta R \cos(\omega_1 t + \theta)] I \cos \omega_2 t \tag{1-68}$$

It is seen that the time-varying resistance causes the amplitude of the voltage $v(t)$ to increase and decrease at a rate and amount determined by

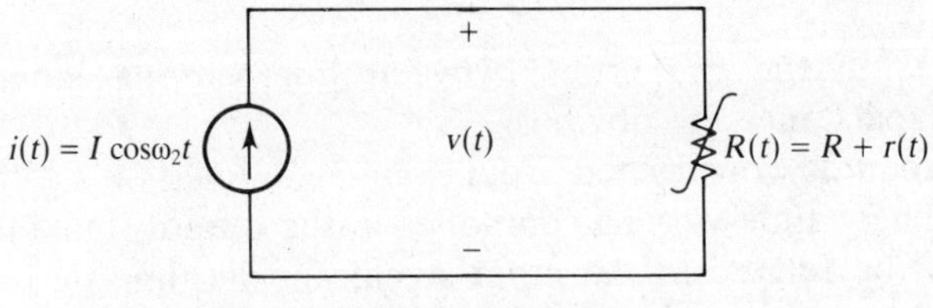

Fig. 1-15 A time-varying-resistance circuit.

how the resistance changes, and that the envelope of the voltage varies in a manner directly proportional to the signal $m(t)$. This is a form of *amplitude modulation* (AM) in which an information-bearing signal $m(t)$ varies the amplitude of the other signal called the *carrier*. Rearrange (1-68) as

$$v(t) = RI[1 + m_a \cos(\omega_1 t + \theta)] \cos \omega_2 t \tag{1-69}$$

where $m_a = \Delta R/R$ represents a *modulation index* for the AM process and is often expressed as a percentage modulation. Further, (1-69) becomes

$$\begin{aligned} v(t) &= RI \cos \omega_2 t + RIm_a \cos \omega_2 t \cos(\omega_1 t + \theta) && \text{(1-70)} \\ &= RI \cos \omega_2 t + \Delta RI \cos \omega_2 t \cos(\omega_1 t + \theta) && \text{(1-70a)} \end{aligned}$$

The first term in (1-70) gives the voltage when the resistance is constant (i.e., when $A = 0$, $m_a = 0$, and $\Delta R = 0$). The second term indicates that multiplication actually occurs between $I \cos \omega_2 t$ and $\Delta R \cos(\omega_1 t + \theta)$ b-ecause of Ohm's law. Consequently, using familiar trigonometric identities to
cast that term into another form will identify particular frequencies that are created by the multiplication process and that are actually present. These new frequencies are called *modulation products*. Therefore (1-70a) becomes

$$\begin{aligned} v(t) = {} & RI \cos \omega_2 t + \tfrac{1}{2} \Delta RI \cos \theta[\cos 2\pi(f_2 + f_1)t + \cos 2\pi(f_2 - f_1)t] \\ & - \tfrac{1}{2} \Delta RI \sin \theta[\sin 2\pi(f_2 + f_1)t - \sin 2\pi(f_2 - f_1)t] \end{aligned} \tag{1-71}$$

Two new frequencies $(f_2 + f_1)$ and $(f_2 - f_1)$ have been created that were not present in the input signals. Linear mixing in a time-invariant circuit does not produce any new frequencies.

The instantaneous power absorbed by $R(t)$ is $p(t) = v(t)\, i(t)$. Thus,

$$\begin{aligned} p(t) = {} & RI^2 \cos^2 \omega_2 t + \tfrac{1}{2} \Delta RI^2 \cos \theta[\cos(\omega_2 + \omega_1)t \cos \omega_2 t \\ & + \cos(\omega_2 - \omega_1)t \cos \omega_2 t] \\ & - \tfrac{1}{2} \Delta RI^2 \sin \theta[\sin(\omega_2 + \omega_1)t \cos \omega_2 t - \sin(\omega_2 - \omega_1)t \cos \omega_2 t] \end{aligned} \tag{1-72}$$

If $\omega_1 \neq 2\omega_2$, the average of all the terms in (1-72) is zero except the $\cos^2 \omega_2 t$ term whose average is 1/2. Therefore, if $\omega_1 \neq 2\omega_2$, the average power absorbed by $R(t)$ is

$$P = \frac{RI^2}{2}$$

But if $\omega_1 = 2\omega_2$, the $\cos(\omega_2 - \omega_1)t \cos \omega_2 t$ term in (1-72) becomes $\cos^2 \omega_2 t$ and this leads to an average power given by

$$P = \frac{RI^2}{2} + \frac{\Delta RI^2 \cos \theta}{4} = \frac{RI^2}{2}\left[1 + \frac{m_a \cos \theta}{2}\right] \tag{1-73}$$

It is seen that the average power absorbed by $R(t)$ depends on θ such that

$$P_{\max} = \frac{RI^2}{2}\left[1 + \frac{m_a}{2}\right] \tag{1-73a}$$

when $\theta = 0$, and

$$P_{\min} = \frac{RI^2}{2}\left[1 - \frac{m_a}{2}\right] \tag{1-73b}$$

when $\theta = \pi$. The average power absorbed by $R(t)$ is always positive.

Additional insight about the nature of time-varying resistance can be obtained after expressing the voltage and current in their phasor representations. First, put (1-71) in standard cosine form for the case when $\omega_1 = 2\omega_2$. Thus,

$$\begin{aligned} v(t) = RI \cos \omega_2 t &+ \tfrac{1}{2}\,\Delta RI \cos \theta[\cos 3\omega_2 t + \cos \omega_2 t] \\ &+ \tfrac{1}{2}\,\Delta RI \sin \theta[\cos(3\omega_2 t + \tfrac{1}{2}\pi) + \cos(\omega_2 t + \tfrac{1}{2}\pi)] \end{aligned} \tag{1-74}$$

There are now only two frequencies involved with $v(t)$, namely, f_2 and $3f_2$. Therefore, at frequency f_2, the phasor voltage is

$$V(f_2) = RI\underline{/0} + \tfrac{1}{2}\,\Delta RI \cos \theta\underline{/0} + \tfrac{1}{2}\,\Delta RI \sin \theta\underline{/\pi/2} \tag{1-75}$$

$$= I(R + \tfrac{1}{2}\,\Delta R \cos \theta + j\,\tfrac{1}{2}\,\Delta R \sin \theta) \tag{1-75a}$$

and at frequency $3f_2$, the phasor voltage is

$$V(3f_2) = \tfrac{1}{2}\,\Delta RI \cos \theta\underline{/0} + \tfrac{1}{2}\,\Delta RI \sin \theta\underline{/\pi/2} \tag{1-76}$$

$$= I(\tfrac{1}{2}\,\Delta R \cos \theta + j\,\tfrac{1}{2}\,\Delta R \sin \theta) \tag{1-76a}$$

Since $i(t) = I \cos \omega_2 t + 0 \cos 3\omega_2 t$, the phasor current at frequency f_2 is

$$I(f_2) = I\underline{/0} \tag{1-77}$$

and the phasor current at frequency $3f_2$ is

$$I(3f_2) = 0\underline{/0} \tag{1-78}$$

The impedance of the time-varying resistance at frequency f_2 is

$$Z(f_2) = \frac{V(f_2)}{I(f_2)} = R + \tfrac{1}{2}\,\Delta R \cos \theta + j\,\tfrac{1}{2}\,\Delta R \sin \theta \tag{1-79}$$

and the impedance at frequency $3f_2$ is

$$Z(3f_2) = \frac{V(3f_2)}{I(3f_2)} = \infty \quad \text{(i.e., an open circuit at } 3f_2) \tag{1-80}$$

Equation (1-79) indicates that the time-varying resistance has an inductive or capacitive reactive component when $\sin\theta \neq 0$, $\theta \neq 0$, or $\theta \neq \pi$. Further, the real part of the impedance $Z(f_2)$ can be constant and greater than R when $\cos\theta > 0$, or constant and less than R (but still always positive) when $\cos\theta < 0$.

1-7.2 Time-varying Capacitance

Assume that a capacitance $C(t)$ is composed of the sum of constant and time-varying components, that is,

$$C(t) = C + c(t) \tag{1-81}$$

Further, let the value of $c(t)$ be directly proportional to a controlling signal $m(t)$, or

$$c(t) = k_c m(t) \tag{1-82}$$

where k_c represents a capacitance deviation constant in farads per unit amplitude of $m(t)$. The numerical value of the parameter k_c depends on the specific mechanism (herein unspecified) whereby the capacitance is made to be time-varying. Then, if $m(t) = A\cos(\omega_1 t + \theta)$,

$$c(t) = k_c A \cos(\omega_1 t + \theta) = \Delta C \cos(\omega_1 t + \theta)$$

in which $\Delta C = k_c A$ represents the maximum deviation of the capacitance from its constant value C. Hence,

$$C(t) = C + \Delta C \cos(\omega_1 t + \theta) \tag{1-83}$$

where $\Delta C \leqslant C$ because negative capacitance is not necessarily produced by this process.

Now apply a voltage source $v(t) = V\cos\omega_2 t$ to the time-varying capacitance as shown in Fig. 1-16. The capacitor current is, from (1-64a) and using familiar trigonometric identities,

$$\begin{aligned} i(t) = -\omega_2 CV \sin\omega_2 t - \tfrac{1}{2}(\omega_1 + \omega_2)\,\Delta CV[&\cos\theta \sin(\omega_1 + \omega_2)t \\ &+ \sin\theta \cos(\omega_1 + \omega_2)t] \\ - \tfrac{1}{2}(\omega_1 - \omega_2)\,\Delta CV[&\cos\theta \sin(\omega_1 - \omega_2)t + \sin\theta\cos(\omega_1 - \omega_2)t] \end{aligned} \tag{1-84}$$

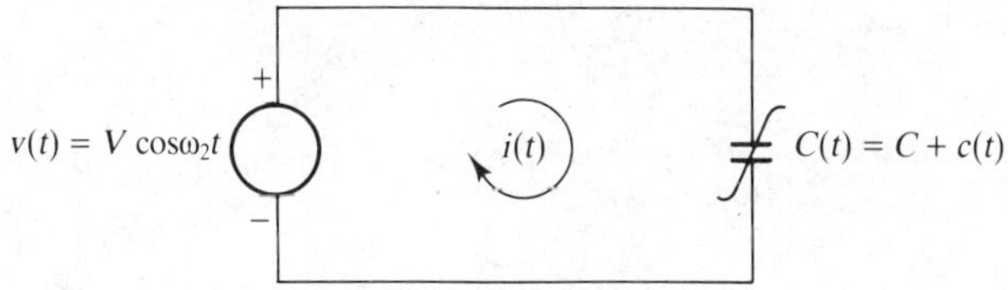

Fig. 1-16 A time-varying-capacitance circuit.

Notice that two new frequencies have been created that were not present in the input signals. But the most significant result is found by examining the average power absorbed by the time-varying capacitance to see whether the minus signs introduced by differentiation cause negative absorbed power. Thus, since the instantaneous power $p(t) = v(t)\,i(t)$, each term in (1-84) is multiplied by $V \cos \omega_2 t$. Then, if $\omega_1 \neq 2\omega_2$, the average power obtained from that expression is zero. However, if $\omega_1 = 2\omega_2$ the instantaneous power is

$$\begin{aligned} p(t) = &-\tfrac{1}{2}\omega_2 CV^2 \sin 2\omega_2 t - \tfrac{1}{2}3\omega_2\,\Delta CV^2[\cos\theta \sin 3\omega_2 t \\ &+ \sin\theta \cos 3\omega_2 t] \cos \omega_2 t \\ &- \tfrac{1}{2}\omega_2\,\Delta CV^2[\cos\theta \sin \omega_2 t + \sin\theta \cos \omega_2 t] \cos \omega_2 t \end{aligned} \tag{1-85}$$

Now, when $\omega_1 = 2\omega_2$, there is a nonzero average power because the last term in (1-85) contains $\cos^2 \omega_2 t$ whose average is $\frac{1}{2}$. Therefore, for $\omega_1 = 2\omega_2$, the average power *absorbed* by the time-varying capacitance is

$$P_c = -\tfrac{1}{4}\omega_2\,\Delta CV^2 \sin\theta \tag{1-86}$$

revealing the fact that the time-varying capacitance is a source of power if $\sin\theta > 0$. It absorbs power if $\sin\theta < 0$. The power given in (1-86) is the most negative when $\sin\theta = 1$ or $\theta = \pi/2$. Hence, the most power that can be *delivered* by the time-varying capacitance is

$$P_{c,\max} = +\tfrac{1}{4}\omega_2\,\Delta CV^2 \tag{1-87}$$

and occurs when

$$C(t) = C - \Delta C \sin 2\omega_2 t \tag{1-88}$$

and with

$$v(t) = V \cos \omega_2 t$$

The signal source that controls the time-varying capacitance is usually called the *pump*, and it is the ultimate origin of any power delivered by the capacitance or the sink of any power actually absorbed by the capacitance.

When $\omega_1 = 2\omega_2$ and $\theta = \pi/2$, the current is, from (1-84),

$$\begin{aligned} i(t) = & -\omega_2 CV \sin \omega_2 t && \text{(from } C\,dv/dt\text{)} \\ & -\tfrac{1}{2}\omega_2\, \Delta CV \cos \omega_2 t && \text{(from the difference frequency term } \cos(\omega_1 - \omega_2)t\text{)} \\ & -\tfrac{1}{2}3\omega_2\, \Delta CV \cos 3\omega_2 t && \text{(from the sum frequency term } \cos(\omega_1 + \omega_2)t\text{)} \end{aligned} \tag{1-89}$$

Express this in standard form in order next to identify phasor currents at the two frequencies f_2 and $3f_2$. Thus,

$$\begin{aligned} i(t) &= \omega_2 CV \cos(\omega_2 t + \pi/2) + \tfrac{1}{2}\omega_2\, \Delta CV \cos(\omega_2 t + \pi) \\ &\quad + \tfrac{1}{2}3\omega_2\, \Delta CV \cos(3\omega_2 t + \pi) \end{aligned} \tag{1-89a}$$

From (1-89a) the phasor current at frequency f_2 is

$$\begin{aligned} I(f_2) &= \omega_2 CV\underline{/\pi/2} + \tfrac{1}{2}\omega_2\, \Delta CV\underline{/\pi} \\ &= V(-\tfrac{1}{2}\omega_2\, \Delta C + j\omega_2 C) \end{aligned} \tag{1-90}$$

and the phasor current at frequency $3f_2$ is

$$I(3f_2) = -\tfrac{1}{2}3\omega_2\, \Delta CV + j0$$

Since the applied voltage is

$$v(t) = V \cos \omega_2 t + 0 \cos 3\omega_2 t$$

the phasor voltage at frequency f_2 is

$$V(f_2) = V\underline{/0}$$

and the phasor voltage at frequency $3f_2$ is

$$V(3f_2) = 0 + j0$$

Therefore, the admittance of the time-varying capacitance at frequency f_2 is

$$Y(f_2) = \frac{I(f_2)}{V(f_2)} = -\tfrac{1}{2}\omega_2\, \Delta C + j\omega_2 C = -G_c + jB_c$$

where $-G_c = -\omega_2\, \Delta C/2$ represents a negative conductance produced by the time-varying capacitance and $B_c = \omega_2 C$ represents the capacitive susceptance of the constant-capacitance portion of the time-varying capacitance.

The admittance of the time-varying capacitance at frequency $3f_2$ is

$$Y(3f_2) = \frac{I(3f_2)}{V(3f_2)} = \infty$$

(i.e., a short circuit at the sum frequency $3f_2$). Therefore, the sinusoidal steady-state equivalent signal circuit of a time-varying capacitance, subject to the conditions described above, can be developed as shown in Fig. 1-17a. The negative conductance $-G_c$ gives rise to the current source $G_c V(f_2)$ and this is interpreted as a voltage-controlled current source having the direction indicated with respect to the polarity of the voltage as shown in Fig. 1-17b. The time-varying capacitance is also a source of current at frequency $3f_2$, but since there is no voltage at the frequency, the equivalent signal circuit must include a short circuit at frequency $3f_2$ (i.e., the sum of the input signal frequency f_2 and the pump frequency $2f_2$). The presence of the controlled source or negative conductance distinguishes an active device from a passive device, and power amplification becomes possible. This is the basic concept that underlies a parametric amplifier. Since the operating mechanism is not temperature-dependent, the parametric amplifier can be cooled to the temperature of liquid helium (4.2 K) to minimize the effects of random noise.

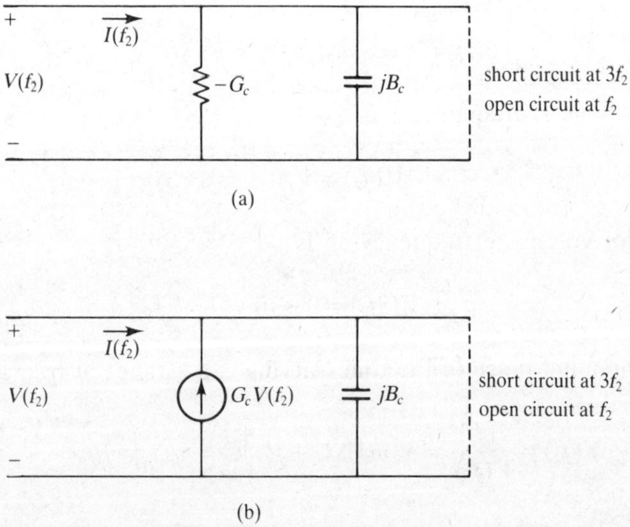

Fig. 1-17 Sinusoidal steady-state equivalent signal circuits of a time-varying capacitance subject to conditions (see text): (a) using negative conductance $-G_c$; (b) using a voltage-controlled current source $G_c V(f_2)$.

Example 1-6

The circuit shown in Fig. 1-18 provides amplification of the signal power delivered to $G_L = 2$ mS at 200 MHz from the signal source. The time-varying capacitance, from (1-88), $C(t) = C - \Delta C \sin 2\omega_2 t$, is assumed to be

$$C(t) = (38.2 - 3.66 \sin 2\pi \times 400 \times 10^6 t) \text{ pF}$$

and it remains unspecified how this $C(t)$ is obtained. It has been shown that the sinusoidal steady-state circuit equivalent of this "pumped" capacitance is the parallel combination of negative conductance $-G_c = -\omega_2 \Delta C/2 = -2.3$ mS and capacitive susceptance $j\omega_2 C = j48$ mS as given in Fig. 1-18. Further, the parallel inductance is sufficient to make $\omega_2 C = 1/(\omega_2 L)$ so that, at $f_2 = 200$ MHz.

$$\begin{aligned} Y_{\text{in}}(f_2) &= (G_1 + G_L - G_c) + j0 \\ &= (0.5 + 2 - 2.3) \text{ mS} + j0 = 0.2 \text{ mS} + j0 \\ &= G_{\text{in}} + j0 \end{aligned}$$

The circuit bandwidth is given by $BW = (G_s + G_{\text{in}})/(2\pi C) = 5$ MHz.

At frequency f_2, the input power from the signal source is

$$P_{\text{in}} = \frac{|I|^2}{2G_{\text{in}}}$$

and the output power delivered to G_L is

$$P_{\text{out}} = \frac{|I|^2 G_L}{2G_{\text{in}}^2}$$

where $|I|$ is the input phasor current magnitude at frequency f_2. The power gain is

$$G_p = \frac{P_{\text{out}}}{P_{\text{in}}} = \frac{G_L}{G_{\text{in}}} = \frac{2 \times 10^{-3}}{0.2 \times 10^{-3}} = 10$$

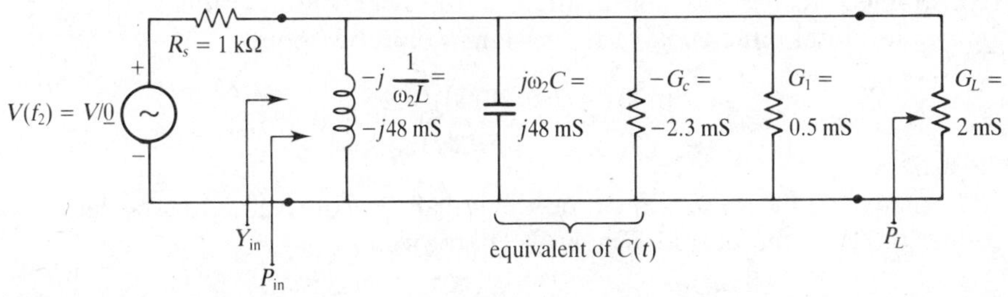

Fig. 1-18 Idealized amplifier employing time-varying capacitance $C(t)$. The series LC combination that provides the short circuit at $3f_2$ is not shown.

This is a respectable amount of power gain at 200 MHz but it is very sensitive to small variations in ΔC around the 3.66 pF value and in the conductance values because $G_{in} = 0.2$ mS is obtained from the difference between two larger numbers (i.e., 2.3 and 2.5). Larger power gains are possible if ΔC and the conductances are changed.

1-8 TIME DELAY, TIME ADVANCE, AND PHASE SHIFT

Rearrange (1-27) into the form

$$x_1(t, z) = A \cos \omega\left(t - \frac{\beta z}{\omega}\right)$$

Since $v_P = \omega/\beta$, this becomes

$$x_1(t, z) = A \cos \omega\left(t - \frac{z}{v_p}\right) \tag{1-91}$$

The quantity z/v_p represents the time required for the signal to arrive at the location z from the location $z = 0$. Thus, z/v_p is interpreted to be a delay time (or a propagation time) between the signal at point z and the signal at point $z = 0$. Further, from (1-27), βz is the phase lag in radians between the signal at point z and the signal at point $z = 0$.

When a sinusoidal signal is applied to a system, it travels through each interconnecting wire, device, component, and structure, with the corresponding phase velocity of each, and then arrives at the output at a later time. Calculating this overall delay time at a single frequency for even a simple system is not pleasant to contemplate because of its obvious complexity, and attempting the calculations for several frequencies with dispersive media (the usual situation) is conducive to marked mental anguish. Fortunately, for many systems, such calculations, even if they could be performed successfully, would not be particularly worthwhile because the physical dimensions may be such that the delay time is insignificant anyway. Hence, if the maximum value of βz in (1-27) is constrained to be no more than about $0.001 \times 2\pi$ radians, then the maximum total dimension of the system would be about

$$z_{max} \leqslant \frac{0.001 \times 2\pi}{\beta} = \frac{0.001 \times 2\pi}{2\pi/\lambda_0} = 0.001\lambda_0$$

Alternatively, the maximum time delay can be constrained to be less than about 0.001 of the period $T_0 = 1/f_0$. Therefore,

$$\frac{z_{max}}{v_p} \leqslant 0.001 T_0$$

or

$$z_{\max} \leqslant 0.001 \frac{v_p}{f_0} = 0.001\lambda_0$$

In both cases λ_0 is taken to be the free-space wavelength. The actual composite wavelength associated with the traveling signal inside the system is less than the free-space wavelength because the various individual phase velocities are less than the velocity of light, owing to intricate physical phenomena such as drift, diffusion, polarization and relaxation of electric dipoles in dielectric materials, magnetization and relaxation of magnetic dipoles and domains in magnetic materials, the effects of abrupt boundaries, fringing and stray fields, and fortuitous couplings, and numerous other processes. The 0.001 value selected above is assumed to be small enough to account for all the consequences of the above processes that are distributed throughout the system. For example, at 30 MHz, $\lambda_0 = 10\text{ m} = 1000\text{ cm}$, so if a system has $z_{\max} \leqslant 0.001 \times 1000\text{ cm} = 1\text{ cm}$, then the propagation time of a 30-MHz signal through that system is insignificant (unless it is known *a priori* that extraordinarily slow physical actions occur). When the propagation time is insignificant, the effect of an input signal is felt simultaneously throughout the system and the components can each be "lumped" or concentrated at their corresponding locations rather than being distributed throughout the system. Systems may contain only lumped components, only distributed components, or both lumped and distributed components. Any distributed component automatically leads to a propagation time, so that the output signal is delayed in time relative to the input signal.

The propagation time is zero in a lumped system, but phase shift can be present whenever the system contains energy-storage elements. There is a time delay or advance associated with the phase shift. Consider a steady-state signal at some point in a lumped system

$$x(t) = A \cos(\omega_0 t + \phi) \tag{1-92}$$

where ϕ represents the phase shift in radians relative to a reference phase at the input. The phase shift may be positive (leading) or negative (lagging) depending on the nature of the energy-storage elements contained in the system. Rearrange (1-92) into the form

$$x(t) = A \cos \omega_0 \left(t + \frac{\phi}{\omega_0}\right) \tag{1-93}$$

The quantity ϕ/ω_0 is the time shift of the signal relative to the reference. A positive sign associated with ϕ/ω_0 denotes a steady-state time advance caused by a phase lead, and a negative sign denotes a steady-state time delay caused by a phase lag. This time delay or time advance in a lumped system, with each of the two being a direct consequence of phase shift, is just as real as the propagation time in a distributed system.†

It is usual to refer to the time shift as the *phase delay* t_p. For t_p to be a positive number representing phase delay, which occurs when ϕ is negative, phase delay is defined as

$$t_p = -\frac{\phi}{\omega_0} \tag{1-94}$$

Consequently, negative phase delay is the same as positive phase advance.

1-8.1 Amplitude and Phase Distortion; Group Delay

Consider two sinusoidal signal voltages whose frequencies are nearly the same. The sum of the two signals is applied to the input of a linear system whose voltage gain (or attenuation) $A(j\omega) = |A(j\omega)|\underline{/\phi(\omega)}$ may be frequency-dependent at the frequencies of the input signals. Let one of the signals be $V_1 \cos \omega t$, represented by the phasor $V_1\underline{/0}$ at frequency ω. Let the other signal be $V_2 \cos(\omega + \Delta\omega)t$, represented by the phasor $V_2\underline{/0}$ at the frequency $\omega + \Delta\omega$. Then, by superposition, the output voltage at frequency ω is $V_1|A(j\omega)|\underline{/\phi(\omega)}$, and the output voltage at frequency $\omega + \Delta\omega$ is $V_2|A(j(\omega + \Delta\omega))|\,\underline{/\phi(\omega + \Delta\omega)}$. If $|A(j\omega)| \neq |A(j(\omega + \Delta\omega))|$, then the ratio of the amplitudes of the output voltages at the two different frequencies is not the same as the initial ratio of the amplitudes of the input voltages, and amplitude distortion is said to occur.

The time shift of the output voltage at freqeuncy ω is found by rewriting

$$V_1|A(j\omega)| \cos(\omega t + \phi(\omega)) \quad \text{as} \quad V_1|A(j\omega)| \cos \omega\left(t + \frac{\phi(\omega)}{\omega}\right)$$

Thus, this time delay is $t_{p_1} = -\phi(\omega)/\omega$.

The time shift of the output voltage at frequency $\omega + \Delta\omega$ is found by rewriting $V_2A(j(\omega + \Delta\omega)) \cos((\omega + \Delta\omega)t + \phi(\omega + \Delta\omega))$ as

$$V_2A(j(\omega + \Delta\omega)) \cos(\omega + \Delta\omega)\left(t + \frac{\phi(\omega + \Delta\omega)}{\omega + \Delta\omega}\right)$$

Thus, the time delay at this frequency is $t_{p_2} = -\phi(\omega + \Delta\omega)/(\omega + \Delta\omega)$. But, since $\Delta\omega$ is small, it is convenient to use the approximations

$$\phi(\omega + \Delta\omega) = \phi(\omega) + \frac{d\phi(\omega)}{d\omega}\Delta\omega$$

† Using a reversing switch to produce the opposite polarity of a signal voltage does not introduce phase shift in the sense intended here, because the reference is changed.

and

$$\frac{1}{\omega + \Delta\omega} = \frac{1}{\omega}\left(\frac{1}{1 + \dfrac{\Delta\omega}{\omega}}\right) \cong \frac{1}{\omega}\left(1 - \frac{\Delta\omega}{\omega}\right)$$

Therefore, this phase delay becomes

$$t_{p_2} \cong -\left[\phi(\omega) + \frac{d\phi(\omega)}{d\omega}\Delta\omega\right]\left(\frac{1}{\omega}\right)\left(1 - \frac{\Delta\omega}{\omega}\right)$$

$$\cong -\frac{\phi(\omega)}{\omega} - \frac{\Delta\omega}{\omega}\left[\frac{d\phi(\omega)}{d\omega} - \frac{\phi(\omega)}{\omega}\right]$$

where the second-order infinitesimal is neglected. The quantity known as group delay t_g is defined as

$$t_g = -\frac{d\phi(\omega)}{d\omega} \tag{1-95}$$

If the group delay is the same as the phase delay, $t_{p_1} = t_{p_2}$, and there is no phase (or time-delay) distortion, because the input signals are both delayed (or advanced) the same amount when passing through the system and they combine at the output with the same relative phase positions that they had at the input.

Thus,

$$\frac{d\phi(\omega)}{d\omega} = \frac{\phi(\omega)}{\omega}$$

Separating variables and integrating yields

$$\ln \phi(\omega) = \ln \omega + \ln k$$

or

$$\phi(\omega) = k\omega$$

where k is a proportionality constant. This demonstrates that the phase shift of the gain (or attenuation) of a linear system should be a linear function of frequency in order for zero phase (or time-delay) distortion to occur. Also, $d\phi(\omega)/d\omega = k$, meaning that the group delay must be constant for zero phase (or time-delay) distortion.

Example 1-7

(a) A 1-MHz sinusoidal signal is shifted -0.2π radians in passing through a linear system. Calculate thc phase delay. (b) Calculate the phase shift required of the same system for zero phase distortion with a second signal at 1.1 MHz.

Solution (a) The amount of phase delay is

$$t_p = \frac{\phi}{2\pi f_0} = -\frac{(-0.2\pi)}{2\pi \times 10^6} = 0.1\ \mu\text{s}$$

The phase delay is positive because the angle is negative.
(b) For zero phase distortion,

$$\frac{\phi_1}{2\pi f_1} = \frac{\phi_2}{2\pi f_2} = t_p = \text{a constant for all frequencies}$$

The phase shift required at 1.1 MHz is

$$-\phi = 2\pi \times t_p \times f_2 = 2\pi \times 0.1 \times 1.1 = 0.22\pi \text{ radians}$$

which shows that the required phase shift is proportional to frequency.

1-8.2 Amplitude Modulation and Group Delay

Additional insight into the meaning of group delay can be obtained by passing an amplitude-modulated signal through a frequency-dependent, linear, and time-invariant system such as a tuned amplifier, a bandpass filter, or other similar transmission channel. No new frequencies are created by such a system, and the output should be an amplitude-scaled, time-shifted replica of the AM input signal.

As shown briefly in Section 1-7.1, when a sinusoidal carrier of radian frequency ω_2 is amplitude-modulated by a sinusoidal modulating signal of radian frequency ω_1, new components having radian frequencies $\omega_2 + \omega_1$ and $\omega_2 - \omega_1$ are present in addition to the carrier. Therefore, if $\omega_1 \ll \omega_2$, the three frequencies are clustered closely together and constitute a frequency group. Of course, the concept of the frequency group is lost when the three frequencies are widely separated.

Using the form of (1-63), let the initial AM signal voltage be

$$v_i(t) = V(1 + m_a \cos \omega_1 t) \cos \omega_2 t \tag{1-96}$$

in which, for simplicity, $\theta = 0$. This is expandable immediately into

$$v_i(t) = V \cos \omega_2 t + \frac{m_a V}{2} \cos (\omega_2 + \omega_1)t + \frac{m_a V}{2} \cos (\omega_2 - \omega_1)t \tag{1-97}$$

If the system through which the AM signal passes is linear, then superposition can be used to determine the output $v_o(t)$ as the sum of the output voltages produced separately at each of the three different frequencies involved.

Assume that the frequency-dependent amplifier gain is $K_2\underline{/\phi(\omega_2)}$ at the carrier frequency, $K_1\underline{/\phi(\omega_2+\omega_1)}$ at the frequency $\omega_2+\omega_1$, and $K_1\underline{/\phi(\omega_2-\omega_1)}$ at the frequency $\omega_2-\omega_1$. Further, for this frequency group, represent the phase angles by the first two terms in their Taylor series expanded about ω_2. Thus,

$$\phi(\omega_2+\omega_1) \cong \phi(\omega_2) + \left.\frac{d\phi(\omega)}{d\omega}\right|_{\omega=\omega_2} \Delta\omega = \phi(\omega_2) + \frac{d\phi(\omega_2)}{d\omega}\,\omega_1$$

in which $\Delta\omega = (\omega_2+\omega_1) - \omega_2 = \omega_1$, and also

$$\phi(\omega_2-\omega_1) \cong \phi(\omega_2) - \left.\frac{d\phi(\omega)}{d\omega}\right|_{\omega=\omega_2} \Delta\omega = \phi(\omega_2) - \frac{d\phi(\omega_2)}{d\omega}\,\omega_1$$

in which $\Delta\omega = \omega_2 - (\omega_2-\omega_1) = \omega_1$. Of course, if $\Delta\omega$ is not small, additional terms in the series expansion will be present and the situation becomes more complicated. Using the above, the output voltage is

$$\begin{aligned} v_o(t) &= K_2 V \cos(\omega_2 t + \phi(\omega_2)) \\ &\quad + \frac{K_1 m_a V}{2} \cos\left[(\omega_2+\omega_1)t + \phi(\omega_2) + \frac{d\phi(\omega_2)}{d\omega}\,\omega_1\right] \\ &\quad + \frac{K_1 m_a V}{2} \cos\left[(\omega_2-\omega_1)t + \phi(\omega_2) - \frac{d\phi(\omega_2)}{d\omega}\,\omega_1\right] \end{aligned}$$

which can be rearranged as

$$\begin{aligned} v_o(t) &= K_2 V \cos(\omega_2 t + \phi(\omega_2)) \\ &\quad + \frac{K_1 m_a V}{2} \cos\left[(\omega_2 t + \phi(\omega_2)) + \left(\omega_1 t + \frac{d\phi(\omega_2)}{d\omega}\,\omega_1\right)\right] \\ &\quad + \frac{K_1 m_a V}{2} \cos\left[(\omega_2 t + \phi(\omega_2)) - \left(\omega_1 t + \frac{d\phi(\omega_2)}{d\omega}\,\omega_1\right)\right] \end{aligned}$$

But $\frac{1}{2}\cos(A+B) + \frac{1}{2}\cos(A-B) = \cos A \cos B$. Hence, the above becomes

$$\begin{aligned} v_o(t) &= K_2 V \cos(\omega_2 t + \phi(\omega_2)) \\ &\quad + K_1 m_a V \cos\left[\omega_1 t + \frac{d\phi(\omega_2)}{d\omega}\,\omega_1\right] \cos[\omega_2 t + \phi(\omega_2)] \\ &= K_2 V \left\{1 + \frac{K_1 m_a}{K_2} \cos\left[\omega_1 t + \frac{d\phi(\omega_2)}{d\omega}\,\omega_1\right]\right\} \cos[\omega_2 t + \phi(\omega_2)] \end{aligned} \tag{1-98}$$

or, finally,

$$v_o(t) = K_2 V\left\{1 + m'_a \cos \omega_1\left[t + \frac{d\phi(\omega_2)}{d\omega}\right]\right\} \cos \omega_2\left[t + \frac{\phi(\omega_2)}{\omega_2}\right] \quad (1\text{-}99)$$

where $m'_a = K_1 m_a / K_2$ is the modulation index of the output voltage. The output voltage expressions (1-98) and (1-99) are in exactly the same forms as the AM input voltage expressions (1-96) and (1-97), but there are some important differences that need to be mentioned.

1. The carrier amplitude is multiplied by the gain constant K_2, and the carrier undergoes a phase delay $t_p = -\phi(\omega_2)/\omega_2$ as expected. However, the steady carrier itself contains no message information.
2. The envelope is subjected to a group delay (or envelope delay) $t_g = -d\phi(\omega_2)/d\omega$. Since the envelope is the information-bearing part of the modulated signal, the group delay is the true signal delay rather than the phase delay.
3. The modulation index $m'_a = K_1 m_a / K_2$ of the output voltage differs from the initial modulation index m_a of the input voltage whenever $K_1 \neq K_2$. Thus, m'_a could be larger or smaller than m_a depending on the shape of the amplifier gain characteristic.

PROBLEMS

1-1 Sketch the resultant waveforms for each of the following.

(a) $x_a(t) = 20 \cos 20\pi t + 5 \cos 400\pi t$

(b) $x_b(t) = 5 \cos 20\pi t + 20 \cos 400\pi t$

(c) $x_c(t) = 5 \cos 20\pi t + 5 \cos 400\pi t$

1-2 For each of the following that is periodic, find the period, the fundamental frequency, and the harmonic numbers.

(a) $x_a(t) = 1.7 \sin (2\pi \times 10t - 23°) - 3.4 \cos (2\pi \times 10.2t + 71°)$

(b) $x_b(t) = 2 \cos 2\pi \times 8.1t - 3 \sin 2\pi \times 9.6t + 4 \cos 2\pi \times 10.4t$

(c) $x_c(t) = -3\pi \sin 2\pi \times 100t + 2 \cos 2\pi(100 + \pi)t$

(d) $x_d(t) = 2 \cos 2\pi^2 \times 10t - 3e \sin 2\pi \times 13t$

(e) $x_e(t) = \sqrt{\pi} \cos 2\pi \times 10et + \cos 2\pi \times 27t$

(f) $x_f(t) = \sqrt{2} \cos 2\pi \times 10\sqrt{3t} + \sqrt{3} \sin (2\pi \times 15\sqrt{3t} - 26°)$

(g) $x_g(t) = 10\pi \sin 2\pi \times 100\sqrt{2t} - 10e \cos 2\pi \times 200\sqrt{3t}$

1-3 (a) Sketch the beats that are produced by each of the following combinations of frequencies for equal-amplitude sinusoidal signals. (b) Determine the beat period, beat frequency, and whether the beat period equals the true period or how many beat periods equals the true period.

(1) $f_2 = 900$ Hz, $f_1 = 800$ Hz
(2) $f_2 = 9.3$ Hz, $f_1 = 9.0$ Hz
(3) $f_2 = 13.14$ kHz, $f_1 = 10$ kHz
(4) $f_2 = 900$ Hz, $f_1 = 850$ Hz
(5) $f_2 = 102$ kHz, $f_1 = 100$ kHz
(6) $f_2 = 456$ kHz, $f_1 = 454$ kHz

1-4 Plot the normalized envelope and the function $F(t)$ over one cycle of the beat frequency if:

(a) $x_a(t) = 4 \cos 2\pi \times 500t + 2 \cos 2\pi \times 450t$

(b) $x_b(t) = \cos 2\pi \times 500t + 10 \cos 2\pi \times 450t$

1-5 (a) Five cosine functions whose maximum amplitudes are each $2V$ and whose frequencies differ by 100 Hz are grouped around an average frequency of 2000 Hz. Sketch the spectrum and plot the envelope $E(t)$. (b) Repeat the above using eight cosines instead of five.

1-6 For each of the following, determine the conditions required for $x(t)_{\max}$ to be $V_1 + V_2$ and for $x(t)_{\min}$ to be $-(V_1 + V_2)$. Show with examples whether $x(t)_{\max} = V_1 + V_2$ and $x(t)_{\min} = -(V_1 + V_2)$ can occur together at their respective instants. Assume that $f_2 > f_1$ and that each $x(t)$ is periodic.

(a) $x_1(t) = V_1 \cos 2\pi f_1 t - V_2 \cos 2\pi f_2 t$

(b) $x_2(t) = V_1 \sin 2\pi f_1 t - V_2 \sin 2\pi f_2 t$

(c) $x_3(t) = V_1 \cos 2\pi f_1 t + V_2 \sin 2\pi f_2 t$

1-7 For each of the following, calculate the period of $x(t)$ and, if $x(t)_{\min} = -x(t)_{\max}$, determine their values and where they occur in a period of $x(t)$.

(a) $x_1(t) = 2 \cos 2\pi 21t + 3 \cos 2\pi 93t + 4 \cos 2\pi 273t$

(b) $x_2(t) = 2 \sin 2\pi 21t + 3 \sin 2\pi 93t + 4 \sin 2\pi 273t$

1-8 A tone synthesizer can give listeners the *impression* that very low-frequency tones are present when in fact such tones may not be physical entities at all. Suppose the loudspeakers of a certain synthesizer can reproduce frequencies down to 40 Hz, but you wish to create the *impression* that tones below 40 Hz are reproduced by the synthesizer. Calculate the fundamental frequency when two equal-amplitude sinusoids are added having frequencies of 40 Hz and:

(a) $\frac{3}{2} \times 40$ Hz

(b) $\frac{6}{5} \times 40$ Hz

(c) $\frac{21}{20} \times 40$ Hz

1-9 Show that $x_2(t, x) = B \cos(\omega t + \beta z)$ can be interpreted as a wave motion in the $-z$ direction.

1-10 The phase velocity of a wave in a certain medium is related to wavelength by

$$v_p = 3(1 - e^{-\lambda/100}) \times 10^8 \text{ m/s}$$

(a) Sketch v_p vs. λ

(b) Calculate the group velocity at $\lambda = 60$ m

1-11 The two voltage waves in (1-48) have $V_2 = -V_1$. Sketch the standing waves for several instants of time and compare with the result of Case A in Section 1-5.2.

1-12 Sketch the envelope produced by the linear mixing of waves in (1-48) when (a) $V_1 = 3V_2$; (b) $V_2 = 4V_1$.

1-13 Design the resistive mixers of Fig. 1-11 for each of the following.

(a) $R_a = R_b = R_L = 50\ \Omega$

(b) $R_a = R_b = R_L = 300\ \Omega$

(c) $R_a = 50\ \Omega,\quad R_b = 90\ \Omega,\quad R_L = 50\ \Omega$

In each design, check the resistive match at each port, calculate the isolation in dB between the signal generators, calculate the attenuation in dB between each input terminal voltage and the corresponding output voltage, and obtain the final expression for the output voltage in terms of each input terminal voltage to show the linear mixing process.

1-14 Design a resistive mixer to combine the outputs of three 50-Ω signal generators in a 50-Ω load resistor. Compare the isolation and attenuation numbers to the results of Problem 1-13(a).

1-15 Repeat Problem 1-14 using four (instead of three) 50-Ω signal generators.

1-16 What do the results of Problem 1-14 and 1-15 suggest about the possible design of resistive mixers having specified isolation and attenuation for mixing outputs of only two 50-Ω signal generators? Give an example.

1-17 Repeat the analysis of the hybrid combiner with the signal generator connected to port *b*.

1-18 Repeat the analysis of the hybrid combiner with the signal generator connected to port *c*.

1-19 The Q of a component is defined from its impedance at a certain frequency to be the ratio of the magnitude of its series reactance to its series resistance. Using (1-79) for the impedance of a time-varying resistance, determine:

(a) the value of θ that maximizes the Q at frequency f_2

(b) the maximum Q when $R = 1\ \text{k}\Omega$ with $\Delta R = 200\ \Omega$ and $\Delta R = 800\ \Omega$.

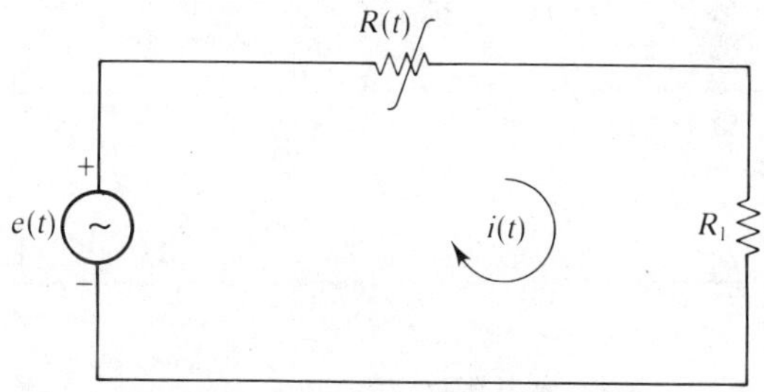

Fig. P1-20 Circuit for Prob. 1-20.

1-20 A time-varying resistance is used in the circuit of Fig. P1-20. Assume that $e(t) = 5 \cos 2\pi \times 10^4 t$ V, $R(t) = 100 + 10 \cos 2\pi \times 10^3 t\ \Omega$, and $R_1 = 100\ \Omega$. Manipulate the expression for the current $i(t)$ so that the approximation

$$\frac{1}{1+x} \cong 1 - x$$

for small x, can be used. Calculate: (a) the average power absorbed by R_1. (b) m_a in percent.

1-21 The value of ΔC produced by the unspecified process in Example 1-5 is directly proportional to its controlling signal $m(t) = A \cos(\omega_1 + \theta)$, and when $A = 0.6$ V, $\Delta C = 3.66$ pF. Calculate a new value of A such that the power gain of Example 1-5 is $G_p = 100$.

1-22 Derive the equations for P_{in}, P_{out}, and G_p used in Example 1-5.

1-23 Perform a development similar to that in Section 1-7.2 but starting with a time-varying inductance $L(t) = L + \ell(t)$ to determine: (a) the conditions for which the average power absorbed by $L(t)$ is negative; (b) the sinusoidal steady-state equivalent signal circuit of $L(t)$ subject to the conditions imposed in part (a).

2

NONLINEAR MIXING

2-0 INTRODUCTION

Nonlinear mixing is a generic name given to the mixing of signals in a nonlinear system or device. It is a fact that the fundamental electrical behavior of electronic devices is intrinsically nonlinear: a degree of nonlinearity is always present to some extent. Linearity is an idealization that is actually unattainable and can only be approximated in physically realizable systems and devices. We must acknowledge the existence of nonlinearity, seek to understand some of its consequences, and learn how to cope with it.

In this chapter we will become acquainted with some types of nonlinear behavior after applying elementary mathematical methods and an effort will be made to resist the temptation of dwelling overly long on the attractions of the underlying mathematical theory itself. The treatment will not be exhaustive and only selected topics having immediate application to the analog radio communications system will be considered.

2-1 SYSTEM CLASSIFICATIONS

An object, device, component, or other operating entity, as well as a process or a mental concept that generates an output signal from a possible input signal can be viewed as a *system*. Hence, a system may be as abbreviated as a single resistor, as complicated as the complete automatic guidance control system of a space vehicle, or an oscillator that generates a periodic output signal but whose separate input signal is zero.

An analog system is one that transforms an analog input signal into an analog output signal. A discrete-time system is one that transforms a discrete-time input signal into a discrete-time output signal. Mixed or hybrid systems that include both are also possible. This book will deal only with analog systems.

A system that operates equally well in either direction (i.e., either from input to output or from output to input) is said to be *bilateral.* A *unilateral* system operates in only one direction, usually taken to be from input to output.

Systems are further classified as to their causality and stability. A system is said to be *causal* if the output at any time is caused only by the input at that present time and past times, and not by any possible future values of the input. A system is said to be *stable* if a finite-amplitude input signal produces a finite-amplitude output signal. If sustained oscillation occurs when the input signal is zero, the system is an oscillator. We will assume that the analog systems discussed in this book are causal and stable unless otherwise noted.

Other important system classifications to be discussed in detail are

1. Fixed vs. time varying
2. Instantaneous vs. dynamic
3. Linear vs. nonlinear

We will let y_T represent the total instantaneous output and x_T represent the total instantaneous input. The quantities y_T and x_T are assumed to be functions of the continuous-time variable denoted by t and they may represent voltage, current, electric charge, force, displacement, or whatever other physical quantities are required by the particular analog system under consideration. Systems are often represented by an equation that specifies how the output y_T is related to the input x_T. Inspection of the system equation yields the needed information about the system classification.

2-1.1 Fixed vs. Time-varying

A system is said to be fixed (or time invariant) if a time delay or advance of the input signal causes the output signal to be time shifted the same amount, otherwise the system is time-varying. This classification is readily determined from the system equation because the system is fixed if the coefficients of y_T and x_T and their time derivatives and integrals are constants. The following are examples.

$y_T = 2x_T$	fixed
$y_T = 2(\sin t)x_T$	time varying because of sin t
$t\dfrac{dy_T}{dt} + y_T^2 = x_T$	time-varying because of the coefficient t
$3\dfrac{dy_T}{dt} = 2\dfrac{dx_T}{dt} + 1$	fixed

2-1.2 Instantaneous vs. Dynamic

A system is said to be *instantaneous* if its output y_T at each instant t depends only on its input x_T at that same instant. The accompanying conditions must be such that the effects of any possible energy-storage elements within the system are zero or insignificant; therefore the system equation does not contain any time derivatives or integrals of y_T or x_T. The input–output equations of several instantaneous systems follow.

$$y_T = 2x_T \qquad y_T = \begin{cases} 10^{-4}x_T^{3/2}, & \text{for } x_T \geq 0 \\ 0, & \text{for } x_T < 0 \end{cases}$$

$$y_T = 2x_T + 3x_T^2$$

$$y_T = 10^{-8}(e^{25x_T} - 1) \qquad y_T = 3x_T + 2$$

A dynamic system is obtained when the effects of energy storage elements within the system must be included. For example, the effect of a constant inductance or capacitance is to include a term such as $L\,di/dt$, $(1/L)\int v\,dt$, $C\,dv/dt$, or $(1/C)\int i\,dt$ in the system equation. This leads to a steady-state input–output relationship that is characterized by frequency-dependent amplitude and phase responses. Consequently, a radio receiver may have the steady-state response of an instantaneous system to an input signal at the frequency to which the receiver is tuned but of a dynamic system to off-tuned input signals.

A plot of the input–output equation for a fixed instantaneous system on rectangular coordinates with the total instantaneous input as abscissa and the total instantaneous output as ordinate is called the *static characteristic* of the system. For a dynamic system, the plot is called the *dynamic characteristic*.

There are two kinds of memory associated with analog systems: *temporal* and *sequential*. Temporal memory applies to dynamic systems that contain energy-storage elements in which the transient nature of the response to a perturbation makes the system "forget" what happened a long time ago, be insensitive to the sequence in which the perturbations occur, and "remember" only an event that occurred in the recent past. Temporal-memory durations depend on time constants. The presence of an energy-storage element accounts for the temporal-memory aspect of a dynamic system. An instantaneous system does not have any temporal memory.

An analog system that is subject to a triggering action is said to have sequential memory. For example, consider an instantaneous system composed of the series connection of a tunnel diode, a resistor, and a variable DC voltage source whose polarity is such that the diode conducts in the forward direction. Let the resistor value be such that the DC load line at a certain location intersects the diode static characteristic at three places simultaneously. As the source voltage is gradually increased from zero, the load line moves along the same branch of the diode static characteristic causing the current at the intersection of the load line to increase initially and then decrease slightly until the load line becomes tangent to the diode static characteristic. At that point, an infinitesimal further increase in the source voltage causes the intersection point to appear on an entirely different branch of the diode static characteristic. The sudden jump of the operating point from one branch to another, which results in a sudden change in the current, is called *triggering*. As the source voltage is still further increased, the current continues to increase in a way determined by the intersection of the load line with the second branch. If the source voltage is now gradually reduced, the current initially decreases, then increases slightly, and finally jumps to a higher value because triggering moves the load-line intersection point back onto the first branch of the diode static characteristic. This system exhibits a "memory" for the sequence whereby voltage was applied; therefore, the system is said to have *sequential memory*. Sequential-memory durations can be for arbitrarily long time intervals. All analog systems subject to triggering have sequential memory.

2-1.3 Linear vs. Nonlinear

A linear system is one that satisfies superposition. Superposition means that when the input is composed of the sum of several amplitude-scaled signals, then the output can be found from the superposition (i.e., the output amplitude-scaled sum) of the individual responses for each of the various input signals when all initial conditions are zero. Thus, if y_{T1} and y_{T2} are the responses to the inputs x_{T1} and x_{T2}, respectively, linearity requires that the following two properties hold.

1. **Homogeneity**

   The system equation must be homogeneous† in the sense that if the system equation is multiplied by a constant, then that constant can be associated with the signal variables in *every* term such that each signal amplitude is scaled by an amount equal to that same constant. Consequently, if the input is ax_{T1} the output is ay_{T1}, and if the input is bx_{T2} the output is by_{T2}.

2. **Additivity**

   It must be possible to add the two system equations that relate x_{T1} to y_{T1} and x_{T2} to y_{T2} and combine terms in that result to show that if the input is $x_{T1}+x_{T2}$ then the output is $y_{T1}+y_{T2}$ because the system equation is satisfied for the additive signals.

† A function $F(x, y)$ is homogeneous of degree n if $F(gx, gy) = g^nF(x, y)$. Thus, if gx and gy are substituted for x and y, respectively, in $F(x, y)$ and then g^n is factored out, the other remaining factor is the original function $F(x, y)$. For example, $F_1(x, y) = x^2 + y^2$ is homogeneous of degree $n = 2$ since $F_1(gx, gy) = (gx)^2 + (gy)^2 = g^2(x^2 + y^2) = g^2F_1(x, y)$. But $F_2(x, y) = x^2 + y^2 + 4$ is not homogeneous.

If either or both of the two properties is violated, superposition does not apply and the system is nonlinear in the variables y_T and x_T.

These two properties of a linear system may be combined into a single requirement as follows:

If the input is $ax_{T1} + bx_{T2}$ **then the output is** $ay_{T1} + by_{T2}$

We can base some general conclusions about system linearity on the system equation.

1. If any term in the system equation is nonlinear in the signal variables or their time derivatives and integrals, then the system defined by the equation will be nonlinear in those variables.
2. If every term in the system equation is linear in the signal variables or their time derivatives and integrals, then the system defined by the equation will be linear in those variables.

Consider the system equation

$$y_T = Ax_T + B(t) \tag{2-1}$$

where A is a constant that could be resistance, transfer conductance, voltage amplification, etc., as determined by the nature of the system, and $B(t)$ is an extra quantity, constant or time-varying, within the system that is independent of the signal variables. Homogeneity of the respective equations yields

$$ay_{T1} = A\,ax_{T1} + aB(t)$$

and

$$by_{T2} = A\,bx_{T2} + bB(t)$$

Add these and combine terms to obtain

$$ay_{T1} + by_{T2} = A(ax_{T1} + bx_{T2}) + (a+b)B(t) \tag{2-2}$$

But the combined homogeneity/additivity requirement applied to (2-1) means that

$$ay_{T1} + by_{T2} = A(ax_{T1} + bx_{T2}) + B(t) \tag{2-3}$$

Since (2-2) and (2-3) are not equal because of the effect of $B(t)$, superposition does not hold and therefore the system defined by (2-1) is nonlinear in the variables y_T and x_T even though the equation itself is linear.

The system defined by (2-1) is (a) time-varying if $B(t)$ is not constant, (b) instantaneous because there are no time derivatives or integrals involving x_T or y_T in the equation, and (c) nonlinear in the variables y_T and x_T because of $B(t)$.

2-2 INCREMENTAL LINEARITY

Many electrical devices require certain constant voltages or currents to establish their desired modes of operation. The word *bias* is appropriately used in connection with the manner whereby the device is predisposed, by a constant offset or bias amount, to respond in a certain way when a signal is applied to it. Terms such as "forward bias," "bias current," etc., apply to the biasing arrangement. The bias values determine the quiescent operating condition known as the Q point.

Let the total instantaneous input be given by

$$x_T = X_Q + x \tag{2-4}$$

where X_Q represents the constant bias at the input and x represents the instantaneous input signal. Also let the total instantaneous output be given by

$$y_T = Y_Q + y \tag{2-5}$$

where Y_Q represents the constant bias at the output and y represents the instantaneous output signal. The signals x and y may have nonzero average values. Substitute (2-4) and (2-5) into (2-1) for the case when $B(t) = B$, a constant. Thus,

$$Y_Q + y = A(X_Q + x) + B \tag{2-6}$$

Since there is no output signal when there is no input signal, $y = 0$ when $x = 0$. Hence, at the Q point

$$Y_Q = AX_Q + B \tag{2-7}$$

so that (2-6) becomes

$$y = Ax \tag{2-8}$$

This equation satisfies the conditions for a linear system in terms of the instantaneous signal variables y and x instead of y_T and x_T. Therefore the system defined by $y_T = Ax_T + B$ is linear in the signal variables y and x even though it is nonlinear in the total instantaneous variables y_T and x_T; the system is *incrementally linear.*

In similar fashion, when $B(t)$ is not constant, the system defined by $y_T = Ax_T + B(t)$ is found to be incrementally linear.

2-3 POWER-LAW NONLINEARITY

Consider an instantaneous unilateral system whose total instantaneous output is related to the total instantaneous input by the power law

$$y_T = \begin{cases} Kx_T^n, & \text{when } x_T \geqslant 0 \\ 0, & \text{when } x_T < 0 \end{cases} \tag{2-9}$$

where K and the exponent n have numerical values that depend on the internal physical properties of the system. Again, let

$$x_T = X_Q + x \qquad \text{and} \qquad y_T = Y_Q + y$$

Substitute these into (2-9) to obtain

$$Y_Q + y = K(X_Q + x)^n = KX_Q^n\left(1 + \frac{x}{X_Q}\right)^n \tag{2-10}$$

So long as $x < X_Q$, the condition $x_T < 0$ can never occur.

Using the binomial expansion,†

$$(1+b)^n = 1 + nb + \frac{n(n-1)}{2!}b^2 + \frac{n(n-1)(n-2)}{3!}b^3 + \cdots + \frac{n!}{(n-k)!\,k!}b^k + \cdots \tag{2-11}$$

for $b^2 < 1$ and the facts that $y = 0$ for $x = 0$, and $Y_Q = KX_Q^n$, (2-10) becomes

$$Y_Q + y = Y_Q\left[1 + n\left(\frac{x}{X_Q}\right) + \frac{n(n-1)}{2!}\left(\frac{x}{X_Q}\right)^2 + \frac{n(n-1)(n-2)}{3!}\left(\frac{x}{X_Q}\right)^3 + \cdots\right] \tag{2-12}$$

† The binomial expansion is

$$(a+b)^n = a^n + na^{n-1}b + \frac{n(n-1)}{2!}a^{n-2}b^2 + \ldots + \frac{n!}{(n-k)!\,k!}a^{n-k}b^k + \cdots$$

This series terminates when n is a positive integer or zero and then holds for all a and b. For all other real values of n, the binomial expansion is a convergent infinite series provided that $b^2 < a^2$.

for $x/x_Q^2 < 1$, which then simplifies to

$$y = Y_Q\left[n\left(\frac{x}{X_Q}\right) + \frac{n(n-1)}{2!}\left(\frac{x}{X_Q}\right)^2 + \frac{n(n-1)(n-2)}{3!}\left(\frac{x}{X_Q}\right)^3 + \cdots\right] \quad (2\text{-}13)$$

This expression relates the instantaneous input signal x to the resulting instantaneous output signal y at a Q point established by the bias values. Further, (2-13) shows a linear term in x, a squared term, a cubed term, etc., and this indicates nonlinearity between increments in y_T and x_T, that is, between the *signal* components y and x, both of which can have nonzero average values.

When the Q point has been established for a particular system, the values of k, n, X_Q, and Y_Q are fixed. Therefore (2-13) becomes the power series

$$y = Ax + Bx^2 + Cx^3 + \cdots \quad (2\text{-}14)$$

where

$$A = \frac{nY_Q}{X_Q} \quad (2\text{-}14a)$$

$$B = \frac{n(n-1)Y_Q}{2!\,X_Q^2} \quad (2\text{-}14b)$$

$$C = \frac{n(n-1)(n-2)Y_Q}{3!\,X_Q^3} \quad (2\text{-}14c)$$

and the general rule for other coefficients is obvious.

2-4 EXPONENTIAL-LAW NONLINEARITY

Some instantaneous systems have an input–output static characteristic expressed by the nonlinear exponential law over their normal operating range. Thus

$$y_T = ae^{bx_T} \quad (2\text{-}15)$$

where a and b are constants whose values depend on the internal physical properties of the system. Again let

$$x_T = X_Q + x \qquad \text{and} \qquad y_T = Y_Q + y$$

Thus, (2–15) becomes

$$Y_Q + y = ae^{b(X_Q+x)} = ae^{bx_Q}e^{bx} \quad (2\text{-}16)$$

Since $y = 0$ when $x = 0$,

$$Y_Q = ae^{bx_Q}$$

so that (2-16) becomes

$$Y_Q + y = Y_Q e^{bx}$$

But the exponential can be expanded into a power series given by

$$e^{bx} = 1 + bx + \frac{b^2x^2}{2!} + \frac{b^3x^3}{3!} + \cdots + \frac{b^kx^k}{k!} + \cdots \tag{2-17}$$

Therefore,

$$Y_Q + y = Y_Q\left(1 + bx + \frac{b^2x^2}{2!} + \frac{b^3x^3}{3!} + \cdots\right)$$

or, canceling terms,

$$y = Y_Q\left(bx + \frac{b^2x^2}{2!} + \frac{b^3x^3}{3!} + \cdots\right) \tag{2-18}$$

When the Q point has been established, the values of a, b, Y_Q, and X_Q are fixed, so that (2-18) becomes the power series

$$y = Ax + Bx^2 + Cx^3 + \cdots \tag{2-19}$$

where $A = bY_Q$, $B = (b^2/2!)Y_Q$, $C = (b^3/3!)Y_Q$ and the general rule for the other coefficients is obvious.

2-5 GENERAL NONLINEARITY AND TAYLOR SERIES

Assume that the nonlinear input–output static characteristic is given in the general functional form

$$y_T = f(x_T) \tag{2-20}$$

Expand this into a Taylor series about the operating point at $y_T = Y_Q$ and $x_T = X_Q$ to obtain

$$y_T = Y_Q + A(x_T - X_Q) + B(x_T - X_Q)^2 + C(x_T - X_Q)^3 + \cdots + N(x_T - X_Q)^n + \cdots \tag{2-21}$$

where as many terms can be used as the amount of curvature demands. The coefficient A is found by differentiating (2-21) with respect to x_T to obtain

$$\frac{dy_T}{dx_T} = 0 + A + 2B(x_T - X_Q) + 3C(x_T - X_Q)^2 + \cdots \tag{2-22}$$

Then evaluate dy_T/dx_T at the Q point where $x_T = X_Q$. Thus,

$$A = \left.\frac{dy_T}{dx_T}\right|_{x_T = X_Q} \tag{2-23}$$

equals the slope of the static characteristic at the operating point.

Coefficient B is found by differentiating (2-22) with respect to x_T, giving

$$\frac{d^2y_T}{dx_T^2} = 0 + 0 + 2B + 2(3)(C)(x_T - X_Q) + \cdots \tag{2-24}$$

which, when evaluated at $x_T = X_Q$ yields

$$B = \left.\frac{1}{2}\frac{d^2y_T}{dx_T^2}\right|_{x_T = X_Q} \tag{2-25}$$

In similar manner, the general coefficient is found to be

$$N = \left.\frac{1}{n!}\frac{d^ny_T}{dx_T^n}\right|_{x_T = X_Q} \tag{2-26}$$

Thus, (2-21) becomes

$$y = Ax + Bx^2 + Cx^3 + \cdots + Nx^n + \cdots \tag{2-27}$$

in which $y = y_T - y_Q$ and $x = x_T - X_Q$ represent the instantaneous signal components.

When $y_T = f(x_T)$ has continuous derivatives with respect to x_T up to the nth order only, and all higher order derivatives are zero, the series (2-27) terminates with the Nx^n term. If continuous derivatives of all orders exist, then (2-27) is an infinite series. Consequently, any nonlinear instantaneous system whose static characteristic yields derivatives evaluated at an operating point will have an output signal component y related to the input signal component x by the power series (2-27), and y can have an average value that depends on x.

Whenever the input–output static characteristic is available as a known mathematical expression, then the numerical values of the various coefficients in (2-27) can be calculated using (2-23)–(2-26).

2-5.1 Linearization

In (2-27), if $|x|$ can purposely be maintained sufficiently small, the higher-power terms will become negligible in amplitude compared to the linear term. Hence, the nonlinearity becomes essentially linear for small signals. To make the amplitude of the squared term less than 1 percent of the amplitude of the linear term requires that

$$\left|\frac{Bx^2}{Ax}\right| < 0.01$$

from which is obtained

$$|x| < 0.01 \frac{|A|}{|B|} \tag{2-28}$$

Substituting the A and B values for the power-law nonlinearity with $n = 1.5$, the squared term is considered to be negligible if $|x| < 0.04X_Q$. The amplitude of the cubed term is then found to be 0.67×10^{-2} percent of the amplitude of the linear term, and so it also is negligible. Consequently, in this case with $n = 1.5$, if $|x| < 0.04X_Q$, the nonlinearity approaches the linear relationship

$$y \cong Ax \tag{2-29}$$

for the signal components and therefore incremental linearity is closely approximated. In fact, any improved degree of approximation to incremental linearity can be achieved merely by limiting the input signal to an even smaller amplitude. This is the basic principle underlying the linear small-signal amplifying property of biased active nonlinear devices. Then, as the device is subjected to increasing amplitudes of input signals, the effects of the nonlinear terms become more significant.

The nonlinear system itself provides the operation of raising x_T to the exponent n in (2-9) or generating the functional form in (2-20). The energy required for that process comes from the sources of the signals and of the biases and from reduced system loss. Therefore, any component produced from the power-series representation (2-12), (2-19), or (2-27), actually exists and can be measured and is not just an abstract mathematic invention.

2-6 STATIC AND DYNAMIC CHARACTERISTICS

Typical static characteristics are shown in Fig. 2-1. In each case the static characteristic is a single-valued function of the total instantaneous input variable x_T. Suppose the system represented by Fig. 2-1c is subjected to a sinusoidal signal x superimposed on the bias X_Q at an operating point Q as

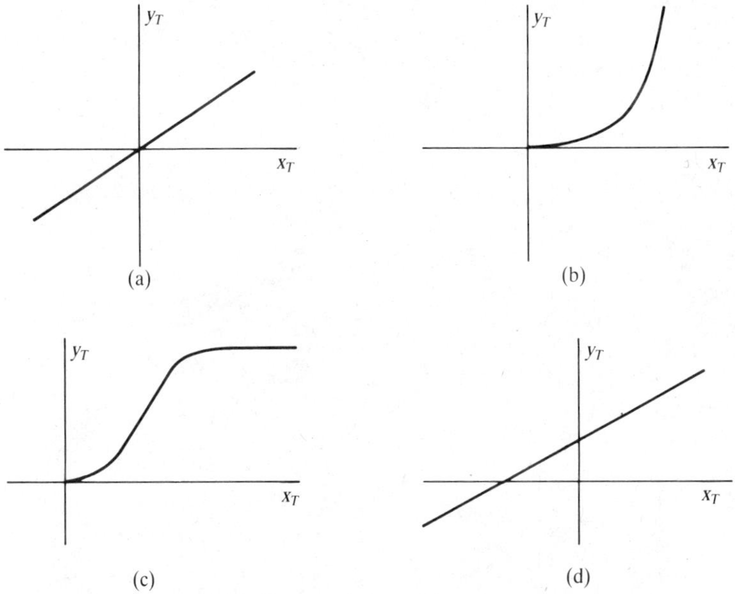

Fig. 2-1 Typical characteristics of instantaneous systems.

in Fig. 2-2a. The origin of the axes for the signals x and y is at the Q point, so that the *signal* operation alone can be depicted as in Fig. 2-2b. Let the input signal be $x = X_0 \cos \omega_0 t$ and divide its time base into segments such as at instants where the instantaneous amplitudes are zero, half-maximum, and maximum magnitudes marked as points 1 through 9 on the input signal waveform and at any other points that may help the graphical constructions. Projecting each of these points upward onto the static characteristic and across yields the corresponding instantaneous points of the output signal waveform. The path of operation of the system is up and down along the static characteristic.

Observe that whenever the portion of the static characteristic upon which the input projections are made is not a straight line, the output signal waveform differs from the input signal waveform. The degree of dissimilarity between the two waveforms is a function of the nonlinearity of the static characteristic about the Q point, and of the amplitude of the input signal.

Now suppose that the steady-state output waveform is delayed by 45° as in Fig. 2-3 because of some energy-storage effect within the dynamic system. What does the dynamic characteristic look like? This question is answered by projecting the amplitudes of both signals at equal instants across and upward until corresponding projections intersect as shown in Fig. 2-3. The closed-loop figure thus obtained is the *dynamic* characteristic for the waveforms shown. In this case, the path of operation is in a counterclock-

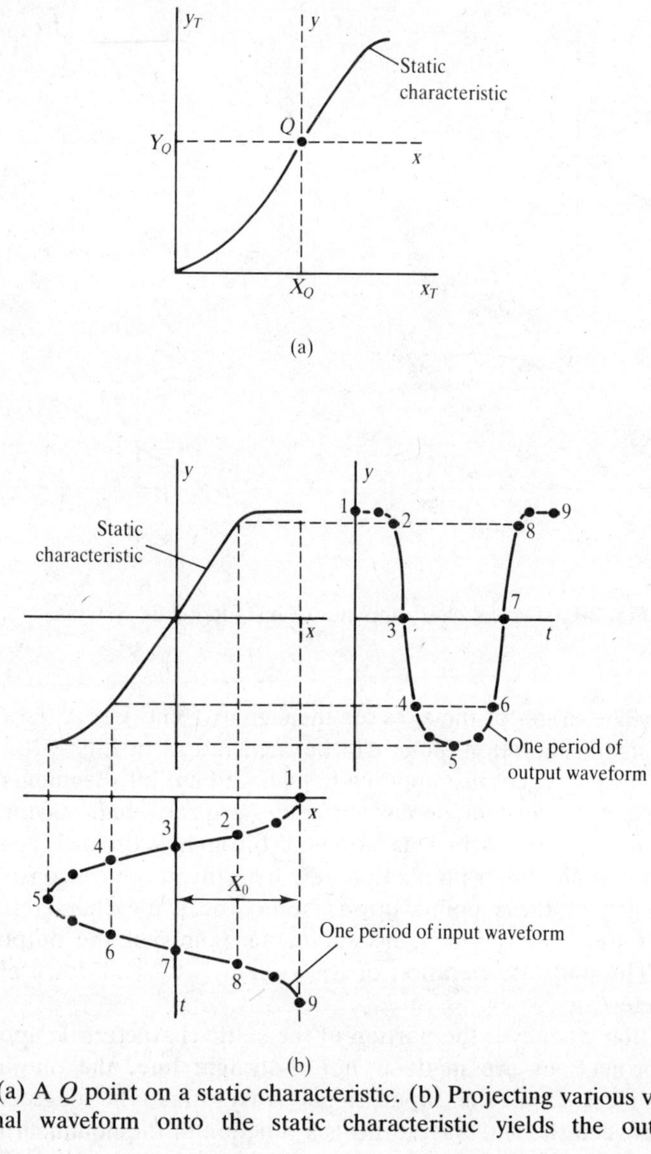

Fig. 2-2 (a) A Q point on a static characteristic. (b) Projecting various values of an input signal waveform onto the static characteristic yields the output signal waveform.

wise direction around the loop. The size and shape of the dynamic characteristic depends on the amplitude and frequency of the input signal and on the amount of system nonlinearity. This process of locating intersecting projections is reminiscent of the graphical procedure used to predict Lissajous figures on an oscilloscope screen.

The variable nature of the dynamic characteristic makes it unsuitable as

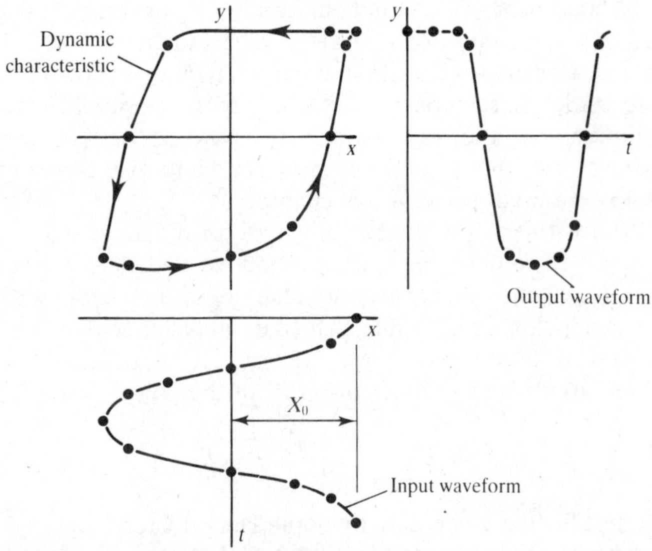

Fig. 2-3 Projecting the signal amplitudes at equal instants across and upward until corresponding projections intersect yields the closed-loop figure known as the dynamic characteristic.

the original point of departure in analyzing system response. It is usually desirable to operate an analog communications system under tuned or resonant conditions so as to effectively eliminate energy-storage actions and render the system instantaneous in its steady-state response. In this case, therefore, the closed-loop dynamic characteristic degenerates into the static characteristic. Thus, the static characteristic becomes the important property that determines whether many significant nonlinear system responses occur.

A dynamic characteristic can be observed on an oscilloscope by applying the input signal to the horizontal amplifier and the output signal to the vertical amplifier. The observed figure can be changed into a static characteristic by inserting a phase-shift network prior to the horizontal amplifier to compensate for the phase shift of the system. The negative of the amount of compensating phase shift required to produce the static characteristic is the same as the phase shift of the system.

2-7 THE METHOD OF LEAST SQUARES

The static characteristics of systems are sometimes specified by the manufacturer in numerical or graphical form on the system data sheets. But, when necessary, the user may obtain information about a particular system from laboratory measurements.

To analyze nonlinear effects mathematically, the experimental, numerical or graphical raw data must be converted to a functional form. When this kind of data is used instead of theoretical analysis to obtain the functional form of the static characteristic, the result is an empirical equation that closely approximates the raw data. The power-series functional form between y and x at the Q point is best suited to the type of nonlinear analysis that is the main topic of this chapter.

We will start with an *odd* number of values for the input signal x_d that are chosen at equal increments in amplitude above and below a zero central value at $x_d = 0$ and the corresponding values y_d of the output such as given in the first two columns of Table 2-1 (the subscript d denotes the given data). Using the method of least squares, it is possible to find the coefficients of any desired number of terms of the power series

$$y = Ax + Bx^2 + Cx^3 + Dx^4 + Ex^5 + \cdots \tag{2-30}$$

such that it best fits the given data by minimizing the total squared error. No separate constant term appears in (2-30) because $y = 0$ when $x = 0$. The number of pairs of original raw data must be greater than the number of unknown coefficients for the method of least squares to be beneficial.

The alleged error at, say, $x_d = 4$, is the difference between $y(4)$ given by the desired terms of the polynomial (2-30) and the given $y_d(4) = 0.667$. Hence, the total squared error is

$$\begin{aligned} S = {} & [y(4) - 0.667)]^2 + [y(3) - 0.6]^2 + [y(2) - 0.5]^2 + [y(1) - 0.333]^2 \\ & + [y(0) - 0]^2 + [y(-1) - (-0.333)]^2 + [y(-2) - (-0.5)]^2 \\ & + [y(-3) - (-0.6)]^2 + (y(-4) - (-0.667)]^2 \end{aligned} \tag{2-31}$$

Thus, using (2-30), this becomes

$$\begin{aligned} S = {} & [A(4) + B(4)^2 + C(4)^3 + \cdots - 0.667]^2 \\ & + [A(3) + B(3)^2 + C(3)^3 + \cdots - 0.6]^2 \\ & + [A(2) + B(2)^2 + C(2)^3 + \cdots - 0.5]^2 \\ & + [A(1) + B(1)^2 + C(1)^3 + \cdots - 0.333]^2 \\ & + [A(0) + B(0)^2 + C(0)^3 + \cdots - 0]^2 \\ & + [A(-1) + B(-1)^2 + C(-1)^3 + \cdots - (-0.333)]^2 \\ & + [A(-2) + B(-2)^2 + C(-2)^3 + \cdots - (-0.5)]^2 \\ & + [A(-3) + B(-3)^2 + C(-3)^3 + \cdots - (-0.6)]^2 \\ & + [A(-4) + B(-4)^2 + C(-4)^3 + \cdots - (-0.667)]^2 \end{aligned} \tag{2-32}$$

TABLE 2-1 Data for the Method of Least Squares

Raw Data		Calculated Values								
x_d	y_d	x_d^2	x_d^4	x_d^6	x_d^8	x_d^{10}	$x_d y_d$	$x_d^2 y_d$	$x_d^3 y_d$	$x_d^5 y_d$
4	0.667	16	256	4096	65,536	1,048,576	2.667	10.667	42.667	683.008
3	0.600	9	81	729	6561	59,049	1.800	5.400	16.200	145.800
2	0.500	4	16	64	256	1024	1.000	2.000	4.000	16.000
1	0.333	1	1	1	1	1	0.333	0.333	0.333	0.333
0	0	0	0	0	0	0	0	0	0	0
−1	−0.333	1	1	1	1	1	0.333	−0.333	0.333	0.333
−2	−0.500	4	16	64	256	1024	1.100	−2.000	4.000	16.000
−3	−0.600	9	81	729	6561	59,049	1.800	−5.400	16.200	145.000
−4	−0.667	16	256	4096	65,536	1,048,576	2.667	−10.667	42.667	683.008
Column sums										
$\sum x_d$	$\sum y_d$	$\sum x_d^2$	$\sum x_d^4$	$\sum x_d^6$	$\sum x_d^8$	$\sum x_d^{10}$	$\sum x_d y_d$	$\sum x_d^2 y_d$	$\sum x_d^3 y_d$	$\sum x_d^5 y_d$
= 0	= 0	= 60	= 708	= 9780	= 144,708	= 2,217,300	= 11.6	= 0	= 126.4	= 1690.282

The minimum total square-error is found by differentiating (2-32) with respect to A, B, C, etc., in sequence and thereafter setting those partial derivatives equal to zero. This gives linear equations in the coefficients A, B, C, etc., which must then be solved simultaneously. Thus,

$$\begin{aligned}\frac{\partial S}{\partial A} &= 2[A(4) + B(4)^2 + C(4)^3 + \cdots - 0.667](4) \\ &\quad + 2[A(3) + B(3)^2 + C(3)^3 + \cdots - 0.6](3) \\ &\quad + 2[A(2) + B(2)^2 + C(2)^3 + \cdots - 0.5](2) \\ &\quad + 2[A(1) + B(1)^2 + C(1)^3 + \cdots - 0.333](1) \\ &\quad + 2[A(0) + B(0)^2 + C(0)^3 + \cdots - 0](0) \\ &\quad + 2[A(-1) + B(-1) + C(-1)^3 + \cdots - (-0.333)](-1) \\ &\quad + 2[A(-2) + B(-2)^2 + C(-2)^3 + \cdots - (-0.5)](-2) \\ &\quad + 2[A(-3) + B(-3)^2 + C(-3)^3 + \cdots - (-0.6)](-3) \\ &\quad + 2[A(-4) + B(-4)^2 + C(-4)^3 + \cdots - (-0.667)](-4) = 0 \qquad (2\text{-}33)\end{aligned}$$

Therefore, canceling the 2s, summing by columns, and generalizing yields

$$A \sum x_d^2 + B \sum x_d^3 + C \sum x_d^4 + \cdots = \sum x_d y_d \qquad (2\text{-}34)$$

where $\sum$ indicates the sum of the data-related values as shown in Table 2-1.

In similar fashion, setting the other partial derivatives equal to zero, combining terms in corresponding columns, and generalizing gives

$$A \sum x_d^3 + B \sum x_d^4 + C \sum x_d^5 + \cdots = \sum x_d^2 y_d \qquad (2\text{-}35)$$

$$A \sum x_d^4 + B \sum x_d^5 + C \sum x_d^6 + \cdots = \sum x_d^3 y_d \qquad (2\text{-}36)$$

and an additional equation is required for each additional coefficient in (2-30) to be included. If only the A, B, and C terms in (2-30) are to be used, then (2-34), (2-35), and (2-36) terminate with the C terms and they form the three equations whose simultaneous solutions yield the optimum values for A, B, and C. If the D term in (2-30) is included, then (2-34) ends with $D \sum x_d^5$, (2-35) ends with $D \sum x_d^6$, (2-36) ends with $D \sum x_d^7$, and the fourth equation is

$$A \sum x_d^5 + B \sum x_d^6 + C \sum x_d^7 + D \sum x_d^8 = \sum x_d^4 y_d \qquad (2\text{-}37)$$

The extensions required to include even more terms in the approximating power series (2-30) should be obvious.

All of the equations thus obtained simplify considerably because $\sum x_d = 0$, $\sum x_d^3 = 0$, $\sum x_d^5 = 0$ (and similarly for *all odd* exponents m, $\sum x_d^m = 0$)

because an *odd* number of values of x_d were originally chosen at equal increments in amplitude about a zero central value. Thus, the equations become

$$A\sum x_d^2 + C\sum x_d^4 + \cdots = \sum x_d y_d \tag{2-38}$$

$$B\sum x_d^4 + \cdots = \sum x_d^2 y_d \tag{2-39}$$

$$A\sum x_d^4 + C\sum x_d^6 + \cdots = \sum x_d^3 y_d \tag{2-40}$$

$$\vdots \qquad\qquad\qquad \vdots$$

Another advantage in choosing an odd number of pairs of raw data in the manner described is that the final simultaneous equations appear as two independent sets having fewer unknowns each, and this makes them easier to solve than just one set of equations containing all of the unknown coefficients.

To use the cubic polynomial (2-30) to approximate the raw data in Table 2-1, the equations (2-38)–(2-40) become

$$60A + 708C = 11.6 \tag{2-41}$$

$$708B = 0 \tag{2-42}$$

$$708A + 9780C = 126.4 \tag{2-43}$$

from which are obtained

$$A = 0.280, \qquad B = 0, \qquad C = -0.00735$$

Therefore, the raw data is closely approximated by the power series

$$y = 0.280x - 0.00735x^3 \tag{2-44}$$

The total squared error (for up to the C term) is found to be $S_C \cong 0.01$. Equation (2-44) is plotted in Fig. 2-4 and compared to the raw data points. The match is reasonably good. It is typical of the method of least squares that the approximating curve tends to be above some data points and below other data points almost in an alternating pattern but in a way that minimizes the total squared error.

Readers should pursue the details sufficiently to convince themselves that if the E term is included in (2-30), the equations for this example become

$$60A + 708C + 9780E = 11.600$$

$$708B + 9780D = 0$$

$$708A + 9780C + 144708E = 126.400$$

$$9780B + 144708D = 0$$

$$9780A + 144708C + 2217300E = 1690.282$$

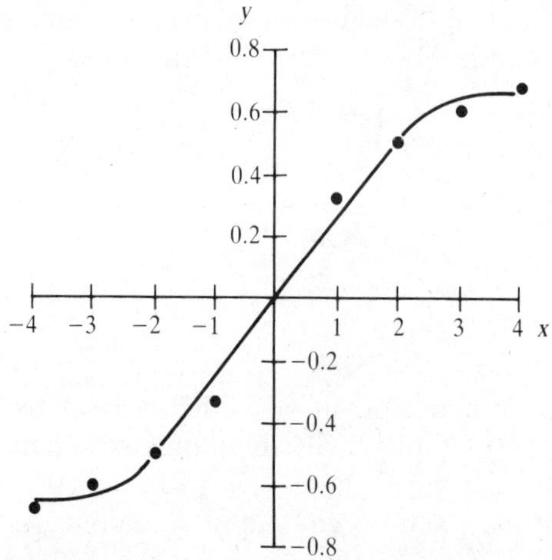

Fig. 2-4 Plot of $y = 0.280x - 0.00735x^3$ found by the method of least squares to approximate the original data shown by the dotted points.

from which are obtained by the Gauss method of successive elimination of unknowns (see Appendix B),

$$A = 0.3371, \qquad B = 0, \qquad C = -0.02209$$
$$D = 0, \qquad E = 0.00768$$

Therefore, the approximating polynomial becomes

$$y = 0.337x - 0.022x^3 + 0.00072x^5 \tag{2-45}$$

and this fits the data points much better than (2-44) does because the total squared error is found to be $S_E \cong 0.002$, which is 5 times smaller than before.

Treat the raw data as exact numbers and carry all computations to enough significant figures to give reliable results from beginning to end.

Unfortunately, the raw data are subject to errors and variations because of measurement uncertainties and such things as drifting of device parameters with age, changes caused by replacing a defective component, instability of Q points caused by an unregulated power-supply voltage, and fluctuating temperatures. It is therefore to be expected that the coefficients in (2-45) should undergo possibly significant and continual irregularities if the original data are not reliable.

2-8 NONLINEARITY PRODUCTS—SINGLE SINUSOIDAL INPUT SIGNAL

Suppose a single sinusoidal signal (sometimes referred to as a *tone* regardless of its audible quality) given by

$$x = X_0 \cos \omega_0 t = X_0 \cos \phi \tag{2-46}$$

where $\phi = 2\pi f_0 t = \omega_0 t$, is applied at the input of a system whose static characteristic is represented by the power series

$$y = Ax + Bx^2 + Cx^3 + \cdots \tag{2-47}$$

Then the output is obtained by substituting (2-46) into (2-47). Thus

$$y = AX_0 \cos\phi + BX_0^2 \cos^2\phi + CX_0^3 \cos^3\phi + \cdots \tag{2-48}$$

At this point a table of trigonometric identities can be used to expand $\cos^n \phi$ into its various harmonic components. Rather than attempt to locate a suitable table, especially easy and instructive expansion procedures will be explained.

2-8.1 Expansion of $\cos^n \phi$

The exponential equivalent† of $\cos\phi$ is

$$\cos\phi = \tfrac{1}{2}(e^{j\phi} + e^{-j\phi})$$

Applying the binomial expansion described in Section 2-3, this becomes, with $n = 2$,

$$\cos^2\phi = \frac{1}{2^2}(e^{j2\phi} + 2e^{j\phi}e^{-j\phi} + e^{-j2\phi}) = \frac{1}{2^{2-1}}\left(\frac{e^{j2\phi} + e^{-j2\phi}}{2} + \frac{2}{2}\right)$$

$$= \frac{1}{2^{2-1}}\left(\cos 2\phi + \frac{2}{2}\right)$$

which is recognized as a form of a familiar trigonometric identity. Still using the binomial expansion, now with $n = 3$:

$$\cos^3\phi = \frac{1}{2^3}(e^{j3\phi} + 3e^{j2\phi}e^{-j\phi} + 3e^{j\phi}e^{-j2\phi} + e^{-j3\phi})$$

$$= \frac{1}{2^{3-1}}\left(\frac{e^{j3\phi} + e^{-j3\phi}}{2} + 3\frac{e^{j\phi} + e^{-j\phi}}{2}\right)$$

$$= \frac{1}{2^{3-1}}(\cos 3\phi + 3\cos\phi)$$

† The sign of ϕ in *both* exponential terms is normally the same as the sign of ϕ in $\cos\phi$. The negative sign in $e^{-j\phi}$ is normally associated with the symbol j. Each exponential can be considered to be a rotating phasor with $e^{j\phi} = e^{j2\pi f_0 t}$ rotating in the counterclockwise (positive) direction and $e^{-j\phi} = e^{-j2\pi f_0 t}$ rotating in the clockwise (negative) direction even when *both* frequency f_0 and time t are positive. The concept of a "double-sided" spectrum places the negative sign with f_0 and this allows the same clockwise rotating phasor because $e^{-j2\pi f_0 t}$ is the same as $e^{j2\pi(-f_0)t}$. This invention of an abstract and physically impossible negative frequency allows significant mathematical simplifications of the exponential Fourier series and the Fourier transform.

As a last example, with $n = 4$:

$$\cos^4 \phi = \frac{1}{2^4}(e^{j4\phi} + 4e^{j3\phi}e^{-j\phi} + 6e^{j2\phi}e^{-j2\phi} + 4e^{j\phi}e^{-j3\phi} + e^{-j4\phi})$$

$$= \frac{1}{2^{4-1}}\left(\frac{e^{j4\phi} + e^{-j4\phi}}{2} + 4\frac{e^{j2\phi} + e^{-j2\phi}}{2} + \frac{6}{2}\right)$$

$$= \frac{1}{2^{4-1}}\left(\cos 4\phi + 4\cos 2\phi + \frac{6}{2}\right)$$

In the final forms of these examples (and in others of Problem 2-9), it is noted that:

1. The multiplier of the trigonometric series expansion is $1/(2^{n-1})$.
2. The highest harmonic number is the same as the exponent n and the harmonic numbers decrease by 2 for each succeeding trigonometric function.
3. When n is even the harmonic numbers are all even, and when n is odd the harmonic numbers are all odd.
4. The coefficients of the harmonic trigonometric functions come from the binomial coefficients.
5. When n is even a constant term is present, given by the central number of the binomical coefficient series divided by 2, but when n is odd the constant is absent.
6. When n is even a fundamenal frequency term is absent, but when n is odd a fundamental frequency term is present.

The binomial coefficients to be used in the expansion of $\cos^n \phi$ are quickly determined by constructing the easily remembered Pascal's triangle shown in Fig. 2-5. Start with the 1 at the top of the triangle. Then each number in every following row is calculated from the sum of the numbers just on either side of it in the preceding row. The numbers in a row are the binomial coefficients for that value of n. A separate column for n is not necessary because the second number in any row is the same as the value of n for that row. The sum of the numbers in a row is 2^n. The sum of the numbers along a diagonal, starting with 1 and ending with a number N, is equal to the number in the next row below N located along the oppositely directed diagonal through N.

When n is odd there is an even number of coefficients in the row, and when n is even there is an odd number of coefficients in the row and the central number in that row (identified by the circles in Fig. 2-5) when divided by 2, gives the constant term in the expansion.

Actually, only the left half of the triangle is necessary because of the way the exponentials combine into the cosines.

1	$n = 0$
1 1	1
1 **2** 1	2
1 3 3 1	3
1 4 **6** 4 1	4
1 5 10 10 5 1	5
1 6 15 **20** 15 6 1	6
1 7 21 35 35 21 7 1	7
1 8 28 56 **70** 56 28 8 1	8
⋮	⋮

Fig. 2-5 Pascal's triangle for $\cos^n \phi$. Each row gives the binomial coefficients for the expansion of $\cos^n \phi$. Each number is the sum of the numbers just on either side of it in the preceeding row. The circled numbers divided by 2 give the constant term when n is even.

Using these procedures the expansion for $\cos^n \phi$ is quickly obtained. Thus, for example,

$$\cos^7 \phi = \frac{1}{2^6}(\cos 7\phi + 7 \cos 5\phi + 21 \cos 3\phi + 35 \cos \phi)$$

and

$$\cos^8 \phi = \frac{1}{2^7}\left(\cos 8\phi + 8 \cos 6\phi + 28 \cos 4\phi + 56 \cos 2\phi + \frac{70}{2}\right)$$

2-8.2 Expansion of $\sin^n \phi$

The exponential equivalent of $\sin \phi$ is

$$\sin \phi = \frac{e^{j\phi} - e^{-j\phi}}{2j}$$

Then, with $n = 2$, the binomial expansion yields

$$\begin{aligned}\sin^2 \phi &= \frac{1}{2^2 j^2}(e^{j2\phi} - 2e^{j\phi}e^{-j\phi} + e^{-j2\phi}) \\ &= \frac{1}{2^{2-1}(-1)}\left(\frac{e^{j2\phi} + e^{-j2\phi}}{2} - \frac{2}{2}\right) \\ &= \frac{1}{2^{2-1}}\left(-\cos \phi + \frac{2}{2}\right)\end{aligned}$$

which is another familiar trigonometric identity. Next, still using the binomial expansion, with $n = 3$,

$$\sin^3 \phi = \frac{1}{2^3 j^3}(e^{j3\phi} - 3e^{j2\phi}e^{-j\phi} + 3e^{j\phi}e^{-j2\phi} - e^{-j3\phi})$$

$$= \frac{1}{2^{3-1}j^2}\left(\frac{e^{j3\phi} - e^{-j3\phi}}{2j} - 3\frac{e^{j\phi} - e^{-j\phi}}{2j}\right)$$

$$= \frac{1}{2^{3-1}}(-\sin 3\phi + 3\sin\phi)$$

Finally, with $n = 4$,

$$\sin^4 \phi = \frac{1}{2^4 j^4}(e^{j4\phi} - 4e^{j3\phi}e^{-j\phi} + 6e^{j2\phi}e^{-j2\phi} - 4e^{j\phi}e^{-j3\phi} + e^{-j4\phi})$$

$$= \frac{1}{2^{4-1}(1)}\left(\frac{e^{j4\phi} + e^{-j4\phi}}{2} - 4\frac{e^{j2\phi} + e^{-j2\phi}}{2} + \frac{6}{2}\right)$$

$$= \frac{1}{2^{4-1}}\left(\cos 4\phi - 4\cos 2\phi + \frac{6}{2}\right)$$

In the final forms of the these examples (and in others of Problem 2-10), it is noted that:

1. The multiplier of the trigonometric series expansion is $1/(2^{n-1})$.
2. The highest harmonic number is the same as the exponent n and the harmonic numbers decrease by 2 for each succeeding trigonometric function.
3. When n is even the harmonic numbers are all even, when n is odd the harmonic numbers are all odd.
4. The coefficients of the harmonic trigonometric functions come from the binomial coefficients but some signs are negative.
5. When n is even the expansion contains cosine terms and a constant term is present given by the central number of the binomial coefficient series divided by 2, but when n is odd the expansion contains only sine terms and the constant is absent.
6. When n is even a fundamental frequency is absent, but when n is odd a fundamental frequency term is present.

The binomial coefficients to be used in the expansion of $\sin^n \phi$ are quickly determined by constructing the easily remembered Pascal's triangle shown in Fig. 2-6. First, form the triangular array of numbers as discussed in Section 2-8.1 for $\cos^n \phi$. Next, starting with the 1 at the top of the triangle,

1	$n = 0$
1 −1	1
−1 (2) −1	2
−1 3 −3 1	3
1 −4 (6) −4 1	4
1 −5 10 −10 5 −1	5
−1 6 −15 (20) −15 6 −1	6
−1 7 −21 35 −35 21 −7 1	7
1 −8 28 −56 (70) −56 28 −8 1	8
⋮	⋮

Fig. 2-6 Pascal's triangle for $\sin^n \phi$. When n is even the expansion contains cosine terms and when n is odd the expansion contains sine terms.

place signs with the ones down the left side of the triangle in the alternating pattern $++--++--++--\cdots$. Then place signs with the numbers in each row in an alternating pattern given by either $+-+-+-+\cdots$ or $-+-+-+-+\cdots$ as determined by the sign of the first 1 in a row previously assigned. The resulting numbers in a row are the binomial coefficients and their proper signs for that value of n in $\sin^n \phi$. A separate column for n is not necessary because the absolute value of the second number in any row is the same as the value of n for that row.

Using these procedures, the expansion for $\sin^n \phi$ is quickly obtained. Thus, for example,

$$\sin^7 \phi = \frac{1}{2^6}(-\sin 7\phi + 7 \sin 5\phi - 21 \sin 3\phi + 35 \sin \phi)$$

and

$$\sin^8 \phi = \frac{1}{2^7}\left(\cos 8\phi - 8 \cos 6\phi + 28 \cos 4\phi - 56 \cos 2\phi + \frac{70}{2}\right)$$

Another way to view these expansions is to note that raising an even function such as $\cos \phi$ to any integer power yields an even function composed only of even functions in its expansion. Raising an odd function such as $\sin \phi$ to an odd integer power yields an odd function composed only of odd functions in its expansion. Raising an odd function to an even integer power yields an even function composed only of even functions in its expansion.

Returning now to the output produced by a single sinusoidal input signal as expressed by (2-48),

$$y = AX_0 \cos\phi + BX_0^2 \cos^2\phi + CX_0^3 \cos^3\phi + DX_0^4 \cos^4\phi + \cdots \qquad (2\text{-}48a)$$

This expands into

$$\begin{aligned} y = AX_0 \cos\phi &+ \frac{BX_0^2}{2}(\cos 2\phi + 1) \\ &+ \frac{CX_0^3}{2^2}(\cos 3\phi + 3\cos\phi) \\ &+ \frac{DX_0^4}{2^3}(\cos 4\phi + 4\cos 2\phi + 3) + \cdots \end{aligned} \qquad (2\text{-}49)$$

Now collect terms of like frequencies:

$$\begin{aligned} y = {} & \frac{BX_0^2}{2} + \frac{3DX_0^4}{8} + \cdots && \begin{cases}\text{zero-frequency components} \\ \text{from all even-exponent terms}\end{cases} \\ & + \left(AX_0 + \frac{3CX_0^3}{4} + \cdots\right)\cos\phi && \begin{cases}\text{fundamental frequency components} \\ \text{from all odd-exponent terms}\end{cases} \\ & + \left(\frac{BX_0^2}{2} + \frac{DX_0^4}{2} + \cdots\right)\cos 2\phi && \begin{cases}\text{2nd harmonic components from} \\ \text{even exponent terms with } n \geqslant 2\end{cases} \\ & + \left(\frac{CX_0^3}{4} + \cdots\right)\cos 3\phi && \begin{cases}\text{3rd harmonic components from} \\ \text{odd-exponent terms with } n \geqslant 3\end{cases} \\ & + \left(\frac{DX_0^4}{8} + \cdots\right)\cos 4\phi && \begin{cases}\text{4th harmonic components from} \\ \text{even-exponent terms with } n \geqslant 4\end{cases} \\ & + \text{higher harmonic terms} \end{aligned} \qquad (2\text{-}50)$$

$$= Y_0 + Y_1 \cos\phi + Y_2 \cos 2\phi + Y_3 \cos 3\phi + \cdots \qquad (2\text{-}50a)$$

Terms that result from even exponents are referred to as even-order terms and those that result from odd exponents are referred to as odd-order terms. It is noticed that (2-50) that the output contains many new harmonic components that were not present at the input. This is a direct consequence of the nonlinearity of the static characteristic. Each even-exponent term in the power series representation of the static characteristic produces a zero-frequency constant component and one or more even harmonics of the original frequency at the input, and each odd-exponent term produces a fundamental frequency component and one or more odd harmonics.

2-8.3 Harmonic Extensions

Each binomial coefficient in Pascal's triangle for $\cos^n \phi$ is given by

$$\frac{n!}{(n-k)!\,k!} \tag{2-51}$$

where n is the integer exponent and k is an integer that denotes the sequential position of a binomial coefficient in a row starting with $k = 0$ on the left end. For each zero-frequency constant component in the center of a row, it is noticed that $k = n/2$ for n even only. Therefore the binomial coefficient for each zero-frequency constant component is, from (2-51) with $k = n/2$,

$$\frac{n!}{\left(\frac{n}{2}\right)!\left(\frac{n}{2}\right)!} \qquad \text{for } n \text{ even}$$

In the expansion of $\cos^n \phi$ each such constant is divided by its respective 2^n.

For each fundamental frequency component, it is noticed that $k = (n - 1)/2$ for n odd only. Therefore the binomial coefficient for each fundamental frequency component is, from (2-51) with $k = (n - 1)/2$,

$$\frac{n!}{\left(\frac{n+1}{2}\right)!\left(\frac{n-1}{2}\right)!} \qquad \text{for } n \geqslant 1 \text{ and odd only}$$

For each second harmonic component, it is noticed that $k = (n - 2)/2$, for $n \geqslant 2$ and even only. Therefore the binomial coefficient for each second harmonic is, from (2-51) with $k = (n - 2)/2$,

$$\frac{n!}{\left(\frac{n+2}{2}\right)!\left(\frac{n-2}{2}\right)!} \qquad \text{for } n \geqslant 2 \text{ and even only}$$

In similar fashion it is found that the binomial coefficient for each pth harmonic component is

$$\frac{n!}{\left(\frac{n+p}{2}\right)!\left(\frac{n-p}{2}\right)!} \tag{2-52}$$

where $n \geqslant p$ and n is even if p if even and n is odd if p is odd. In the

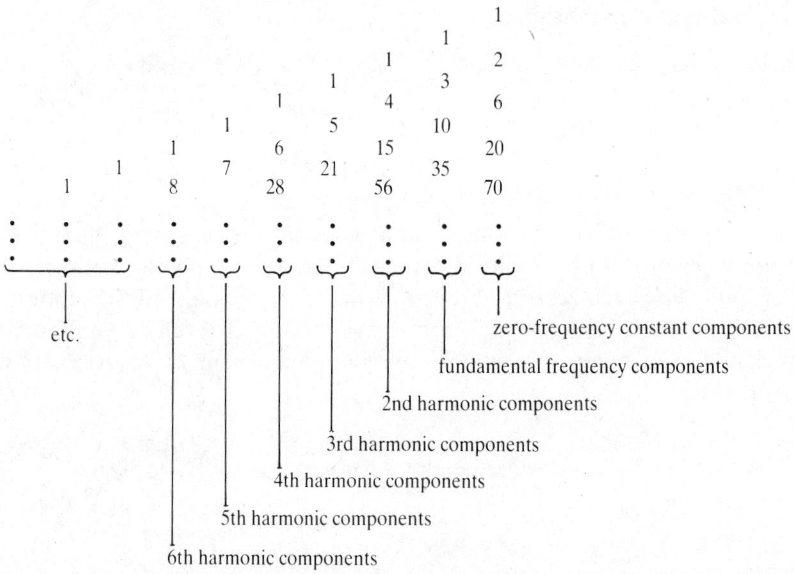

Fig. 2-7 The numbers in a column yield the binomial coefficients for components having the same frequencies that are produced by different exponents.

expansion of $\cos^n \phi$, each pth harmonic component is divided by 2^{n-1} except when $p = 0$ divide by 2^n. But (2-52) represents the binomial coefficients in a particular *column* of Pascal's triangle. The central column gives the binomial coefficients for the zero-frequency constant components ($p = 0$); the first column to the left of the central one is for the fundamental frequency components ($p = 1$); the second column to the left of the central one is for the second harmonic components ($p = 2$), and so forth for the other columns. Several harmonic components are identified column by column in the Pascal's triangle of Fig. 2-7. All of the binomial coefficients in a particular column of Pascal's triangle for $\sin^n \phi$ have the same sign.

Now we are able to extend the use of Pascal's triangle to determine the output amplitudes of all harmonic components having the same frequency but produced by different exponents. For example, if (2-47) is extended to

$$y = Ax + Bx^2 + Cx^3 + Dx^4 + Ex^5 + Fx^6 + Gx^7 + Hx^8 + Ix^9 + Jx^{10} + \cdots \tag{2-53}$$

(when all 26 letters of the alphabet are used, start over with $A_1x^{27} + B_1x^{28} + \cdots$ if necessary), then from the above considerations, the zero-frequency output component produced by all even exponents in (2-53) with $x = X_0 \cos \phi$, is

$$Y_0 = \frac{2BX_0^2}{4} + \frac{6DX_0^4}{16} + \frac{20FX_0^6}{64} + \frac{70HX_0^8}{256} + \frac{252JX_0^{10}}{1024} + \cdots \tag{2-54}$$

which can be simplified somewhat by dividing out the common factors in each term.

The fundamental frequency component produced by all odd exponents in (2-53) with $x = X_0 \cos \phi$ is

$$Y_1 \cos \phi = \left(AX_0 + \frac{3CX_0^3}{4} + \frac{10EX_0^5}{16} + \frac{35GX_0^7}{64} + \frac{126IX_0^9}{256} + \cdots \right) \cos \phi \quad (2\text{-}55)$$

The second harmonic component is

$$Y_2 \cos 2\phi = \left(\frac{BX_0^2}{2} + \frac{4DX_0^4}{8} + \frac{15FX_0^6}{32} + \frac{56HX_0^8}{128} + \frac{210JX_0^{10}}{512} + \cdots \right) \cos 2\phi \quad (2\text{-}56)$$

The third harmonic component is

$$Y_3 \cos 3\phi = \left(\frac{CX_0^3}{4} + \frac{5EX_0^5}{16} + \frac{21GX_0^7}{64} + \frac{84IX_0^9}{256} + \cdots \right) \cos 3\phi \quad (2\text{-}57)$$

and similarly for higher-order harmonic components.

2-8.4 Instantaneous and rms Characteristics

Equation (2-55) indicates that the amplitude of the fundamental output depends on odd-order nonlinearity of the instantaneous static characteristic and not on the even-order nonlinearity. If, for practical reasons, measurements are made of rms input $X_r = X_0/\sqrt{2}$ and rms fundamental output $Y_{1r} = Y_1/\sqrt{2}$, the pairs of values can be approximated in the least-squares sense by the power series

$$Y_{1r} = A'X_r + C'X_r^3 + E'X_r^5 + G'X_r^7 + \cdots \quad (2\text{-}58)$$

where $A', C', E', G', \cdots$ are the appropriate coefficients for the series (2-58) to approach the measured data.

In terms of rms input X_r and rms fundamental output Y_{1r}, (2-55) yields

$$Y_{1r} = AX_r + \frac{3}{2} CX_r^3 + \frac{5EX_r^5}{2} + \frac{35GX_r^7}{8} + \cdots \quad (2\text{-}59)$$

It is evident that the rms input–rms fundamental output characteristic defined by (2-59) is not the same as the instantaneous static characteristic given by (2-53), because of the additional fractions and because the even-order terms are missing.

When the coefficeints $A', C', E', \ldots$ are determined from measurements for the *rms* characteristic, the odd-order constants for the *instantaneous* characteristic can be found by equating like-order coefficients in (2-58) and (2-59) to obtain

$$A = A', \qquad C = \tfrac{2}{3}C', \qquad E = \tfrac{2}{5}E', \qquad G = \tfrac{8}{35}G', \qquad \ldots \tag{2-60}$$

Even-order terms in the instantaneous static characteristic have been lost.

The even-order terms can be recovered if additional measurements are made of a quantity that depends on the even exponents. It is seen in (2-54) and (2-56) that both the zero-frequency and the second-harmonic components depend on all even-order terms. Therefore, if the variation of the zero-frequency component is measured over a range of rms input values, the approximating power series is

$$Y_0 = B'X_r^2 + D'X_r^4 + F'X_r^6 + H'X_r^8 + \cdots \tag{2-61}$$

where $B', D', F', H', \ldots$ are the appropriate coefficients for the series to approach the measured data.

In terms of the rms input $X_r = X_0/\sqrt{2}$, (2-54) yields

$$Y_0 = \frac{BX_r^2}{1} + \frac{3DX_r^4}{2} + \frac{5FX_r^6}{2} + \frac{35HX_r^8}{8} + \cdots \tag{2-62}$$

Equating like-order coefficients in (2-61) and (2-62) results in

$$B = B', \qquad D = \tfrac{2}{3}D', \qquad F = \tfrac{2}{5}F', \qquad H = \tfrac{8}{35}H', \qquad \ldots \tag{2-63}$$

showing that the even-order coefficients can be recovered.

Consequently, the odd-order terms of the instantaneous static characteristic can be determined from measurements of rms input and rms fundamental output, whereas the even-order terms can be determined from the rms input and zero-frequency output measurements after applying the method of least squares in each case.

2-8.5 Coefficients of the Trigonometric Fourier Series

The amplitudes of the various harmonic components can also be determined by calculating the coefficients of the corresponding trigonometric Fourier series provided that the nonlinearity is not too complicated for the required integrations to be performed. For example, consider the static characteristic shown in Fig. 2-8, which is a typical idealization of an instantaneous system whose output amplitude is limited symmetrically by combination of the saturation and cutoff processes. The static characteristic is initially linear with a slope $m_1 = y_1/x_1$ and the output amplitude remains constant at $\pm y_1$ when $|x| > |x_1|$. Let the input signal be

$$x(t) = X_0 \cos \omega_0 t$$

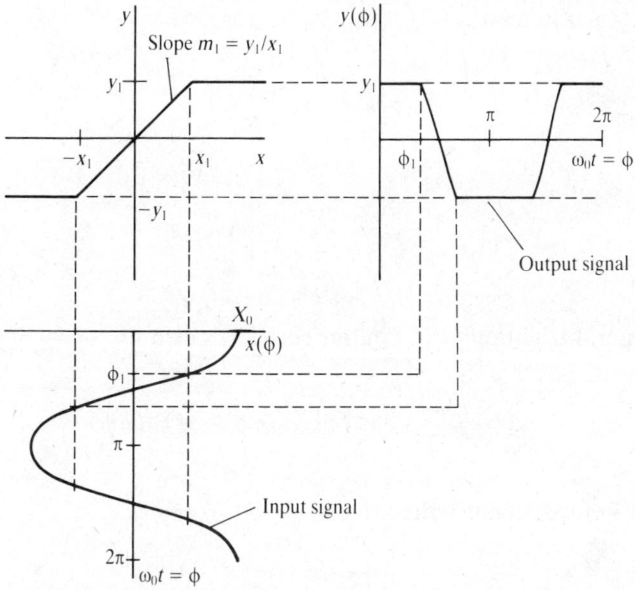

Fig. 2-8 Construction showing how a nonlinear static characteristic can cause distortion of the output signal.

or

$$x(\phi) = X_0 \cos \phi \tag{2-64}$$

where $\phi = \omega_0 t$ radians. Then the output signal is the function

$$y(t) = y(X_0 \cos \omega_0 t)$$

or

$$y(\phi) = y(X_0 \cos \phi) \tag{2-65}$$

The angle ϕ_1 that occurs at a break in the characteristic is determined at time t_1 from

$$x_1 = X_0 \cos \omega_0 t_1 = X_0 \cos \phi_1$$

where $\phi_1 = \omega_0 t_1$. Therefore

$$\cos \phi_1 = \frac{x_1}{X_0} \tag{2-66}$$

from which is obtained

$$\phi_1 = \cos^{-1}\frac{x_1}{X_0} \tag{2-67a}$$

and

$$\frac{\pi}{2} - \phi_1 = \sin^{-1}\frac{x_1}{X_0} \tag{2-67b}$$

The general trigonometric Fourier series is given by

$$y(\phi) = \frac{a_0}{2} + \sum_{n=1}^{\infty} (a_n \cos n\phi + b_n \sin n\phi) \tag{2-68}$$

where the Fourier coefficients are calculated from

$$a_n = \frac{1}{\pi}\int_{-\pi}^{\pi} y(\phi)\cos n\phi \, d\phi \tag{2-69a}$$

and

$$b_n = \frac{1}{\pi}\int_{-\pi}^{\pi} y(\phi)\sin n\phi \, d\phi \tag{2-69b}$$

Three properties of the output waveform in Fig. 2-8 simplify the Fourier series and its coefficient evaluations:

1. $y(\phi)$ has zero average, therefore $a_0 = 0$.
2. $y(\phi)$ is an even function of ϕ, therefore $b_n = 0$ and the sine terms are absent.
3. $y(\phi)$ has odd half-wave symmetry, therefore only odd harmonics are present.

Because of these properties, the Fourier series simplifies in this case to

$$y(\phi) = \sum_{n=1}^{\infty} a_n \cos n\phi, \qquad n \text{ odd} \tag{2-70}$$

and the remaining coefficients are found from

$$a_n = \frac{4}{\pi}\int_{0}^{\pi/2} y(\phi)\cos n\phi \, d\phi, \qquad n \text{ odd} \tag{2-71}$$

Therefore, for $X_0 \geqslant x_1$,

$$a_n = \frac{4}{\pi}\left[\int_0^{\phi_1} m_1 x_1 \cos n\phi \, d\phi + \int_{\phi_1}^{\pi/2} m_1 X_0 \cos \phi \cos n\phi \, d\phi\right], \qquad n \text{ odd} \quad (2\text{-}72)$$

Straightforward integrations but tedious simplifications lead to the result that, for $n = 1$ only, the output fundamental amplitude is

$$a_1 = m_1 X_0 f\left(\frac{x_1}{X_0}\right) \quad (2\text{-}73)$$

where

$$f\left(\frac{x_1}{X_0}\right) = \frac{2}{\pi}\left[\sin^{-1}\frac{x_1}{X_0} + \frac{x_1}{X_0}\sqrt{1 - \left(\frac{x_1}{X_0}\right)^2}\right], \qquad X_0 \geqslant x_1 \quad (2\text{-}73a)$$

$$= 1, \qquad X_0 \leqslant x_1 \quad (2\text{-}73b)$$

When $X_0 \leqslant x_1$ so that $a_1 = m_1 X_0$ because the system is operating in its linear region, all harmonics are zero and $f(x_1/X_0)$ equals 1 as in (2-73b).

For $n > 1$ and odd, the higher-order harmonic amplitudes are given by

$$a_n = \frac{2m_1 X_0}{\pi n}\left[\frac{\sin (n+1)\phi_1}{n+1} - \frac{\sin (n-1)\phi_1}{n-1}\right], \qquad n = 3, 5, 7, \ldots \quad (2\text{-}74)$$

2-8.5.1 Harmonic Distortion Distortion is said to occur when the waveform of the output signal differs from the waveform of the input signal. A measure of the effect of each harmonic alone caused by the nonlinear static characteristic is given by its harmonic distortion defined as

$$n\text{th harmonic distortion} = \frac{a_n}{a_1} \times 100\% \quad (2\text{-}75)$$

In practice, a wave analyzer (i.e., a narrow-bandwidth tunable voltmeter) is used to measure separately the rms amplitudes of the fundamental and of each harmonic. Thus, if measurements yield $a_1 = 0.5$ V rms, $a_2 = 0.02$ V rms, and $a_3 = 0.015$ V rms, then

$$\text{2nd harmonic distortion} = \frac{0.02}{0.5} \times 100\% = 4\%$$

and

$$\text{3rd harmonic distortion} = \frac{0.015}{0.5} \times 100\% = 3\%$$

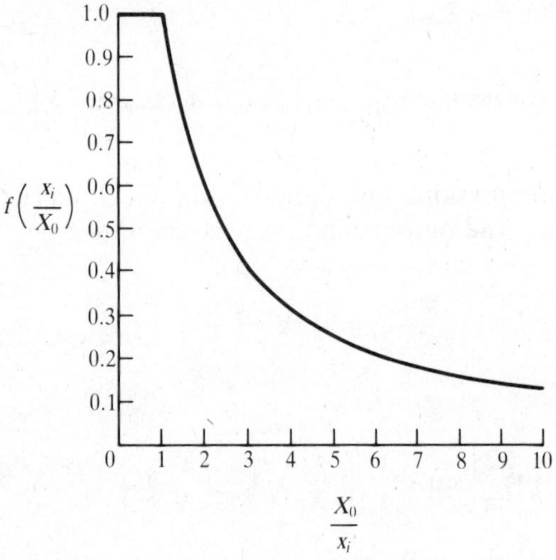

Fig. 2-9 The compression function $f(x_i/X_0)$.

The total harmonic distortion is the square root of the sum of the squares of all higher-order harmonic amplitudes divided by the total rms amplitude, expressed as a percentage. In the above example, if all harmonics above the third have negligible amplitudes, the total harmonic distortion is

$$\frac{\sqrt{0.02^2 + 0.015^2} \times 100\%}{\sqrt{0.5^2 + 0.02^2 + 0.015^2}} = \frac{\sqrt{0.02^2 + 0.015^2} \times 100\%}{0.5006} \cong \sqrt{(\%2\text{nd})^2 + (\%3\text{rd})^2}$$

The percentage of total harmonic distortion is given approximately by the square root of the sum of the squares of the individual harmonic percentages only when the total harmonic distortion is small (less than about 10%).

2-8.5.2 Compression The function $f(x_1/X_0)$, defined by (2-73a) and plotted in Fig. 2-9 from values given in Table 2-2, represents the factor whereby the fundamental output amplitude increases at a continually decreasing rate as the input amplitude X_0 is increased beyond x_1. This phenomenon is known as *compression.* The same compression effect is also observed from the harmonic extensions for the fundamental output given by (2-55) when the higher-order coefficients are negative and this means that the fundamental output amplitude increases in less than a linear manner. The inverse process to compression is called *expansion*.

The degree of compression produced by the nonlinearity of Fig. 2-8 is

TABLE 2-2 Some Calculated Values For the Function

$$f\left(\frac{x_i}{X_0}\right) = \frac{2}{\pi}\left[\sin^{-1}\frac{x_i}{X_0} + \frac{x_i}{X_0}\sqrt{1-\left(\frac{x_i}{X_0}\right)^2}\right]$$

$\dfrac{X_0}{x_i}$	$f\left(\dfrac{x_i}{X_0}\right)$	$\dfrac{X_0}{x_i}$	$f\left(\dfrac{x_i}{X_0}\right)$
1.0	1.000	5.0	0.253
1.5	0.781	5.5	0.230
2.0	0.609	6.0	0.211
2.5	0.495	7.0	0.181
3.0	0.416	8.0	0.159
3.5	0.359	9.0	0.141
4.0	0.315	10.0	0.127
4.5	0.281		

given by a *compression ratio* CR defined as

$$\mathrm{CR} = \frac{\text{actual fundamental output}}{\text{fundamental output if linear}} = \frac{\dfrac{y_1 X_0}{x_1} f\left(\dfrac{x_1}{X_0}\right)}{\dfrac{y_1 X_0}{x_1}} = f\left(\frac{x_1}{X_0}\right) \tag{2-76a}$$

Consequently, the function $f(x_1/X_0)$ is a compression function but is also called a saturation function.

It is usual to express the compression ratio in decibels. Thus,

$$\mathrm{CR}_{\mathrm{dB}} = 20 \log \mathrm{CR} = 20 \log f\left(\frac{x_1}{X_0}\right) \tag{2-76b}$$

Since $f(x_1/X_0) \leqslant 1$, the compression ratio becomes negative decibels. Numerical analysis shows that when $X_0 = 1.26x_1$ the compression ratio becomes -1 dB. Popular usage refers to this as $+1$ dB of compression.

2-8.6 A General Nonlinearity

Analysis of the general piecewise nonlinearity shown in Fig. 2-10 yields results that are applicable to other piecewise nonlinear static characteristics. It is assumed that $y(x)$ is an odd function of x. Therefore, the characteristic actually extends into the first and third quadrants but only its first quadrant

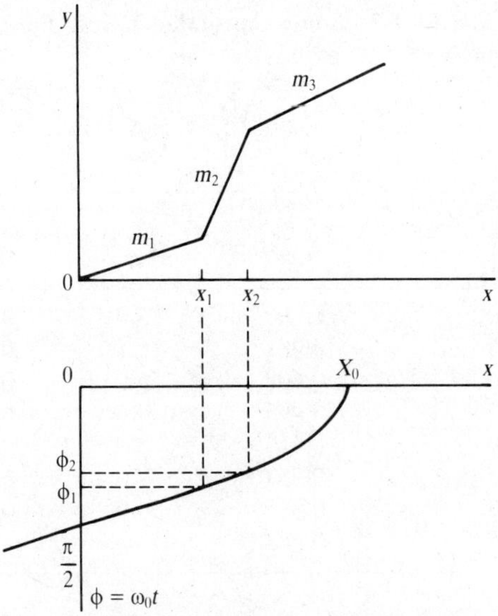

Fig. 2-10 One quadrant of a general piecewise odd nonlinearity.

and the first quarter-period of the input signal are shown for simplicity. The output for each piecewise-linear segment is

$$y = m_1 x, \qquad 0 \leqslant x \leqslant x_1 \tag{2-77}$$

$$y = (m_1 - m_2)x_1 + m_2 x, \qquad x_1 \leqslant x \leqslant x_2 \tag{2-78}$$

$$y = (m_1 - m_2)x_1 + (m_2 - m_3)x_2 + m_3 x, \qquad x_2 \leqslant x \tag{2-79}$$

These can bc cxtended very quickly by induction to include any number of additional segments.

If the input signal is $x = X_0 \cos \phi$, where $\phi = \omega_0 t$, then

$$\cos \phi_1 = \frac{x_1}{X_0}$$

$$\cos \phi_2 = \frac{x_2}{X_0}$$

or

$$\cos \phi_i = \frac{x_i}{X_0}, \qquad i = 1, 2, 3, \ldots \tag{2-80}$$

and also

$$\frac{\pi}{2} - \phi_i = \sin^{-1} \frac{x_i}{X_0} \tag{2-81}$$

The Fourier series is given by (2-70) and the coefficients are given by (2-71). Thus, for $n = 1$, (2-71) becomes

$$a_1 = \frac{4}{\pi}\left\{\int_0^{\phi_2} [(m_1 - m_2)x_1 + (m_2 - m_3)x_2 + m_3 X_0 \cos\phi] \cos\phi \, d\phi \right.$$
$$\left. + \int_{\phi_2}^{\phi_1} \left[(m_1 - m_2)x_1 + m_2 X_0 \cos\phi\right] \cos\phi \, d\phi + \int_{\phi_1}^{\pi/2} m_1 X_0 \cos\phi \cos\phi \, d\phi \right\} \tag{2-82}$$

from which is obtained, after integration and simplification,

$$a_1 = \frac{2X_0}{\pi}\left[(m_1 - m_2)(\cos\phi_1 \sin\phi_1 - \phi_1) \right.$$
$$\left. + (m_2 - m_3)(\cos\phi_2 \sin\phi_2 - \phi_2) + m_1 \frac{\pi}{2}\right] \tag{2-83}$$

Now add and subtract $m_2\,\pi/2$ and $m_3\,\pi/2$ inside the bracketed term and then combine to obtain

$$a_1 = \frac{2X_0}{\pi}\left[(m_1 - m_2)\left(\cos\phi_1 \sin\phi_1 + \frac{\pi}{2} - \phi_1\right) \right.$$
$$\left. + (m_2 - m_3)\left(\cos\phi_2 \sin\phi_2 + \frac{\pi}{2} - \phi_2\right) + m_3 \frac{\pi}{2}\right] \tag{2-84}$$

which becomes

$$a_1 = X_0\left[(m_1 - m_2) f\left(\frac{x_1}{X_0}\right) + (m_2 - m_3) f\left(\frac{x_2}{X_0}\right) + m_3\right] \tag{2-85}$$

where

$$f\left(\frac{x_i}{X_0}\right) = \frac{2}{\pi}\left[\sin^{-1}\frac{x_i}{X_0} + \frac{x_i}{X_0}\sqrt{1 - \left(\frac{x_i}{X_0}\right)^2}\right], \qquad i = 1, 2, 3, \ldots \tag{2-86}$$

Equation (2-85) can be extended very quickly by induction to include additional segments in a piecewise-linear static characteristic. For example, if another segment having slope $+m_4$ starting at x_3 is included when $X_0 > x_3$, (2-85) extends into

$$a_1 = X_0\left[(m_1 - m_2) f\left(\frac{x_1}{X_0}\right) + (m_2 - m_3) f\left(\frac{x_2}{X_0}\right) + (m_3 - m_4) f\left(\frac{x_3}{X_0}\right) + m_4\right] \tag{2-87}$$

When the phasor that represents the fundamental output component is divided by the phasor that represents the fundamental input sinusoid, the resulting quantity is called the *describing function* of the nonlinear system for a single sinusoidal input.

For $n = 3, 5, 7, \ldots$, (2-71) becomes

$$a_n = \frac{4}{\pi}\left\{\int_0^{\phi_2} [(m_1 - m_2)x_1 + (m_2 - m_3)x_2 + m_3 X_0 \cos\phi] \cos n\phi \, d\phi \right.$$
$$+ \int_{\phi_2}^{\phi_1} [(m_1 - m_2)x_1 + m_2 X_0 \cos\phi] \cos n\phi \, d\phi$$
$$\left. + \int_{\phi_1}^{\pi/2} m_1 X_0 \cos\phi \cos n\phi \, d\phi \right\} \tag{2-88}$$

from which is obtained, after integration and simplification,

$$a_n = \frac{2X_0}{\pi n}\left\{(m_1 - m_2)\left[\frac{\sin(n+1)\phi_1}{n+1} - \frac{\sin(n-1)\phi_1}{n-1}\right]\right.$$
$$\left. + (m_2 - m_3)\left[\frac{\sin(n+1)\phi_2}{n+1} - \frac{\sin(n-1)\phi_2}{n-1}\right]\right\} \tag{2-89}$$

Equation (2-89) can be extended very quickly by induction to include additional segments in a piecewise-linear static characteristic. For example, if another segment having slope $+m_4$ starting at x_3 is included when $X_0 > x_3$, the term

$$(m_3 - m_4)\left[\frac{\sin(n+1)\phi_3}{n+1} - \frac{\sin(n-1)\phi_3}{n-1}\right]$$

is added inside the braces of (2-89). In all cases where equations are extended by induction, it is assumed that $x_1 < x_2 < x_3 < x_4$, etc.

Example 2-1

One quadrant of the idealized static characteristic of a nonlinear system is shown in Fig. 2-11. The input signal is $x(t) = X_0 \cos \omega_0 t$ or $x(\phi) = X_0 \cos\phi$. Calculate the amplitudes of the fundamental a_1 and the third harmonic a_3 (in terms of the slope m and the input amplitude X_0) that appear in the output of the system when $x_2 = 1.2x_1$ and $X_0 = 2x_2$.

Solution The slopes of the general piecewise nonlinearity in Fig. 2-10 become $m_1 = 0$ for the region $0 \leqslant x \leqslant x_1$, $m_2 = m$ for the region $x_1 \leqslant x \leqslant x_2$, and $m_3 = 0$ for the region $x \geqslant x_2$. Also

$$\cos\phi_1 = \frac{x_1}{X_0} = \frac{x_2}{1.2X_0} = \frac{x_2}{1.2(2)x_2} = \frac{1}{2.4}$$

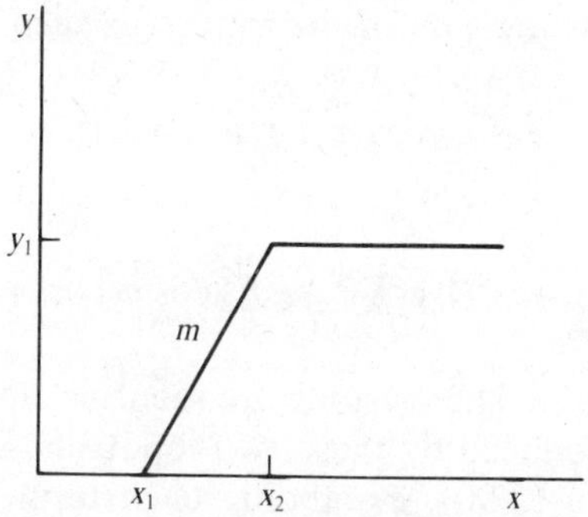

Fig. 2-11 One quadrant of the static characteristic for Example 2-1.

from which is obtained $\phi_1 = 65.376°$. Then

$$\cos \phi_2 = \frac{x_2}{X_0} = \frac{x_2}{2x_2} = \frac{1}{2}$$

from which is obtained $\phi_2 = 60°$. Thus (2-85) yields

$$a_1 = X_0\left[-mf\left(\frac{x_1}{X_0}\right) + mf\left(\frac{x_2}{X_0}\right)\right] = mX_0\left[-f\left(\frac{1}{2.4}\right) + f\left(\frac{1}{2}\right)\right]$$
$$= mX_0(-0.515 + 0.609) = 0.094mX_0$$

and (2-89) yields

$$a_3 = \frac{2X_0}{3\pi}\left\{-m\left[\frac{\sin 4(65.376°)}{4} - \frac{\sin 2(65.376°)}{2}\right]\right.$$
$$\left. + m\left[\frac{\sin 4(60°)}{4} - \frac{\sin 2(60°)}{2}\right]\right\}$$
$$= -0.005mX_0$$

The minus sign indicates simply that the phase of the third harmonic has shifted by 180°.

2-9 NONLINEARITY PRODUCTS—TWO SINUSOIDAL INPUT SIGNALS

Consider the effect of the static characteristic given by the power series

$$y = Ax + Bx^2 + Cx^3 + Dx^4 + Ex^5 + Fx^6 + \cdots \tag{2-90}$$

when the input signal is the sum of two sinusoids such as

$$x = E_1 \cos \omega_1 t + E_2 \cos \omega_2 t$$

or

$$x = E_1 \cos \phi_1 + E_2 \cos \phi_2 \tag{2-91}$$

where $\phi_i = \omega_i t,\ i = 1, 2$. The various frequencies and amplitudes of the output components produced by these two input sinusoids is determined by substituting (2-91) into (2-90), expanding each term, and then combining those components having the same frequency. We will do this for each term separately in (2-90). Thus, the linear term is

$$Ax = AE_1 \cos \phi_1 + AE_2 \cos \phi_2 \tag{2-92}$$

which shows that each input sinusoid is simply multipled by the constant A. The squared or second-order term is

$$\begin{aligned}
Bx^2 &= B(E_1 \cos \phi_1 + E_2 \cos \phi_2)^2 \\
&= B(E_1^2 \cos \phi_1 + 2E_1E_2 \cos \phi_1 \cos \phi_2 + E_2^2 \cos^2 \phi_2) \\
&= \frac{BE_1^2}{2} + \frac{BE_2^2}{2} && \Big\}\text{zero-frequency constant terms} \\
&\quad + \frac{BE_1^2}{2} \cos 2\phi_1 + \frac{BE_2^2}{2} \cos 2\phi_2 && \Big\}\text{2nd harmonics} \\
&\quad + BE_1E_2[\cos (\phi_1 + \phi_2) + \cos (\phi_1 - \phi_2)] && \Big\}\text{second-order IM products}
\end{aligned} \tag{2-93}$$

The second-order nonlinear term produces, in addition to the constants and 2nd harmonics of both input frequencies, two equal-amplitude components whose frequencies are given by the sum and the difference of the two input frequencies. These new sum and difference frequencies are generally not harmonically related to either of the input frequencies and they are called *intermodulation* (IM) products. The second-order nonlinearity produces second-order IM products whose amplitudes are linearly related to both E_1 and E_2.

The process of mixing two sinusoidal signals having different frequencies in a nonlinear system such that the sum and difference (or beat) frequency components actually exists is called *heterodyning*. Heterodyning does not occur in a linear system because no new frequencies can be created in a linear system, but beating can occur. Heterodyning makes the beat frequency a physical reality.

The cubic or third-order term is

$$
\begin{aligned}
Cx^3 &= C(E_1 \cos \phi_1 + E_2 \cos \phi_2)^3 \\
&= C(E_1^3 \cos^3 \phi_1 + 3E_1^2 E_2 \cos^2 \phi_1 \cos \phi_2 \\
&\quad + 3E_1 E_2^2 \cos \phi_1 \cos^2 \phi_2 + E_2^3 \cos^3 \phi_2)
\end{aligned}
$$

Expand each term with the aid of Pascal's triangle, use the identity

$$2 \cos A \cos B = \cos (A + B) + \cos (A - B)$$

and then collect terms to obtain

$$
\begin{aligned}
Cx^3 = {} & \frac{3C}{2}\left(E_1 E_2^2 + \frac{E_1^3}{2}\right) \cos \phi_1 + \frac{3C}{2}\left(E_1^2 E_2 + \frac{E_2^3}{2}\right) \cos \phi_2 && \Big\}\text{fundamentals} \\
& + \frac{CE_1^3}{4} \cos 3\phi_1 + \frac{CE_2^3}{4} \cos 3\phi_2 && \Big\}\text{3rd harmonics} \\
& + \frac{3CE_1^2 E_2}{4} [\cos (2\phi_1 + \phi_2) + \cos (2\phi_1 - \phi_2)] && \Big\}\text{third-order} \\
& + \frac{3CE_1 E_2^2}{4} [\cos (2\phi_2 + \phi_1) + \cos (2\phi_2 - \phi_1)] && \Big\}\text{IM products}
\end{aligned}
\qquad (2\text{-}94)
$$

It is seen that the third-order nonlinearity produces the 3rd harmonics and the fundamentals of both input frequencies, and third order IM products whose frequencies are $2f_1 + f_2$ and $2f_2 + f_1$. Notice that the sum of the harmonic numbers involved in the IM frequencies is the same as the order of the nonlinear term.

Although it is relatively easy to continue this procedure mathematically, it soon becomes tedious and lengthy to perform completely, especially when higher-order terms are considered. Therefore, only the results are given for two more terms, Dx^4 and Ex^5.

$$
\begin{aligned}
Dx^4 = {} & D\left(\frac{3E_1^4}{8} + \frac{3E_1^2 E_2^2}{2} + \frac{3E_2^4}{8}\right) && \Big\}\text{zero-frequency constant terms} \\
& + \frac{DE_1^4}{8} \cos 4\phi_1 + D\left(\frac{E_1^4}{2} + \frac{3}{2} E_1^2 E_2^2\right) \cos 2\phi_1 && \Big\}\text{fourth and second} \\
& + \frac{DE_2^4}{8} \cos 4\phi_2 + D\left(\frac{E_2^4}{2} + \frac{3}{2} E_1^2 E_2^2\right) \cos 2\phi_2 && \Big\}\text{harmonics}
\end{aligned}
$$

$$
\left.\begin{aligned}
&+\frac{DE_1^3E_2}{2}[\cos(3\phi_1+\phi_2)+\cos(3\phi_1-\phi_2)]\\
&+\frac{3DE_1^2E_2^2}{2}[\cos(2\phi_1+2\phi_2)+\cos(2\phi_1-2\phi_2)]\\
&+\frac{DE_1E_2^3}{2}[\cos(\phi_1+3\phi_2)+\cos(\phi_1-3\phi_2)]
\end{aligned}\right\}\text{fourth-order IM products}
$$

$$
+D\left(\frac{3E_1^3E_2}{2}+\frac{3E_1E_2^3}{2}\right)[\cos(\phi_1+\phi_2)+\cos(\phi_1-\phi_2)]\quad\}\text{ second-order IM products} \tag{2-95}
$$

$$
\begin{aligned}
Ex^5 &= \left.E\left(\frac{5}{8}E_1^5+\frac{15E_1^3E_2^2}{4}+\frac{15E_1E_2^4}{8}\right)\cos\phi_1 + E\left(\frac{5}{8}E_2^5+\frac{15E_1^2E_2^3}{4}+\frac{15E_1^4E_2}{8}\right)\cos\phi_2\right\}\text{fundamentals}\\
&\left.+\frac{EE_1^5}{16}\cos 5\phi_1+E\left(\frac{5E_1^5}{16}+\frac{5E_1^3E_2^2}{4}\right)\cos 3\phi_1 +\frac{EE_2^5}{16}\cos 5\phi_2+E\left(\frac{5E_2^5}{16}+\frac{5E_1^2E_2^3}{4}\right)\cos 3\phi_2\right\}\text{fifth and third harmonics}\\
&\left.+\frac{5EE_1^4E_2}{16}\cos(\pm4\phi_1\pm\phi_2)+\frac{5EE_1^3E_2^2}{8}\cos(\pm3\phi_1\pm2\phi_2) +\frac{5EE_1^2E_2^3}{8}\cos(\pm2\phi_1\pm3\phi_2)+\frac{5EE_1E_2^4}{16}\cos(\pm\phi_1\pm4\phi_2)\right\}\text{fifth-order IM products}\\
&\left.+\left(\frac{5EE_1^4E_2}{4}+\frac{15EE_1^2E_2^3}{8}\right)\cos(\pm2\phi_1\pm\phi_2) +\left(\frac{5EE_1E_2^4}{4}+\frac{15EE_1^3E_2^2}{8}\right)\cos(\pm\phi_1\pm2\phi_2)\right\}\text{third-order IM products}
\end{aligned} \tag{2-96}
$$

Basing the conclusions on these results (and on other expansions such as Fx^6 that readers may wish to pursue for their own satisfaction) some general observations can be made.

1. Harmonics of each of the input frequencies are produced such that:
   (a) The highest harmonic number of each is the same as the exponent n of the particular term involved and the harmonic numbers decrease by 2 for each succeeding harmonic frequency.
   (b) When n is even the harmonic numbers are all even, and when n is odd the harmonic numbers are all odd. Thus, when n is even a fundamental frequency component is absent, but when n is odd a fundamental frequency component is present.

(c) When n is even constant terms are present.

2. Various IM frequencies are produced such that:
   (a) The highest-order IM frequency is the same as the exponent n and the order decreases by 2 for each succeeding IM frequency produced by the same exponent.
   (b) When n is even the order of all IM products is even, and when n is odd the order of all IM frequencies is odd.
   (c) There are no zero-order or first-order IM frequencies because such components actually are the zero-frequency (or constant) and fundamental components, respectively.
   (d) If $|\alpha_1|$ and $|\alpha_2|$ are the harmonic numbers, the IM frequencies can be written as $\pm\alpha_1 f_1 \pm \alpha_2 f_2$ for simplicity, where $|\alpha_1| + |\alpha_2|$ is the order. The meaning attached to this notation is that, when actual numerical values are specified: (i) the sum frequency is obtained by using both + signs together, (ii) the difference frequency is obtained by using whichever combination of + and − signs yields a positive value for the frequency, and (iii) both negative signs are never to be used simultaneously.
   (e) In all IM products of any order, the amplitude of a sum frequency is the same as the amplitude of a difference frequency when the same set of harmonic numbers is involved.

There is a particularly easy method, illustrated in Table 2-3, whereby all frequencies produced can be identified and listed merely by inspection. For example, suppose the power series (2-90) terminates with the Fx^6 term when the two input frequencies are f_1 and f_2. The Fx^6 term alone produces not only the 6th harmonics and the sixth-order IM frequencies, but also the 4th and 2nd harmonics, the fourth and second order IM frequencies, and a constant, all of which simply add to the components produced by the Bx^2 and Dx^4 terms. The Ex^5 term alone produces not only the 5th harmonics and fifth-order IM frequencies but also the 3rd harmonics and fundamental components, all of which simply add to the components produced by the Ax and Cx^3 terms. Consequently, in general, the highest even-order term and the highest odd-order term determine *all* the different *frequencies* appearing in the output. The effect on the output frequency content is as though only those two highest-order terms were in the power series.

Table 2-3 shows that to obtain the various IM frequencies having a certain order O, start with the harmonic number $\alpha_1 = \text{O} - 1$ for f_1 and the lowest harmonic number $\alpha_2 = 1$ for f_2. Then that particular entry is $\pm(\text{O} - 1)f_1 + f_2$. For each succeeding IM frequency, decrease α_1 by 1 and increase α_2 by 1 until finally $\alpha_1 = 1$ and $\alpha_2 = \text{O} - 1$. Thus, there will be $\text{O} - 1$ entries in a column for a certain order O, with each entry representing two equal-amplitude IM frequencies.

TABLE 2-3 Frequencies Produced by a Power Series Terminating in Fx^6 When $x = E_1 \cos 2\pi f_1 t + E_2 \cos 2\pi f_2 t$

			IM Frequencies (by order)				
Constant	Fundamentals	Harmonics	6th	5th	4th	3rd	2nd
Present	f_1, f_2	$6f_1, 6f_2$	$\pm 5f_1 \pm f_2$	$\pm 4f_1 \pm f_2$	$\pm 3f_1 \pm f_2$	$\pm 2f_1 \pm f_2$	$\pm f_1 \pm f_2$
		$5f_1, 5f_2$	$\pm 4f_1 \pm 2f_2$	$\pm 3f_1 \pm 2f_2$	$\pm 2f_1 \pm 2f_2$	$\pm f_1 \pm 2f_2$	
		$4f_1, 4f_2$	$\pm 3f_1 \pm 3f_2$	$\pm 2f_1 \pm 3f_2$	$\pm f_1 \pm 3f_2$		
		$3f_1, 3f_2$	$\pm 2f_1 \pm 4f_2$	$\pm f_1 \pm 4f_2$			
		$2f_1, 2f_2$	$\pm f_1 \pm 5f_2$				

If $f_1 = 0.5$ MHz and $f_2 = 1.5$ MHz, a frequency such as $\pm 3f_1 \pm 2f_2 = \pm 1.5\text{ MHz} \pm 3\text{ MHz}$ means that the sum frequency $[(+1.5 + 3)\text{ MHz} = +4.5\text{ MHz}]$ and the difference frequency $[(-1.5 + 3)\text{ MHz} = +1.5\text{ MHz}]$ are present as IM frequencies. Choose the signs such that the difference frequency is positive. Both negative signs are never to be used simultaneously.

2-9.1 Compression, Desensitization, and Blocking

Consider the odd-order static characteristic in Fig. 2-12 in which

$$y = \begin{cases} Ax + Cx^3 = 12x - x^3, & |x| \leqslant 2 \\ 16, & |x| > 2 \end{cases} \tag{2-97}$$

where $A = 12$ and $C = -1$. Let $E_1 \cos \phi_1$ represent a desired input signal and let $E_2 \cos \phi_2$ be another and possibly undesired interfering signal. Then $x = E_1 \cos \phi_1 + E_2 \cos \phi_2$ will be the total input. The desired output signal at frequency f_1 is found from (2-92) and (2-94) to be

$$\begin{aligned} Y_1 &= AE_1 \cos \phi_1 + \frac{3C}{2}\left(E_1E_2^2 + \frac{E_1^3}{2}\right) \cos \phi_1 \\ &= 12E_1\left(1 - \frac{E_1^2}{16} - \frac{E_2^2}{8}\right) \cos \phi_1 \end{aligned} \tag{2-98}$$

provided that operation is restricted to the range $|x| \leqslant 2$. If the other signal has negligible amplitude,

$$Y_1 \cong 12E_1\left(1 - \frac{E_1^2}{16}\right) \cos \phi_1 \tag{2-99}$$

provided that $|E_1| < 2$. The compression ratio for this region of the static characteristic is

$$\mathrm{CR} = \frac{12E_1\left(1 - \frac{E_1^2}{16}\right)}{12E_1} = 1 - \frac{E_1^2}{16} \tag{2-100}$$

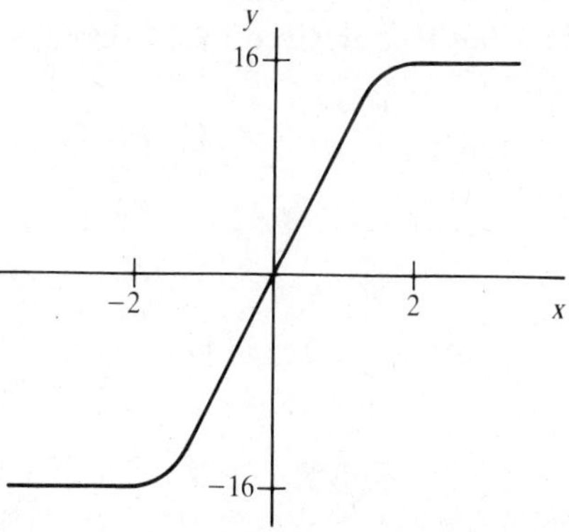

Fig. 2-12 An odd-order static characteristic for which $y = 12x - x^3$ when $|x| \leqslant 2$, and $y = 16$ when $|x| > 2$.

or

$$\mathrm{CR_{dB}} = 20 \log \left(1 - \frac{E_1^2}{16}\right) \tag{2-101}$$

For 1 dB of compression,

$$-1 = 20 \log \left(1 - \frac{E_1^2}{16}\right)$$

from which is obtained

$$E_1 = 4(1 - 10^{-1/20})^{1/2} = 1.32 \text{ V}$$

This result should be valid because it is less than 2 V.

Further, if $E_1^2/16 < 0.01$ or $E_1 < 0.4V$, linear operation is closely approximated so that $Y_1 \cong 12E_1$ and the system functions essentially as a linear small-signal amplifier with negligible compression less than 0.09 dB.

Now let E_1 be negligible (i.e., $E_1 < 0.4V$) while E_2 is increased. Then (2-98) becomes

$$Y_1 \cong 12E_1\left(1 - \frac{E_2^2}{8}\right) \cos \phi_1 \tag{2-102}$$

It is observed that the amplitude E_2 of the other signal causes the amplitude of the desired output to be smaller than expected of a linear system and that the action is similar to that caused by compression. But when it is the *other* signal that produces compression of the desired output, the effect is called *desensitization.* Hence a desensitization ratio can be defined as

$$\mathrm{DS} = \frac{\left|\begin{array}{l}\text{desired output due to a desired input signal (for negligible} \\ \text{compression) in the presence of another signal}\end{array}\right|}{\left|\begin{array}{l}\text{desired output due to a desired input signal only (for negligible} \\ \text{compression}\end{array}\right|}$$

$$= 1 - \frac{E_2^2}{8} \tag{2-103}$$

and

$$\mathrm{DS_{dB}} = 20 \log \left(1 - \frac{E_2^2}{8}\right)$$

Thus, continuing the above example, if $E_1 < 0.4V$ and $E_2 = 2V$, $\mathrm{DR_{dB}} = -6$ dB and a desensitization of 6 dB is said to occur. This represents a change in the desired output voltage by a factor of 1/2 and a change in the desired output power by a factor of 1/4.

The desensitization phenomenon can occur in any odd-order system subject to the limiting natures of the saturation and cutoff processes, but it is most generally applied to radio receiving systems. The usual observed action is that the receiver output level for a desired and possibly weak signal suddenly decreases when a strong interfering signal, strong perhaps merely because of its proximity, also suddenly appears at the input. This undesirable response is directly attributable to a deficiency in the receiver caused by the nonlinearity, and not to a defect in either transmitter.

Because of the minus sign in (2-102) it appears that the desired output Y_1 will be compressed to zero amplitude and therefore completely blocked when

$$1-\frac{E_2^2}{8}=0$$

from which is obtained $E_2=2.83V$. But this value is not reliable because $E_2=2.83V>2V$ and operation is forced into the region outside of the range for which (2-102) is valid. In any event, however, the blocking effect is indicated. Blocking is the extreme case of desensitization in which the desired output is reduced to zero amplitude by a strong interfering signal.

The desired signal cannot cause blocking of itself, but it can cause compression of itself. This can contribute to the desensitization and ultimate blocking that results from the presence of a strong interfering signal.

Showing theoretically that a strong interfering signal actually does block a weaker desired signal is an exceptionally complicated exercise. It will not be pursued further here because it can involve developing the Fourier series in two variables (because there are two signals), delineating appropriate values of the variables that occur simultaneously over the range of operation of the nonlinearity, reducing the integrals encountered to a form of one of the elliptic integrals, and familiarizing ourselves with elliptic integrals, their identities, and tabulated values. For a taste of what is involved, consult the references by Bennett (see the Bibliography).

2-10 NONLINEARITY PRODUCTS—MULTIPLE SINUSOIDAL INPUT SIGNALS

The binomial theorem is useful when expanding the sum of *two* signals raised to an integer exponent, like the term $(x_1+x_2)^n$. A binomial coefficient is a coefficient in the expansion of $(x_1+x_2)^n$. When the sum of more than two signals is raised to an integer exponent, such as in a term of a power series representation of a static characteristic like $(x_1+x_2+\cdots x_M)^n$, a *multinomial* occurs. This expansion is aided by the multinomial theorem. The multinomial theorem is

$$(x_1+x_2+\cdots+x_M)^n = \sum_{n_1,\, n_2,\, \ldots,\, n_M}\left(\frac{n!}{n_1!\, n_2! \cdots n_M!}\right)x_1^{n_1}x_2^{n_2}\cdots x_M^{n_M} \quad (2\text{-}104)$$

in which the sum is taken over all possible ways whereby the nonnegative values of $n_1, n_2, \ldots, n_M$ can combine such that $n_1 + n_2 + \cdots + n_M = n$. The quantity in parentheses in (2-104) is called a *multinomial coefficient.* An example best illustrates what is involved.

Example 2-2

Use the multinomial theorem to expand $(x_1 + x_2 + x_3)^2$.

Solution Since $n = 2$, then $n_1 + n_2 + n_3 = 2$. The various ways this sum can be obtained are kept track of by preparing a table of the numbers. The sum of the numbers in each row equals 2.

n_1	n_2	n_3
2	0	0
1	1	0
0	2	0
1	0	1
0	0	2
0	1	1

Each multinomial coefficient is found from the numbers in a row. Thus

$$\text{Row 1 yields } \frac{2!}{2!\,0!\,0!} x_1^2 x_2^0 x_3^0 = x_1^2$$

$$\text{Row 2 yields } \frac{2!}{1!\,1!\,0!} x_1^1 x_2^1 x_3^0 = 2x_1x_2$$

$$\text{Row 3 yields } \frac{2!}{0!\,2!\,0!} x_1^0 x_2^2 x_3^0 = x_2^2$$

$$\text{Row 4 yields } \frac{2!}{1!\,0!\,1!} x_1^1 x_2^0 x_3^1 = 2x_1x_3$$

$$\text{Row 5 yields } \frac{2!}{0!\,0!\,2!} x_1^0 x_2^0 x_3^2 = x_3^2$$

$$\text{Row 6 yields } \frac{2!}{0!\,1!\,1!} x_1^0 x_2^1 x_3^1 = 2x_2x_3$$

Adding these six results together to effect the summation required in (2-104) shows that

$$(x_1 + x_2 + x_3)^2 = x_1^2 + 2x_1x_2 + x_2^2 + 2x_1x_3 + 2x_2x_3 + x_3^2$$

as obtained from the multinomial theorem. That the result is correct is readily shown by multiplying $x_1 + x_2 + x_3$ by itself.

Using the multinomial theorem to expand the binomial $(x_1 + x_2)^n$ yields the same results as does using the binomial theorem.

The following development, which parallels that given by Sea (in the reference given in the Bibliography) yields a formula whereby the amplitude of any frequency component can be calculated without performing the entire expansion.

Let each input signal be represented as $x_i = E_i \cos \phi_i$, where $\phi_i = \omega_i t = 2\pi f_i t$. The product part of the numerator of (2-104) is

$$x_1^{n_1} x_2^{n_2} \cdots x_M^{n_M} = \prod_{i=1}^{M} (E_i \cos \phi_i)^{n_i} = \left[\prod_{i=1}^{M} E_i^{n_i}\right]\left[\prod_{i=1}^{M} (\cos \phi_i)^{n_i}\right] \tag{2-105}$$

The expansion of $(\cos \phi_i)^{n_i}$ in (2-105) is carried out as described earlier. Thus,

$$(\cos \phi_i)^{n_i} = \frac{1}{2^{n_i - 1}} \sum_{k_i} \frac{n_i!}{k_i!\,(n_i - k_i)!} \cos (n_i - 2k_i)\phi_i \tag{2-106}$$

where the sum is taken over k_i from 0 to $n_i/2$ when n_i is even or to $(n_i - 1)/2$ when n_i is odd. Further, the constant term for $k_i = n_i/2$ must be divided by 2. Only the positive physically realizable frequencies occur and the single-sided spectrum is obtained. But $\cos \phi = \frac{1}{2}(e^{j\phi} + e^{-j\phi})$, so (2-106) becomes

$$(\cos \phi_i)^{n_i} = \frac{1}{2^{n_i}} \sum_{k_i=0}^{n_i} \frac{n_i!}{k_i!\,(n_i - k_i)!} e^{j(n_i - 2k_i)\phi_i} \tag{2-107}$$

Both positive and negative frequencies are involved in the exponential form, so that the double-sided spectrum in obtained. Therefore the product term in (2-105) becomes

$$\prod_{i=1}^{M} (\cos \phi_i)^{n_i} = \prod_{i=1}^{M} \frac{1}{2^{n_i}} \sum_{k_i=0}^{n_i} \frac{n_i!}{k_i!\,(n_i - k_i)!} e^{j(n_i - 2k_i)\phi_i} \tag{2-108}$$

This represents the product of the exponential expansions of each $\cos \phi$ raised to its own exponent with $i = 1, 2, \ldots, M$. The result obtained from (2-108) is a form having the sum of various products. Thus, (2-108) can be written as

$$\prod_{i=1}^{M} (\cos \phi_i)^{n_i} = \frac{1}{2^n} \sum_{k_1=0}^{n_1} \cdots \sum_{k_M=0}^{n_M} \left[\prod_{p=1}^{M} \frac{n_p}{k_p!\,(n_p - k_p)!}\right] e^{j[\sum_{i=1}^{M} (n_i - 2k_i)\phi_i]} \tag{2-109}$$

in which $n_1 + n_2 + \cdots + n_M = n$. An advantage of using the exponentials instead of the cosines is the ease with which the angles are added when exponential products are required. A disadvantage is that twice as many exponential terms are obtained in the expansion as are really needed. Fortunately, conversion back to the cosines is performed before the various amplitudes are calculated.

Consequently, (2-104) becomes

$$(x_1 + x_2 + \cdots + x_M)^n = \sum_{n_1, \ldots, n_M} \frac{n!\, E_1^{n_1} \cdots E_M^{n_M}}{2^n} \sum_{k_1=0}^{n_1} \cdots \sum_{k_M=0}^{M} \left[\prod_{p=1}^{M} \frac{1}{k_p!\,(n_p - k_p)!} \right] e^{j[\sum_{i=1}^{M} (n_i - 2k_i)\phi_i]} \tag{2-110}$$

in which $n_1!\, n_2! \cdots n_M!$ in the denominator of (2-104) cancels with $\prod_{p=1}^{M} n_p!$ in the numerator of (2-109).

The frequencies contained in (2-110) are found from the sum

$$\phi = (n_1 - 2k_1)\phi_1 + (n_2 - 2k_2)\phi_2 + \cdots + (n_M - 2k_M)\phi_M \tag{2-111}$$

This has the form

$$\phi = \alpha_1\phi_1 + \alpha_2\phi_2 + \cdots + \alpha_M\phi_M \tag{2-112}$$

where each α_p is a positive or negative integer whose magnitude represents a harmonic number. Thus,

$$\pm|\alpha_p| = n_p - 2k_p, \qquad n = 1, 2, \ldots, M$$

from which is obtained

$$2k_p = n_p \mp |\alpha_p| \tag{2-113}$$

If the amplitude V_n of only one frequency component from only one term in the power series having exponent n is desired, the entire expansion (2-110) is not necessary. After selecting the values of α_p for the desired frequency, the only terms required in (2-110) will involve $(e^{j\phi} + e^{-j\phi})/2$, which is $\cos\phi$ at the desired frequency. Therefore the remainder of the multiple summations is not needed in this situation and so all of those terms can be ignored. Hence, substituting k_p from (2-113) into the inner product of (2-110) and simplifying, the voltage at the desired frequency is given by

$$V_n = \sum_{n_1, \ldots, n_M} \frac{n!\, E_1^{n_1} \cdots E_M^{n_M}}{2^{n-1}} \prod_{p=1}^{M} \frac{1}{\left(\dfrac{n_p + |\alpha_p|}{2}\right)! \left(\dfrac{n_p - |\alpha_p|}{2}\right)!} \tag{2-114}$$

This must be divided by 2 if the zero-frequency component for $\phi = 0$ with $\alpha_p = 0$ is the desired component. The value given by (2-114) must be multipled by the appropriate coefficient $A, B, C, \ldots$ in the power series to determine the final output amplitude at the desired frequency.

After choosing a particular frequency component by selecting α_p values, the only terms in (2-114) that contribute to that frequency are those for which $n_p = \alpha_p + 2q_p$, where $q_j = 1, 2, 3, \ldots,$ a nonnegative integer index. Define the order of a particular desired frequency component as given by the capital Greek letter omicron

$$\mathrm{O} = |\alpha_1| + |\alpha_2| + \cdots + |\alpha_M| \tag{2-115}$$

Then the summation rule for (2-114) becomes

$$\begin{aligned} n &= n_1 + n_2 + \cdots + n_M = |\alpha_1| + 2q_1 + |\alpha_2| + 2q_2 + \cdots + |\alpha_M| + 2q_M \\ &= \mathrm{O} + 2(q_1 + q_2 + \cdots + q_M) \end{aligned}$$

from which is obtained

$$q_1 + q_2 + \cdots + q_M = \frac{n - \mathrm{O}}{2} = L \tag{2-116}$$

Equation (2-116) becomes the new summation rule when n_p is eliminated from (2-114) by substituting $n_p = |\alpha_p| + 2q_p$. Thus, the final formula becomes

$$V_n = \sum_{q_1,\, q_2,\, \ldots,\, q_M} \frac{n!}{2^{n-1}} \prod_{p=1}^{M} \frac{E_p^{|\alpha_p| + 2q_p}}{(|\alpha_p| + q_p)!\, q_p!} \tag{2-117}$$

Example 2-3

Use Eq. (2-117) to calculate the output voltage produced by the power-series term $C(E_1 \cos \phi_1 + E_2 \cos \phi_2)^3$ at the frequency ϕ given by: (a) $2\phi_1 - \phi_2$, (b) $3\phi_1$, (c) ϕ_2.

Solution It is seen that $n = 3$ and $M = 2$.

(a) $\mathrm{O} = 3$, $\alpha_1 = 2$ and $\alpha_2 = -1$. Therefore, $L = (n - \mathrm{O})/2 = 0 = q_1 + q_2$. Hence, $q_1 = 0$ and $q_2 = 0$. From (2-117), the output voltage is

$$V_{(\mathrm{a})} = C\left(\frac{3!}{2^2}\right) \prod_{p=1}^{2} \frac{E_1^2 E_2^1}{|\alpha_p|!\, 0!} = C \frac{6E_1^2 E_2}{4} \frac{1}{|\alpha_1|!} \frac{1}{|\alpha_2|!} = \frac{C6E_1^2 E_2}{8} = \frac{C3E_1^2 E_2}{4}$$

(b) $\alpha_1 = 3$, $\alpha_2 = 0$, $\mathrm{O} = 3$, $q_1 + q_2 = L = 0$, so $q_1 = q_2 = 0$;

$$V_{(\mathrm{b})} = C\left(\frac{3!}{2^2}\right) \prod_{p=1}^{2} \frac{E_1^3 E_2^0}{|\alpha_p|!\, 0!} = C \frac{3!}{4} \frac{E_1^3}{|\alpha_1|!} \frac{1}{|\alpha_2|!} = \frac{C6E_1^3}{4 \cdot 6 \cdot 1} = \frac{CE_1^3}{4}$$

(c) $\alpha_1 = 0$, $\alpha_2 = 1$, $O = 1$, $L = (3-1)/2 = 1$. Thus there are two sets of q_1 and q_2 given by $q_1 = 1$ and $q_2 = 0$, and $q_1 = 0$ and $q_2 = 1$.

$$V_{(c)} = C\left[\frac{3!\,E_1^{0+2}E_2^{1+0}}{2^2(0+1)!\,0!} + \frac{3!\,E_1^{0+0}E_2^{1+2}}{2^2(0+0)!\,0!\,(1+1)!\,1!}\right] = C\left(\frac{3E_1^2E_2}{2} + \frac{3E_2^3}{4}\right)$$

The amplitudes calculated for all parts agree with the results obtained in (2-94).

Example 2-4

A static characteristic is represented by the power series

$$y = Ax + Bx^2 + Cx^3 + Dx^4 + Ex^5$$

Use (2-117) to calculate the total output produced by $x = E_1 \cos\phi_1 + E_2 \cos\phi_2$ at the IM frequency given by $2\phi - \phi_2$.

Solution The desired frequency has $O = 3$ with $\alpha_1 = 2$ and $\alpha_2 = -1$. Here, third-order IM products are produced *only* by the Cx^3 and Ex^5 terms. The effect of the Cx^3 term was determined in part (a) of Exercise 2-3 to be $3CE_1^2E_2/4$. We need to add to that amount the result produced by the Ex^5 term. Thus, for this term, $n = 5$, $O = 3$, and $L = (n - O)/2 = (5-3)/2 = 1$. Since $L = q_1 + q_2 = 1$, there are two sets of values of q_1 and q_2 given by $q_1 = 1$ and $q_2 = 0$, and $q_1 = 0$ and $q_2 = 1$. From (2-117) the output at $2\phi_1 - \phi_2$ produced by Ex^5 is

$$E\left(\frac{5!}{2^4}\right)\left[\frac{E_1^{2+2}}{(2+1)!\,1!}\frac{E_2^{1+0}}{(1+0)!\,0!} + \frac{E_1^{2+0}}{(2+0)!\,0!}\frac{E_2^{1+2}}{(1+1)!\,1!}\right]$$

$$= E\left(\frac{5E_1^4E_2}{4} + \frac{15E_1^2E_2^3}{8}\right)$$

This result agrees with that of (2-96) and is certainly more readily obtained. The total output at the desired frequency $2\phi_1 - \phi_2$ is

$$\frac{3CE_1^2E_2}{4} + \frac{5EE_1^4E_2}{4} + \frac{15EE_1^2E_2^3}{8}$$

If the original power series is such that additional higher-order odd-exponent terms are included, each of those odd-order terms will produce a third-order IM frequency $2\phi_1 - \phi_2$ (as well as other components too) that can be calculated separately by (2-117) and added to the above result.

PROBLEMS

2-1 Input–output equations for several systems are given below; the total instantaneous input is x_T and the total instantaneous output is y_T. Classify each system as fixed or time-varying, instantaneous or dynamic, linear or nonlinear in x_T and y_T, and state which are incrementally linear in the signal variables x and y.

(a) $2y_T = 5x_T + 7$ (b) $2y_T = 5x_T^2 + x_T$

(c) $2ty_T = 5x_T + 7$ (d) $\dfrac{dy_T}{dt} + 2y_T = 3x_T$

(e) $\dfrac{d^2y_T}{dt^2} + 3y_T\dfrac{dy_T}{dt} = 5x_T$

2-2 An instantaneous system is represented by

$$y_T = \begin{cases} 0.01x_T^4, & x_T \geq 0 \\ 0, & x_T < 0 \end{cases}$$

If $x_T = 8 + x$ and $y_T = Y_Q + y$, evaluate the required coefficients and write the equivalent power series for the signal components x and y such that $|x| < 8$. Does the series terminate?

2-3 Repeat Problem 2-2 for $y_T = 0.01x_T^{3/2}$, $x_T \geq 0$.

2-4 The current through a certain semiconductor diode at room temperature is approximated by

$$i_T = I_s(e^{v_T/40} - 1), A$$

where v_T is in miltivolts. Forward current exists when v_T is positive and reverse current exists when v_T is negative. The current I_s is the reverse saturation current when v_T is a large negative voltage. (a) A measurement shows that $i_T = 1$ mA when $v_T = 700$ mV. Calculate I_s. (b) The diode is now operated such that $v_T = 700 + v$, mV and $i_T = I_Q + i$, where v and i represent signal components. Evaluate the required coefficients and write several terms of the power series that relates i to v.

2-5 (a) Expand $y_T = \sqrt{x_T}$ into a Taylor series about an operating point at $x_T = 4$. (b) If now $x_T = 4 + x$ and $y_T = Y_Q + y$, calculate the amplitude of signal x to insure that the amplitude of the squared term is less than 1 percent of the amplitude of the linear term. (c) For the conditions of part (b), calculate the amplitudes of the third and fourth power terms.

2-6 The device represented by the raw data in Table 2-1 is subjected to a higher operating temperature. For the same values of x_d, this causes each positive value of y_d to become more positive by 5 percent of each tabulated amount, and each negative value of y_d becomes more negative by 5 percent of each tabulated amount.

Use the method of least squares and the Gauss method of successive elimination to best represent this new raw data by a polynomial through the Ex_d^5 term. Plot both the new raw data and the approximating polynomial on the same graph for comparison.

2-7 On rectangular coordinates with x as abscissa and y as ordinate, plot the function $y = x^n$ for $0 \leq x \leq 1$ using $n = 1, 2, 5, 9, 14$, and 20. This shows how the amount of curvature in a static characteristic is related to the exponent in a power series.

2-8 Use a separate piece of paper for this problem and save the results for possible future reference in other problems.

(a) Prepare the left half of Pascal's triangle to expand $\cos^n \phi$ for positive integers $n \leq 20$. Check the accuracy of your numbers by adding the numbers in the entire last row for $n = 20$ (using the missing right half too in the addition) to see whether they total to 2^{20}. If not, spot check the sum of other rows to see whether each one adds to the appropriate 2^n sum.

(b) In a contrasting color or size of mark, indicate the signs required to expand $\sin^n \phi$.

2-9 Use Pascal's triangle developed in Problem 2-8(a) to expand:

(a) $\cos^8 \phi$ (d) $\cos^{14} \phi$
(b) $\cos^9 \phi$ (e) $\cos^{18} \phi$
(c) $\cos^{13} \phi$ (f) $\cos^{19} \phi$

2-10 Use Pascal's triangle developed in Problem 2-8(b) to expand:

(a) $\sin^8 \phi$ (d) $\sin^{14} \phi$
(b) $\sin^9 \phi$ (e) $\sin^{18} \phi$
(c) $\sin^{13} \phi$ (f) $\sin^{19} \phi$

2-11 Any selected number in Pascal's triangle represents the number of possible downward-directed zigzag paths (never vertical or horizontal) from the apex through other entries of the triangle that lie above the selected number and which then end at the selected number (whew!). Test this fact using the entire Pascal's triangle for $n = 6$ for several selected scattered numbers including the number 20 in the center of the row for $n = 6$.

2-12 The number of combinations of n things taken k at a time is

$$\binom{n}{k} = \frac{n!}{(n-k)!\,k!}$$

This is the same as the kth binomial coefficient corresponding to the nth row, starting with $k = 0$ for the first element of a row for a given n. Then,

$$\binom{n}{k+1} = \frac{n!}{(n-k-1)!\,(k+1)!}$$

and

$$\binom{n+1}{k+1} = \frac{(n+1)!}{(n+1-k-1)!\,(k+1)!} = \frac{(n+1)!}{(n-k)!\,(k+1)!}$$

By manipulating the factorials, show that

$$\binom{n+1}{k+1} = \binom{n}{k} + \binom{n}{k+1}$$

which is the rule by which elements of the $(n+1)$th row of Pascal's triangle are calculated from the elements of the nth row.

2-13 Extend Eq. (2-50) by including the term $EX_0^5 \cos^5 \phi$ in (2-48).

2-14 Use Pascal's triangle and Eq. (2-53) with $x = X_0 \cos \phi$ to determine the output components at (a) the 4th harmonic, and (b) the 5th harmonic.

2–15 A certain system has the ideal static characteristic shown in Fig. 2-8, where $y_1 = 0.1$ V and $x_1 = 1$ V. (a) Using Eq. (2-73), plot a_1 versus X_0 for $0 \leq X_0 \leq 10x_1$. (b) Use numerical analysis to calculate X_0 for a compression of 2 dB.

2-16 Using Eqs. (2-73) and (2-74), plot the 3rd, 5th, and 7th harmonic percentage distortions versus X_0/x_1 for $1 \leq X_0/x_1 \leq 8$. Pay particular attention to values where a curve reaches a maximum or goes to zero. Mark each asymptotic distortion value for $X_0/x_1 \rightarrow \infty$.

2-17 Repeat Example 2-1 for $x_2 = 1.3x_1$ and $X_0 = 3x_2$.

2-18 Use the results in Section 2-8.6 for a general piecewise nonlinearity to obtain expressions for the output amplitudes a_1 and a_3 in each of the cases shown in Fig. P2-18.

2-19 (a) Using Pascal's triangle, expand Fx^6 with $x = E_1 \cos \phi_1 + E_2 \cos \phi_2$ to determine the amplitude of *only* the second-order IM product $\cos(\phi_1 - \phi_2)$ produced by Fx^6 alone. (b) Add the results of part (a) to those of Section 2-9 to determine the total amplitude of the second-order IM product $\cos(\phi_1 - \phi_2)$ produced by the power series in (2-90).

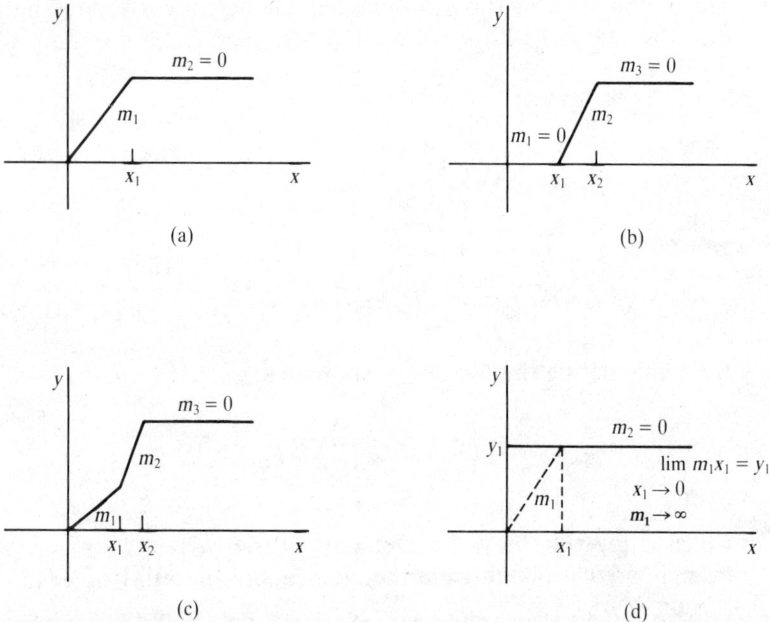

Fig. P2-18 One quadrant only for each of four piecewise nonlinear static characteristics for Problem 2-18. Each static characteristic is an odd function of x.

2-20 (a) Tabulate, as in Table 2-3, all *frequencies* created by a power series terminating in Hx^8 when $x = E_1 \cos 2\pi f_1 t + E_2 \cos 2\pi f_2 t$. (b) Encircle those frequencies produced by the Hx^8 term alone, and underline those frequencies produced by the Gx^7 term alone.

2-21 Show that the expansion of $(x_1 + x_2)^4$ by the multinomial theorem gives the same result as expansion by the binomial theorem.

2-22 Use Eq. (2-117) to calculate the output produced by the power series term $D(E_1 \cos \phi_1 + E_2 \cos \phi_2)^4$ at (a) $\phi_1 - \phi_2$, (b) $2\phi_1$, (c) zero frequency.

2-23 Use (2-117) to calculate the output produced by the power series term $F(E_1 \cos \phi_1 + E_2 \cos \phi_2)^6$ at $(\phi_1 - \phi_2)$ and compare the result to that of Problem 2-19(a).

2-24 The signal current i of a certain device is related to its signal voltage v by

$$i = 0.32 \times 10^{-3}\, v + 3.2 \times 10^{-6}\, v^2, \text{ A}$$

If $v = A \cos(2000\pi t)$ V, determine the amplitude A such that there is 5 percent second-harmonic content in the current.

2-25 A certain amplifier has an output voltage that is related to the input voltage by

$$y = \begin{cases} 12x - x^3, & |x| \leq 2 \\ 16, & |x| > 2 \end{cases}$$

The input signal is represented by $x = 0.5 \cos 2\pi f_1 t + 1.5 \cos 2\pi f_2 t$, V. Calculate the amplitude of the output component whose frequency is:
(a) f_1 (b) $2f_2 - f_1$.

3

NOISE I

3-0 INTRODUCTION

The topic of electrical noise is both fascinating and frustrating to the electronic system designer or user. It is fascinating because of the intellectual challenges it continually offers in discovering its various origins and characteristics and, at the same time, frustrating because of the subtleties and nuances that lead to a lack of sufficient understanding about how best to deal with it in spite of the wealth of pertinent literature.

It is not the purpose here to become involved with a classification of everything that can be regarded as noise mechanisms or with any subjective implications of the fact that what can be considered by one person to be an annoying noise disturbance under certain conditions may be enjoyed by another person as beautiful music, and vice versa.

At the microscopic level, the basic discrete constituents of matter undergo continual temperature-dependent random tiny motions of mechanical and quantum-mechanical origins. Merely interconnecting components to form an operating system will not cause these fundamental chaotic motions to cease. Any physical quantity that depends ultimately on how the essential constituents of matter react to some external stimulus and to each other must fluctuate spontaneously about its average, nominal, or macroscopic value.

In what follows, the word *noise* is intended to mean the spontaneous fluctuation of an electrical quantity (e.g., current, voltage, resistance, capacitance, or inductance) in a random way that depends originally on the discrete nature of electric charge and on its motions. We acknowledge its inevitable presence with a sigh of resignation. The net nuisance effect of noise is to render a desired signal more ambiguous or to obscure it partially or completely by limiting the useful sensitivity of any amplifier, or to cause an undesirable response from an electronic system, or to degrade its performance by reducing the certainty with which observation or measurement can be made. All signals are immersed in an omnipresent electronic bedlam and weak signals particularly must suffer in their hostile environments.

3-1 PORTS AND TRANSFER FUNCTIONS

Many components have only a pair of accessible terminals whereby their electrical connections are made. A pair of terminals is always necessary to provide for a useful circuit and each such pair is called a *port*. All two-terminal devices or a series–parallel interconnection of them can be represented as one-port† networks, as in Fig. 3-1a.

A voltage obtained from an independent voltage source is the independent variable and the resulting current is the dependent variable. If the current is obtained from an independent current source it is the independent variable and the resulting voltage is the dependent variable. The independent variable is variously called the forcing function, stimulus, driving function, excitation, input signal, or, simply, the input.‡ The dependent variable is variously called the reaction, response, output signal, or, simply, the output.‡

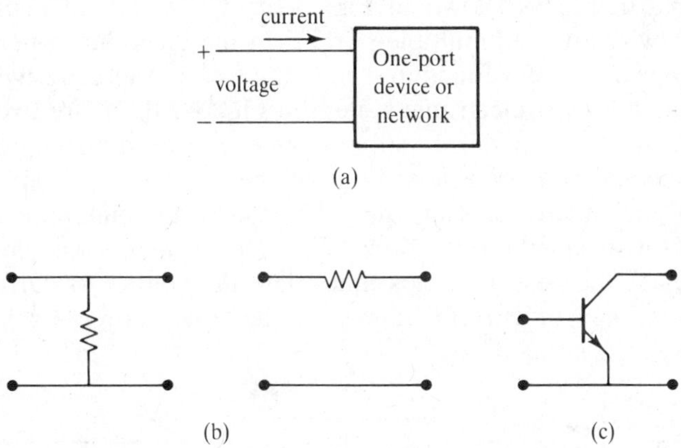

Fig. 3-1 (a) A one-port device or network having only one pair of terminals. (b) Some two-port networks made from one-ports. (c) A three-terminal device connected as a two-port.

† The attributive adjective "one-port" is often used as a noun so that a two-terminal device is sometimes called a *one-port*.

‡ Using a word as either an adjective or a noun is often meaningful and expedient.

The voltage and current at a port can be related to each other in either the time domain or the frequency domain. In the time domain the voltage $v(t)$ and the current $i(t)$ are usually related by a differential equation. In the frequency domain there are two possibilities. In one choice the system can be analyzed or measurements made using steady-state AC theory and procedures. Since steady-state is assumed, initial conditions are not necessary because by definition their effects disappear before steady-state is reached. The circuit elements become the impedances R, $j2\pi fL$, and $1/j2\pi fC$; further, $i(t)$ and $v(t)$ at the port become the phasors $I(f)$ and $V(f)$ respectively. Then $Z(f) = R(f) + jX(f) = V(f)/I(f)$ is the driving-point impedance of the one-port or $Y(f) = I(f)/V(f) = G(f) + jB(f)$ is the driving-point admittance of the one-port.

The other choice is to transform the system into the complex frequency domain represented by the complex variable $s = \sigma + j2\pi f$ of the Laplace transform. Again initial conditions are taken to be zero. The circuit elements become the transformed impedances R, sL, and $1/sC$, and $i(t)$ and $v(t)$ become $I(s)$ and $V(s)$ respectively. Then $Z(s) = V(s)/I(s)$ or $Y(s) = I(s)/V(s)$ become the transformed driving-point impedance or admittance, respectively, of the one-port. Further, $Z(f)$ or $Y(f)$ may be obtained from $Z(s)$ or $Y(s)$ simply by replacing s with $j2\pi f$.

A single two-terminal device can become a two-port device by adding terminals and showing how the two ports are obtained as indicated in Fig. 3-1b. A single three-terminal device becomes a two-port† by choosing a common terminal to be paired with one or the other of the two remaining terminals. Each pair of terminals is a port and two pairs are possible for a three-terminal device as shown in Fig. 3-1c. There are, of course, more complicated two-ports and multiple-ports than the ones shown here.

The concept of transfer function is identified when there are two or more ports. For the two-port electronic system in Fig. 3-2 there are two terminal voltages and two terminal currents that can be related to one another in the frequency domain by their ratios. Four of these ratios, each involving the voltage and current at the same port, yield driving-point impedance and admittances for specified terminations. *Transfer function* is the generic name given to each of the possible ratios that relate the voltage or current at any port to the voltage or current at any of the other ports. Each of these

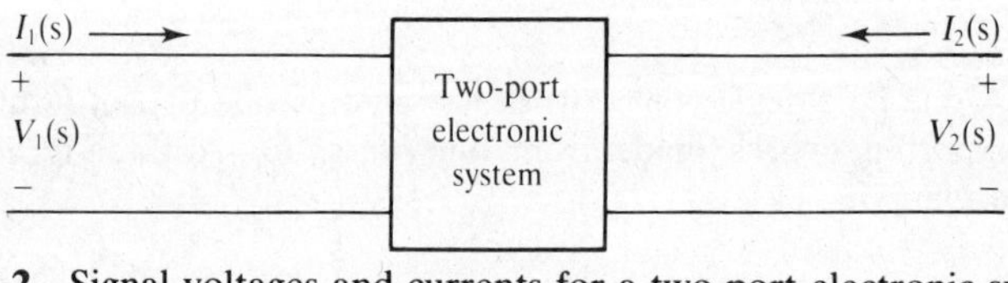

Fig. 3-2 Signal voltages and currents for a two-port electronic system.

† The attributive adjective "two-port" is used here as a noun.

particular ratios describes a "trans-port" action that is of fundamental importance because it indicates how a voltage or current at one port is transferred to another port. Each ratio may also have some other more popular name.

A transfer function is a characteristic of the system, including its terminations, and is not influenced by the actual input or output signals provided that the system remains linear and time-invariant, and it is always determined for zero initial conditions. We will use $H(f)$ to designate a transfer function obtained by steady-state AC circuit theory and $H(s)$ to designate a transfer function obtained from Laplace transform procedures. Either one can be found from the other by interchanging s and $j2\pi f$.

In Fig. 3-2, a transfer function can be the phasor ratio of each of the following:

$$H(f) = \frac{V_2(f)}{V_1(f)} = A_v(f), \quad \text{a voltage-gain function}$$

$$= \frac{I_2(f)}{I_1(f)} = A_i(f), \quad \text{a current-gain function}$$

$$= \frac{V_2(f)}{I_1(f)}, \quad \text{a forward-direction transfer impedance}$$

$$= \frac{I_2(f)}{V_1(f)}, \quad \text{a forward-direction transfer admittance}$$

In each ratio the denominator quantity is the independent variable, and the various ratios are not necessarily equal. Of course, other ratios can be defined for reverse-direction transfer functions. In the laboratory the voltage gain $|A_v(f)|$ is probably the best to measure because it is usually easier and more accurate to measure voltage than current.

In all cases it is assumed that the system itself contains no internal independent sources but it may contain dependent sources. Each internal independent source, such as a noise source possibly, can be associated with its own port which can then be considered removed from internal to the system. For example, a two-port having one internal independent noise source can be considered as a three-port having *no* internal independent sources. Then numerous transfer functions can be identified and determined by straightforward application of the superposition principle.

3-2 NOISE SPECTRA

Noise is a random signal and we like to think of all signals, random or otherwise, in terms of the relative amplitudes and phases of their frequency components. Signals such as noise are known as *power signals* (as contrasted to energy signals) because they are assumed to exist over the entire infinite time interval. Therefore, noise can be characterized by a frequency-dependent *spectral density*† $W(f)$ that can be associated with a power signal. Energy signals are characterized by their energy-density spectra. Even though noise is completely without regularity in its detailed instantaneous amplitudes and cannot be described analytically, its spectral density is usually definite and determinable in spite of the fact that the amplitudes and phases of the frequency components of noise are distributed completely at random. The spectral density is a real and even function of frequency but we are interested in only the physically realizable frequency range $0<f<\infty$. We will not be concerned with interpretations about how averages are obtained to derive the spectral density of a random signal or how spectral density is related to the Fourier transform.

The distribution of the spectral density over some portion of the frequency range $0<f<\infty$ is used to identify and name certain types of noise. Thus, *white noise* is the name given to a noise whose spectral density $W(f)$ is constant. Noise that has its $W(f)$ proportional to frequency is known as *blue noise.* Noise having its $W(f)$ inversely proportional to frequency is called *pink noise.* When a wide frequency range is considered, it is possible for a noise to be pink noise at low frequencies, white noise at mid-range frequencies, and blue noise at high frequencies.

The spectral density $W_x(f)$ of a noise $x_n(t)$ is related to the mean-square value of the noise by

$$\overline{x_n^2(t)} = \int_0^\infty W_x(f)\, df \tag{3-1}$$

where the overbar represents that a long-time average has been taken. The rms value of the noise is

$$X_n = \sqrt{\overline{x_n^2(t)}} \tag{3-2a}$$

Therefore

$$X_n^2 = \overline{x_n^2(t)} \tag{3-2b}$$

The noise $x(t)$ can be either a voltage or a current depending on how $W_x(f)$ is specified.

When noise is applied to the input of a linear system, the frequency components of the noise are transmitted through the system in a manner that depends on a transfer function $H(f)$ of that system as represented by

$$W_y(f) = |H(f)|^2\, W_x(f) \tag{3-3}$$

† It is also known as power spectrum, power-density spectrum, power spectral density, spectral intensity, and noise spectrum. The symbol $S(f)$ is often used.

where $W_x(f)$ is the input noise spectral density and $W_y(f)$ is the output noise spectral density caused by the input noise. Therefore noise will still be noise after passing through a linear transmission system having any transfer function whatever. The particular transfer function needed for a given situation depends on whether $W_x(f)$ is the spectral density of a noise current or of a noise voltage and whether $W_y(f)$ is the spectral density of a noise current or of a noise voltage. The mean-square value of the output noise caused by the input noise is given by

$$\overline{y_n^2(t)} = \int_0^\infty W_y(f)\, df \tag{3-4}$$

and the rms value of the output noise is

$$Y_n = \sqrt{\overline{y_n^2(t)}} \tag{3-5a}$$

Therefore,

$$Y_n^2 = \overline{y_n^2(t)} \tag{3-5b}$$

Unfortunately, the system itself usually contains internal independent noise sources of its own and they cause additional noise, represented by a spectral density $W_s(f)$, to be delivered to the output. If these internal noise mechanisms are independent of or uncorrelated with the noise applied to the input, as they usually are, the total output noise spectral density $W_o(f)$ is obtained from

$$W_o(f) = W_y(f) + W_s(f) \tag{3-6}$$

The mean-square value of the total output noise $y_o(t)$ is

$$\overline{y_o^2(t)} = \int_0^\infty W_o(f)\, df \tag{3-7}$$

and the rms value of the total output noise is

$$Y_o = \sqrt{\overline{y_o^2(t)}} \tag{3-8a}$$

Therefore

$$Y_o^2 = \overline{y_o^2(t)} \tag{3-8b}$$

Not only do the internal noise sources deliver some noise to the output port but they also deliver some noise to anything that is connected to any other port or to intermediate test points. This action sometimes can cause certain noise measurements to differ from what is expected unless the effect is recognized and accounted for.

If the noise voltage is across or the noise current is through a 1-Ω resistance, the mean-square value given by (3-1), (3-4), or (3-7) is also numerically equal to the average power absorbed by the 1-Ω resistance. It is when this average power is a nonzero, finite, positive value that the signal represented by the noise voltage or noise current is called a power signal. If the resistance is any nonzero, finite, positive constant value R, the average noise power is obtained from V_n^2/R or $I_n^2 R$.

When the resistance component of impedance $Z(f) = R(f) + jX(f)$ is a function $R(f)$ of frequency f, the average noise power is given by

$$N = \int_0^\infty R(f)\, W_i(f)\, df \tag{3-9}$$

where $W_i(f)$ is the spectral density of the noise current through $R(f)$.

When the conductance component of admittance $Y(f) = G(f) + jB(f)$ is a function $G(f)$ of frequency f, the average noise power is given by

$$N = \int_0^\infty G(f)\, W_v(f)\, df \tag{3-10}$$

where $W_v(f)$ is the spectral density of the noise voltage across $G(f)$. The inappropriateness of calling $W_i(f)$ and $W_v(f)$ power-density spectra is evident from inspection of (3-9) and (3-10), where it is seen that the products $R(f)W_i(f)$ and $G(f)W_v(f)$ are actually power-density spectra. It is only for the special cases when $R(f) = G(f) = 1$ that $W_i(f)$ and $W_v(f)$ appear to represent power-density spectra. This illogical lack of uniformity in terminology and an inadequate standardization of even some basic definitions contribute to the frustrations of many who have to deal with noise and its effects.

3-3 THERMAL NOISE

Ohmic resistance (i.e., resistance that is dissipative by nature) is one of the most often encountered pernicious sources of noise. Small instantaneous currents are caused by the chaotic Brownian motions† of electrons in an ohmic resistance at thermal equilibrium.‡ These Brownian motions occur in resistance materials in response to the random thermal motions of the

† In 1828 Robert Brown, a Scottish botanist, published a paper that described the irregular movements of sufficiently small solid particles suspended in a liquid. According to kinetic theory, electrons in a conductor are subject to the irregular Brownian motions.

‡ Thermal equilibrium requires that the resistance receives the same amount of energy that it delivers, otherwise the temperature of the resistance would change, thus violating the requirement of thermal equilibrium.

molecules about their normal locations in the lattice structure. The small instantaneous currents give rise to the fluctuations called *thermal noise* and also known as Johnson noise, thermal agitation, and resistance noise. The spectral density of thermal-noise voltage has been found to be given by

$$W_v(f) = \frac{4hfR(f)}{e^{hf/kT} - 1} \tag{3-11}$$

where T = absolute temperature, K(=°Celsius + 273)
k = Boltzmann's constant = 1.38×10^{-23} J/K
h = Plank's constant = 6.62×10^{-34} J s
f = frequency, Hz
$R(f)$ = ohmic component of the Thévenin equivalent impedance $Z(f)$.

This complicated expression can be simplified by noting that $e^x \cong 1 + x$ for small x (i.e., $x \leq 0.2$). Thus for $x = hf/kT$ and $T = 290$ K, the highest frequency for which $hf/kT \leq 0.2$ is found to be approximately $f_h =$ 1200 GHz, which is in the near infrared region. Therefore, for frequencies up to about 1200 GHz, a range that includes a most useful portion of the entire frequency spectrum, (3-11) becomes

$$W_v(f) = 4kTR(f) \tag{3-12}$$

Using (3-12) in (3-1) and (3-2) we find

$$\overline{v_n^2(t)} = 4kT \int_0^\infty R(f)\, df \tag{3-13}$$

and from (3-2), V_n is the square root of the value found in (3-13). Of course, the upper limit in (3-13) is not ∞ but one or two thousand gigahertz for only a few percent error.

Example 3-1

Calculate the rms thermal-noise voltage that appears across the open terminals of a parallel combination of a constant capacitance $C = 1000$ pF and a constant resistance $R = 10$ kΩ at $T = 17$°C = 290 K.

Solution The impedance of the parallel combination is $Z(f) = (R/j\omega C)/(R + 1/j\omega C)$. After some manipulation this becomes $Z(f) = (R - j\omega R^2C)/(1 + \omega^2R^2C^2)$, whose real part is $R(f) = R/(1 + \omega^2R^2C^2)$. The 3-dB frequency is defined to be $f_c = 1/(2\pi RC)$, therefore $R(f) = Rf_c^2/(f^2 + f_c^2)$. Thus (3-13) becomes

$$\overline{v_n^2(t)} = 4kT \int_0^\infty \frac{Rf_c^2}{f^2 + f_c^2}\, df = 4kTRf_c \tan^{-1} \frac{f}{f_c}\bigg|_0^\infty = 4kTR\left(\frac{\pi}{2} f_c\right) = \frac{kT}{C}$$

Using numbers, it is found that $\overline{v_n^2(t)} = 4 \times 10^{-12}\ \text{V}^2$. The rms thermal-noise voltage is found by taking the square root to obtain $V_n = 2\ \mu\text{V}$.

If the ohmic resistance component $R(f)$ is constant, that is $R(f) = R$, the spectral density given by(3-12) becomes

$$W_v(f) = 4kTR \tag{3-14}$$

and the mean-square thermal-noise voltage of (3-13) becomes

$$\overline{v_n^2(t)} = 4kTR \int_0^\infty df \tag{3-15}$$

Equation (3-15) indicates that the mean-square thermal-noise voltage becomes infinite because of the upper limit of the integral. This actually does not occur because (3-14) is valid only up to frequency $f_h \cong 1200$ GHz. Above this frequency, thermal noise decreases rapidly and quantum noise that is proportional to frequency predominates. Therefore, the infinite upper limit in (3-15) does not apply. But since (3-14) shows that thermal noise is white noise having equal contributions for any identical frequency increments up to about f_h, it is appropriate to consider (3-15) to be

$$\overline{v_n^2(t)} = 4kTR \int_{f_1}^{f_2} df = 4kTR(f_2 - f_1) = 4kTRB\ \text{V}^2 \tag{3-16}$$

and (3-16) gives the mean-square thermal-noise voltage caused by noise components in a noise bandwidth $B = f_1 - f_1$ when $f_2 > f_1$ and $f_2 < 1200$ GHz.

Thermal noise in resistance R can be represented by an rms thermal-noise voltage V_n in series with noiseless resistance R as shown by the Thévenin equivalent circuit in Fig. 3-3a and it represents the thermal-noise open-circuit rms voltage across the terminals of R. Alternatively, thermal noise can be represented by an rms thermal-noise current source I_n in parallel with noiseless resistance R as shown in the Norton equivalent circuit in Fig. 3-3b and it represents the thermal-noise short-circuit rms current through the shorted terminals of R. Thus,

$$W_i(f) = \frac{4kT}{R} = 4kTG \tag{3-17}$$

where $G = 1/R$, and

$$\overline{i_n^2(t)} = \frac{4kTB}{R} = 4kTGB \tag{3-18}$$

Also, the rms noise current is the square root of the value given by (3-18).

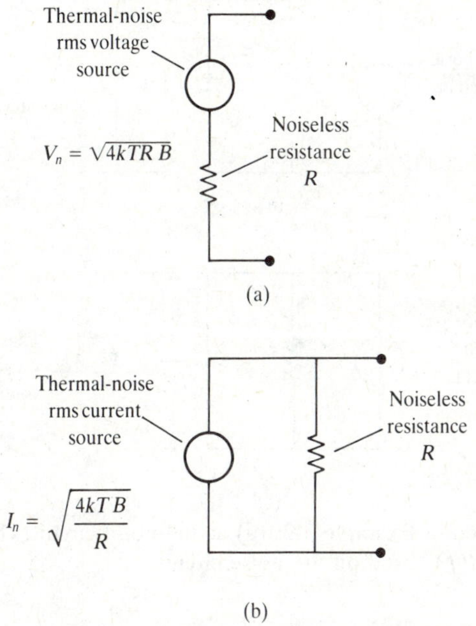

Fig. 3-3 Circuit representations for thermal noise in resistance R. (a) Series rms thermal-noise voltage source. (b) Parallel rms thermal-noise current source.

Thermal noise generated by metallic resistors follows (3-16) and (3-18) closely, but additional noise is present in carbon-composition resistors ("crazy-contact" noise), metal-film, and semiconductor resistors that may also depend on a superimposed current. Thermal noise in metallic resistors is independent of current if the heating effect of the current is negligible. Thermal noise in lossy dielectrics is caused by the random thermal excitations of polarizable molecules that form small fluctuating dipoles.

The static dielectric constant of a dielectric material depends on the average of the thermally induced dipole-moment fluctuations, and the instantaneous dielectric constant depends on the instantaneous polarizations thereby induced.

Example 3-2

Repeat the calculations of Example 3-1 using input and output spectral densities with the transfer function $H(f)$.

Solution The circuit needed to find $H(f)$ is based on the actual noisy circuit shown in Fig. 3-4a. The input is a noise current and the output is a noise voltage. Therefore, the input spectral density if found from (3-17)

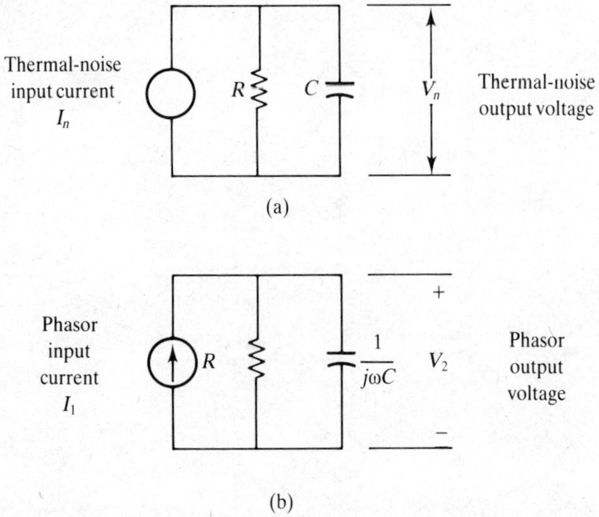

Fig. 3-4 Circuits used in Example 3-2: (a) actual noise circuit, (b) steady-state AC circuit used to find $H(f)$ based on the noise circuit.

to be

$$W_i(f) = \frac{4kT}{R}$$

which is constant and independent of frequency. The transfer function $H(f)$ is found from the equivalent circuit in the frequency domain shown in Fig. 3-4b. Thus,

$$H(f) = \frac{V_2(f)}{I_1(f)} = \frac{R/j\omega C}{R + 1/j\omega C} = \frac{R}{1 + j\omega RC} = \frac{Rf_c}{f_c + jf} \tag{3-19}$$

where $f_c = 1/(2\pi RC)$. Then

$$|H(f)|^2 = \frac{R^2 f_c^2}{f_c^2 + f^2} \tag{3-20}$$

Using (3-3), the spectral density of the output noise is

$$W_y(f) = |H(f)|^2\, W_i(f) = \frac{R^2 f_c^2}{f_c^2 + f^2}\,\frac{4kT}{R} = \frac{4kTRf_c^2}{f_c^2 + f^2} \tag{3-21}$$

Finally, from (3-4) and (3-5) for the output noise,

$$\overline{v_n^2(t)} = \int_0^\infty \frac{4kTRf_c^2}{f_c^2+f^2}\,df = 4kTRf_c \tan^{-1}\frac{f}{f_c}\bigg|_0^\infty = 4kTR\left(\frac{\pi}{2}f_c\right) = \frac{kT}{C}$$

which is the same result as in Example 3-1.

Example 3-3

The spectral density of the noise voltage across the parallel RC circuit of Example 3-2 is applied to an ideal lossless filter having a transfer function $|H(f)| = |A_v|$ (i.e., a positive constant voltage gain) for frequencies between f_1 and f_2 ($f_2 > f_1$) and $|H(f)|$ is zero at all other frequencies. Find the mean-square output noise voltage from the filter. Assume that the input impedance of the filter is so high that adding the filter does not affect $W_v(f)$.

Solution From Example 3-2,

$$W_v(f) = \frac{4kTRf_c^2}{f^2+f_c^2}$$

Therefore (3-3) yields

$$\overline{v_n^2(t)} = \int_{f_1}^{f_2} |A_v|^2 \frac{4kTRf_c^2}{f^2+f_c^2}\,df = |A_v|^2\,4kTRf_c\left(\tan^{-1}\frac{f_2}{f_c} - \tan^{-1}\frac{f_1}{f_c}\right)$$

or

$$V_n^2 = |A_v|^2\,4kTRf_c \tan^{-1}\left(\frac{(f_2-f_1)/f_c}{1+(f_1f_2/f_c^2)}\right) \tag{3-22}$$

This complicated expression can be simplified somewhat if $f_2 - f_1$ is less than about $0.1f_c$, because then it is assured that $\tan x \approx x$. Consequently,

$$V_n^2 \cong |A_v|^2 \frac{4kTR(f_2-f_1)}{1+(f_1f_2/f_c^2)} \tag{3-23}$$

If $f_1 = f_s - \Delta f/2$ and $f_2 = f_s + \Delta f/2$, where $\Delta f \ll f_s$ for narrow bandwidth, then (3-23) becomes

$$V_n^2 \cong |A_v|^2 \frac{4kTR\,\Delta f}{1+(f_s/f_c)^2}$$

or

$$\frac{V_n^2}{\Delta f} \cong |A_v|^2 \frac{4kTR}{1+(f_s/f_c)^2} \tag{3-24}$$

It is seen that the mean-square thermal-noise voltage depends on the filter bandwidths $B = \Delta f$ and also on where the specified frequency f_s is located relative to f_c.

3-4 SHOT NOISE

When discrete charge carriers move at random times across an energy barrier or energy discontinuity, such as when there is current through a semiconductor junction or when there is emission of electrons from a photosurface or from a temperature-limited thermionic cathode, the resulting minute fluctuations are called *shot noise*. The spectral density of the shot-noise current is given by

$$W_i(f) = 2qI_{dc} \tag{3-25}$$

where q = electronic charge = 1.6×10^{-19} C

I_{dc} = average current magnitude across the energy barrier, A.

Therefore, the mean-square shot-noise component of the direct current becomes

$$\overline{i_n^2(t)} = \int_0^\infty 2qI_{dc}\, df \tag{3-26}$$

This indicates that $\overline{i_n^2(t)}$ becomes infinite because of the upper limit. This actually does not happen because (3-25) is valid only for frequencies below $f_\tau \cong 1/\tau$, where τ is the transit time of the charge carriers for the particular device. The effect of transit time occurs when the frequency of the signal is so high that some of the charge carriers that are injected into the energy-barrier region by the signal do not have time to continue all the way across the barrier before the signal changes polarity and forces some or all of the charge carriers to turn around and return to their points of origin. The frequency f_τ depends on the kind of device and on its physical construction. Generally the smaller the device is, the higher f_τ becomes. Therefore the upper limit of infinity in (3-26) does not apply. But since (3-25) shows that shot noise is white noise having equal contributions for any identical frequency increments up to f_τ, it is appropriate to consider (3-26) to be

$$\overline{i_n^2(t)} = \int_{f_1}^{f_2} 2qI_{dc}\, df = 2qI_{dc}(f_2 - f_1) = 2qI_{dc}B \tag{3-27}$$

and (3-27) gives the mean-square shot-noise current caused by noise components in a noise bandwidth $B = f_2 - f_1$ if $f_2 > f_1$ and $f_2 < f_\tau$. The rms shot-noise current I_n is the square root of the value given by (3-27).

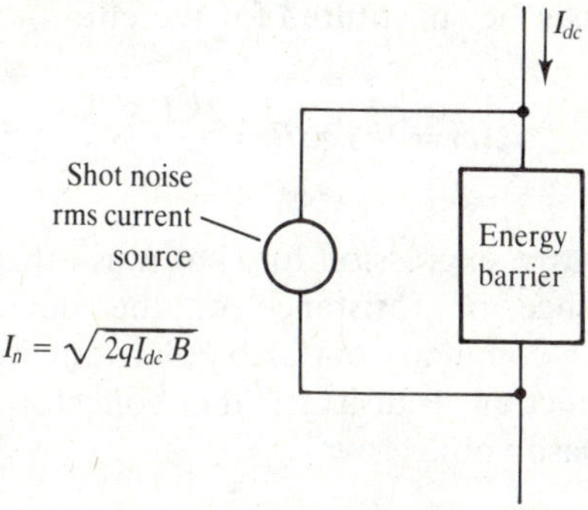

Fig. 3-5 Circuit representation for shot noise.

Shot noise is represented by an rms pure current source I_n connected across the terminals of the device having the energy barrier through which the current I_{dc} exists, as shown in Fig. 3-5. Energy must be supplied continually to the charge carriers for them to move across the energy barrier, so the system is not in thermal equilibrium as in the case of thermal noise.

Equation (3-27) gives what is known as full shot noise. It must be modified by a significant reduction factor when the current becomes space-charge limited instead of temperature limited. The velocity distributions and transit times of the charge carriers contribute to reducing shot noise.

When the volt–ampere relationship for the device is known, the mean-square shot-noise current given by (3-27) can be expressed another way. For example, the base current of a bipolar junction transistor in the common-emitter configuration is approximated by

$$I_B \cong \frac{I_s}{\beta} e^{qV_B/kT}$$

where V_B = base-emitter voltage for forward bias
I_s = emitter saturation current
β = DC current-gain ratio.

The slope of the base-current characteristic represents a signal-circuit conductance given by

$$g_\pi = \frac{dI_B}{dV_B} = \frac{q}{kT} I_B$$

Solving this for the current yields

$$I_B = \frac{kTg_\pi}{q} = \frac{kT}{qr_\pi}$$

where $g_\pi = 1/r_\pi$. This can be substituted for the current in (3-27) to obtain

$$\overline{i_n^2(t)} = 2kTg_\pi B = \frac{2kTB}{r_\pi} \tag{3-28}$$

and (3-28) is an alternative expression for shot noise that is useful when the signal-circuit conductance or resistance of the device is used in the signal-circuit model of a common-emitter bipolar junction transistor.

Similarly, for a junction transistor, the collector current is related approximately to the base voltage by

$$I_C \cong I_s e^{qV_B/kT}$$

then

$$g_m = \frac{dI_C}{dV_B} = \frac{q}{kT} I_C$$

describes a signal-circuit forward transfer function called mutual conductance g_m. Therefore, a mean-square shot-noise current in the collector circuit is

$$\overline{i_n^2(t)} = 2qI_C B = 2kTg_m B \tag{3-29}$$

3-5 OTHER NOISE SOURCES

Avalanche or ionization noise is produced when intense electric fields excite electrons into conduction or accelerate charge carriers sufficiently to dislodge other charge carriers by physical impact. Zener diodes and gas diodes operating in their breakdown regions are notoriously noisy, as well as operating neon signs, sodium vapor lamps, and fluorescent lights.

Flicker noise is a type of noise that increases as frequency is decreased below a few kilohertz, hence the name $1/f$ noise is sometimes used. Also known as excess noise and as semiconductor noise, it has been attributed to various mechanisms such as the generation-recombination random processes within the depletion region (or space-charge layer) of a semiconductor junction, possible continuing diffusion and structural rearrangements within and on the surfaces of the materials used, structural imperfections of the crystal, the trapping and subsequent release of charge carriers, electrode contacts, surface reverse-leakage currents, and various modulation effects.

Partition or interception noise is the result of the random division of current between electrodes when that partition possibility exists. Other noises are aptly described as popcorn noise and burst noise. The magnetization and demagnetization of magnetic materials produces what is called Barkhausen noise.

Noisy components can cause both amplitude and angle modulation of a desired signal. Such fortuitous modulation creates noise sidebands that accompany the output signal from an amplifier or oscillator. Unduly large noise sidebands of the local oscillator(s) in a superheterodyne radio receiver can seriously impair that receiver's performance. These noise effects also cause results known as amplitude jitter, frequency jitter, and phase jitter.

The antenna of a radio receiver picks up not only some of the electromagnetic radiation of a desired signal (with interfering signals from undesired stations too, perhaps) but also black-body radiation (the electromagnetic equivalent of thermal noise); the noisy radiation created naturally by lightning and electrical discharges between clouds in thunderstorms (producing what is sometimes called static) and by solar flares on the sun and violent outbursts on other stars in the far reaches of galactic space; and man-made radiation and coupling from such things as power lines, motors, neon signs, light flashers and dimmers, vehicle ignition, medical apparatus, industrial equipment, switching power supplies, and computers. Anything run by electricity is a potential source of noise.

The amount of noise at any instant at some location depends on how many of the above noise sources there are, their proximity, the status of their intensity (such as during a sunspot cycle), and the frequency. The general trend is that noise intensity increases as the cause of the noise occurs less frequently. For example, more energy can be expended in a bolt of lightning than in an automotive system, or more energy can be expended in an earthquake than in vibrations caused by hourly class changes in a school. More energy was released in the "big bang" than in the Mount St. Helen's volcanic eruption. It has been proposed that part of the present background level of radiation is the remnant of the vast amount caused eons ago by the event referred to as the "big bang."

Noise from the various natural and man-made sources can be coupled or induced into the circuitry, including ground returns, test points, and connections to measuring instruments. These problems are usually dealt with by appropriate shielding, lead dress, and ground-circuit design. Generally, noise introduced by the internal components of an amplifier can be reduced by lowering the temperature, the bias currents, the resistance, and the bandwidth, until limited by other considerations (e.g., minimum bandwidth required by a desired signal).

Originally, the spectral densities of the various noises are broadband in nature. But when broadband noise is applied to a system having selectivity, the spectrum of the output noise is narrowed by the selectivity characteristic. Broadband noise and narrowband noise can exist simultaneously, however, and it is sometimes desirable to know their relative amounts. Noise measurements are such that what may appear to be minor errors can become severely exaggerated in low-noise situations.

Some undesirable signals are nonrandom in nature and have a characteristic regularity in their detailed properties. Examples are the over-the-horizon radar "woodpecker," power-line-frequency "hum" in electronic equipment, "microphonics" caused by mechanical vibrations, the periodic "wow" due to some irregularity in the rotation of mechanical apparatus, and various effects known as "howl" and "motorboating" caused by feedback. In theory, undesired nonrandom signals can be eliminated if the characteristics of the interfering signals can be identified, if the sources of the interference can be located, and if the proper suppression techniques can be devised and applied as close to the sources as possible. Indeed, if noise were not random such disturbances would cease to be a problem.

3-6 NOISE BANDWIDTH

When white noise having $W_x(f) = W_x$ (i.e., a frequency-independent constant) is applied to a linear system having a transfer function $H(f)$, the mean-square value of the output noise caused by the input noise is, from (3-3) and (3-4)

$$\overline{y_n^2(t)} = W_x \int_0^\infty |H(f)|^2 \, df \tag{3-30}$$

The integral in (3-30) represents the area under the $|H(f)|^2$ curve and the integral converges when the transfer function represents a lowpass or a bandpass frequency characteristic. It is convenient to think of that area as being equal to the area of a rectangle in Fig. 3-6 whose height is the maximum of $|H(f)|^2$ at some frequency f_0 and whose width B is called *noise bandwidth*. Therefore, for equal areas,

$$\int_0^\infty |H(f)|^2 \, df = |H(f_0)|^2 B$$

from which the noise bandwidth is found to be, by definition,

$$B = \frac{1}{|H(f_0)|^2} \int_0^\infty |H(f)|^2 \, df \tag{3-31}$$

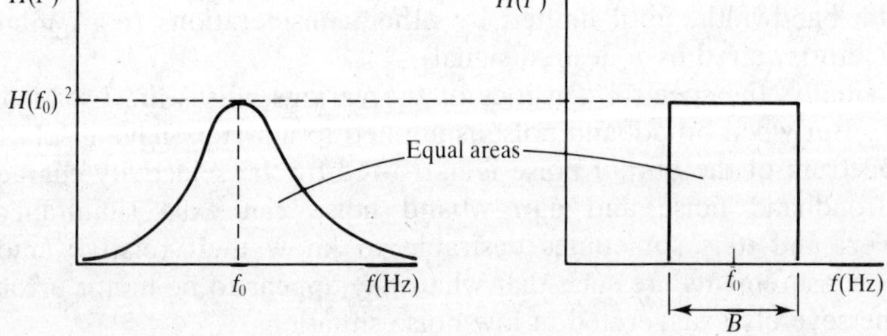

Fig. 3-6 (a) A curve for $|H(f)|^2$. (b) A rectangle having the same area as (a) used to define noise bandwidth B.

Hence (3-30) becomes

$$\overline{y_n^2(t)} = W_x\,|H(f_0)|^2\,B \tag{3-32}$$

The utility of noise bandwidth now becomes clear. When white noise is applied to a fictitious circuit whose transfer function is constant at $H(f_0)$ over the noise bandwidth B, the mean-square value of the output noise caused by the input noise is simply the product of W_x times $|H(f_0)|^2$ times B. Therefore, instead of evaluating the integral in (3-30), it is necessary to determine the maximum value of $|H(f)|^2$ and also the noise bandwidth. But the noise bandwidth defined by (3-31) involves exactly the same integral as before, so it does not appear that any simplification has actually been achieved. The simplification that does result occurs because the noise bandwidths based on the maximum of $|H(f)|^2$ at some frequency f_0 have already been calculated for many systems and tabulated for posterity. Thus, for example, a maximally flat single RC-coupled or high-Q tuned amplifier stage has $B = 1.57BW$, the same as for a single-time-constant RL or RC circuit (as in Example 3-4). A 2-stage RC-coupled amplifier or a 2-stage synchronously tuned cascade has $B = 1.22BW$. A 3-stage synchronously tuned cascade has $B = 1.16BW$. A single stagger-tuned pair or a 4-pole Butterworth filter has $B = 1.11BW$. A 6-pole Butterworth filter has $B = 1.05BW$. The noise bandwidth B is seen to be somewhat larger than the 3-dB bandwidth BW of lowpass and bandpass circuits.

It is common practice to substitute BW for B whenever multistage amplifiers are under consideration because of the simplicity thereby achieved with small residual error. Sometimes the ratio B/BW is simply approximated depending on the complexity of the system, in which case experience becomes the guide.

If there is more than one peak in the $|H(f)|^2$ curve, more than one B may need to be determined. For example, in a radio receiver where both signal and image frequencies are being accepted. $|H(f)|^2$ may show two widely separated peaks, and so there may be two different noise bandwidths that depend on the nature of the two responses.

Example 3-4

Determine the noise bandwidth of the parallel circuit of Example 3-2.

Solution From Example 3-2 it is seen that the maximum value of $H(f)$ occurs at $f = f_0 = 0$. Thus $|H(f_0)|^2 = R^2$ and the transfer-function magnitude squared is

$$|H(f)|^2 = \frac{R^2 f_c^2}{f^2 + f_c^2}$$

where $f_c = 1/(2\pi RC)$ is also the same as the 3-dB bandwidth BW of this lowpass circuit. Then using (3-31) the noise bandwidth is found to be

$$B = \int_0^\infty \frac{f_c^2}{f^2 + f_c^2}\, df = f_c \tan^{-1}\frac{f}{f_c}\bigg|_0^\infty = \frac{\pi}{2} f_c = 1.57 f_c = 1.57 BW \qquad (3\text{-}33)$$

Note that $(\pi/2)f_c$ is the same as the quantity that occurred in the results of the previous examples.

When the mathematical representation of the transfer function is not known or is too unwieldy to consider obtaining by direct analysis, the transfer function of the operating system can be measured at a series of frequencies and squared; then numerical integration can be performed to approximate the area needed to determine the noise bandwidth. One possibility is illustrated in Fig. 3-7. Measurements are made at several frequencies separated by k equal (for convenience) increments over the frequency range required and the area under the $|H(f)|^2$ curve is approximated by the sum of the several smaller rectangular areas. Thus,

$$B \cong \frac{1}{|H(f_0)|}\left[|H(f_1)|^2 + |H(f_2)|^2 + \cdots + |H(f_i)|^2 + \cdots |H(f_k)|^2\right] \Delta f \qquad (3\text{-}34)$$

where Δf = frequency increment selected
f_i = center frequency of each Δf at which each measurement is made
$|H(f_i)|$ = transfer-function magnitude at each f_i.
Other numerical integration procedures, such as Simpson's rule for irregular areas may be useful or even preferred.

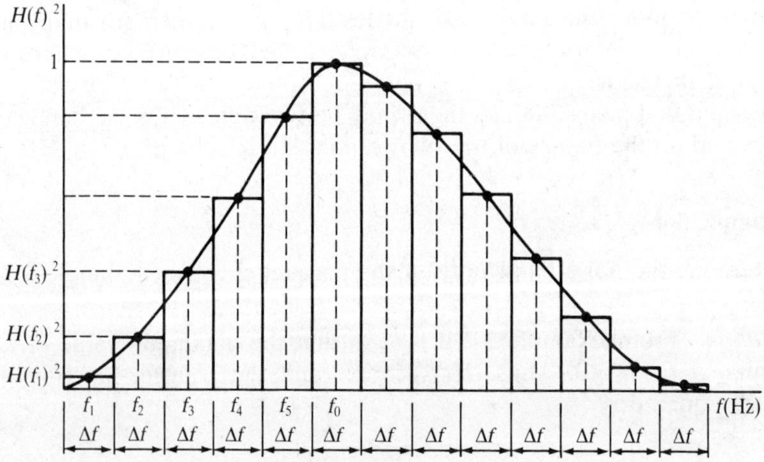

Fig. 3-7 Measurements at equally separated frequencies allow the sum of the small rectangular areas to approximate the actual area under the $|H(f)|^2$ curve.

Plots of $|H(f)|^2$ can be contrived that are composed of simple geometric figures. This allows the area under the curve to be calculated exactly by inspection from the sum of the areas of individual triangles, rectangles, or other well-known shapes that comprise the area under the curve rather than by actually performing the integration of (3-31). Several such plots are included in the problems for practice in calculating B and BW.

To see that there can be a difference in noise bandwidth if the transfer-function-magnitude squared curve is found from voltage-gain measurements or from power-gain measurements, consider the following. Suppose $H(f)$ is determined from voltage-gain measurements. Then

$$H(f) = A_v(f) = \frac{V_2(f)}{V_1(f)}$$

and the noise bandwidth is found very accurately from (3-31) as

$$B = \int_0^\infty \frac{|A_v(f)|^2}{|A_v(f_0)|^2}\, df$$

Input power $P_i(f)$ and output power $P_L(f)$ are related to the voltages by

$$P_i(f) = |V_1(f)|^2\, G_i(f) \tag{3-35}$$

and

$$P_L(f) = |V_2(f)|^2\, G_L(f) \tag{3-36}$$

where $G_i(f)$ = the real part of the input admittance across which $V_1(f)$ occurs
$G_L(f)$ = the real part of the load admittance across which $V_2(f)$ occurs
Therefore the operating power gain is

$$G_p(f) = \frac{P_L(f)}{P_i(f)} = |A_v(f)|^2 \frac{G_L(f)}{G_i(f)}$$

from which is obtained

$$|A_v(f)|^2 = G_p(f)\frac{G_i(f)}{G_L(f)} \tag{3-37}$$

It is clear that $|A_v(f)|^2 = G_p(f)$ only if $G_i(f) = G_L(f)$. If the maximum voltage-gain-magnitude squared occurs at $f = f_0$, then

$$|A_v(f_0)|^2 = \left[G_p(f)\frac{G_i(f)}{G_L(f)}\right]_{\text{maximized at } f = f_0}$$

It is seen that the maximum voltage-gain-magnitude squared does not necessarily occur at the same frequency for which maximum power gain alone occurs, because of the possible skewing effect of the ratio $G_i(f)/G_L(f)$. If that ratio has negligible skewing effect, then

$$|A_v(f_0)|^2 = G_p(f_0)\frac{G_i(f_0)}{G_L(f_0)} \tag{3-38}$$

and maximum power gain and maximum voltage-gain-magnitude squared both occur at $f = f_0$. The noise bandwidth becomes

$$B = \int_0^\infty \frac{G_p(f)}{G_p(f_0)} \frac{G_i(f)/G_i(f_0)}{G_L(f)/G_L(f_0)} df \tag{3-39}$$

Observe that the noise bandwidth obtained from the voltage-gain-magnitude squared curve will be the same as the noise bandwidth obtained from the power-gain curve provided that

$$\frac{G_i(f)}{G_i(f_0)} = \frac{G_L(f)}{G_L(f_0)} \tag{3-40}$$

But this equality of the normalized input and load conductance variations with frequency is seldom if ever a design objective in practice. Therefore it is less accurate to use power-gain measurements instead of voltage-gain measurements to calculate noise bandwidth even if the skewing effect of $G_i(f)/G_L(f)$ is negligible.

3-7 INTEGRAL EVALUATIONS

Originally the concept of a fictitious noise bandwidth was introduced primarily to avoid the integration required in (3-30). But there is nothing intrinsic about the area of the equivalent rectangle that compels its amplitude to equal the maximum of $|H(f)|^2$ at some frequency f_0. It is the area that is important and it is equally valid to select any finite positive number to be the value of $|H(f_0)|^2$ and then adjust the noise bandwidth B to suit that choice. Or the noise bandwidth could be chosen equal to the 3-dB bandwidth and then the amplitude of the equivalent rectangle adjusted to obtain equal areas.

Fortunately a table has been developed that allows integrals of the form

$$I_n = \frac{1}{j2\pi}\int_{-j\infty}^{j\infty} \left| \frac{c_{n-1}s^{n-1} + \cdots + c_0}{d_n s^n + \cdots + d_0} \right|^2 ds \tag{3-41}$$

to be evaluated for $1 \leq n \leq 10$, where s is the Laplace-transform complex-frequency variable. The denominator polynomial must have its zeros in the left half-plane only. Use the Routh criterion if there is any question about whether there are any zeros in the right half-plane. Since $H(f) = c(f)/d(f)$ is the result of analyzing the system using steady-state AC circuit theory, replace $(j2\pi f)$ by the variable s to obtain

$$\int_0^\infty |H(f)|^2\, df = A = \frac{1}{j2\pi}\int_0^{j\infty} \left|\frac{c(s)}{d(s)}\right|^2 ds = \frac{1}{2}\left[\frac{1}{j2\pi}\int_{-j\infty}^{j\infty} \left|\frac{c(s)}{d(s)}\right|^2 ds\right]$$

or

$$A = \tfrac{1}{2} I_n \tag{3-42}$$

where I_n is given by (3-41).

It is convenient to determine the transfer function $H(s) = c(s)/d(s)$ directly from the transformed system in terms of s rather than by using AC circuit theory and then converting that result by replacing $j2\pi f$ with s. If the system is lumped, linear, and time-invariant, $H(s)$ will turn out to be a ratio of polynomials in s having the proper form as required in (3-41). Then the area A is calculated from (3-42) using the table of integrals in Appendix C to find I_n. Finally, using (3-30),

$$\overline{y_n^2(t)} = W_x A \tag{3-43}$$

Example 3-5

A parallel GLC circuit is shown in Fig. 3-8a: (a) Calculate the rms noise voltage V_n caused by the thermal-noise current source I_n of the conductance; (b) calculate the noise bandwidth.

Solution (a) From (3-17), the spectral density of the thermal-noise current of conductance G is

$$W_i(f) = 4kTG$$

The transformed circuit is in Fig. 3-8b and the transfer function is

$$H(s) = \frac{V(s)}{I(s)} = \frac{sL}{s^2LC + sGL + 1}$$

Since $n = 2$, the table of integrals in Appendix C yields

$$I_2 = \frac{c_1^2 d_0 + c_0^2 d_2}{2d_0 d_1 d_2}$$

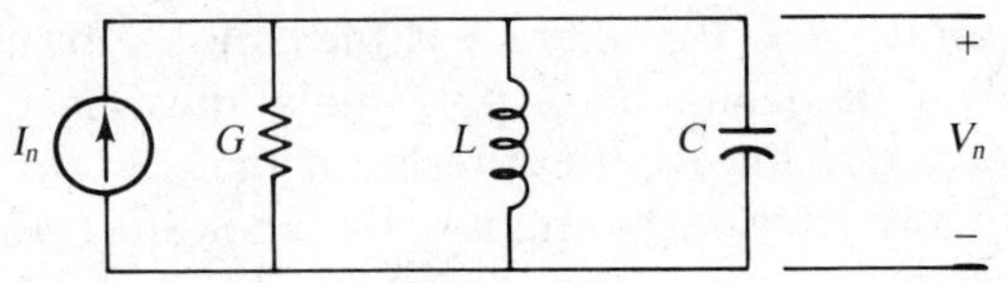

(a)

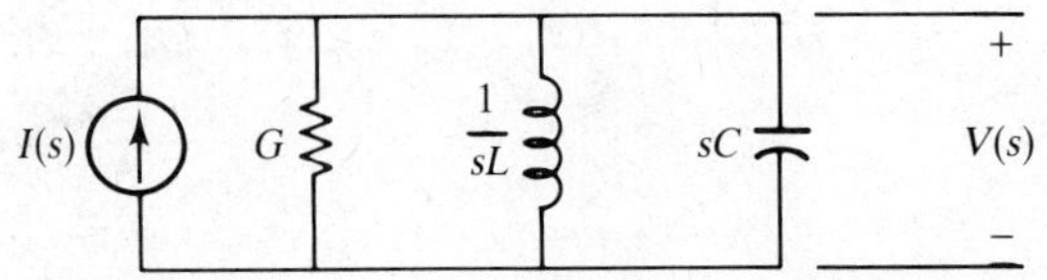

(b)

Fig. 3-8 (a) A parallel *GLC* circuit with the thermal-noise current source of conductance *G*. (b) Transformed-circuit admittances used to determine the transfer function $H(s) = V(s)/I(s)$.

in which $c_1 = L$, $c_0 = 0$, $d_0 = 1$, $d_1 = GL$, and $d_2 = LC$. Hence,

$$I_2 = \frac{1}{2GC}$$

so that

$$A = \frac{1}{2} I_2 = \frac{1}{4GC}$$

Therefore from (3-43),

$$V_n^2 = 4kTG\left(\frac{1}{4GC}\right) = \frac{kT}{C}$$

and the rms thermal-noise voltage is

$$V_n = \sqrt{\frac{kT}{C}}$$

(b)

$$H(f) = H(s)\Big|_{s=j\omega} = \frac{j\omega L}{1 - \omega^2 LC + j\omega GL}$$

Therefore

$$|H(f)|^2 = \frac{(\omega L)^2}{(1-\omega^2 LC)^2 + (\omega GL)^2}$$

This has a maximum value $|H(f_0)|^2 = 1/G^2$ when the frequency is $f_0 = 1/(2\pi\sqrt{LC})$. Equating the area of the rectangle $|H(f_0)|^2 B$ to the area A found above and solving for B yields

$$B = \frac{A}{|H(f_0)|^2} = \frac{G}{4C}$$

Since the 3-dB bandwidth is $BW = G/(2\pi C)$, then $B = (\pi/2)BW$.

3-8 AMPLIFIER NOISE CALCULATIONS

The circuit of a typical audiofrequency transistor amplifier is shown in Fig. 3-9a and a simplified small-signal equivalent circuit for it is in Fig. 3-9b. The

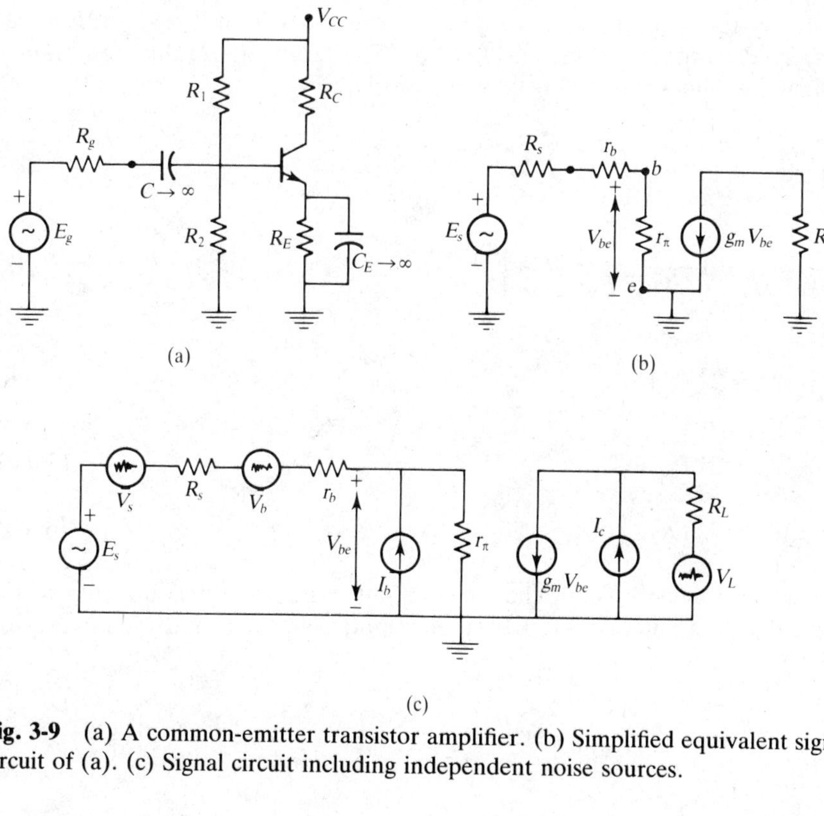

Fig. 3-9 (a) A common-emitter transistor amplifier. (b) Simplified equivalent signal circuit of (a). (c) Signal circuit including independent noise sources.

capacitances are assumed to be sufficiently large that they are signal short-circuits and the supply-voltage terminal is taken to be a signal ground. Thévenin resistance R_s is the parallel combination of R_g, R_1, and R_2, and the Thévenin voltage E_s is the open-circuit voltage across R_2 (i.e., $E_s = E_g R_1 \| R_2/(R_g + R_1 \| R_2)$. For simplicity, we assume that R_1 and R_2 are sufficiently large that $R_s \cong R_g$ and $E_s \cong E_g$.

Several white-noise sources can be identified as follows.

1. Thermal-noise voltage V_s in R_s having spectral density $W_1 = 4kTR_s$
2. Thermal-noise voltage V_b in r_b having spectral density $W_2 = 4kTr_b$
3. Base current shot noise I_b having spectral density $W_3 = 2kT/r_\pi$
4. Collector current shot noise I_c having spectral density $W_4 = 2kTg_m$
5. Thermal-noise voltage V_L in R_L having spectral density $W_5 = 4kTR_L$

All five noise sources are assumed to be uncorrelated for simplicity. All other noise sources are ignored here. When these five noise sources are inserted into the equivalent circuit, Fig. 3-9c is obtained. To determine the spectral density of the total output noise voltage across R_L, it is convenient to use superposition with voltage and current division to calculate the transfer function and the spectral density of the output noise voltage separately for each of the above five noise sources from (3-3) and then add those spectral-density values using (3-6). The following results are obtained for the five white-noise sources, respectively:

1. $$W_{o1} = 4kTR_s\left[\frac{g_m r_\pi R_L}{r_b + r_\pi + R_s}\right]^2 \tag{3-44a}$$

2. $$W_{o2} = 4kTr_b\left[\frac{g_m r_\pi R_L}{r_b + r_\pi + R_s}\right]^2 \tag{3-44b}$$

3. $$W_{o3} = 2kTr_\pi\left[\frac{g_m (r_b + R_s) R_L}{r_b + r_\pi + R_s}\right]^2 \tag{3-44c}$$

4. $$W_{o4} = 2kTg_m R_L^2 \tag{3-44d}$$

5. $$W_{o5} = 4kTR_L \tag{3-44e}$$

The spectral density W_o of the total output noise voltage is the sum of the above five components. After some manipulation, the sum can be expressed as

$$W_o = 4kTR_L\left\{1 + \frac{g_m R_L}{2}\left(1 + g_m r_\pi\left[1 - \frac{r_\pi^2}{(r_b + r_\pi + R_s)^2}\right]\right)\right\} \tag{3-45}$$

Equation (3-45) shows that the spectral density of the total output noise voltage is minimized as T, R_L, B, g_m (or I_c), r_b, or R_s is decreased as much as possible. This helps explain, for example, why a source resistance of 50 Ω results in less output noise than for a source resistance of 300 Ω. Specifically, if $T = 290$ K, $r_\pi = 2.5$ kΩ, $R_L = 10$ kΩ, $g_m = 40$ mS, $r_b = 100$ Ω, the spectral density of the output noise voltage from each of the five noise sources can be calculated from (3-44a)–(3-44e). The results, for several values of R_s, are tabulated as follows, in which all entries should be multiplied by 10^{-12}.

	R_s	1 kΩ	300 Ω	50 Ω	0
	W_{o1}	1.24	0.57	0.11	0
	W_{o2}	0.12	0.19	0.23	0.24
	W_{o3}	0.30	0.061	0.01	0.005
	W_{o4}	0.032	0.032	0.032	0.032
	W_{o5}	0.0002	0.0002	0.002	0.0002
Total	W_o	1.69	0.85	0.38	0.28

It is noticed that the spectral density of the total output noise voltage decreases as R_s decreases even though the component W_{o2} increases because V_{be} increases as R_s decreases as a consequence of voltage division. Of course, W_o is obtained quickly using (3-45). In all cases considered here, the output spectral density W_{o5} produced by the thermal noise in R_L is negligible compared to the other components. Therefore (3-45) becomes

$$W_o = 2kTg_mR_L^2\left(1 + g_mr_\pi\left[1 - \frac{r_\pi^2}{(r_b + r_\pi + R_s)^2}\right]\right) \tag{3-46}$$

The spectral density given by (3-46) represents white noise and is valid up to f_h or f_τ, whichever is lower. This means that the rms value squared of the total output noise voltage is

$$V_o^2 = W_oB = 2kTg_mR_L^2B\left(1 + g_mr_\pi\left[1 - \frac{r_\pi^2}{(r_b + r_\pi + R_s)^2}\right]\right) \tag{3-47}$$

PROBLEMS

3-1 By three different methods, show that the total possible mean-square current in a series RL circuit produced by the thermal-noise of the constant resistance is

$$\overline{i^2(t)} = \frac{kT}{L}$$

3-2 Noise current is produced in a series *RL* circuit by the thermal-noise voltage of the constant resistance. (a) Derive a general expression for the mean-square current for a bandwidth that extends from f_a to f_b. (b) Show that the result of Problem 3-1 is obtained from $f_a = 0$ at $f_b \to \infty$.

3-3 Constant resistances R_1 and R_2 are in parallel with capacitance C. Calculate the total possible noise power delivered to: (a) R_2 by the thermal-noise voltage of R_1; (b) R_1 by the thermal-noise voltage of R_2. Comment on whether you expected these results if the circuit is in thermal equilibrium, i.e., everything is at the same temperature that remains constant.

3-4 A 4 kΩ and a 45 kΩ resistors are in parallel. The temperature is 17°C. Calculate the thermal noise power per hertz of bandwidth delivered *by*:

(a) the 5 kΩ *to* the 45 kΩ; (b) the 45 kΩ *to* 5 kΩ

3-5 Constant resistor R_1, maintained at temperature T_1, is in parallel with constant resistor R_2 which is maintained at temperature T_2. (a) Derive an expression for the total mean-square thermal-noise voltage across the parallel combination for a noise bandwidth B when the thermal noise in each resistor is represented by its own current source. (b) Calculate the rms noise voltage across the circuit when $R_1 = 20$ kΩ, $T_2 = 100$°C, and $B = 50$ kHz. (c) Repeat (b) when $T_1 = T_2 = 17$°C.

3-6 (a) Calculate the maximum possible thermal-noise power per hertz that the thermal noise of resistor R can deliver to a variable resistor $0 \leq R_L \leq \infty$ when both are at temperature $T_0 = 290$ K. (b) Express the numerical answer in (a) in dBm. The dBm unit is a logarithmic quantity that expresses a power P_1 relative to a reference power of 1 mW. Thus $\text{dBm} = 10 \log_{10}(P_1/10^{-3})$.

3-7 A 50-Ω signal source is connected to a noiseless 100-Ω resistor by the two-port network N composed of the two 70.711-Ω resistors as shown in Fig. P3-7. For a temperature of 290 K and a bandwidth of 1 Hz,

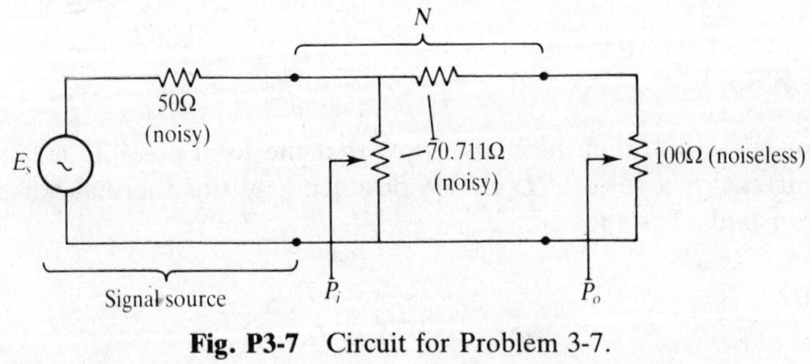

Fig. P3-7 Circuit for Problem 3-7.

calculate: (a) the thermal-noise power delivered to the 100-Ω resistor by the 50-Ω source resistor; (b) the total thermal-noise power delivered to the 100-Ω resistor; (c) the ratio of the input signal power P_i to the output signal power P_o; (d) the ratio of (b) to (a).

3-8 For the fictitious gain-squared functions shown in Fig. P3-8, calculate (a) the signal bandwidth BW, (b) the noise bandwidth B.

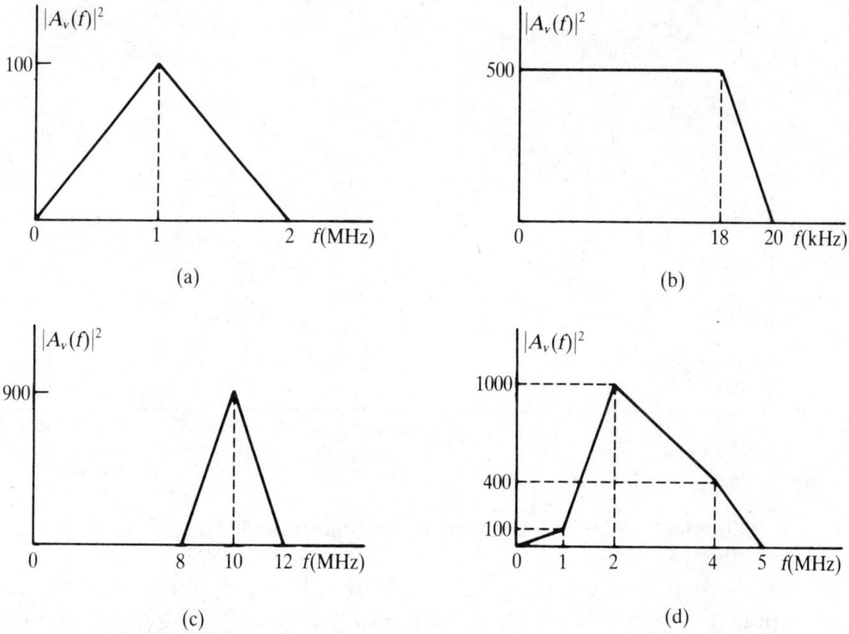

Fig. P3-8 Fictitious gain-squared functions for Problem 3-8.

3-9 Plot the following gain-squared functions and calculate the signal bandwidth BW and the noise bandwidth B.

(a) $$|A_v(f)|^2 = \begin{cases} 50(1 - \cos 2\pi \times 10^{-6}f), & 0 \leq f \leq 1.0\ \text{MHz} \\ 0, & \text{elsewhere} \end{cases}$$

(b) $$|A_v(f)|^2 = \begin{cases} 100(1 - \cos \pi \times 10^{-6}f), & 4\ \text{MHz} \leq f \leq 6\ \text{MHz} \\ 0, & \text{elsewhere} \end{cases}$$

(c) $$|A_v(f)|^2 = \begin{cases} e^{1.1513 \times 10^{-6}f} & 0 \leq f \leq 4\ \text{MHz} \\ 95 + 5 \cos 3\pi \times 10^{-6}f, & 4\ \text{MHz} \leq f \leq 6\ \text{MHz} \\ 100e^{-(f - 6\times 10^6/3\times 10^6)}, & f \geq 6\ \text{MHz} \end{cases}$$

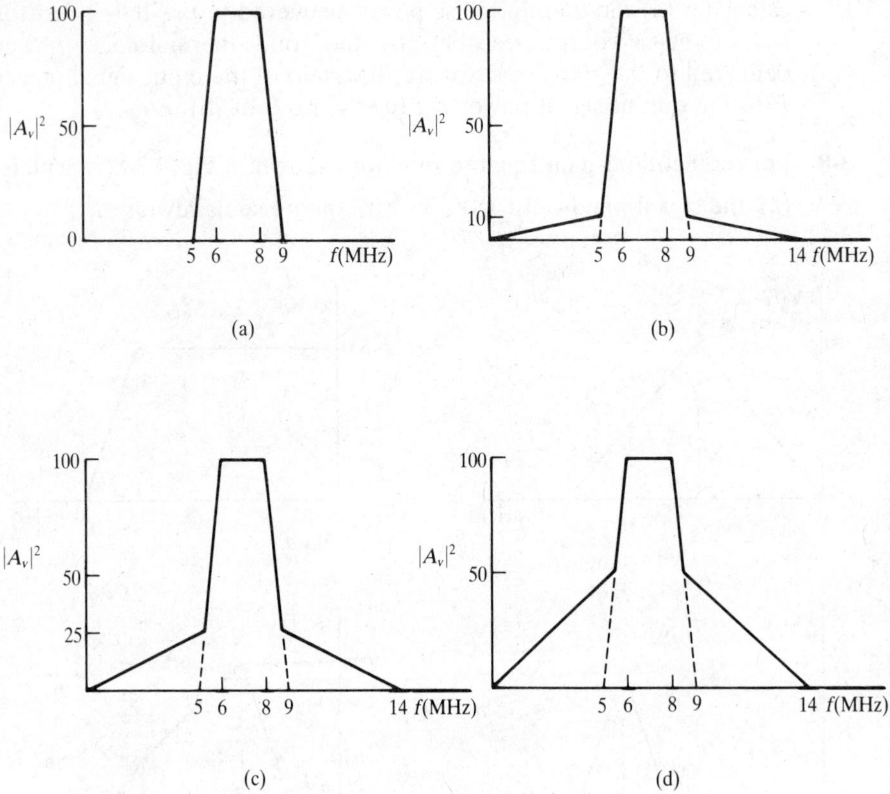

Fig. P3-11 Fictitious gain-squared functions for Problem 3-11.

3-10 Measurements of the overall voltage gain of a radio receiver arc made every 500 Hz with the following results. Calculate the noise bandwidth and the signal bandwidth.

f(kHz)	$\|A_v(f)\|$	f(kHz)	$\|A_v(f)\|$	f(kHz)	$\|A_v(f)\|$
994.0	0.25	998.0	8.7	1002.0	5.0
994.5	0.50	998.5	11.5	1002.5	2.5
995.0	0.8	999.0	15.0	1003.0	1.5
995.5	1.4	999.5	17.3	1003.5	1.0
996.0	2.0	1000.0	17.5	1004.0	0.6
996.5	2.8	1000.5	17.3	1004.5	0.4
997.0	4.1	1001.0	15.0	1005.0	0.25
997.5	6.2	1001.5	10.0	1005.5	0.1
				1006.0	0.04

3-11 A sequence of fictitious gain-squared functions is shown in Fig. P3-11, wherein the "skirt" shapes change progressively. (a) Calculate the noise bandwidth B of each function. (b) Explain why these skirt shapes affect the noise bandwidths but not the signal bandwidths.

4

SPECIAL FUNCTIONS AND FILTERS

PART I Special Functions

Before proceeding with a study of modulation processes, it will be instructive to review some of the mathematical quantities that have intrinsic value in numerous applications. Specifically, the quantity $e^{x \cos \omega t}$ needs to be investigated. This quantity occurs when the voltage across a semiconductor diode is $X_0 \cos \omega t$. The resulting current, as given by the diode equation, is also periodic and contains a term like

$$e^{(qX_0/kT) \cos \omega t} = e^{x \cos \omega t}$$

where $x = qX_0/kT$. Similarly, the collector current of a transistor is exponentially related to the base–emitter voltage.

Further, in angle modulation, the strange-looking functions $\cos(\beta \sin \omega t)$ and $\sin(\beta \sin \omega t)$ need to be interpreted and expanded.

All of these functions are periodic because their arguments repeat with the period $2\pi/\omega$, and any single-valued function of a periodic argument must also repeat itself with the same period. The periodicity of all of these functions suggests that they can be represented by their trigonometric Fourier series. It is in the context of evaluation of the associated Fourier coefficients that *Bessel functions* arise.

Bessel functions involve factorials, and the *Gamma function* is the generalized factorial function that provides a convenient method of interpolating betwen the ordinary factorials of positive integers. In addition, the Gamma function allows the factorials of negative nonintegers to be defined and evaluated.

4-1 GAMMA FUNCTIONS

The Gamma function of the variable n, denoted by $\Gamma(n)$, was discovered by the eighteenth-century Swiss mathcmatician Leonard Euler. It is defined by the continuous function

$$\Gamma(n) = \int_0^\infty t^{n-1} e^{-t}\, dt \tag{4-1}$$

when $n > 0$. Values of $\Gamma(n)$ for the range $1 < n < 2$ are given in Table 4-1. It follows from (4-1) that

$$\Gamma(n+1) = \int_0^\infty t^n e^{-t}\, dt \tag{4-2}$$

Note that this is the same as the Laplace transform of t^n with $s = 1$. Now integrate (4-2) by parts with $u = t^n$ and $dv = e^{-t}\, dt$. Thus,

$$\Gamma(n+1) = -t^n e^{-t}\Big|_0^\infty - (-1)\int_0^\infty nt^{n-1} e^{-t}\, dt = n\int_0^\infty t^{n-1} e^{-t}\, dt$$

or

$$\Gamma(n+1) = n\Gamma(n) \tag{4-3}$$

Since from (4-1),

$$\Gamma(1) = \int_0^\infty e^{-t}\, dt = 1$$

repeatedly setting the integers $n = 1, 2, 3, \ldots,$ into (4-3), we obtain

$$\begin{aligned}
\Gamma(2) &= 1 \cdot \Gamma(1) = 1 = 1! \\
\Gamma(3) &= 2 \cdot \Gamma(2) = 2 \cdot 1 = 2! \\
\Gamma(4) &= 3 \cdot \Gamma(3) = 3 \cdot 2 \cdot 1 = 3! \\
&\vdots \qquad \vdots \qquad \vdots
\end{aligned}$$

TABLE 4-1 Values of $\Gamma(n) = (n-1)!$

n	0	1	2	3	4	5	6	7	8	9
1.0	1.0000	0.9943	0.9888	0.9835	0.9784	0.9735	0.9687	0.9642	0.9597	0.9555
0.1	0.9514	0.9474	0.9436	0.9399	0.9364	0.9330	0.9298	0.9267	0.9237	0.9209
0.2	0.9182	0.9156	0.9131	0.9108	0.9085	0.9064	0.9044	0.9025	0.9007	0.8990
0.3	0.8975	0.8960	0.8946	0.8934	0.8922	0.8912	0.8902	0.8893	0.8885	0.8879
0.4	0.8873	0.8868	0.8864	0.8860	0.8858	0.8857	0.8856	0.8856	0.8857	0.8859
0.5	0.8862	0.8866	0.8870	0.8876	0.8882	0.8889	0.8896	0.8905	0.8914	0.8924
0.6	0.8935	0.8947	0.8959	0.8972	0.8986	0.9001	0.9017	0.9033	0.9050	0.9068
0.7	0.9086	0.9106	0.9126	0.9147	0.9168	0.9191	0.9214	0.9238	0.9262	0.9288
0.8	0.9314	0.9341	0.9368	0.9397	0.9426	0.9456	0.9487	0.9518	0.9551	0.9584
0.9	0.9618	0.9652	0.9688	0.9724	0.9761	0.9799	0.9837	0.9877	0.9917	0.9958

Therefore, in general,

$$\Gamma(n) = (n-1)! \tag{4-4a}$$

$$\Gamma(n+1) = n! \tag{4-4b}$$

This result provides a generalization of the factorial. One consequence of either (4-4a) or (4-4b) is the customary agreement that $\Gamma(1) = 1 = 0!$ which is consistent with $(n-1)! = n!/n$ when $n = 1$. Another important consequence of (4-4b) is that it serves as the definition of factorials of fractional numbers. Since values of $\Gamma(n)$ are given in Table 4-1 for the range $1 < n < 2$, then those values are defined by (4-4a) to be factorials of the corresponding fractional numbers in the range between zero and 1. Hence, $\Gamma(3/2) = \frac{1}{2}! = 0.8862 = \frac{1}{2}\sqrt{\pi}$. Further, with the aid of (4-3), values of the Gamma function can be calculated for the range $2 < n < 3$, and they yield the corresponding fractional factorials in the range between 1 and 2. The process is readily extended to other ranges. Thus, $\Gamma(5/2) = (3/2)! = \frac{3}{2}(0.8862) = 1.3293 = \frac{3}{4}\sqrt{\pi}$.

Also, from (4-3),

$$\Gamma(n) = \frac{\Gamma(n+1)}{n} \tag{4-5}$$

This allows values of $\Gamma(n)$, when $0 < n < 1$, to be calculated from the Gamma functions in the range $1 < n < 2$. Thus, $\Gamma(\frac{1}{2}) = 2\Gamma(3/2) = 1.7724 = \sqrt{\pi}$.

Since

$$\lim_{n\to 0+} \Gamma(n) = \lim_{n\to 0+} \frac{\Gamma(n+1)}{n} = +\infty$$

and

$$\lim_{n\to 0-} \Gamma(n) = \lim_{n\to 0-} \frac{\Gamma(n+1)}{n} = -\infty$$

it is seen that $\Gamma(n)$ is discontinuous at $n = 0$. Also, from (4-5),

$$\Gamma(-n) = \frac{\Gamma(-n+1)}{-n} \tag{4-6}$$

Thus $\Gamma(-n)$ is discontinuous at $n = +1$ because $\Gamma(-1+1)$ is discontinuous as shown above. Further, $\Gamma(-n)$ is discontinuous at $n = +2$ because $\Gamma(-2+1) = \Gamma(-1)$ is discontinuous. Similarly it is seen that $\Gamma(-n)$ is discontinuous whenever n is a positive integer. However, values of $\Gamma(-n)$ are finite for noninteger positive values of n and they can be calculated from (4-6) using known values of the Gamma function for a range greater than $-n$ by unity. Thus, $\Gamma(-\frac{1}{2}) = -2\Gamma(\frac{1}{2}) = -2(1.7724) = -3.5448 = -2\sqrt{\pi}$.

4-2 BESSEL FUNCTIONS

F. W. Bessel (1784–1846) was a distinguished German mathematician and astronomer and director of a German observatory. In a study of planetary motion he encountered an important second-order differential equation that now bears his name, one form of which is

$$x^2\frac{d^2y}{dx^2}+x\frac{dy}{dx}+(x^2-n^2)y=0 \tag{4-7}$$

Bessel originated and developed the solutions of this equation in the forms of power series. The solutions are known as Bessel functions of nth order, which refers only to the real term n and not to the actual order of the differential equation itself. A solution of any order is obtained by letting n be the positive or negative number of that order.

Bessel functions are indispensable in the study of electric currents, propagation of waves, electromagnetic fields, and fluid flow in cylindrical geometry, the vibration of circular membranes, and in angle modulation.

There are two independent solutions of any order of Bessel's equation. The first of these is called a Bessel function of the first kind of order n. For real, positive, integer values of n, it is given by

$$J_n(x)=\frac{x^n}{2^n}\left[\frac{1}{n!}-\frac{x^2}{2^2 1!\,(n+1)!}+\frac{x^4}{2^4 2!\,(n+2)!}-\frac{x^6}{2^6 3!\,(n+3)!}+\cdots\right] \tag{4-8a}$$

$$=\sum_{k=0}^{\infty}\frac{(-1)^k x^{2k+n}}{2^{2k+n}\,k!\,(k+n)!} \tag{4-8b}$$

Note that $J_n(x)$ is an even function of x when n is even, odd when n is odd.

When n is positive integer, the term $(k+n)!$ in (4-8b) becomes $n!$ for $k=0$, $n!\,(n+1)$ for $k=1$, $n!\,(n+1)(n+2)$ for $k=2$, etc. Thus (4-8a) becomes

$$J_n(x)=\frac{x^n}{2^n n!}\left[1-\frac{x^2}{2^2 1!\,(n+1)}+\frac{x^4}{2^4 2!\,(n+1)(n+2)}\right.$$
$$\left.-\frac{x^6}{2^6 3!\,(n+1)(n+2)(n+3)}+\cdots\right] \tag{4-9}$$

Graphs of several Bessel functions of the first kind are shown in Fig. 4-1 and some tabulated values are given in Tables 4-2 and 4-3.

4-2.1 A Recurrence Formula

The series (4-8a) can be used to deduce a recurrence formula between Bessel functions of the first kind of orders $n-1$, n, and $n+1$. First, obtain

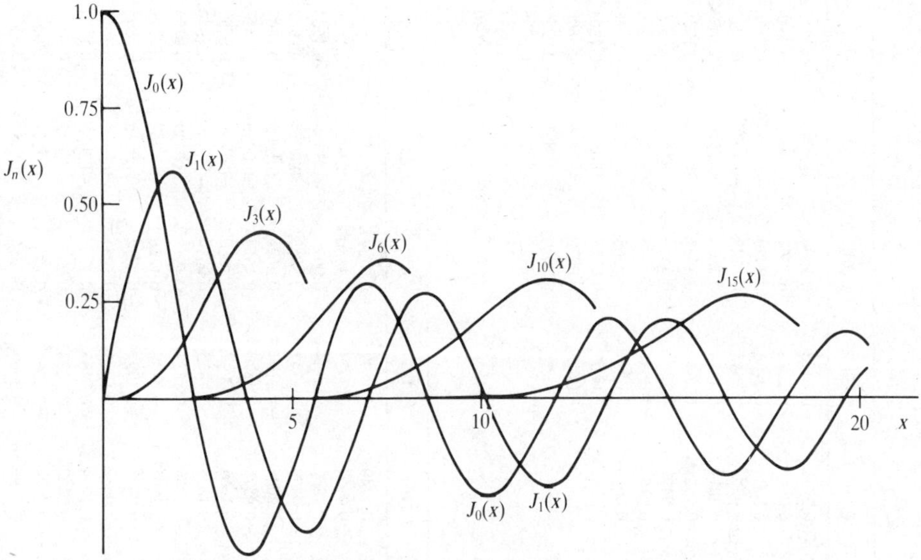

Fig. 4-1 Partial graphs of several Bessel functions.

several terms of the series for $J_{n-1}(x)$ by simply replacing n with $n-1$ in (4-8a). Similarly, obtain several terms of the series for $J_{n+1}(x)$, divide the initial factor $x^{n+1}/2^{n+1}$ of this series by $x^2/2^2$ to obtain the same factor as for the $J_{n-1}(x)$ series. Then multiply each of the other terms of this series by $x^2/2^2$. Now add the two separate expressions for $J_{n-1}(x)$ and $J_{n+1}(x)$, combine terms in like powers of x, factor, and simplify. When that result is

TABLE 4-2 Some Roots of Bessel Functions

	$J_n(\beta_p) = 0$		
p	$n = 0$	$n = 1$	$n = 2$
1	2.4048	3.8317	5.1356
2	5.5201	7.0156	8.4172
3	8.6537	10.1735	11.6198
4	11.7915	13.3237	14.7960
	$J_n'(\beta_p) = 0$		
p	$n = 0$	$n = 1$	$n = 2$
1	0.0000	1.8412	3.0542
2	3.8317	5.3314	6.7061
3	7.0156	8.5363	9.9695
4	10.1735	11.7060	13.1704

TABLE 4-3 Bessel Functions

n	$J_n(0.1)$	$J_n(0.3)$	$J_n(0.5)$	$J_n(0.7)$	$J_n(1)$	$J_n(2)$	$J_n(3)$	$J_n(4)$	$J_n(5)$	$J_n(6)$	$J_n(7)$	$J_n(8)$
0	0.9975	0.9776	0.9385	0.8812	0.7652	0.2239	−0.2601	−0.3971	−0.1776	0.1506	0.3001	0.1717
1	0.0499	0.1483	0.2423	0.3290	0.4401	0.5767	0.3391	−0.0660	−0.3276	−0.2767	−0.0047	0.2346
2	0.0012	0.0112	0.0306	0.0589	0.1149	0.3528	0.4861	0.3641	0.0466	−0.2429	−0.3014	−0.1130
3		0.0006	0.0029	0.0073	0.0196	0.1289	0.3091	0.4302	0.3648	0.1148	−0.1676	−0.2911
4				0.0007	0.0025	0.0340	0.1320	0.2811	0.3912	0.3576	0.1578	−0.1054
5						0.0070	0.0430	0.1321	0.2611	0.3621	0.3479	0.1858
6						0.0012	0.0114	0.0491	0.1311	0.2458	0.3392	0.3376
7							0.0025	0.0152	0.0534	0.1296	0.2336	0.3206
8								0.0040	0.0184	0.0565	0.1280	0.2235
9									0.0055	0.0212	0.0589	0.1263
10									0.0015	0.0070	0.0235	0.0608
11										0.0020	0.0083	0.0256
12											0.0027	0.0096
13												0.0033
14												0.0010

n	$J_n(9)$	$J_n(10)$	$J_n(11)$	$J_n(12)$	$J_n(13)$	$J_n(14)$	$J_n(15)$	$J_n(16)$	$J_n(17)$	$J_n(18)$	$J_n(19)$	$J_n(20)$
0	−0.0903	−0.2459	−0.1712	0.0477	0.2069	0.1711	−0.0142	−0.1749	−0.1699	−0.0134	0.1466	0.1670
1	0.2453	0.0435	−0.1768	−0.2234	−0.0703	0.1334	0.2051	0.0904	−0.0977	−0.1880	−0.1057	0.0668
2	0.1448	0.2546	0.1390	−0.0849	−0.2177	−0.1520	0.0416	0.1862	0.1584	−0.0075	−0.1578	−0.1603
3	−0.1809	0.0584	0.2273	0.1951	0.0033	−0.1768	−0.1940	−0.0439	0.1349	0.1863	0.0725	−0.0989
4	−0.2655	−0.2196	−0.0150	0.1825	0.2193	0.0762	−0.1192	−0.2026	−0.1107	0.0696	0.1806	0.1307
5	−0.0550	−0.2341	−0.2383	−0.0735	0.1316	0.2204	0.1305	−0.0575	−0.1870	−0.1554	0.0036	0.1512
6	0.2043	−0.0145	−0.2016	−0.2437	−0.1180	0.0812	0.2061	0.1667	0.0007	−0.1560	−0.1788	−0.0550
7	0.3275	0.2167	0.0184	−0.1703	−0.2406	−0.1508	0.0345	0.1825	0.1875	0.0514	−0.1165	−0.1842
8	0.3051	0.3179	0.2250	0.0451	−0.1410	−0.2320	−0.1740	−0.0070	0.1537	0.1959	0.0929	−0.0739
9	0.2149	0.2919	0.3089	0.2304	0.0670	−0.1143	−0.2200	−0.1895	−0.0429	0.1228	0.1947	0.1251
10	0.1247	0.2075	0.2804	0.3005	0.2338	0.0850	−0.0901	−0.2062	−0.1991	−0.0732	0.0916	0.1865
11	0.0622	0.1231	0.2010	0.2704	0.2927	0.2357	0.0999	−0.0682	−0.1914	−0.2041	−0.0984	0.0614
12	0.0274	0.0634	0.1216	0.1953	0.2615	0.2855	0.2367	0.1124	−0.0486	−0.1762	−0.2055	−0.1190
13	0.0108	0.0290	0.0643	0.1201	0.1901	0.2536	0.2787	0.2368	0.1228	−0.0309	−0.1612	−0.2041
14	0.0039	0.0120	0.0304	0.0650	0.1188	0.1855	0.2464	0.2724	0.2364	0.1316	−0.0151	−0.1464
15	0.0013	0.0045	0.0130	0.0316	0.0656	0.1174	0.1813	0.2399	0.2666	0.2356	0.1389	−0.0008

TABLE 4-3 (continued)

n	$J_n(10)$	$J_n(11)$	$J_n(12)$	$J_n(13)$	$J_n(14)$	$J_n(15)$	$J_n(16)$	$J_n(17)$	$J_n(18)$	$J_n(19)$	$J_n(20)$
16	0.0016	0.0051	0.0140	0.0327	0.0661	0.1162	0.1775	0.2340	0.2611	0.2345	0.1452
17		0.0019	0.0057	0.0149	0.0337	0.0665	0.1150	0.1739	0.2286	0.2559	0.2331
18			0.0021	0.0063	0.0158	0.0346	0.0669	0.1138	0.1706	0.2235	0.2511
19				0.0025	0.0068	0.0166	0.0354	0.0671	0.1127	0.1676	0.2189
20					0.0028	0.0074	0.0173	0.0362	0.0673	0.1116	0.1647
21					0.0010	0.0030	0.0079	0.0180	0.0369	0.0674	0.1106
22						0.0012	0.0034	0.0084	0.0187	0.0375	0.0676
23							0.0013	0.0036	0.0089	0.0193	0.0381
24								0.0015	0.0039	0.0093	0.0199
25									0.0017	0.0042	0.0098
26										0.0018	0.0045
27											0.0019

compared term by term with the series for $(2n/x)J_n(x)$, it is noted that

$$J_{n-1}(x) + J_{n+1}(x) = \frac{2n}{x} J_n(x) \tag{4-10a}$$

from which is obtained a recurrence formula

$$J_{n+1}(x) = \frac{2n}{x} J_n(x) - J_{n-1}(x) \tag{4-10b}$$

Not only can this recurrence formula be used to extend a table of Bessel functions, but values for any integral order can be calculated step by step in terms of $J_1(x)$ and $J_0(x)$. Thus, $J_2(x) = (2/x)J_1(x) - J_0(x)$, and $J_3(x) = (4/x)J_2(x) - J_1(x) = ((8/x^2) - 1)J_1(x) - (4/x)J_0(x)$.

Another useful application of this recurrence formula occurs when x in $J_n(x)$ is a fractional value not found Table 4-3. Then, instead of trying to interpolate between adjacent columns to find $J_n(3.5)$, for example, use the following procedure. According to Table 4-3, $J_n(3.5)$ certainly vanishes to at least two decimal places for $n > 8$. Now arbitrarily select the values 0 for $J_{10}(3.5)$, using for n 2 more than the 8, and 1 for $J_9(3.5)$. Then use the recurrence formula to calculate $J_8(3.5)$ and all other entries in the second column of Table 4-4, in descending order of n, rounding off to the nearest integer at each step. Thus, for example,

$$J_{n-1}(3.5) = \left(\frac{2n}{3.5}\right) J_n(3.5) - J_{n+1}(3.5)$$

TABLE 4-4 Calculations for $J_n(3.5)$

n	Calculated from Recurrence Formula	Second Column Normalized $J_n(3.5)$
10	0 } arbitrarily	0.0000
9	1 } chosen	0.0003
8	5	0.0015
7	22	0.0067
6	83	0.0254
5	263	0.0805
4	668	0.2044
3	1264	0.3868
2	1499	0.4587
1	449	0.1374
0	−1242	−0.3800

Therefore

$$J_8(3.5) = \left(\frac{2\times 9}{3.5}\right)J_9(3.5) - J_{10}(3.5)$$

$$= \frac{18}{3.5}\times 1 - 0 = 5$$

Notice that $n = 9$ in this example. When calculating $J_5(3.5)$, $n = 6$. The central number in each triad of numbers involved at each step becomes the value of n to use at that step. Also, the calculated values depend on what numbers were arbitrarily chosen initially and on how large n was. However, since

$$J_0(x) + 2J_2(x) + 2J_4(x) + 2J_6(x) + \cdots = 1$$

it is necessary to normalize the calculated values by dividing each by the sum

$$-1242 + 2(1499) + 2(668) + 2(83) + 2(5) = 3268$$

When this is accomplished, the values in the third column are obtained. These are the required values for $J_n(3.5)$ and the number of correct significant figures in those values is the same as the number of digits in the respective calculated values in the second column. Accuracy is lost when the starting value of n is too small because then the calculated values may have too few digits.

4-2.2 Bessel Functions of Negative Order

Because of n^2 in (4-7), negative values of n can be significant. Thus, replace n in (4-8a) by $-n$ to obtain the Bessel function of the first kind of negative order given by

$$J_{-n}(x) = \frac{x^{-n}}{2^{-n}}\left[\frac{1}{(-n)!} - \frac{x^2}{2^2 1!\,(1-n)!} + \frac{x^4}{2^4 2!\,(2-n)!} - \frac{x^6}{2^6 3!\,(3-n)!} + \cdots\right] \tag{4-11a}$$

$$= \sum_{k=0}^{\infty} \frac{(-1)^k x^{2k+n}}{2^{2k+n} k!\,(k-n)!} \tag{4-11b}$$

When n is not an integer, replace $(k-n)!$ by $\Gamma(k-n+1)$ to evaluate $J_{-n}(x)$. This means that a linear combination of $J_n(x)$ and $J_{-n}(x)$ is the complete solution of Bessel's equation when $|n|$ is not an integer or zero. If $n = 0$ the two solutions are identical.

When n is an integer, all of the factors $1/(k-n)!$ in (4-11a) are seen to be zero when $k<n$ because $\Gamma(k-n+1)=(k-n)!$ is not finite when $(k-n)$ is a negative integer. Hence

$$J_{-n}(x)=\sum_{k=n}^{\infty}\frac{(-1)^k x^{2k-n}}{2^{2k-n}k!\,(k-n)!}$$

But we can let the index of the summation be something else if convenient. Therefore, replace k by $k+n$ to obtain

$$J_{-n}(x)=\sum_{k+n=n}^{\infty}\frac{(-1)^{k+n}x^{2k+2n-n}}{2^{2k+2n-n}(k+n)!\,(k+n-n)!}$$

which is the same as

$$J_{-n}(x)=(-1)^n\sum_{k=0}^{\infty}\frac{(-1)^k x^{2k+n}}{2^{2k+n}(k+n)!\,k!}$$

Finally, using (4-11b),

$$J_{-n}(x)=(-1)^n J_n(x) \tag{4-12}$$

when n is an integer. Consequently, $J_n(x)$ and $J_{-n}(x)$ are linearly related when n is an integer. This means that a linear combination of $J_n(x)$ and $J_{-n}(x)$ is *not* the complete solution of Bessel's equation when n is an integer including zero. In this case, the other, and, as it turns out, more complicated independent solution $Y_n(x)$ is called a Bessel function of the second kind of order n. It is sometimes denoted by $N_n(x)$ and called a Neumann function. It is infinite at $x=0$ and cannot be used to express physically finite quantities as x approaches zero, and we will not need it.

Bessel functions of the third kind, or more generally, Hankel functions of the first and second kind, are obtained from the complex functions

$$H_n^{(1)}(x)=J_n(x)+jY_n(x)$$
$$H_n^{(2)}(x)=J_n(x)-jY_n(x)$$

These functions are encountered in certain types of wave propagation and aerodynamic flutter analysis.

4-2.3 Modified Bessel Functions

The second-order differential equation

$$x^2\frac{d^2y}{dx^2}+x\frac{dy}{dx}-(x^2+n^2)y=0 \tag{4-13}$$

differs from Bessel's equation only in the sign of x^2 in the coefficient of y. Change the variable using $w = jx$, to obtain the form of Bessel's equation

$$w^2 \frac{d^2y}{dw^2} + w\frac{dy}{dw} + (w^2 - n^2)y = 0$$

whose solution involves $J_n(w)$, $J_{-n}(w)$, etc., or, in terms of the original variable, $J_n(jx)$, $J_{-n}(jx)$, etc. Thus, for example,

$$J_n(jx) = j^n \sum_{k=0}^{\infty} \frac{x^{2k+n}}{2^{2k+n}k!\,(k+n)!}$$

But this function is not always real because of j^n. A real solution (4-13) is given by the function

$$I_n(x) = j^{-n}J_n(jx) = \sum_{k=0}^{\infty} \frac{x^{2k+n}}{2^{2k+n}k!\,(k+n)!} \tag{4-14}$$

This is known as the modified Bessel function of the first kind of order n. All of the terms in this series are positive and do not alternate in sign as in the $J_n(x)$ series. There will also be modified Bessel functions of negative order, and modified Bessel functions of the second kind of order n. $I_0(x)$ and $I_1(x)$ are plotted in Fig. 4-2. All of the modified Bessel functions increase monotonically with x.

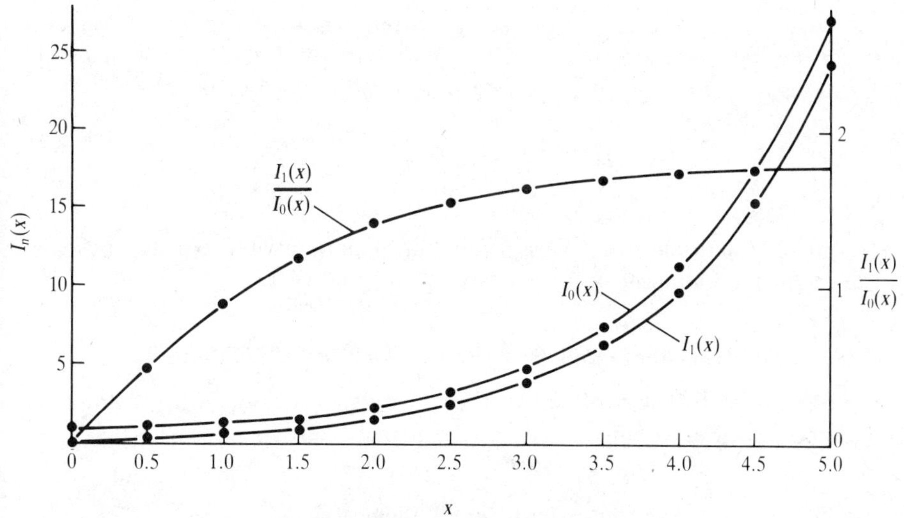

Fig. 4-2 Modified Bessel functions.

4-2.4 Bessel Functions of Fractional Order

When n is any positive number or any negative noninteger number, the generalized factorial, $\Gamma(k+n+1)$, replaces $(k+n)!$ in (4-8b). This allows the factorial of any positive number and of any negative noninteger number to be evaluated by the methods described earlier. Consequently, Bessel functions of the first kind and of fractional orders can be computed by summing the series.

However, when the fractional order is half an odd integer, the Bessel functions can be evaluated in closed form in terms of elementary functions. To show this, first let $n = \frac{1}{2}$ in (4-8b) and replace $(k+n)!$ by $\Gamma(k+n+1)$ to obtain

$$J_{1/2}(x) = \sum_{k=0}^{\infty} \frac{(-1)^k x^{2k+1/2}}{2^{2k+1/2} k!\, \Gamma\left(\dfrac{2k+3}{2}\right)} \tag{4-15}$$

Now expand (4-15) into several terms, use $\Gamma(\frac{1}{2}) = \sqrt{\pi}$ and $\Gamma(n+1) = n\Gamma(n)$, and then simplify the various terms to identify a familiar series for $\sin x$. Thus, (4-15) reduces to

$$J_{1/2}(x) = \sqrt{\frac{2}{\pi x}} \sin x \tag{4-16}$$

It is not difficult to show by a similar procedure that

$$J_{-1/2}(x) = \sqrt{\frac{2}{\pi x}} \cos x \tag{4-17}$$

A recurrence formula for Bessel functions whose order is half an odd integer can be obtained from (4-10a) or (4-10b) by replacing n with $p - \frac{1}{2}$, where p is a positive or negative integer. Thus (4-10b) becomes

$$J_{p+1/2}(x) = \frac{2p-1}{x} J_{p-1/2}(x) - J_{p-3/2}(x) \tag{4-18}$$

Thus, these fractional-order Bessel functions can be expressed, step-by-step, in terms of the elementary functions in (4-16) and (4-17).

4-2.5 Bessel Real (ber) and Bessel Imaginary (bei) Functions

In the study of the skin effect on the steady-state current on a cylindrical conductor, a differential equation of the form

$$r^2 \frac{d^2y}{dr^2} + r\frac{dy}{dr} - r^2(jp)y = 0$$

is obtained, where p is a constant whose value depends of the permeability and resistivity of the conductor, and on the frequency of the sinusoidal current. If we let $v = \sqrt{jp}\, r$, we obtain Bessel's modified equation of zero order (similar to (4-13))

$$v^2 \frac{d^2y}{dv^2} + v\frac{dy}{dv} - v^2 y = 0$$

whose solution is proportional to the modified Bessel function of the first kind and zero order denoted by $I_0(v)$. Then, in terms of the original variable, $I_0(v) = I_0(\sqrt{jp}\, r) = I_0(\sqrt{j}\, x)$, where $x = \sqrt{p}\, r$. Thus, using (2-18),

$$I_0(\sqrt{j}\, x) = \sum_{k=0}^{\infty} \frac{j^k x^{2k}}{2^{2k}(k!)^2} \tag{4-19}$$

This can be split into two series, one that is real when k is even, and one that is imaginary when k is odd. The real series is denoted by $\mathrm{ber}_0\,(x)$ and the imaginary series by $\mathrm{bei}_0\,(x)$, so that

$$I_0(\sqrt{j}\, x) = \mathrm{ber}_0\,(x) + j\,\mathrm{bei}_0\,(x)$$

Thus, from (4-19),

$$\mathrm{ber}_0\,(x) = 1 - \frac{x^4}{2^4(2!)^2} + \frac{x^8}{2^8(4!)^2} - \cdots$$

$$\mathrm{bei}_0\,(x) = \frac{x^2}{2^2} - \frac{x^6}{2^6(3!)^2} + \frac{x^{10}}{2^{10}(5!)^2} - \cdots$$

4-2.6 Bessel Polynomials

The second-order differential equation

$$x^2\frac{d^2y}{dx^2} + (2x+2)\frac{dy}{dx} - n(n+1) = 0 \tag{4-20}$$

where n is a positive integer, is a distant relative of Bessel's equation. One of its solutions is in the form of polynomials of order n, denoted by $y = D_n(x)$, and known as Bessel polynomials. Several of the Bessel polynomials are:

$$D_0(x) = 1$$
$$D_1(x) = 1 + x$$
$$D_2(x) = 1 + 3x + 3x^2$$
$$D_3(x) = 1 + 6x + 15x^2 + 15x^3$$
$$D_4(x) = 1 + 10x + 45x^2 + 105x^3 + 105x^4$$
$$D_5(x) = 1 + 15x + 105x^2 + 420x^3 + 945x^4 + 945x^5$$

It can be shown that any of these Bessel polynomials is a solution of (4-20) by simply substituting it into the differential equation. Their general form is

$$D_n(x) = \sum_{k=0}^{n} \frac{(n+k)!\, x^k}{2^k (n-k)!\, k!} \tag{4-21}$$

A recursion formula for the Bessel polynomials is

$$D_{n+1}(x) = (2n+1) \times D_n(x) + D_{n-1}(x)$$

The other solution of (4-20) is

$$y = e^{2/x} D_n(-x)$$

which is infinite at $x = 0$. Bessel polynomials can be expressed in terms of Bessel functions of half-integral order. They occur in the theory of traveling spherical waves and certain types of electric filters whose group delay is to be maintained as constant as possible. In the filter application, the polynomials that are encountered are known as the delay polynomials $D_n(p)$. They are obtained from the Bessel polynomials by replacing the variable x in (4-21) with $1/p$ and multiplying the result by p^n. Thus, (4-21) becomes

$$D_n(p) = \sum_{k=0}^{n} \frac{(n+k)!\, p^{n-k}}{2^k (n-k)!\, k!}$$

This is put into a little more convenient form by replacing $n - k$ by k, so that $n + k = 2n - k$ and $k = n - k$. Therefore, since the range for the sum is the same.

$$D_n(p) = \sum_{k=0}^{n} \frac{(2n-k)!\, p^k}{2^{n-k} k!\, (n-k)!} \tag{4-22}$$

Several delay polynomials are:

$$D_0(p) = 1$$
$$D_1(p) = 1 + p$$
$$D_2(p) = 3 + 3p + p^2$$
$$D_3(p) = 15 + 15p + 6p^2 + p^3$$
$$D_4(p) = 105 + 105p + 45p^2 + 10p^3 + p^4$$
$$D_5(p) = 945 + 945p + 420p^2 + 105p^3 + 15p^4 + p^5$$

A recursion formula for the delay polynomials is

$$D_{n+1}(p) = (2n+1)D_n(p) + p^2 D_{n-1}(p)$$

Notice that the coefficients of the delay polynomials are reversed from the coefficients of the corresponding Bessel polynomials.

This concludes our survey of Gamma and Bessel functions. In what follows, we will need only the Bessel functions of the first kind of order n, the modified Bessel functions of the first kind of order n, and the delay polynomials.

4-3 EXPANSIONS OF $\cos(\beta \sin \omega t)$ AND $\sin(\beta \sin \omega t)$

Instead of evaluating the coefficients of the trigonometric Fourier series by the standard integral formulas, an alternative procedure will be employed. It begins with the power series

$$\cos x = 1 - \frac{x^2}{2!} + \frac{x^4}{4!} - \frac{x^6}{6!} + \cdots$$

and

$$\sin x = x - \frac{x^3}{3!} + \frac{x^5}{5!} - \frac{x^7}{7!} + \cdots$$

where $x = \beta \sin \omega t$. Therefore, using the procedures involving Pascal's triangle for expanding $\sin^n \omega t$ described in Chapter 2, we find that

$$\begin{aligned} \cos(\beta \sin \omega t) = 1 &- \frac{\beta^2}{2!}\tfrac{1}{2}(1 - \cos 2\omega t) + \frac{\beta^4}{4!}\frac{1}{2^3}(\cos 4\omega t - 4\cos 2\omega t + 3) \\ &- \frac{\beta^6}{6!}\frac{1}{2^5}(-\cos 6\omega t + 6\cos 4\omega t - 15\cos \omega t + 10) \\ &+ \frac{\beta^8}{8!}\frac{1}{2^7}(\cos 8\omega t - 8\cos 6\omega t + 28\cos 4\omega t - 56\cos 2\omega t + 35) \\ &- \cdots \end{aligned} \tag{4-23}$$

Selecting from (4-23) the sum of all the constant terms only results in

$$\begin{aligned} \sum \text{constants} &= 1 - \frac{\beta^2}{2!\,2} + \frac{\beta^4 \cdot 3}{4!\,2^3} - \frac{\beta^6 \cdot 10}{6!\,2^5} + \frac{\beta^8 \cdot 35}{8!\,2^7} - \cdots \\ &= 1 - \frac{\beta^2}{2^2} + \frac{\beta^4}{2^4 2!\,2} - \frac{\beta^6}{2^6 3!\,2 \cdot 3} + \frac{\beta^8}{2^8 4!\,2 \cdot 3 \cdot 4} - \cdots \\ &= J_0(\beta) \end{aligned} \tag{4-24}$$

But the infinite series (4-24) is exactly the same as the series for the Bessel function of the first kind of zero order, identified as $J_0(\beta)$. Fortunately, these functions have already been calculated. A brief tabulation of $J_0(\beta)$ is given in Table 4-3.

Returning to (4-23), next group together all of the $\cos 2\omega t$ terms and factor $\cos 2\omega t$ from each term to obtain the coefficient in the form of an infinite series. Thus,

$$\begin{aligned}\cos 2\omega t \text{ coefficient} &= \frac{\beta^2}{2!\,2} - \frac{\beta^4 4}{4!\,2^3} + \frac{\beta^6 15}{6!\,2^5} - \frac{\beta^8 56\cdot 4}{8!\,2^7} + \cdots \\ &= \frac{\beta^2}{2^2}\left[1 - \frac{\beta^2}{2^2 3} + \frac{\beta^4}{2^4 2!\,3\cdot 4} - \frac{\beta^6}{2^6 3!\,3\cdot 4\cdot 5} + \cdots\right] \\ &= 2J_2(\beta) \end{aligned} \tag{4-25}$$

It turns out that the series (4-25) is exactly twice the series known as the Bessel function of the first kind of second order, identified as $J_2(\beta)$. Some tabulated values are given in Table 4-3. Notice that the order of the Bessel function is the same as the integer harmonic number.

In similar fashion, the coefficient of $\cos 4\omega t$ will be an infinite series. Thus, from (4-23),

$$\cos 4\omega t \text{ coefficient} = \frac{\beta^4}{2^3 4!}\left[1 - \frac{\beta^2}{2^2 5} + \frac{\beta^4}{2^4 2\cdot 5\cdot 6} - \cdots\right] = 2J_4(\beta) \tag{4-26}$$

When this process is continued for as many other coefficients as desired, it is found that

$$\begin{aligned}\cos(\beta \sin \omega t) &= J_0(\beta) + 2J_2(\beta)\cos 2\omega t + 2J_4(\beta)\cos 4\omega t + \cdots \\ &\quad + 2J_n(\beta)\cos n\omega t + \cdots \\ &= J_0(\beta) + \sum_{\text{even } n=2}^{\infty} 2J_n(\beta)\cos n\omega t \end{aligned} \tag{4-27}$$

The expansion of $\sin(\beta \sin \omega t)$ proceeds in similar manner, details of which are reserved for the Problems at the end of this chapter. Consequently,

$$\sin(\beta \sin \omega t) = \sum_{\text{odd } n=1}^{\infty} 2J_n(\beta)\cos n\omega t \tag{4-28}$$

Tabulated values of several Bessel functions are given in Table 4-3.

4-4 EXPANSION OF $e^{x \cos \omega t}$

The series representation of an exponential is

$$e^a = 1 + \frac{a}{1!} + \frac{a^2}{2!} + \frac{a^3}{3!} + \cdots + \frac{a^n}{n!} + \cdots \tag{4-29}$$

provided that a^2 is finite. Therefore,

$$e^{x \cos \omega t} = 1 + (x \cos \omega t) + \frac{(x \cos \omega t)^2}{2!} + \frac{(x \cos \omega t)^3}{3!} + \cdots \tag{4-30}$$

Now, using the procedures involving Pascal's triangle for expanding $\cos^n \omega t$ described earlier, we find that a few terms are

$$e^{x \cos \omega t} = 1 + x \cos \omega t + \frac{x^2}{4}(\cos 2\omega t + 1) + \frac{x^3}{24}(\cos 3\omega t + 3 \cos \omega t)$$
$$+ \frac{x^4}{192}(\cos 4\omega t + 4 \cos 2\omega t + 3) + \cdots \tag{4-31}$$

Next, group together those terms having the same frequency. Thus,

$$e^{x \cos \omega t} = \left(1 + \frac{x^2}{4} + \frac{x^4}{64} + \cdots\right) + \left(x + \frac{x^3}{8} + \cdots\right) \cos \omega t$$
$$+ \left(\frac{x^2}{4} + \frac{x^4}{48} + \cdots\right) \cos 2\omega t + \cdots \tag{4-32}$$

This represents the trigonometric Fourier series of $e^{x \cos \omega t}$. Each Fourier coefficient is an infinite series. To see that these Fourier coefficients are actually in the same forms of familiar series, compare them with some terms of a few modified Bessel functions. Thus,

$$I_0(x) = 1 + \frac{x^2}{4} + \frac{x^4}{64} + \cdots \tag{4-33a}$$

$$I_1(x) = \frac{x}{2} + \frac{x^3}{2 \cdot 8} + \cdots \tag{4-33b}$$

$$I_2(x) = \frac{x^2}{2 \cdot 4} + \frac{x^4}{2 \cdot 48} + \cdots \tag{4-33c}$$

and the pattern now becomes obvious: the Fourier coefficients are related to the modified Bessel functions. Therefore,

$$e^{x \cos \omega t} = I_0(x) + 2I_1(x) \cos \omega t + 2I_2(x) \cos 2\omega t + \cdots$$
$$= I_0(x) + \sum_{n=1}^{\infty} 2I_n(x) \cos n\omega t \tag{4-34}$$

An important application of this expansion pertains to circuits like the ones shown in Fig. 4-3. In Fig. 4-3a the reverse-biasing voltage V_R is

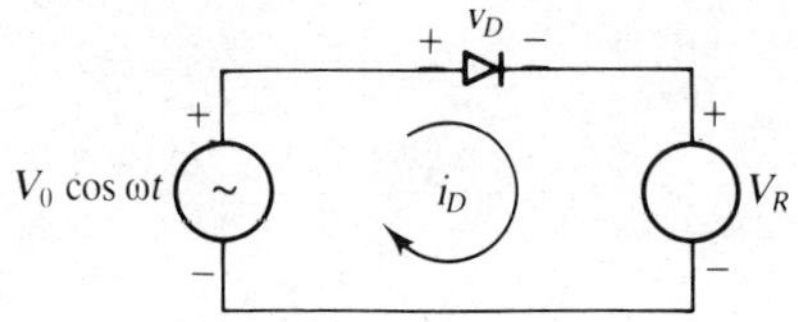

(a)

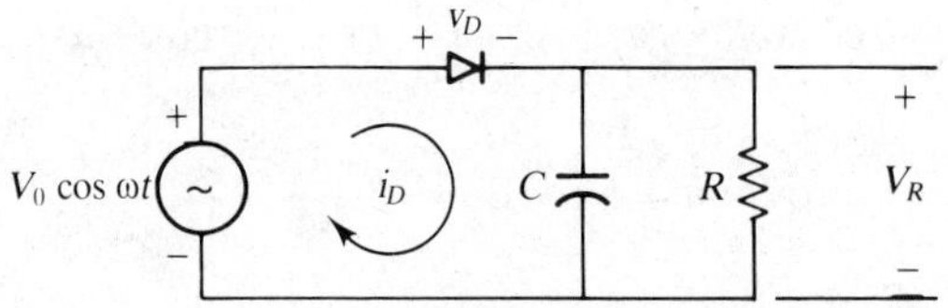

(b)

Fig. 4-3 Diode rectifiers. (a) A DC voltage source produces a constant reverse-biasing voltage V_R that is independent of the current. (b) A long RC time constant produces a reverse-biasing voltage that depends on the current.

constant and independent of the current. The total current is, from the diode equation,

$$i_D = I_s(e^{qv_D/kT} - 1) = I_s(e^{v_D/V_T} - 1)$$

where I_s is the reverse saturation current, v_D is the total diode voltage, and V_T involves Boltzmann's constant k, temperature T in kelvin, and the electronic charge q in coulombs ($V_T = 25$ mV at 290 K). Since $v_D = V_0 \cos \omega t - V_R$, the current becomes

$$i_D = I_s e^{-V_R/V_T} e^{V_0 \cos \omega t/V_T} - I_s = I_s e^{-V_R/V_T} e^{x \cos \omega t} - I_s$$

where $x = V_0/V_T$. Substituting its expansion for $e^{x \cos \omega t}$, this becomes

$$i_D = [I_s e^{-V_R/V_T} I_0(x) - I_s] + 2I_s e^{-V_R/V_T} \sum_{n=1}^{\infty} I_n(x) \cos n\omega t \tag{4-35}$$

The first term (in square brackets) represents the average current and the second term, which involves the summation, represents the sum of the various harmonic components contained in the diode current. Since zero resistance is shown in Fig. 4-3a, the only reverse voltage acting is V_R, which is constant and independent of the current. Consequently, as the amplitude V_0 of the input sinusoid is increased, the value of x increases, therefore $I_0(x)$ and $I_n(x)$ both increase, so that, in turn, the average current and the amplitudes of all harmonic components also increase continually.

A similar development can be carried out for a bipolar junction transistor by approximating its operation by the Ebers–Moll equations. The collector and emitter currents are both exponentially related to the base–emitter voltage in a form parallel to that of the diode equation, therefore the currents behave similarly to i_D above.

Of course, actual operating circuits are not ideal as was assumed in Fig. 4-3a. The signal source has its own output resistance R_o, the diode (or other device) has some internal resistance R_i, and generally there is other external resistance R across which the output voltage is obtained. Usually, however, $R \gg R_o + R_i$, and the small values are omitted in Fig. 4-3b. Now, any current produces a constant voltage component in the reverse direction given by $V_R = RI_D$, where I_D is the average current. The voltage V_R is no longer constant as before but increases as the amplitude V_0 of the input sinusoid increases. When $V_R \ll V_T = 25$ V, the exponentials in (4-35) are essentially unity, and the currents vary with V_0 as though $V_R = 0$. But as V_0 increases, i_D and I_D increase, which increases V_R, and this, in turn, makes the exponentials smaller. Thus, the currents are not as large as expected for constant $V_R = 0$. When applied to any of the harmonic components, this is a manifestation of compression.

The compression effect is more clearly in evidence with a bias-stabilized junction transistor, where feedback prevents any significant change in the average collector current. It is as though the average current in (4-35) tends to be held to a constant value I_Q. Thus,

$$I_s e^{-V_R/V_T} I_0(x) - I_s = I_Q$$

from which is obtained

$$I_s e^{-V_R/V_T} = \frac{I_Q + I_s}{I_0(x)} \tag{4-36}$$

Solving (4-36) for V_R reveals how it must change, when V_0 changes, to maintain I_Q constant. When the above substituted into (4-35), we obtain

$$i_D = I_Q + 2(I_Q + I_s) \sum_{n=1}^{\infty} \frac{I_n(x)}{I_0(x)} \cos n\omega t \tag{4-37}$$

Consequently, if the average current in (4-35) is held constant, the amplitudes of the harmonic components now depend on the ratio $I_n(x)/I_0(x)$ instead of on just $I_n(x)$. The ratio $I_1(x)/I_0(x)$ for the fundamental component is plotted in Fig. 4-2, and its shape demonstrates the compression phenomenon caused by the feedback mechanism.

PART 2 Filters

4-5 STEADY-STATE AMPLITUDE RESPONSE

Consider a transfer function like

$$H(s)=\frac{1+b_1s+b_2s^2+b_3s^3+\cdots+b_ms^m}{1+a_1s+a_2s^2+a_3s^3+\cdots+a_ns^n} \tag{4-38}$$

in which $n>m$. Then, using $s=j\omega$, the sinusoidal steady-state transfer function becomes

$$H(j\omega)=\frac{(1-b_2\omega^2+b_4\omega^4-b_6\omega^6+\cdots)+j(b_1\omega-b_3\omega^3+b_5\omega^5-\cdots)}{(1-a_2\omega^2+a_4\omega^4-a_6\omega^6+\cdots)+j(a_1\omega-a_3\omega^3+a_5\omega^5-\cdots)} \tag{4-39}$$

Notice that the signs alternate in each of the terms in (4-39), that the index of each coefficient is the same as the exponent of the corresponding ω factor, that the real parts are even functions of frequency, and that the imaginary parts are odd functions of frequency. Obviously, $H(j\omega)$ is a complex quantity having magnitude $|H(j\omega)|$ and an angle $\phi(\omega)$, or $H(j\omega)=|H(j\omega)|\underline{/\phi(\omega)}$. The variations of $|H(j\omega)|$ and $\phi(\omega)$ with frequency are called the *amplitude response* and the *phase response* respectively. It is seen that $|H(j\omega)|$ involves the square root of the sum of the squares of the real and imaginary parts of the numerator and the denominator, and that $\phi(\omega)$ depends on the difference between the inverse tangents of the imaginary parts divided by the real parts.

The square roots that occur in $|H(j\omega)|$ are eliminated by squaring. Therefore, after expanding and collecting terms, we obtain

$$|H(j\omega)|^2=\frac{\begin{array}{l}1+(b_1^2-2b_2)\omega^2+(b_2^2-2b_1b_3+2b_4)\omega^4\\ \qquad +(b_3^2-2b_2b_4+2b_1b_5-2b_6)\omega^6+\cdots+b_m^2\omega^{2m}\end{array}}{\begin{array}{l}1+(a_1^2-2a_2)\omega^2+(a_2^2-2a_1a_3+2a_4)\omega^4\\ \qquad +(a_3^2-2a_2a_4+2a_1a_5-2a_6)\omega^6+\cdots+a_n^2\omega^{2n}\end{array}} \tag{4-40}$$

For the amplitude response to decrease as ω becomes larger, to obtain a lowpass filter characteristic, it is necessary that $n>m$, or the highest power of ω in the denominator must be larger than the highest power of ω in the numerator.

It is helpful to deduce a general pattern whereby the calculated coefficients in (4-40) can be quickly evaluated, rather than by going through the tedious algebraic manipulations required to obtain (4-40) for higher values of n. That pattern, obtained by inspection, is as follows:

1. The signs alternate in each calculated coefficient, starting with +.
2. The first and positive term in the calculated coefficient of ω^{2i} comes from a_i or b_i in (4-38). Thus, a_3s^3 in (4-38) leads to $a_3^2\omega^6$.
3. The second term is -2 times the product of the original coefficients in (4-38) on either side once removed from a_i or b_i. Since a_2 and a_4 are on either side once removed from a_3, that second term is $-2a_2a_4$.
4. The third term is $+2$ times the product of the original coefficients in (4-38) on either side twice removed from a_i or b_i. Thus, a third term is $+2a_1a_5$.
5. Continue the same process for each calculated coefficient one at a time until no more can be calculated. Notice that if an original polynomial in s contains an even number of terms, then the central calculated coefficient has the most terms to calculate, but if the original polynomial contains an odd number of terms, then the central *two* calculated coefficients both have the most terms.

Example 4-1

If $A(s) = 1 + a_1s + a_2s^2 + a_3s^3 + a_4s^4 + a_5s^5$, determine $|A(j\omega)|^2$ by inspection.

Solution Using the pattern just described, $|A(j\omega)|^2$ is written immediately. Thus,

$$|A(j\omega)|^2 = 1 + (a_1^2 - 2a_2)\omega^2 + (a_2^2 - 2a_1a_3 + 2a_4)\omega^4$$
$$+ (a_3^2 - 2a_2a_4 + 2a_1a_5)\omega^6 + (a_4^2 - 2a_3a_5)\omega^8 + a_5^2\omega^{10}$$

Observe that products involving a_6, a_7, a_8, etc., are zero because $a_6 = a_7 = a_8 = a_9 = a_{10} = 0$, otherwise $A(s)$ would not end with a_5s^5.

4-6 MAXIMALLY-FLAT AMPLITUDE RESPONSE

Observe that negative signs occur in the calculated coefficients in (4-40). This means that the original values of $a_1, a_2, \ldots, a_n$, and $b_1, b_2, \ldots, b_m$ in (4-38) determine whether some of the coefficients in (4-40) are negative, zero, or positive. Negative coefficients can lead to ripple in the amplitude response, rather than a flat type of response. Ripple does not occur when the numerator and denominator terms track one another.

It is often desired that the amplitude response be as flat as possible. The greatest degree of flatness is obtained when all of the coefficients in (4-40) are zero. Then, $|H(j\omega)| = 1$, which is as flat as can be obtained and all frequencies are passed equally. For a lowpass response to take place, however, one or more positive coefficients must predominate at high frequencies in the denominator. Onc measure of flatness pertains to how many derivatives of $|H(j\omega)|^2$ equal zero at $\omega = 0$. Then, if the slope of the slope of the slope, etc., is zero at $\omega = 0$, $|H(j\omega)|^2$ remains essentially flat over a large frequency range as ω increases, and then drops off rapidly. To investigate these derivatives, it is convenient to express (4-40) as

$$|H(j\omega)|^2 = \frac{1 + N_2\omega^2 + N_4\omega^4 + N_6\omega^6 + \cdots + N_{2m}\omega^{2m}}{1 + D_2\omega^2 + D_4\omega^4 + D_6\omega^6 + \cdots + D_{2n}\omega^{2n}} \tag{4-41}$$

where $n > m$ for a lowpass response. Then, by long division, (4-41) becomes

$$\begin{aligned}|H(j\omega)|^2 = 1&\\ &+ (N_2 - D_2)\omega^2\\ &+ [(N_4 - D_4) - (N_2 - D_2)]\omega^4\\ &+ [(N_6 - D_6) - (N_4 - D_4)D_2 + (N_2 - D_2)(D_2^2 - D_4)]\omega^6\\ &+ [(N_8 - D_8) - (N_6 - D_6)D_2 + (N_4 - D_4)(D_2^2 - D_4)\\ &- (N_2 - D_2)(D_2^3 - 2D_2D_4 + D_6)]\omega^8\\ &+ [(N_{10} - D_{10}) - (N_8 - D_8)D_2 + (N_6 - D_6)(D_2^2 - D_4)\\ &- (N_4 - D_4)(D_2^3 - 2D_2D_4 + D_6) + (N_2 - D_2)(D_2^4 - 3D_2^2D_4\\ &+ 2D_2D_6 - D_8 + D_4^2)]\omega^{10}\\ &+ \text{higher power terms in } \omega^2\end{aligned} \tag{4-42}$$

Although certain patterns can be discerned among the various terms in (4-42), they will not be needed in what follows and so they will not be pursued here. Notice, however, that the numerator coefficients N_{2i} occur only with the denominator coefficients D_{2i} in the differences $(N_{2i} - D_{2i})$, and that those differences are factors in every term of (4-42).

The reason for using long division in (4-41) is that the result in (4-42) is easy to differentiate with respect to ω and to evaluate the derivatives at $\omega = 0$. Thus, with $m = 2$ and $n = 3$, for example, (4-42) becomes

$$\begin{aligned}\frac{d\,|H|^2}{d\omega} &= 2(N_2 - D_2)\omega + 4[(N_4 - D_4) - (N_2 - D_2)D_2]\omega^3\\ &\quad + 6[-D_6 - (N_4 - D_4)D_2 + (N_2 - D_2)(D_2^2 - D_4)]\omega^5 + \cdots\end{aligned}$$

which is zero at $\omega = 0$. Then, differentiating again,

$$\frac{d^2|H|^2}{d\omega^2} = 2(N_2 - D_2) + 12[(N_4 - D_4) - (N_2 - D_2)D_2]\omega^2$$
$$+ 30[-D_6 - (N_4 - D_4)D_2 + (N_2 - D_2)(D_2^2 - D_4)]\omega^4 + \cdots$$

which is zero at $\omega = 0$ provided that $N_2 = D_2$. Consequently,

$$\frac{d^2|H|^2}{d\omega^2} = 0 + 12(N_4 - D_4)\omega^2 + 30[-D_6 - (N_4 - D_4)D_2]\omega^4 + \cdots$$

Thus, the next derivative is

$$\frac{d^3|H|^2}{d\omega^3} = 24(N_4 - D_4)\omega + 120[-D_6 - (N_4 - D_4)D_2]\omega^3 + \cdots$$

which is zero at $\omega = 0$. Continuing,

$$\frac{d^4|H|^2}{d\omega^4} = 24(N_4 - D_4) + 360[-D_6 - (N_4 - D_4)D_2]\omega^2 + \cdots$$

which is zero at $\omega = 0$ provided that $N_4 = D_4$. Hence,

$$\frac{d^4|H|^2}{d\omega} = 360(-D_6)\omega^2 + \cdots$$

which is zero at $\omega = 0$. Further,

$$\frac{d^5|H|^2}{d\omega^5} = -720D_6\omega + \cdots$$

which is zero at $\omega = 0$. Finally,

$$\frac{d^6|H|^2}{d\omega^6} = -720D_6 + \cdots$$

which cannot be equal to zero because of D_6. In general, the maximum number of derivatives that can be made equal to zero at $\omega = 0$ is given by $(2n - 1)$, or 5 in this example. That there is a maximum number of derivatives, each made equal to zero at $\omega = 0$, is the origin of the terminology *maximally flat* amplitude response. As shown above, this requires $N_{2i} = D_{2i}$, $i = 1, 2, 3, \ldots, m$, that is, the numerator and denominator coefficients of like powers in ω^2 must be equal to D_{2n} in (4-41). The term "maximally flat" was coined in 1941 by Landon.

When the ratio of polynomials in ω^2 expressed by (4-41) represents an amplitude-squared function, the response is said to be maximally flat amplitude. If the ratio of polynomials represents group delay, the response is said to be maximally flat group delay, which also means maximally *linear* phase response.

Example 4-2

$$H(s) = \frac{1 + 2s + 3s^2}{1 + a_1 s + a_2 s^2 + a_3 s^3}$$

(a) Write $|H(j\omega)|^2$ by inspection.
(b) For $|H(j\omega)|^2$ to be maximally flat amplitude and have the value $\frac{1}{2}$ at $\omega = 1$, calculate a_1, a_2, and a_3.
(c) Plot $|H(j\omega)|^2$ for $0 \leqslant \omega \leqslant 1.1$ in steps of 0.1.

Solution

(a)
$$|H(j\omega)|^2 = \frac{1 - 2\omega^2 + 9\omega^4}{1 + (a_1^2 - 2a_2)\omega^2 + (a_2^2 - 2a_1 a_3)\omega^4 + a_3^2 \omega^6}$$

(b) Equating coefficients of like powers of ω^2 in the above yields

$$a_1^2 - 2a_2 = -2 \tag{4-43}$$

$$a_2^2 - 2a_1 a_3 = 9 \tag{4-44}$$

Then $|H(j\omega)|^2 = 1$ at $\omega = 1$ results in

$$1 + (a_1^2 - 2a_2) + (a_2^2 - 2a_1 a_3) + a_3^2 = 2 - 4 + 18 = 16 \tag{4-45}$$

Using (4-43) and (4-44) in (4-45), it is found that

$$a_3 = \sqrt{8} = 2.8284$$

Substituting this into (4-43) and (4-44) and eliminating a_2 results in

$$a_1^4 + 4a_1^2 - 22.6274a_1 - 32 = 0$$

Routh's criterion indicates that there is one root with a positive real part. Numerical analysis shows that

$$a_1 = 2.8284$$

Consequently, from (4-43),

$$a_2 = \frac{a_1^2 + 2}{2} = 5$$

TABLE 4-5 Calculated Values for Example 4-2

ω	0	0.1	0.2	0.3	0.4	0.5	0.6	0.7	0.8	0.9	1.0	1.1
$\lvert H(j\omega)\rvert^2$	1	1	0.99	0.99	0.97	0.89	0.79	0.70	0.62	0.55	0.5	0.45

(c) Substituting the values for a_1, a_2, and a_3 into $|H(j\omega)|^2$ results in

$$|H(j\omega)|^2 = \frac{1 - 2\omega^2 + 9\omega^4}{1 - 2\omega^2 + 9\omega^4 + 8\omega^6}$$

Notice that the numerator terms are tracked exactly by the first three terms in the denominator as expected for maximally flat amplitude response. The last term, $8\omega^6$, in the denominator, becomes more effective as the frequency is increased. Calculations yield the values of Table 4-5. These values are plotted in Fig. 4-4 where the maximally flat lowpass filter characteristic is seen.

4-6.1 Butterworth Functions

The maximally flat approximation provides an essentially constant amplitude response over a large portion of the desired passband. The response at high frequencies decreases more rapidly when the highest power of ω^2 in the numerator of (4-40) is made ever smaller than the highest power of ω^2 in the denominator. The best that can be done in this regard is to allow the

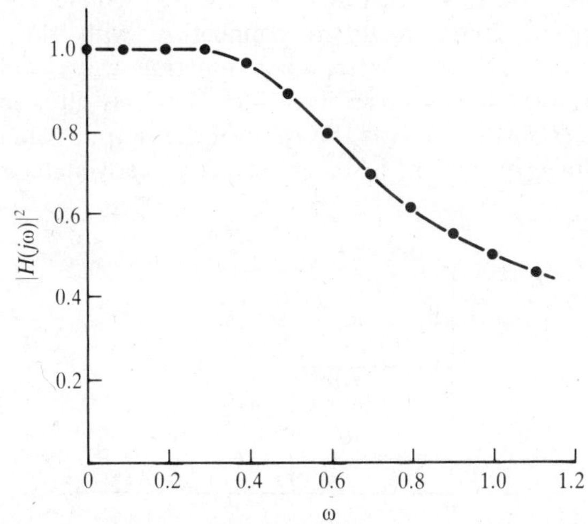

Fig. 4-4 Amplitude response (squared) of filter in Example 4-2.

numerator to contain *no* ω^{2i} terms. Thus, let

$$H(s) = \frac{1}{1 + a_1 s + a_2 s^2 + \cdots + a_n s^n} \tag{4-46}$$

so that, by inspection,

$$|H(j\omega)|^2 = \frac{1}{1 + (a_1^2 - 2a_2)\omega^2 + (a_2^2 - 2a_1 a_3 + 2a_4)\omega^4 + \cdots + a_n^2 \omega^{2n}} \tag{4-47}$$

For this to be maximally flat and equal to $\frac{1}{2}$ at $\omega = 1$, for any n, requires that

$$\begin{aligned} a_n &= 1 \\ a_1^2 - 2a_2 &= 0 \\ a_2^2 - 2a_1 a_3 + 2a_4 &= 0 \\ a_3^2 - 2a_2 a_4 + 2a_1 a_5 - 2a_6 &= 0 \\ \vdots \quad & \quad \vdots \end{aligned}$$

The resulting maximally flat function becomes

$$|H(j\omega)| = \frac{1}{\sqrt{1 + \omega^{2n}}} \tag{4-48}$$

This is known as a Butterworth function. It was originated in 1930 by the British engineer S. Butterworth in connection with the response of vacuum-tube amplifiers. The Butterworth function was developed to approximate the amplitude response of an ideal lowpass filter represented in Fig. 4-5. There, $H(j\omega)$ represents the ratio of the output voltage or current to the input voltage or current in the sinusoidal steady-state condition as a

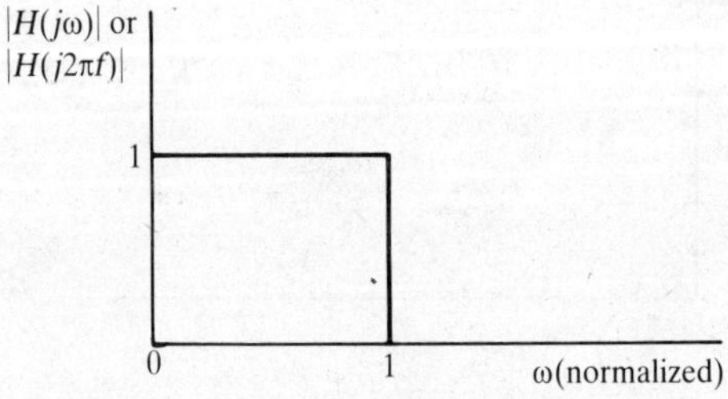

Fig. 4-5 Normalized ideal lowpass filter.

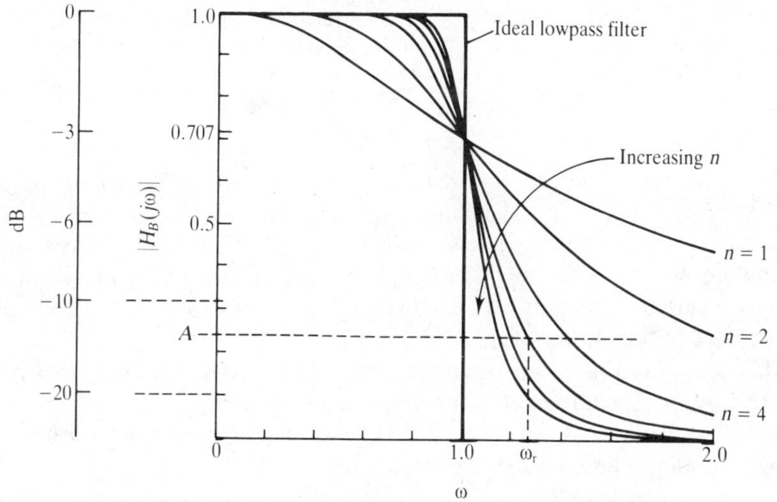

Fig. 4-6 The Butterworth functions (normalized ω).

function of frequency. The Butterworth function is also characterized by the fact that $(2n - 1)$ derivatives with respect to ω are equal to zero at $\omega = 0$.

The ideal lowpass filter allows all frequencies f up to a certain cutoff frequency f_c to pass through the filter circuitry but prevents all frequencies above the cutoff frequency from passing through. Frequencies are said to be normalized when all frequencies are divided by the cutoff frequency f_c. When this is done, the ideal lowpass filter is called a normalized lowpass filter, and it is shown in Fig. 4-5. Since most of the filter functions that we will be using will be normalized, we will let f represent normalized frequency (i.e., actual frequency divided by the cutoff frequency) and make a special point of the matter when reference is made to actual frequency.

The normalized ideal lowpass filter is approximated by the Butterworth function given in (4-48), where n is a positive integer whose value represents the order of the filter and $\omega = 2\pi f$ is the normalized radian frequency. The function expressed by (4-48) approaches unity as ω approaches zero, and it decreases rapidly as ω increases above 1. This response is shown in Fig. 4-6 for several values of the integer n. Observe that the Butterworth function is equal to 0.707 or -3 dB at $\omega = 1$ regardless of the value of n and that it more nearly approximates the ideal shape as n gets progressively larger.

Notice also that the function becomes flatter over a greater range of frequencies as n gets larger.

4-6.2 Rolloff and Attenuation

At high frequencies, the rate at which the Butterworth function decreases in magnitude, for the various values of n, is called "roll off." It is determined by expressing (4-48) in dB when the frequencies are high, that is, $\omega \gg 1$. Then,

$$|H(j\omega)|_{\text{dB}} \cong -20n \log \omega$$

If now ω is varied over an octave (i.e., the frequency is changed by a factor of 2) the corresponding dB change will be $-6n$ dB. If ω is varied over a decade (i.e., the frequency is changed by a factor of 10), the corresponding dB change will be $-20n$ dB. The amplitude response is said to roll off (i.e. decrease) at the rate of $6n$ dB/octave or $20n$ dB/decade. Roll off is determined by the value of n.

Also shown in Fig. 4-6 is the intersection of a line for a certain magnitude A with any of the curves at a (normalized) frequency ω_r as determined by the value of n. Since $A < 1$, then $20 \log A$ yields negative dB, or attenuation instead of gain. Thus, (4-48) becomes

$$A = \frac{1}{\sqrt{1 + \omega_r^{2n}}}$$

from which is obtained

$$n = \frac{\log\left(\dfrac{1}{A^2} - 1\right)}{2 \log \omega_r} \tag{4-49}$$

When the calculated value of n is not an integer, the value should be rounded to the next larger integer (i.e., 4.06 becomes 5). Equation (4-49) yields the order of a Butterworth filter to produce an attenuation of at least A at a specified ω_r when, if necessary, the calculated value of n is rounded to the next larger integer.

Example 4-3

Determine the value of n required for a Butterworth function to yield $A \leq 0.1$ at $\omega_r = 2$.

Solution Substitute the given values into (4-49) to obtain $n = 3.32$, which when rounded to the next larger integer, means that $n = 4$ is required. This rounding to the next larger integer assures that $A < 0.1$ at $\omega_r = 2$, and that the value of n obtained is the smallest usable value.

4-6.3 The Poles of the Butterworth Function

Knowing the value of n from the specifications of A and ω_r does not help much directly with the design of the (normalized) circuit. It is helpful, however, if the transfer function $H(s)$ of the filter to be built from passive components can be determined. This will be done in steps, starting with (4-48).

The square root in (4-48) is eliminated by squaring. Hence,

$$|H(j\omega)|^2 = H(j\omega) \cdot H^*(j\omega) = H(j\omega) \cdot H(-j\omega) = \frac{1}{1+\omega^{2n}} \tag{4-50}$$

where * indicates complex conjugate, and it is assumed that the coefficients in $H(j\omega)$ are real and that $H^*(j\omega) = H(-j\omega)$. For the sinusoidal steady-state condition, use the complex frequency $s = j\omega$ and $\omega = s/j$ in (4-50). Thus,

$$H(s) \cdot H(-s) = \frac{1}{1+(s/j)^{2n}} \tag{4-51}$$

The function on the right-hand side of (4-51) has poles with both negative and positive real parts, as will be shown below. Those poles with negative real parts will be associated with $H(s)$, which is a transfer function that can be realized with the desired passive components. Those poles having positive real parts are assigned to the function $H(-s)$ and discarded, because a transfer function with positive real parts cannot be realized with passive components.

The poles of the function on the right-hand side of (4-51) are found from the values of s for which the denominator equals zero. Hence,

$$\begin{aligned} \left(\frac{s}{j}\right)^{2n} &= -1 + j0 \\ &= \cos\,(2k-1)\pi + j\sin\,(2k-1)\pi, \qquad k = 1, 2, 3, \ldots \\ &= e^{j(2k-1)\pi} \end{aligned}$$

Then,

$$\begin{aligned} \frac{s}{j} &= e^{j(2k-1)\pi/2n} \\ &= \cos\frac{(2k-1)\pi}{2n} + j\sin\frac{(2k-1)\pi}{2n} \end{aligned}$$

and, finally,

$$s = -\sin\frac{(2k-1)90^\circ}{n} + j\cos\frac{(2k-1)90^\circ}{n} \tag{4-52}$$

Since only negative real parts are realizable here, the sine itself in (4-52) must be positive because of the preceding negative sign, and its angle must be in the first or second quadrants, or

$$\frac{(2k-1)90°}{n} < 180°$$

Thus,

$$k < \frac{2n+1}{2} = n + \tfrac{1}{2}$$

But k and n must be integers, hence

$$k \leq n \tag{4-53}$$

assures that the real parts of s are negative. If $n = 2$, then k takes the values 1 and 2 in (4-52). If $n = 3$, then k takes the values 1, 2, and 3.

When (4-52) is expressed in polar form, that is, $s = |s|\underline{/\phi}$, it is seen that $|s| = 1$ and that $\phi = 90° + |\theta|$ as shown in Fig. 4-7. Therefore, the poles of $H(s)$ lie on the arc of a unit circle in the left half of the s-plane. But it is also noticed in Fig. 4-7 that

$$|\theta| = \tan^{-1} \frac{\sin \dfrac{(2k-1)90°}{n}}{\cos \dfrac{(2k-1)90°}{n}} = \tan^{-1} \tan \frac{(2k-1)90°}{n}$$

$$= \frac{(2k-1)90°}{n}$$

Hence,

$$\phi = 90° + (2k-1)\frac{90°}{n} \tag{4-54}$$

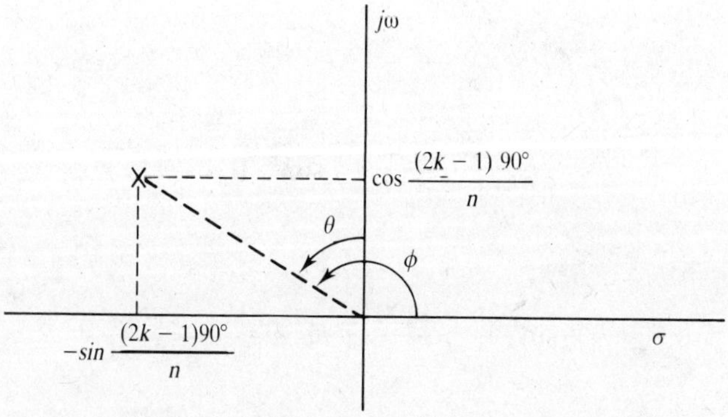

Fig. 4-7 Construction showing the location of a pole of a Butterworth filter.

A pole is at $-1 + j0 = 1/\underline{180°}$ when $\phi = 90° + 90°$, obtained for $2k - 1 = n$ or $k = (n + 1)/2$. Thus, when n is an odd integer, there is an allowable value for k which makes $\phi = 180°$ and also satisfies the requirement that $k \leqslant n$. Further, no poles are located at $\phi = 90°$ or at $\phi = 270°$ because $(2k - 1)/n$ cannot be zero, nor can it be any even multiple of 90°.

The locations in the s-plane of the useful poles (useful because they have negative real parts) of several Butterworth filters are shown by the crosses in Fig. 4-8. The unused poles (unused because they have positive real parts are shown by the encircled crosses.

Observe that all of the poles are equally spaced around the unit circle, that the angle between the radii to adjacent poles is $180°/n$, and that when n is an odd integer a useful pole always occurs at $s = -1$. These facts allow the poles to be evaluated quickly. Thus, for $n = 4$ in Fig. 4-8, the poles

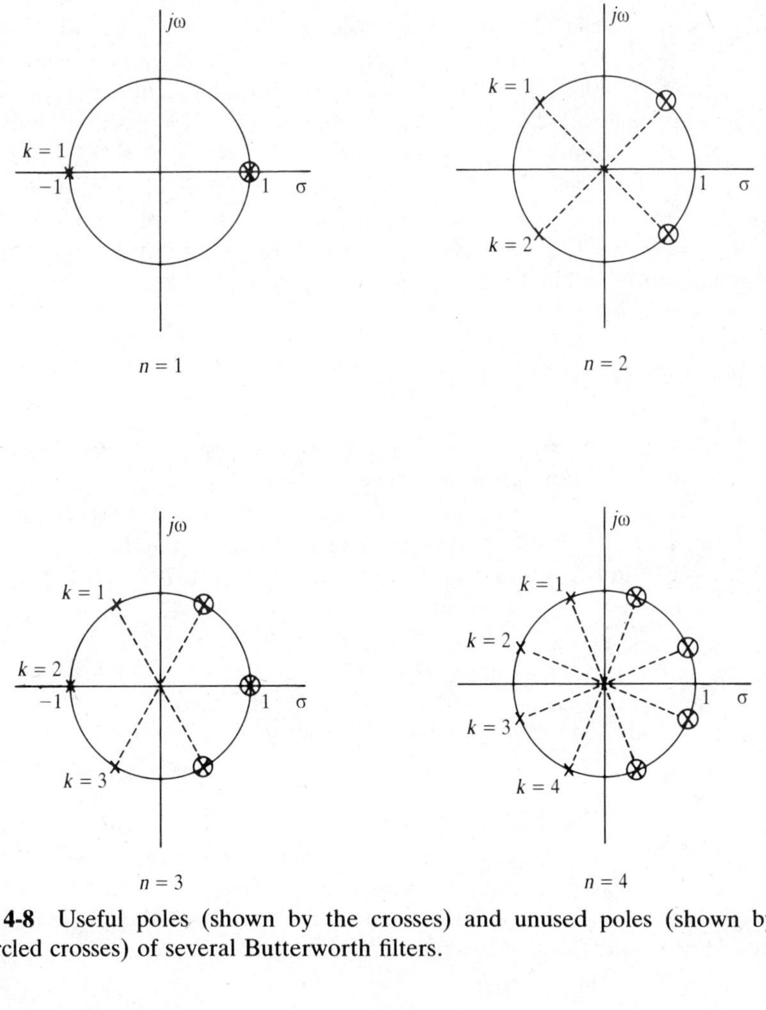

Fig. 4-8 Useful poles (shown by the crosses) and unused poles (shown by the encircled crosses) of several Butterworth filters.

TABLE 4-6 Butterworth Polynomials

n	
1	$s+1$
2	$s^2+\sqrt{2}\,s+1$
3	$(s+1)(s^2+s+1)=s^3+2s^2+2s+1$
4	$(s^2+0.7654s+1)(s^2+1.8478s+1)$
	$=s^4+2.6132s^3+3.4143s^2+2.6132s+1$
5	$(s+1)(s^2+0.6180s+1)(s^2+1.6180s+1)$
	$=s^5+3.2360s^4+5.2360s^3+5.2360s^2+3.2360s+1$

are separated by $180°/4=45°$. The pole marked $k=1$ has the value $1\angle 90°+22.5°=-0.3827+j0.9239$. The pole marked $k=4$ has the value $1\angle -112.5°=-0.3827-j0.9239$. The corresponding factors that produce these poles are $(s+0.3827-j0.9339)$ $(s+0.3927+j0.9239)$ which becomes $(s^2+0.7654s+1)$. Similarly, the poles marked $k=2$ and $k=3$ have the values $1\angle 180°-22.5°=-0.9239+j0.3827$ and $1\angle 180°+22.5°=-0.9239-j0.3827$, respectively. Their corresponding factors are $(s+0.9239-j0.3827)(s+0.9239+j0.3827)=(s^2+1.8478s+1)$. These quadratics agree with those given in Table 4-6 for $n=4$.

4-6.4 The Butterworth Polynomials

The transfer function $H(s)$ is immediately known from the pole locations shown in Fig. 4-8. For example, when $n=1$, the pole is at $s=-1$. Therefore, $H(s)=1/(s+1)$ because the factor $s+1$ has a root at $s=-1$. The quantity $(s+1)$ is the first-order Butterworth polynomial.

When $n=2$, the two poles are located at $1\angle 135°=-0.707+j0.707$ and at $1\angle 225°=-0.707-j0.707$. Therefore,

$$H(s)=\frac{1}{(s+0.707-j0.707)(s+0.707+j0.707)}$$

$$=\frac{1}{s^2+\sqrt{2}\,s+1}$$

The quantity $(s^2+\sqrt{2}\,s+1)$ is the second-order Butterworth polynomial.

When $n=3$, the three poles are $1\angle 120°=-0.5+j0.866$, $-1+j0$, and

$1\underline{/240°} = -0.5 - j0.866$. Therefore,

$$H(s) = \frac{1}{(s+1)(s+0.5-j0.866)(s+0.5+j0.866)}$$

$$= \frac{1}{(s+1)(s^2+s+1)}$$

$$= \frac{1}{s^3+2s^2+2s+1}$$

The quantity $(s^3 + 2s^2 + 2s + 1)$ is the third-order Butterworth polynomial, but it may sometimes be more convenient to use it in the form of its two factors in the preceding line. The other forms of the Butterworth polynomials are determined in similar fashion, and several of the results are given in Table 4-6.

4-6.5 Shape Factor (SF)

The shape factor is defined to be the ratio of the bandwidth of the filter when its amplitude response is at the −60 dB level to the bandwidth of the filter when its amplitude response is at the −6 dB level (sometimes the −3 dB level is used instead of the −6 dB level). The shape factor is a measure of how closely the filter response approaches that of an ideal filter whose shape factor is equal to unity.

For the Butterworth filter, −6 dB corresponds to $|H| = 1/2$ at $\omega = \omega_6$, and −60 dB corresponds to $|H| = 1/1000$ at $\omega = \omega_{60}$. Therefore,

$$\frac{1}{\sqrt{1+\omega_{60}^{2n}}} = \frac{1}{1000} \quad \text{and} \quad \frac{1}{\sqrt{1+\omega_6^{2n}}} = \frac{1}{2}$$

from both of which it is found that $\omega_{60} = (10^6 - 1)^{1/2n} \cong (10^6)^{1/2n}$ and $\omega_6 = 3^{1/2n}$. Therefore, the shape factor for Butterworth filters is

$$\text{SF} = \frac{\omega_{60}}{\omega_6} = (3.33 \times 10^5)^{1/2n} = (577.35)^{1/n} \tag{4-55}$$

When $n = 3$, SF = 8.33, and when $n = 6$, SF = 2.89.

Example 4-4

Calculate the values of L_1, L_2, and C for the lowpass filter shown in Fig. 4-9 so that the amplitude response $|V_2(j\omega)|/|V_1(j\omega)|$ is maximally flat. Let $R = 1\ \text{k}\Omega$ and $f_o = 20$ kHz.

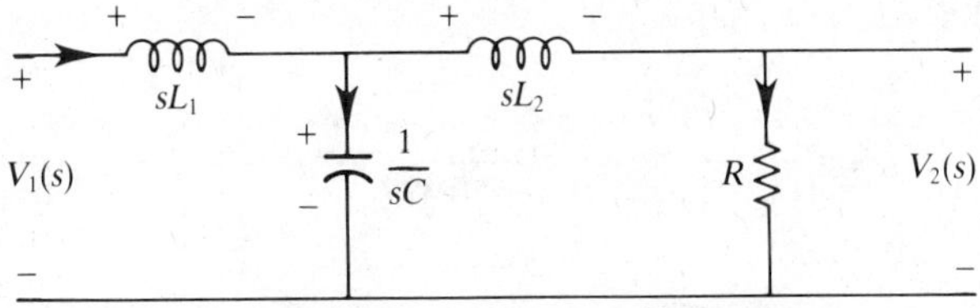

Fig. 4-9 The lowpass filter of Example 4-4.

Solution There are three independent storage elements in the filter circuit, therefore the transfer function has three poles with negative real parts, and the third-order Butterworth polynomial given in Table 4-6 must be used. The transfer function $H(s) = V_2(s)/V_1(s)$ will be calculated. Although $H(s)$ can be found by using series and parallel combinations of sL_1, sL_2, $1/sC$, and R, another informative method will be demonstrated. Let $V_2(s) = 1$ (transformed volts, of course). Then $V_1(s)$ will be calculated which makes $V_2(s) = 1$. Consequently, $H(s) = 1/V_1(s)$.

First, it is seen that the current in both R and L_2 is $1/R$. The voltage across L_2 is sL_2/R. Thus, the voltage across C is $1 + sL_2/R = (R + sL_2)/R$, and the current through C is $sC(R + sL_2)/R$. Then, the current through L_1 is the sum of the currents through C and L_2, or $(s^2CL_2 + sCR + 1)/R$. Multiplying this current by sL_1 yields the voltage across L_1, which is $(s^3L_1L_2C + s^2L_1CR + sL_1)/R$. Finally, $V_1(s)$ is the sum of the voltages across L_1 and C. Thus,

$$V_1(s) = \frac{s^3L_1L_2C + s^2L_1CR + s(L_1 + L_2) + R}{R}$$

Therefore,

$$H(s) = \frac{1}{s^3\dfrac{L_1L_2C}{R} + s^2L_1C + s\dfrac{(L_1 + L_2)}{R} + 1}$$

We notice that the denominator of this $H(s)$ is a third-order polynomial, which is as it should be. Further, nothing about this transfer function is normalized, so it is assumed that actual frequencies are to be used. In order for the amplitude response to be maximally flat as required, this polynomial must be made equal to the third-order Butterworth polynomial whose frequencies are normalized. Thus, the frequencies in the Butterworth polynomial must be expressed in terms of the actual frequencies divided by the cutoff frequency. Consequently, equating the coefficients of like powers of s results in the three equations:

$$\text{(a)}\ \frac{1}{\omega_c^3} = \frac{L_1L_2C}{R} \qquad \text{(b)}\ \frac{2}{\omega_c^2} = L_1C \qquad \text{(c)}\ \frac{2}{\omega_c} = \frac{L_1 + L_2}{R}$$

These three equations must be solved simultaneously for the three unknowns L_1, L_2, and C. It is helpful to obtain the ratio of (a) to (b) and solve for L_2 to obtain $L_2 = R/(2\omega_c)$. Substitute this into (c) to find that $L_1 = 3R/(2\omega_c)$. Then, substitute L_1 into (b) to find that $C = 4/(3\omega_c R)$. Finally, using the given values of R and ω_c in these three formulas, we find that $L_1 = 11.94\,\text{mH}$, $L_2 = 3.98\,\text{mH}$, and $C = 0.011\,\mu\text{F}$. Observe also that the values of the three components are inversely proportional to the cutoff frequency. If the cutoff frequency is decreased by a factor of 1000 to 20 Hz, for, perhaps, the filter for the output from a full-wave rectifier operated from the 60-Hz power line, then $L_1 = 11.94\,\text{H}$, $L_2 = 3.98\,\text{H}$, and $C = 11\,\mu\text{F}$.

4-7 THE CHEBYSHEV APPROXIMATION

Let $|G(j\omega)|$ be an ideal normalized transfer function in an interval from $\omega = -1$ to $\omega = +1$. Also let $|H(j\omega)|$ be a rational approximating function whose coefficients are the adjustable parameters. Then, for a certain set of coefficients and polynomial order in the denominator of $|H(j\omega)|$, the value of $|H(j\omega)|$ may oscillate about $|G(j\omega)|$ as in Fig. 4-10a. A measure of the closeness of approximation is the error $E(j\omega) = |G(j\omega)| - |H(j\omega)|$. The error function can have both positive and negative peaks, some probably larger in magnitude than others, as in Fig. 4-10b. Then, by a trial-and-error step-by-step numerical procedure, the coefficients of the approximating function $|H(j\omega)|$ can be adjusted as found to be necessary so that the largest peak of the error is decreased at each step.

In this process, the zeros of $E(j\omega)$ will shift because the coefficients of the polynomial are changed. The procedure is continued until all of the peak errors become equal in magnitude at the least possible value for the particular polynomial involved and also equal to the magnitudes of the errors at the end points at $\omega = -1$ and $\omega = +1$, as in Fig. 4-10c. The equality of the peak errors over the interval is the reason why the name "equal ripple" is also used for this kind of approximation.

When the peak errors are normalized to 1, the Chebyshev approximation $C(\omega)$ is obtained. In Fig. 4-10d, $C(\omega)$ must be represented by an eighth-order polynomial in ω because 8 zeros are shown over the interval.

A general expression for $C(\omega)$ will now be derived by means of a rather subtle mathematical development. Form the function $[1 - C^2(\omega)]$ shown in Fig. 4-10e. This function has simple zeros at $\omega = \pm 1$ because $C(\omega) = \pm 1$ at those points. This means that $(1 - \omega)$ and $(1 + \omega)$ are factors of $[1 - C^2(\omega)]$. In addition, this function has double zeros at several internal points where it is zero for two reasons: (1) $C(\omega) = \pm 1$ at those points, and (2) the derivative of $[1 - C^2(\omega)] = -2C(\omega)\,dC(\omega)/d\omega$ is zero because $dC(\omega)/d\omega$ is zero at those same points. Therefore, $[1 - C^2(\omega)]$ has $[dC(\omega)/d\omega]^2$ as a

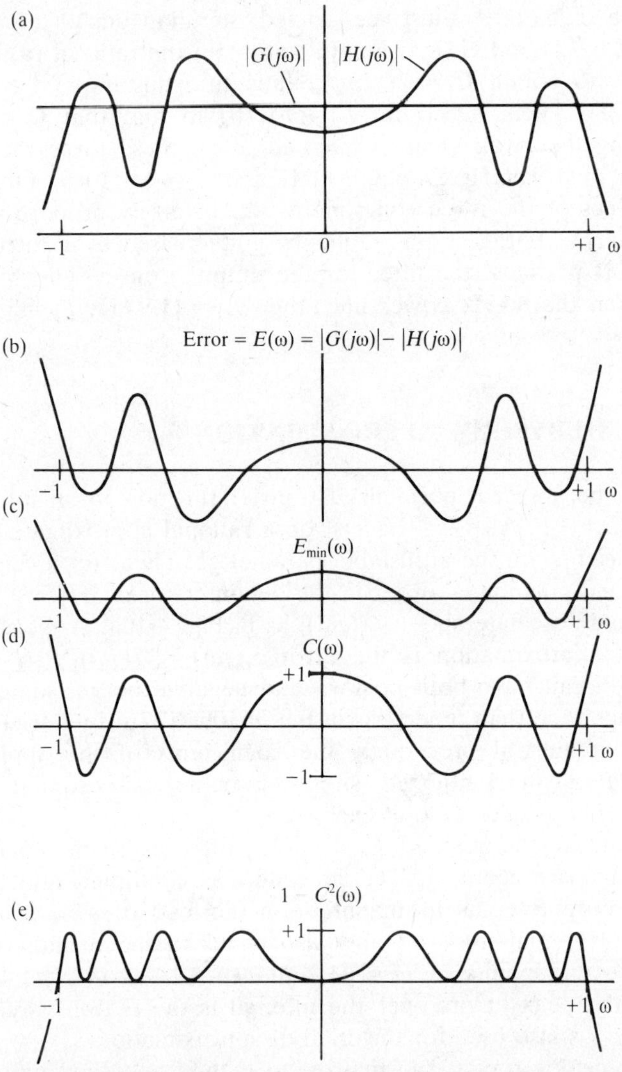

Fig. 4-10 (a) Ideal and approximating functions. (b) The error function. (c) The peak errors minimized and made equal in magnitudes. (d) The normalized error function. (e) The function $1 - C^2(\omega)$.

double factor. Consequently,

$$1 - C^2(\omega) = K^2(1-\omega)(1+\omega)\left(\frac{dC(\omega)}{d\omega}\right)^2$$

$$= K^2(1-\omega^2)\left(\frac{dC(\omega)}{d\omega}\right)^2$$

where K is a constant. Then, take the square root and separate the variables to obtain

$$\frac{dC(\omega)}{\sqrt{1-C^2(\omega)}} = \frac{\pm 1}{K} \frac{d\omega}{\sqrt{1-\omega^2}}$$

which, upon integrating after selecting the + or − sign, yields

$$\cos^{-1} C(\omega) = \frac{1}{K} \cos^{-1} \omega, \qquad \omega^2 < 1, \qquad C^2(\omega) < 1 \qquad (4\text{-}56)$$

and

$$\cosh^{-1} C(\omega) = \frac{1}{K} \cosh^{-1} \omega, \qquad \omega^2 > 1, \qquad C^2(\omega) > 1 \qquad (4\text{-}57)$$

In Fig. 4-10d, it is seen that $C(\omega)$ oscillates n times ($n = 4$ complete "cycles" in Fig. 4-10d) between ± 1 as ω varies over the frequency interval ± 1. Hence, $1/K = n$, so that (4-56) and (4-57) become

$$C_n(\omega) = \cos(n \cos^{-1} \omega), \qquad \omega^2 < 1, \qquad C^2(\omega) < 1 \qquad (4\text{-}58)$$

and

$$= \cosh(n \cosh^{-1} \omega), \qquad \omega^2 > 1, \qquad C^2(\omega) > 1 \qquad (4\text{-}59)$$

where the subscript n has been added to show that $C_n(\omega)$ also depends on n.

That $C_n(\omega)$ can be represented by polynomials is readily shown with the aid of a recursion formula for calculating the various terms in the polynomials. Let $\omega = \cos x$, or $x = \cos^{-1} \omega$. Then

$$C_n(x) = \cos nx$$

Therefore,

$$C_{n+1}(x) = \cos(n+1)x = \cos x \cos nx - \sin x \sin nx$$

and

$$C_{n-1}(x) = \cos(n-1)x = \cos x \cos nx + \sin x \sin nx$$

Add these two expressions to obtain

$$C_{n+1}(x) + C_{n-1}(x) = 2 \cos x \cos nx$$

Thus, returning to ω as the variable, this yields

$$C_{n+1}(\omega) = 2\omega C_n(\omega) - C_{n-1}(\omega) \qquad (4\text{-}60)$$

Equation (4-60) is the recursion formula used for calculating the Chebyshev polynomials. Since, from (4-58), $C_0(\omega) = 1$ and $C_1(\omega) = \omega$, then, using (4-60),

$$C_2(\omega) = 2\omega C_1(\omega) - C_0(\omega) = 2\omega^2 - 1 = 2(\omega^2 - \tfrac{1}{2})$$

$$\begin{aligned} C_3(\omega) &= 2\omega C_2(\omega) - C_1(\omega) = 4\omega^3 - 2\omega - \omega \\ &= 4\omega^3 - 3\omega = 4(\omega^3 - \tfrac{3}{4}\omega) \end{aligned}$$

$$\begin{aligned} C_4(\omega) &= 2\omega C_3(\omega) - C_2(\omega) = 8\omega^4 - 6\omega^2 - 2\omega^2 + 1 \\ &= 8\omega^4 - 8\omega^2 + 1 = 8(\omega^4 - \omega^2 + \tfrac{1}{8}) \end{aligned}$$

In general, the Chebyshev polynomials can be expressed as the series

$$\begin{aligned} C_n(\omega) = 2^{n-1}\Bigg[\omega^n &- \frac{n}{1!\,2^2}\,\omega^{n-2} + \frac{n(n-3)}{2!\,2^4}\,\omega^{n-4} - \frac{n(n-4)(n-5)}{3!\,2^6}\,\omega^{n-6} \\ &+ \frac{n(n-5)(n-6)(n-7)}{4!\,2^8}\,\omega^{n-8} \\ &- \frac{n(n-6)(n-7)(n-8)(n-9)}{5!\,2^{10}}\,\omega^{n-10} + \cdots\Bigg] \end{aligned}$$

The summation is stopped when the exponents of ω are no longer positive.

Thus, the various $C_n(\omega)$ are polynomials. They were originally defined in 1899 by the Russian mathematician P. L. Chebyshev in studies relating to the behavior of mechanical linkages used in steam engines. Several of the Chebyshev polynomials are listed in Table 4-7 and plotted in Fig. 4-11, where the equal-ripple variation over the range $\omega = \pm 1$ is clearly evident.

TABLE 4-7 Chebyshev Polynomials

n	$C_n(\omega)$
0	1
1	ω
2	$2\omega^2 - 1 = 2(\omega^2 - \frac{1}{2})$
3	$4\omega^3 - 3\omega = 4(\omega^3 - \frac{3}{4}\omega)$
4	$8\omega^4 - 8\omega^2 + 1 = 8(\omega^4 - \omega^2 + \frac{1}{8})$
5	$16\omega^5 - 20\omega^3 + 5\omega = 16(\omega^5 - \frac{5}{4}\omega^3 + \frac{5}{16}\omega)$
6	$32\omega^6 - 48\omega^4 + 18\omega^2 - 1 = 32(\omega^6 - \frac{3}{2}\omega^4 + \frac{9}{16}\omega^2 - \frac{1}{32})$

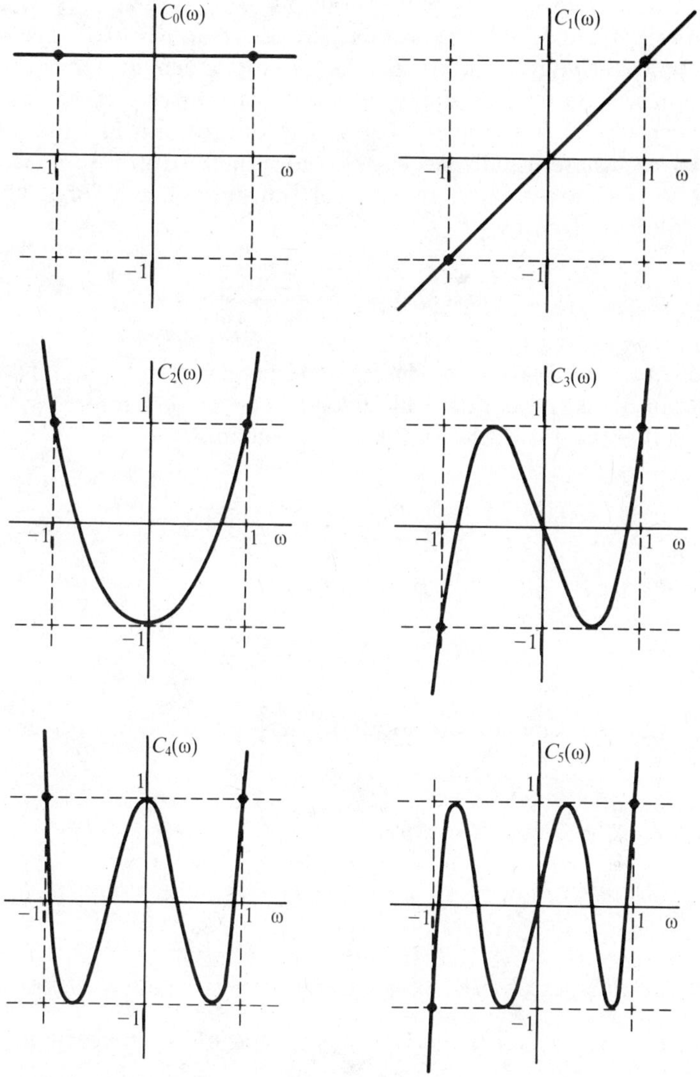

Fig. 4-11 Sketches of several Chebyshev polynomials.

4-7.1 The Poles of the Chebyshev (Equal-ripple) Filter Function

To use the Chebyshev polynomials $C_n(\omega)$ in the transfer function of a lowpass filter with the equal-ripple feature in the passband, let

$$|H(j\omega)| = \frac{1}{\sqrt{1 + C_n^2(\omega)}}$$

represent the required transfer function. Since $C_n(\omega)$ varies between 0 and 1 over the passband, $|H(j\omega)|$ varies between 1 and 0.707. This is a 3-dB equal ripple variation, and it may be far too much in a particular filter application. To obtain less ripple, it is necessary to use less of the possible variation in $C_n^2(\omega)$ over the passband. This is accomplished by multiplying $C_n^2(\omega)$ by ϵ^2, where usually $0 < \epsilon^2 < 1$. More than 3 dB of ripple is obtained when $\epsilon^2 > 1$. Consequently, the typical transfer function of a Chebyshev lowpass filter is given by

$$|H(j\omega)| = \frac{1}{\sqrt{1 + \epsilon^2 C_n^2(\omega)}} \tag{4-61}$$

Sketches of $|H(j\omega)|$ are shown in Fig. 4-12 for $n = 4$ and $n = 5$. The amount of ripple in dB is the negative of $20 \log 1/\sqrt{1 + \epsilon^2}$. Therefore, in a manner similar to that used with the Butterworth functions,

$$|H(j\omega)|^2 = H(j\omega) \cdot H(-j\omega) = \frac{1}{1 + \epsilon^2 C_n^2(\omega)}$$

Then, using $s = j\omega$ and $\omega = -js$, this becomes

$$H(s) \cdot H(-s) = \frac{1}{1 + \epsilon^2 C_n^2(-js)}$$

The poles of this function are located where the denominator equals zero. Thus

$$C_n(-js) = \pm j \frac{1}{\epsilon}$$

But, from (4-58),

$$C_n(-js) = \cos(n \cos^{-1}(-js))$$

To evaluate this, let $U + jV = \cos^{-1}(-js)$, or $-js = \cos(U + jV)$. Then,

$$\begin{aligned} C_n(-js) &= \cos(nU + jnV) = \cos nU \cos jnV - \sin nU \sin jnV \\ &= \cos nU \cosh nV - j \sinh nU \sinh nV = 0 \pm j\frac{1}{\epsilon} \end{aligned}$$

Equating the real parts on each side,

$$\cos nU \cosh nV = 0$$

But since $\cosh nV$ cannot be zero, then $\cos nU = 0$. Therefore, the angles nU must be odd multiples of $\pi/2$ radians or of 90°. Hence

$$U = \frac{90°}{n}, \ldots, \frac{(2k-1)90°}{n}, \qquad k = 1, 2, 3, \ldots \tag{4-62}$$

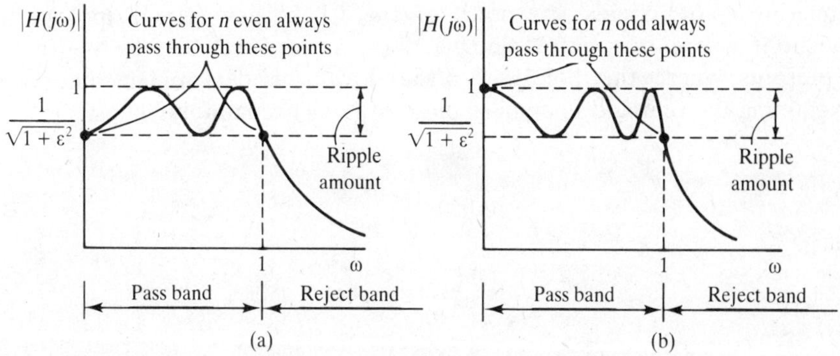

Fig. 4-12 Sketches of $|H(j\omega)|$ for normalized Chebyshev filters: (a) $n = 4$, (b) $n = 5$. Notice that n is the same as the number of variations between maxima and minima of the ripple in the pass band.

These are the values of U that assure that $C_n(-js)$ has no real part. For these values of U,

$$\sin nU = \sin 90°, \sin 270°, \ldots, \sin (2k-1)90° = \pm 1$$

Therefore, in the imaginary part of $C_n(-js)$,

$$\pm \sinh nV = \pm \frac{1}{\epsilon}$$

or

$$nV = \sinh^{-1} \frac{1}{\epsilon}$$

and

$$V = \frac{1}{n} \sinh^{-1} \frac{1}{\epsilon} \tag{4-63}$$

This is the value of V that, along with the associated values of U from (4-62), assure that $C_n(-js) = \pm j/\epsilon$ as required. But

$$\begin{aligned} -js &= \cos (U + jV) = \cos U \cos jV - \sin U \sin jV \\ &= \cosh V \cos U - j \sinh V \sin U \end{aligned}$$

So that, finally,

$$\begin{aligned} s &= \sinh V \sin U + j \cosh V \cos U \\ &= \sigma_k + j\omega_k \end{aligned} \tag{4-64}$$

Equation (4-64) yields the poles of the Chebyshev (equal-ripple) filter function, where $\sigma_k = \sinh V \sin U$ and $\omega_k = \cosh V \cos U$. The figure that represents where the poles are located can be determined quickly by examining the real and imaginary parts of (4-64) separately. Thus,

$$\frac{\sigma_k^2}{\sinh^2 V} = \sin^2 U$$

and

$$\frac{\omega_k^2}{\cosh^2 V} = \cos^2 U$$

Upon adding, we find that

$$\frac{\sigma_k^2}{\sinh^2 V} + \frac{\omega_k^2}{\cosh^2 V} = 1 \tag{4-65}$$

Equation (4-65) is recognized as a general form of an ellipse in the s plane whose major semiaxis is $\cosh V$ and whose minor semiaxis is $\sinh V$, since $\cosh V > \sinh V$. Recall that the poles of the Butterworth filter function are located on a circle having unit radius.

Since the poles of an actual filter must have negative real parts, then, from (4-64), $\sinh V \sin U < 0$. But $\sinh V > 0$, therefore $\sin U < 0$. This means that the angle U must be in the third and fourth quadrants. Hence,

$$180° < (2k - 1)\frac{90°}{n} < 360°$$

from which is obtained

$$n + 1 \leqslant k \leqslant 2n \tag{4-66}$$

Thus, for a second-order Chebyshev filter, $3 \leqslant k \leqslant 4$, meaning that k takes on only the values 3 and 4. For a third-order Chebyshev filter, $4 \leqslant k \leqslant 6$, meaning that k takes on the values 4, 5, and 6. The values of k specified by (4-66) are the only values for which the real parts of the corresponding poles are negative.

4-7.2 Rolloff and Attenuation

At high frequencies, the rate at which the Chebyshev function decreases in magnitude, for the various values of n, is determined by using the leading factor in the general series for the Chebyshev polynomials. Thus, at high frequencies,

$$C_n(\omega) \cong 2^{n-1}\omega^n$$

Therefore,

$$|H(j\omega)| = \frac{1}{\sqrt{1+\epsilon^2 C_n^2(\omega)}} \cong \frac{1}{\epsilon 2^{n-1}\omega^n}$$

Expressing this in dB yields

$$20 \log |H(j\omega)| = -20n \log \omega - 6(n-1) - 20 \log \epsilon \quad \text{dB} \tag{4-67}$$

For a certain filter with n and ϵ constant, the last two terms on the right-hand side of (4-67) are fixed and independent of frequency, whereas the first term indicates that the rolloff is at the rate of $-20n$ dB/decade or $-6n$ dB/octave.

Assume that at a certain normalized radian frequency $\omega_r > 1$, $|H(j\omega_r)|$ of a Chebyshev filter is equal to or less than the amount A. Thus,

$$|H(j\omega_r)| = \frac{1}{\sqrt{1+\epsilon^2 C_n^2(\omega_r)}} = \frac{1}{\sqrt{1+\epsilon^2 \cosh^2(n \cosh^{-1} \omega_r)}} \leq A$$

which, when solved for n, yields

$$n \geq \frac{\cosh^{-1}\left(\frac{1}{\epsilon}\sqrt{\frac{1}{A^2}-1}\right)}{\cosh^{-1} \omega_r} \tag{4-68}$$

Just as with the Butterworth filter, when the calculated value of n is not an integer, the value should be rounded to the next larger integer to produce an attenuation of at least A at the specified ω_r.

4-7.3 Hyperbolic Function Calculations

Some of the currently popular hand-held calculators do not feature the hyperbolic functions, but they do have the exponential and logarithmic functions. Consequently, $\sinh x$ and $\cosh x$ can be calculated using their exponential definitions. Thus,

$$\sinh x = \frac{e^x - e^{-x}}{2} \quad \text{and} \quad \cosh y = \frac{e^y + e^{-y}}{2} \tag{4-69}$$

The inverse hyperbolic functions can be calculated as follows. Let

$$d = \sinh x = \frac{e^x - e^{-x}}{2} \quad \text{and} \quad b = \cosh y = \frac{e^y + e^{-y}}{2}$$

Then

$$e^{2x} - 2de^x - 1 = 0 \qquad \text{and} \qquad e^{2y} - 2be^y + 1 = 0$$

from which are obtained, using the quadratic formula,

$$e^x = \frac{2d + \sqrt{4d^2 + 4}}{2} \qquad \text{and} \qquad e^y = \frac{2b + \sqrt{4b^2 - 4}}{2}$$

Finally, taking the natural logarithm, we obtain

$$x = \sinh^{-1} d = \ln\,(d + \sqrt{d^2 + 1}) \tag{4-70}$$

and

$$y = \cosh^{-1} b = \ln\,(b + \sqrt{b^2 - 1}) \tag{4-71}$$

Notice that the positive square root is required in (4-70) and (4-71).

4-7.4 Shape Factor (SF)

Proceeding in the same manner as for Butterworth filters, it is seen that $|H| = 1/1000$ at ω_{60}, and that $|H| = 1/2$ at ω_6. Hence,

$$\frac{1}{\sqrt{1 + \epsilon^2 C_n^2(\omega_{60})}} = \frac{1}{10^3} \qquad \text{and} \qquad \frac{1}{\sqrt{1 + \epsilon^2 C_n^2(\omega_6)}} = \frac{1}{2}$$

from which is found

$$C_n(\omega_{60}) = \cosh\,(n \cosh^{-1} \omega_{60}) = \sqrt{\frac{10^6 - 1}{\epsilon^2}} \cong \frac{10^3}{\epsilon}$$

and

$$C_n(\omega_6) = \cosh\,(n \cosh^{-1} \omega_6) = \sqrt{\frac{4 - 1}{\epsilon^2}} = \frac{\sqrt{3}}{\epsilon}$$

Then, solving these for ω_{60} and ω_6, the shape factor of the Chebyshev filter is found to be

$$\text{SF} = \frac{\omega_{60}}{\omega_6} = \frac{\cosh\left(\dfrac{1}{n} \cosh^{-1} \dfrac{10^3}{\epsilon}\right)}{\cosh\left(\dfrac{1}{n} \cosh^{-1} \dfrac{\sqrt{3}}{\epsilon}\right)} \tag{4-72}$$

Example 4-5

Design a Chebyshev normalized (i.e., $\omega_c = 1$ rad/s) lowpass filter having 0.5 dB ripple in the passband and whose transfer function magnitude is attenuated by at least 18 dB for $\omega \geqslant 2$ rad/s.

Solution The amount of ripple is related to ϵ by

$$0.5 = -20 \log \frac{1}{\sqrt{1+\epsilon^2}}$$

from which is obtained $\epsilon = 0.349$. Further, $-18\text{ dB} = 20 \log A$, so that $A = 0.126$. The required filter order is given by

$$n \geqslant \frac{\cosh^{-1}[(1/0.349)\sqrt{(1/0.126^2) - 1}]}{\cosh^{-1} 2} = \frac{\cosh^{-1} 22.56}{\cosh^{-1} 2}$$

For calculators without the hyperbolic functions, it is helpful to use

$$n \geqslant \frac{\ln(22.56 + \sqrt{22.56^2 - 1})}{\ln(2 + \sqrt{2^2 - 1})} = 2.892$$

This must be rounded to $n = 3$. Hence, from (4-63),

$$\begin{aligned} V &= \tfrac{1}{3} \sinh^{-1} \frac{1}{0.349} = \tfrac{1}{3} \sinh^{-1} 2.865 \\ &= \tfrac{1}{3} \ln(2.865 + \sqrt{2.865^2 + 1}) = 0.592 \end{aligned}$$

From (4-64), the poles of $H(s)$ are located at

$$s = \sinh 0.592 \sin(2k-1)30° + j \cosh 0.592 \cos(2k-1)30°$$

where $n + 1 \leqslant k \leqslant 2n$, or $4 \leqslant k \leqslant 6$. Therefore, for $k = 4$,

$$s_4 = 0.627 \sin 210° + j1.180 \cos 210° = -0.313 - j1.022$$

For $k = 5$, $s_5 = 0.627 \sin 270° + j1.180 \cos 270° = -0.627 + j0$, and, for $k = 6$, $s_6 = 0.627 \sin 330° + j1.180 \cos 330° = -0.313 + j1.022$. As expected for this third-order filter, all of the poles have negative real parts, a pair of poles are complex conjugates, and one pole is real. Consequently,

$$\begin{aligned} H(s) &= 1/[(s + 0.627)(s + 0.313 + j1.022)(s + 0.313 - j1.022)] \\ &= 1/[s^3 + 1.243s^2 + 1.528s + 0.716] \\ &= 1/[(s + 0.627)(s^2 + 0.626s + 1.142)] \\ &= \left(\frac{1}{1.595s + 1}\right)(1.595 \times 0.876)\left(\frac{1}{0.876s^2 + 0.548s + 1}\right) \end{aligned} \tag{4-73}$$

in all of which s is a normalized quantity. It was demonstrated in Example 4-4 how the component values of a third-order Butterworth filter were calculated from the coefficients in the denominator of the transfer function. That same procedure could also be applied here, but another one will be shown instead because of its application to active filters. Express the desired $H(s)$ in the product form given as (4-73). This product form can be obtained in practice by cascading a first-order filter and a second-order filter. Then the two separate filters are rendered noninteractive in the cascade connection by a central active device whose input impedance is very high, whose output impedance is very low, and whose amplification is set at the product $1.595 \times 0.876 = 1.397$ to achieve an overall amplification of unity at $\omega = 0$. However, if the amplification is set equal to 13.97 instead of 1.397, then the overall amplification at $\omega = 0$ is 10 instead of 1. Then, the overall transfer function is equal to the product of the three individual transfer functions.

Continuing the example, consider the simple first-order lowpass filter shown in Fig. 4-13a. It has a voltage transfer function given by

$$\frac{V_2(s)}{V_1(s)} = \frac{1.595}{sRC + 1} \tag{4-74}$$

in which s is not normalized but is the actual value. This circuit is used to produce the term $1/(1.595s + 1)$ in (4-73), but in this term s is normalized. Recalling that normalized s is the actual s divided by the cutoff frequency ω_c, the latter term becomes

$$\frac{1.595}{1.595\dfrac{s}{\omega_c} + 1} \tag{4-75}$$

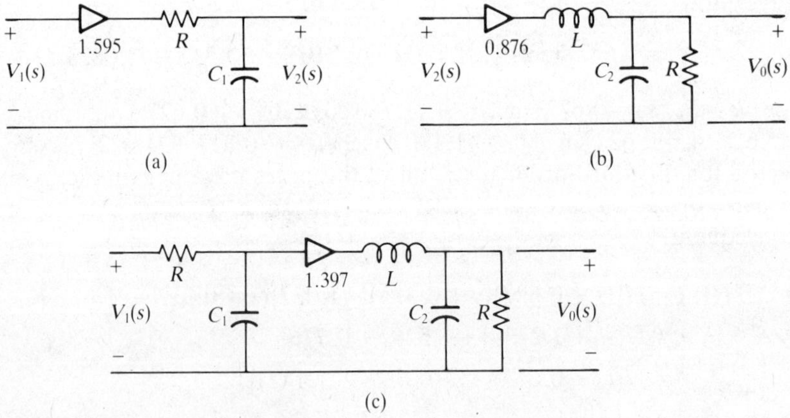

Fig. 4-13 Circuits for Example 4-5. (a) The first-order filter section. (b) The second-order filter section. (c) The complete third-order filter.

in which s is the actual value. Comparing (4-74) and (4-75), it is seen that

$$\frac{1.595}{\omega_c} = RC_1 \tag{4-76}$$

If values for R and the cutoff frequency are specified, then C_1 can be calculated from (4-76).

Now consider the second-order lowpass filter of Fig. 4-13b. It has a voltage transfer function given by

$$\frac{V_o(s)}{V_i(s)} = \frac{0.876}{s^2 LC_2 + s\dfrac{L}{R} + 1} \tag{4-77}$$

in which s is the actual value. This circuit is used to produce the quadratic term in (4-73), but in that term s is normalized. So, since dividing actual s by ω_c is the normalizing method, the quadratic term in (4-73) becomes

$$0.876\left(\frac{s}{\omega_c}\right)^2 + 0.548\left(\frac{s}{\omega_c}\right) + 1 \tag{4-78}$$

Now, comparing the denominator of (4-77) with (4-78), it is found that

$$LC_2 = \frac{0.876}{\omega_c^2} \tag{4-79a}$$

and

$$\frac{L}{R} = \frac{0.548}{\omega_c} \tag{4-79b}$$

Again, if values for R and the cutoff frequency are specified, then L can be calculated from (4-79b), and then C_2 can be calculated from (4-79a). Finally, placing the individual filters and the central amplifier in cascade, Fig. 4-13c results. The overall voltage transfer function, therefore, is that given by (4-73). Specifically, if $R = 500\ \Omega$ and $f_c = 20$ kHz, the above procedure shows that $C_1 = 0.0254\ \mu$F, $L = 21.18$ mH, and $C_2 = 0.0254\ \mu$F.

4-8 FILTER TRANSFORMATIONS

Lowpass filters, whether of the Butterworth or Chebyshev types, are not the only kinds of filters that are needed. There are times when a bandpass filter is desired, or a bandreject type, or, perhaps, a highpass filter. The kind of filter needed is determined by the particular application. Fortunately, however, it is possible to design any of these latter filters from a normalized lowpass filter which is considered to be the prototype.

Let us represent the transfer function of the prototype as $H(p)$, (instead of $H(s)$), and its sinusoidal steady-state transfer function as $H(j\Omega)$, where Ω represents the normalized frequency of the prototype. Then, the transfer function of the actual filter will be given by $H(s)$, and its sinusoidal steady-state transfer function will be $H(j\omega)$, where ω represents actual frequency, not normalized frequency as in the prototype.

4-8.1 The Lowpass-to-Bandpass Transformation

Suppose we wish to design a bandpass filter having a response characteristic whose geometric center is at ω_0 and whose bandwidth is ω_b (to be defined in what follows). The prototype can be transformed into the desired bandpass filter by replacing the variable p of the prototype with $(s^2 + \omega_0^2)/(s\omega_b)$. When $s = j\omega$, p is imaginary, meaning that the j axis in the s plane maps into the j axis of the p plane, and vice versa. Consequently, if $p = j\Omega$ and $s = j\omega$, then

$$\Omega = \frac{\omega^2 - \omega_0^2}{\omega\omega_b} \tag{4-80}$$

A graph of this transformation is shown in Fig. 4-14. Notice that at $\Omega = -1$, $\omega = \omega_L$, and that at $\Omega = +1$, $\omega = \omega_H$. Further, the bandwidth ω_b is shown there to be equal to $\omega_H - \omega_L$. To demonstrate how this transformation is used, consider the lowpass filter of Fig. 4-15a, where the amplitude response extends over the range of Ω from $-\infty$ to $+\infty$. It is seen from the curve in Fig. 4-14, that as Ω varies from $-\infty$ to -1, ω varies from 0 to ω_L. The same general shape of the prototype response in this frequency range is squeezed

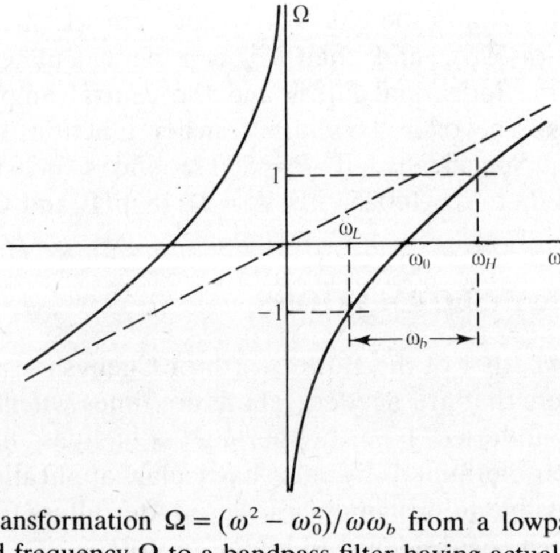

Fig. 4-14 The transformation $\Omega = (\omega^2 - \omega_0^2)/\omega\omega_b$ from a lowpass prototype filter having normalized frequency Ω to a bandpass filter having actual frequency ω.

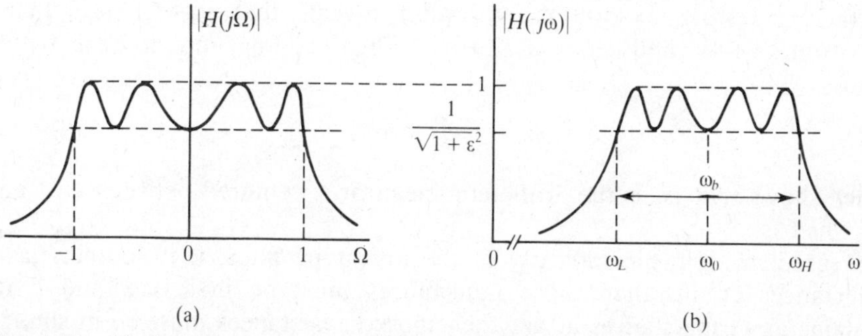

Fig. 4-15 (a) Amplitude response of a Chebyshev prototype filter ($n = 4$). (b) Amplitude response of (a) after transforming into a bandpass filter.

into the corresponding finite range of ω. Then, as Ω varies from -1 to $+1$, ω varies from ω_L, through ω_0, to ω_H. The same general shape of the prototype response in this frequency range is repeated in the corresponding range of ω, and if there are ripples here in the response of the prototype filter, there will be ripples in the corresponding response of the bandpass filter. The prototype response for $-1 \leq \Omega \leq +1$ becomes the response for $\omega_L \leq \omega \leq \omega_H$. Finally, as Ω varies from $+1$ to ∞, ω varies from ω_H to ∞, and the same general shape of the prototype response in this frequency range is repeated over the corresponding range of ω. In this way, the normalized lowpass prototype filter is transformed into the bandpass filter.

Other relationships between the various frequencies can be obtained readily. From Fig. 4-14, at $\Omega = +1$, $\omega = \omega_H$. Substituting these into (4-80) and solving for ω_H yields

$$\omega_H = \frac{\omega_b}{2} + \sqrt{\left(\frac{\omega_b}{2}\right)^2 + \omega_0^2} \tag{4-81}$$

Also, at $\Omega = -1$, $\omega = \omega_L$, so that, using (4-80) again,

$$\omega_L = -\frac{\omega_b}{2} + \sqrt{\left(\frac{\omega_b}{2}\right)^2 + \omega_0^2} \tag{4-82}$$

Now, multiplying (4-81) and (4-82) and simplifying, yields

$$\omega_0 = \sqrt{\omega_L \omega_H} \tag{4-83}$$

which indicates that ω_0 is the geometric mean of ω_L and ω_H. Also, subtracting, we find that the bandwidth is

$$\omega_b = \omega_H - \omega_L \tag{4-84}$$

Further, referring to (4-81) and (4-82) reveals that if $\omega_0 \gg \omega_b$, then $\omega_L = -\omega_b/2 + \omega_0$ and $\omega_H = \omega_b/2 + \omega_0$. Then, adding and dividing by 2 yields

$$\omega_0 = \tfrac{1}{2}(\omega_L + \omega_H) \tag{4-85}$$

which shows that ω_0 is the arithmetic mean or is centered between ω_L and ω_H when $\omega_0 \gg \omega_b$.

Regardless of the complexity of the prototype filter, if it contains any inductances at all, then those inductances must be in series, and if it contains any capacitances at all, then those capacitances must be in shunt. Calculation of the values of the components for the bandpass filter is aided by replacing p of the prototype filter with $(s^2 + \omega_0^2)/(s\omega_b)$. Therefore, each series impedance pL in the prototype becomes the series impedance

$$\frac{s^2 + \omega_0^2}{s\omega_b} L = s\left(\frac{L}{\omega_b}\right) + \frac{1}{s\left(\dfrac{\omega_b}{\omega_0^2 L}\right)} \tag{4-86}$$

which is recognized as being the series connection of an inductor whose value is L/ω_b and a capacitor whose value is $\omega_b/\omega_0^2 L$. The resonant frequency of each such series combination is ω_0.

The shunt admittance of each capacitor pC in the prototype becomes the shunt admittance

$$\frac{s^2 + \omega_0^2}{s\omega_b} C = s\left(\frac{C}{\omega_b}\right) + \frac{1}{s\left(\dfrac{\omega_b}{\omega_0^2 C}\right)} \tag{4-87}$$

which is recognized as being the parallel connection of an inductor whose value is $\omega_b/\omega_0^2 C$ and a capacitor whose value is C/ω_b. The resonant frequency of each such parallel combination is ω_0.

4-8.2 The Lowpass-to-Bandreject Transformation

To transform the prototype lowpass filter into a bandreject filter, replace p by $s\omega_b/(s^2 + \omega_0^2)$. Again, when $s = j\omega$, p is imaginary, and vice versa. Therefore,

$$\Omega = \frac{\omega\omega_b}{\omega_0^2 - \omega^2} \tag{4-88}$$

which is plotted in Fig. 4-16. With this transformation, the prototype response of Fig. 4-15a becomes a bandreject response. To show this, observe in Fig. 4-16 that for $0 \leq \Omega \leq 1$, ω varies from 0 to ω_L; that for $1 \leq \Omega < \infty$, ω varies from ω_L to ω_0; that for $-\infty < \Omega \leq -1$, ω varies from ω_0

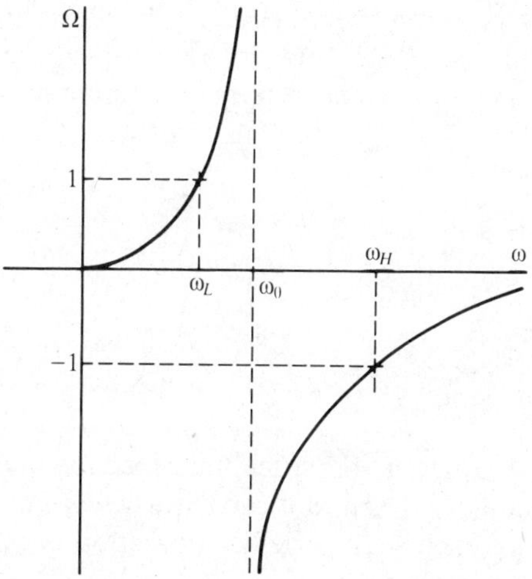

Fig. 4-16 The transformation $\Omega = \omega\omega_b/(\omega_0^2 - \omega^2)$ from a lowpass prototype filter having normalized frequency Ω to a band reject filter having actual frequency ω.

to ω_H; and that for $-1 \leq \Omega \leq 0$, ω varies from ω_H to ∞. If there are ripples in the prototype response in the range $-1 \leq \Omega \leq +1$, then ripples will appear in the ranges $0 \leq \omega \leq \omega_L$ and $\omega_H \leq \omega < \infty$, as shown in Fig. 4-17. Again, it is seen that the bandwidth of the bandreject filter is given by $\omega_b = \omega_H - \omega_L$, and that ω_0 is the geometric mean of ω_L and ω_H.

To calculate the values of the components in the bandreject filter, replace p by $\omega_b s/(s^2 + \omega_0^2)$. Each series impedance pL of the prototype is replaced with the impedance

$$\frac{\omega_b s}{s^2 + \omega_0^2} L = \frac{1}{s\left(\dfrac{1}{\omega_b L}\right) + \dfrac{1}{s\left(\dfrac{\omega_b C}{\omega_0^2}\right)}} \tag{4-89}$$

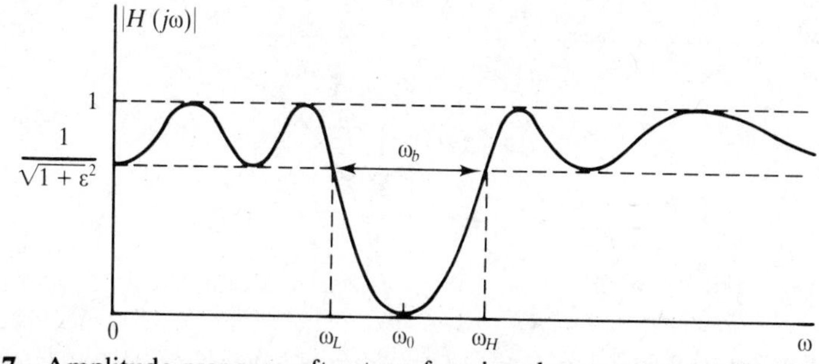

Fig. 4-17 Amplitude response after transforming the prototype in Fig. 4-15a into a band reject filter.

which is recognized as being the impedance of a parallel connection of a capacitor whose value is $1/\omega_b L$ and an inductor whose value is $\omega_b L/\omega_0^2$. The resonant frequency of each such parallel combination is ω_0.

Each shunt susceptance pC of the prototype is replaced with the susceptance

$$\frac{\omega_b s}{s^2+\omega_0^2}C = \frac{1}{s\left(\dfrac{1}{\omega_b C}\right)+\dfrac{1}{s\left(\dfrac{\omega_b C}{\omega_0^2}\right)}} \tag{4-90}$$

which is recognized as being the susceptance of a series connection of an inductor whose value is $1/\omega_b C$ and a capacitor whose value is $\omega_b C/\omega_0^2$. The resonant frequency of each such series combination is ω_0.

4-8.3 The Lowpass-to-Highpass Transformation

This transformation requires replacement of p with ω_0/s. Again, when $s = j\omega$, p is imaginary, and vice versa. Hence, $\Omega = -\omega_0/\omega$. This is plotted in Fig. 4-18. When the prototype filter response is as shown in Fig. 4-15a, this transformation leads to the highpass filter response of Fig. 4-19 (see Problem 4-28). Each series inductor of the prototype is replaced by a capacitor whose value is $1/\omega_0 L$ and each shunt capacitor is replaced by an inductor whose value is $1/\omega_0 C$ (see Problem 4-28).

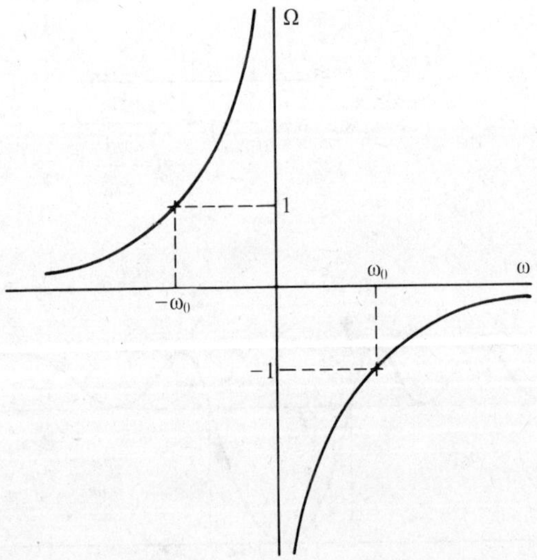

Fig. 4-18 The transformation $\Omega = -\omega_0/\omega$ from a lowpass prototype filter having normalized frequency Ω to a highpass filter having actual frequency ω.

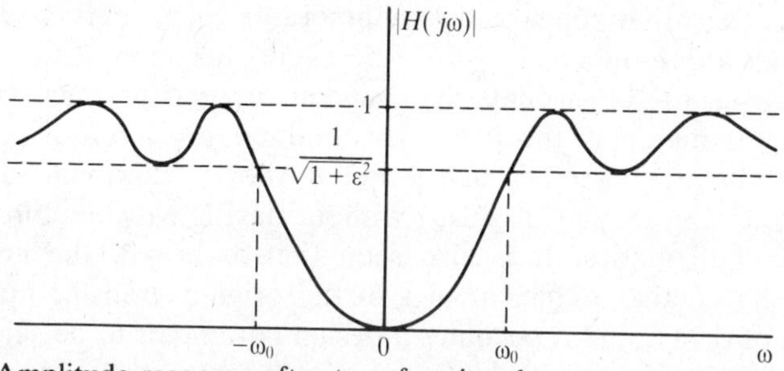

Fig. 4-19 Amplitude response after transforming the prototype in Fig. 4-15a into a highpass filter.

4-9 LINEAR PHASE-RESPONSE FILTERS

Consider a filter whose transfer function is

$$H(s) = \frac{1}{1 + a_1 s + a_2 s^2 + \cdots a_n s^n} \tag{4-91}$$

The sinusoidal steady-state response of the filter is, by inspection,

$$H(j\omega) = \frac{1}{(1 - a_2\omega^2 + a_4\omega^4 - a_6\omega^6 + \cdots) + j(a_1\omega - a_3\omega^3 + a_5\omega^5 - a_7\omega^7 + \cdots)}$$

The phase response of this filter is given by the inverse tangent function. Thus,

$$\phi(\omega) = -\tan^{-1} \frac{a_1\omega - a_3\omega^3 + a_5\omega^5 - a_7\omega^7 + \cdots}{1 - a_2\omega^2 + a_4\omega^4 - a_6\omega^6 + \cdots}$$

Linearizing the phase response by manipulating the inverse tangent function is equivalent to making the group delay $T_g(\omega)$ maximally flat. Since

$$T_g(\omega) = -\frac{d\phi(\omega)}{d\omega} \qquad \text{and} \qquad \frac{d\left(-\tan^{-1}\dfrac{x}{a}\right)}{dx} = \frac{-a}{a^2 + x^2}$$

it will be more tractable mathematically to make the group delay maximally flat than to attempt to linearize the $\tan^{-1}$ function. Consequently, differentiating the $\tan^{-1}$ function with respect to ω, simplifying, and collecting terms, yields

$$T_g(\omega) = \frac{\begin{aligned} a_1 + (a_1a_2 - 3a_3)\omega^2 + (a_2a_3 - 3a_1a_4 + 5a_5)\omega^4 \\ + (a_3a_4 - 3a_2a_5 + 5a_1a_6 - 7a_7)\omega^6 + \cdots \end{aligned}}{\begin{aligned} 1 + (a_1^2 - 2a_2)\omega^2 + (a_2^2 - 2a_1a_3 + 2a_4)\omega^4 \\ + (a_3^2 - 2a_2a_4 + 2a_1a_5 - 2a_6)\omega^6 + \cdots \end{aligned}} \tag{4-92}$$

Notice the pattern that appears in the numerator coefficients: how the signs alternate, how the numbers $1, 3, 5, 7, \ldots$ occur, and how the $a_i a_j$ products are formed from the coefficients that bracket the first product. Notice also that the denominator is the same as obtained for $|H(j\omega)|^2$ previously. Recognizing the existence of such patterns allows additional terms to be included in the expression for $T_g(\omega)$ without having to work through all of the tedious mathematics. It is also seen that at $\omega = 0$, the group delay $T_g(\omega) = a_1$. Thus, the coefficient of s in the original transfer function sets the amount of delay, and it becomes a design parameter to be selected. It is convenient in what follows to let $a_1 = 1$. Then some other value can easily be accommodated in a transfer function by replacing s with $a_1 s$. Consequently,

$$T'_g(\omega) = \frac{\begin{aligned}1 + (a_2 - 3a_3)\omega^2 + (a_2 a_3 - 3a_4 + 5a_5)\omega^4 \\ + (a_3 a_4 - 3a_2 a_5 + 5a_6 - 7a_7)\omega^6 + \cdots\end{aligned}}{\begin{aligned}1 + (1 - 2a_2)\omega^2 + (a_2^2 - 2a_3 + 2a_5)\omega^4 \\ + (a_3^2 - 2a_2 a_4 + 2a_5 - 2a_6)\omega^6 + \cdots\end{aligned}} \tag{4-93}$$

where the prime (′) denotes the group delay when $a_1 = 1$. To make the normalized group delay $T'_g(\omega)$ maximally flat requires that the coefficients of like powers of ω in the numerator and the denominator of (4-93) be equal for the specified value of n. This yields equations that must be solved simultaneously for the values of the coefficients a_2, a_3, a_4, etc. The process will be illustrated here for $n = 3$ only. Thus, $a_4 = a_5 = a_6 = 0$ as well as all other a_i values. For this case, then,

$$T'_g(\omega) = \frac{1 + (a_2 - 3a_3)\omega^2 + (a_2 a_3)\omega^4}{1 + (1 - 2a_2)\omega^2 + (a_2^2 - 2a_3)\omega^4 + (a_3^2)\omega^6} \tag{4-94}$$

Therefore, for maximally flat group delay,

$$1 - 2a_2 = a_2 - 3a_3$$

and

$$a_2^2 - 2a_3 = a_2 a_3$$

Solving simultaneously yields $a_2 = \frac{2}{5}$ and $a_3 = \frac{1}{15}$ (for $a_1 = 1$). Hence, from (4-94),

$$T'_g(\omega) = \frac{1 + \frac{1}{5}\omega^2 + \frac{2}{75}\omega^4}{1 + \frac{1}{5}\omega^2 + \frac{2}{75}\omega^4 + \frac{1}{225}\omega^6} \tag{4-95}$$

Also, from (4-91) with $a_1 = 1$,

$$H'(s) = \frac{1}{1 + s + \frac{2}{5}s^2 + \frac{1}{15}s^3} = \frac{15}{15 + 15s + 6s^2 + s^3} \tag{4-96}$$

Then, finally, replacing each s with $a_1 s = p$ yields

$$H(p) = \frac{15}{15 + 15p + 6p^2 + p^3} \tag{4-97}$$

The polynomial in the denominator of (4-97) is the same as the delay polynomial $D_3(p)$ obtained earlier from the Bessel polynomial $D_3(x)$. Notice also that the constant in a Bessel polynomial is the same as the constant that appears in the numerator of $H(p)$. Further, when $a_1 \neq 1$, the new value of a_2 is equal to the old value multiplied by a_1^2, and the new value of a_3 is the old value multiplied by a_1^3. Thus, for example, if $a_1 = 1$ ms is desired instead of $a_1 = 1$, then the new value of a_2 is $\frac{2}{5} \times 10^{-6}$ and the new value of a_3 is $\frac{1}{15} \times 10^{-9}$. Then the transfer function of this third order linear phase-response filter is, from (4-91),

$$H(s) = \frac{1}{1 + 10^{-3}s + \frac{2}{5} \times 10^{-6}s^2 + \frac{1}{15} \times 10^{-9}s^3}$$

The fact that the delay polynomials $D_n(p)$ are related to the Bessel polynomials is the reason that maximally linear phase-response filters are also known as Bessel filters. It is fortunate that there is a useful recursion formula for the delay polynomials, given by

$$D_{n+1}(p) = (2n + 1)D_n(p) + p^2 D_{n-1}(p) \tag{4-98}$$

This means that only two of the simpler polynomials need to be worked out in detail, and then all others can be found merely by applying the recursion formula. For example, since we have just found $D_3(p) = p^3 + 6p^2 + 15p + 15$, and since it is obvious that $D_1(p) = p + 1$, the recursion formula can be used to determine $D_2(p)$. Thus,

$$(2n + 1)D_n(p) = D_{n+1}(p) - p^2 D_{n-1}(p)$$

Then, for $n = 2$,

$$\begin{aligned}(2 \times 2 + 1)D_2(p) &= D_3(p) - p^2 D_1(p) \\ &= (p^3 + 6p^2 + 15p + 15) - p^2(p + 1) \\ D_2(p) &= \frac{5p^2 + 15p + 15}{5} = p^2 + 3p + 3\end{aligned}$$

Several delay polynomials and their factors are listed in Table 4-8.

TABLE 4-8 Delay Polynomials $D_n(p)$

n	$D_n(p)$
1	$p+1$
2	p^2+3p+3
3	$p^3+6p^2+15p+15=(p+2.3222)(p^2+3.6778p+6.4595)$
4	$p^4+10p^3+45p^2+105p+105$
	$=(p^2+5.7924p+9.1401)(p^2+4.2076p+11.4877)$
5	$p^5+15p^2+105p^3+420p^2+945p+945$
	$=(p+3.6467)(p^2+6.7039p+14.2725)(p^2+4.6493p+18.1563)$
6	$p^6+21p^5+201p^4+1260p^3+4725p^2+10395p+10395$
	$=(p^2+8.4967p+18.8011)(p^2+7.4714p+20.8528)(p^2+5.0319p+26.5140)$

PROBLEMS

4-1 Evaluate (a) $\Gamma(2.7)$, (b) $\Gamma(-1.3)$, (c) $(1.6)!$, (d) $(-1.3)!$, (e) show that $\Gamma(7/2)=(5/2)!=15\sqrt{\pi}/8$.

4-2 Demonstrate that $J_0'(x)=-J_1(x)$.

4-3 (a) Use the recurrence formula to express $J_7(2)$ and $J_8(2)$ in terms of $J_0(2)$ and $J_1(2)$ to obtain $J_7(2)=191J_1(2)-492J_0(2)$ and $J_8(2)=1304J_1(2)-3359J_0(2)$.
(b) If $J_0(2)=0.22389$ and $J_1(2)=0.57672$, calculate $J_7(2)$ and $J_8(2)$.

4-4 Use the recurrence formula to calculate $J_n(4.5)$.

4-5 Using Eq. (4-12) and a table of Bessel functions, determine the various values for $J_{-n}(8)$.

4-6 Use the recurrence formula (4-18) for Bessel functions, whose order is half an odd integer, in several steps to demonstrate that

$$J_{-5/2}(x)=\left(\frac{3}{x^2}-1\right)J_{-1/2}(x)+\frac{3}{x}J_{1/2}(x)$$

4-7 Use the recurrence formula for the delay polynomials to determine $D_6(p)$.

4-8 Demonstrate that

$$\sin(\beta\sin\omega t)=\sum_{\text{odd } n=1}^{\infty}2J_n(\beta)\sin n\omega t$$

by using at least the first four terms in the power series for $\sin x$ with $x = \beta \sin \omega t$.

4-9 Demonstrate that

$$\cos(\beta \cos x) = J_0(\beta) + 2 \sum_{k=1}^{\infty} (-1)^k J_{2k}(\beta) \cos 2kx$$

4-10 Demonstrate that

$$\sin(\beta \cos x) = 2 \sum_{k=0}^{\infty} (-1)^k J_{2k+1}(\beta) \cos(2k+1)x$$

4-11 Derive Eq. (4-50).

4-12 Calculate the values of n required for the Butterworth function to yield:

(a) $A \leq 0.10$ at $\omega_r = 1.5$

(b) $A \leq 0.02$ at $\omega_r = 1.5$

(c) $A \leq 0.01$ at $\omega_r = 2.0$

4-13 (a) Solve Eq. (4-49) for ω_r in terms of A and n.

(b) Using the results of part (a), calculate ω_r when:

(1) $A = 0.1$ and $n = 4$

(2) $A = 0.1$ and $n = 8$

4-14 Calculate values of (normalized) ω_r at which the attenuation of a Butterworth filter is 26 dB when $n = 1, 2, 3, 4$, and 5, one at a time.

4-15 Calculate the high-frequency rolloff in dB/decade and in dB/octave of an nth-order Butterworth filter.

4-16 Calculate the required integral order of a maximally flat lowpass filter if $|H(j\omega)| \leq 0.1$ at a frequency: (a) $2f_C$, (b) $1.5f_C$.

4-17 (a) In separate s-plane figures, sketch the locations of the useful zeros of $1 + (s/j)^{2n} = 0$ for: (1) $n = 6$ and (2) $n = 7$.

(b) Using the results of part (a), write the corresponding Butterworth polynomials in their quadratic-product forms.

4-18 Given the function

$$|H(j\omega)|^2 = \frac{1 + a_2\omega^2}{1 + b_2\omega^2 + b_4\omega^4 + b_6\omega^6}$$

where $a_2 \neq 0$. Determine the relationships between and for the coefficients such the $|H(j\omega)|^2$ is maximally flat (i.e., each of its first five derivatives with respect to ω is zero at $\omega = 0$), and has the value 1/2 at $\omega = 1$. Do not arbitrarily assume initially that any of the coefficients are zero. Hint: First perform a long division and then differentiate.

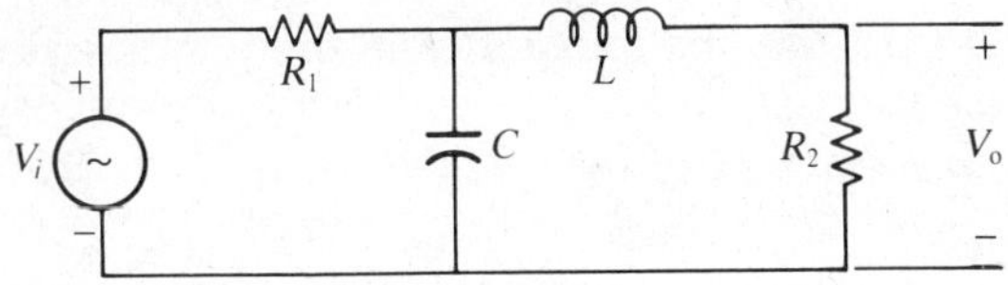

Fig. P4-19 Circuit for Problem 4-19.

4-19 (a) For the network shown in Fig. P4-19, determine L and C in terms of R_1, R_2, and ω_c such that $|H(j\omega)| = |V_0(j\omega)|/|V_i(j\omega)|$ is maximally flat.

(b) If $R_1 = R_2 = 1\,\text{k}\Omega$ and $f_C = 20\,\text{kHz}$, calculate L and C.

(c) Calculate the $-60\,\text{dB}/-6\,\text{dB}$ shape factor.

4-20 Calculate the $-60\,\text{dB}/-6\,\text{dB}$ shape factor and the values of C_1, C_2, and L for the network shown in Fig. P4-20 to be maximally flat when $R_1 = R_2 = 1\,\text{k}\Omega$ and $f_C = 20\,\text{kHz}$.

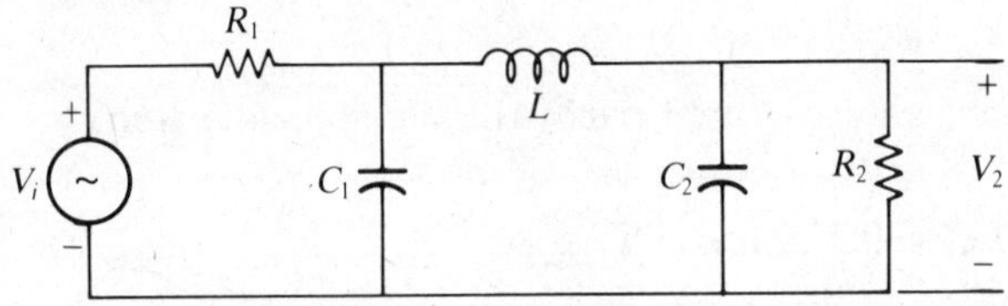

Fig. P4-20 Circuit for Problem 4-20.

4-21 Consider m identical nth-order Butterworth filters in cascade. Let each filter be separated from its neighbors by unity-gain buffer amplifiers to prevent interactions between individual filters. The overall transfer function $|H_o(j\omega)|$ is given by the product of the m individual mth-order transfer functions. Thus,

$$|H_o(j\omega)| = \frac{1}{(1 + \omega^{2n})^{m/2}}$$

The overall response is at the 0.707 or $-3\,\text{dB}$ level at some (normalized) frequency ω_3 called the overall cutoff frequency, which also happens to be the bandwidth of the filter.

(a) Solve the above for the (normalized) frequency ω_3 in terms of n and m.

(b) Demonstrate that the (normalized) ω_3 in part (a) is equal to or less than 1.

(c) Let the overall response be down to the level A_0 at (normalized)

ω_{r0}. What range of values for A_0 (in terms of m) assures that $\omega_{r0} \leq 1$?

(d) Calculate the high-frequency rolloff rate in dB/decade and in dB/octave of this cascaded system.

4-22 Four identical fifth-order Butterworth filters are cascaded as described in Problem 4-21. The actual overall cutoff frequency of the entire cascade is 20 kHz.

(a) Calculate the actual cutoff frequency required of each individual filter.

(b) The overall response of the cascaded system is down to $A_0 = 0.05$ at the normalized frequency ω_{r0}. Calculate ω_{r0} and the ratio ω_{r0}/ω_3.

(c) Calculate the shape factor of the cascaded system, based on the frequencies at which the responses are at -60 dB and -6 dB.

4-23 Using the Chebyshev polynomial $C_4(\omega)$, show that the function $1 - C_4^2(\omega)$ has the factors $(1-\omega)$, $(1+\omega)$, and $(dC_4(\omega)/d\omega)$ *and only those factors* (except for a constant).

4-24 (a) Determine what values of ω make $C_n(\omega) = \cos(n \cos^{-1} \omega) = 0$ by showing that

$$C_n(\omega) = 2^{n-1}\left(\omega - \cos\frac{\pi}{2n}\right)\left(\omega - \cos\frac{3\pi}{2n}\right)(\cdots)\left(\omega - \cos\frac{(2n-1)\pi}{2n}\right)$$

where 2^{n-1} is obtained from Table 4-7.

(b) Use the recursion formula to write out the Chebyshev polynomials $C_5(\omega)$ and $C_6(\omega)$ as the sum of several terms.

(c) Use the result of part (a) to write $C_5(\omega)$ and $C_6(\omega)$ as the product of the several factors.

(d) Make sketches of $C_5(\omega)$ and $C_6(\omega)$ in separate figures. Show the numerical values of the zero points.

4-25 Demonstrate that $C_n^2(\omega) = \frac{1}{2}[1 + C_{2n}(\omega)]$ for $|\omega| < 1$ and $|\omega| > 1$.

4-26 The quantity ϵ is a measure of ripple width in Chebyshev filters. Calculate values of ϵ required for several filters whose ripple varies from 0.5 dB to 4 dB in 0.5 dB steps, and tabulate the results.

4-27 At a specified normalized radian frequency $\omega_r > 1$, $|H(j\omega)|$ of a Chebyshev filter must be equal to or less than the amount A.

(a) Show that the required filter order is given by the next larger integer value to

$$n = \frac{\cosh^{-1}\left(\dfrac{1}{\epsilon}\sqrt{\dfrac{1}{A^2} - 1}\right)}{\cosh^{-1}\omega_r}$$

(b) Calculate the minimum integral order n required for an equal-ripple filter to have $A = 0.1$ at $\omega_r = 1.5$ when the ripple is 1 dB.

(c) Calculate the minimum integral order n required for an equal-ripple filter to have $A = 0.05$ at $\omega_r = 1.4$ when the ripple is 2 dB.

4-28 (a) Demonstrate that when the prototype filter response of Fig. 4-15a is subject to the lowpass-to-highpass transformation, the highpass filter response of Fig. 4-19 is obtained.

(b) By replacing p with ω_0/s, show that each series inductor of the prototype is replaced by a capacitor whose value is $1/\omega_0 L$ and each shunt capacitor is replaced by an inductor whose value is $1/\omega_0 C$.

5

AMPLITUDE-MODULATION PROCESSES

5-1 MODULATION

Suppose you try to converse with someone across town by opening a window and shouting. You soon discover that your voice does not carry far enough, so perhaps you move to the roof or climb a tree and repeat the effort, but still without success. Reasoning that you need more power than your vocal cords can supply,† you propose to use an audio amplifier to control a pneumatically operated vibrating cone of a directional high-powered loud-speaker system, duplicating the apparatus at the other site to provide for 2-way conversations. By so doing you could make yourself heard across town by brute force. But even if such an endeavor proved successful for your purposes alone, it would not be long before numerous unwilling listeners would swamp you with entirely justified complaints about all that abominable racket. And the inevitable lawsuits by neighbors seeking redress for burst eardrums, shattered windows, and disintegrated pet canaries would not be pleasant either. Imagine the scenario and cacophony if everybody acquired and used equipment of like design!

Now suppose you are an astronaut being zipped around untethered in the vacuum of space by your jetpack and you need to communicate with the commander in the space shuttle. Simply talking will not do the job. Sound waves do not exist in a vacuum because there are no atmospheric particles to carry the vibrational energy, therefore no sound waves can be propagated.

† The average power of conversational speech is about 10 μW, becoming about 1 mW when shouting as loudly as possible, and about 10 nW for a faint whisper.

What is needed in both of these situations is a type of inaudible signal whose radiation and propagation characteristics are convenient and harmless and suited to the intervening medium which, in one case, is the Earth's atmosphere, and, in the other case, the vacuum of space. One such desirable signal is a high-frequency transverse electromagnetic wave that can be radiated from a physical structure of appropriate size, called an *antenna* or aerial. An electromagnetic wave does not require a particulate-type environment for its propagation; it exists in either the Earth's atmosphere or the vacuum of space because of the continual exchange of energy between the wave's interdependent electric and magnetic fields. A portion of the radiated electromagnetic wave can be intercepted by a receiving antenna at the distant location and the resulting signal delivered to the receiver for processing to recover the original message. This is the essence of a radio system.

In order for this radiated wave to be useful for communications, at least one of its properties must be varied in accordance with the instantaneous amplitude of the message to be conveyed. This is accomplished by varying some property of the original sinusoidal voltage or current that created the electromagnetic wave. The name given to this process is *modulation*. The system in which modulation takes place is called a *modulator*, and the act itself is described by some form of the verb *to modulate*. When used as a modifier, it sometimes appears as *modulation* techniques, *modulating* signal, and *modulated* waveform. Sometimes the modulating signal itself is referred to as *the modulation*. In all cases the meaning should be clear from the context.

Consider the sinusoidal voltage $x_c(t) = V(t) \cos \phi(t)$. The amplitude $V(t)$ or the angle $\phi(t)$ or both can be varied according to some functional relationship with the modulating signal $m(t)$. Usually, it is desirable to have a linear relationship between $m(t)$ and the amount of change in the quantity undergoing modulation. Then, if $m(t)$ is represented by its various spectral components, the total change in the modulated quantity can be found by superposition of the changes produced by the spectral components individually. Even though this particular linear relationship exists, it does not necessarily give rise to what is called linear modulation.

In *linear modulation*, the *spectrum* of the modulated signal can be found by superposition of the output spectral terms produced individually by the spectral components of $m(t)$. Included under the general heading of linear modulation is the process known as amplitude modulation, which is subdivided into the important categories called double-sideband, single-sideband, and vestigial-sideband modulation.

Amplitude modulation is said to occur when the amplitude $V(t)$ is caused to vary by an amount determined (usually linearly) by the instantaneous amplitudes of the modulating signal $m(t)$, and the angle $\phi(t) = 2\pi f_c t$, where f_c is constant.

The unmodulated sinusoid is called the *carrier*† because its modulated form will carry the information to be conveyed. Thus, reference will be made, for example, to such things as carrier amplitude, carrier power, and carrier frequency. In all cases, the carrier frequency is chosen such that the radiation and propagation of electromagnetic waves are convenient, useful, and legal.

The modulating signal $m(t)$, sometimes also called the baseband signal, is the electrical equivalent of voice, music, temperature, or pressure variations, or some other physical quantity for which an electrical transducer is available. The long-time average value of some signals, such as voice or music, is zero. The spectrum of the baseband signal is sometimes called the baseband.

5-2 DOUBLE-SIDEBAND MODULATION

Perhaps the simplest process, conceptually, whereby the amplitude of a carrier $A_c \cos \omega_c t$ can be varied in a manner that is linearly related to a modulating signal $m(t)$ is by means of the ideal multiplier shown in Fig. 5-1a. The multiplier is a classic example of a linear system having time-varying coefficients. The output voltage of the multiplier is simply the product of the two input voltages, or

$$x_c(t) = A_c m(t) \cos \omega_c t \tag{5-1}$$

Specifically, represent $m(t)$ by its spectral components. Thus

$$m(t) = m_1(t) + m_2(t) + \cdots = A_1 \cos \omega_1 t + A_2 \cos \omega_2 t + \cdots \tag{5-2}$$

Then

$$x_c(t) = A_c A_1 \cos \omega_1 t \cos \omega_c t + A_c A_2 \cos \omega_2 t \cos \omega_c t + \cdots \tag{5-3}$$

each term of which has a waveform as shown in Fig. 5-1b that is like the waveform obtained in the linear mixing of two sinusoids discussed in Chapter 1. Equation (5-3) represents a waveform whose spectrum can be found by using the identity

$$\cos A \cos B = \tfrac{1}{2} \cos (A + B) + \tfrac{1}{2} \cos (A - B)$$

Thus,

$$\begin{aligned} x_c(t) = {} & \tfrac{1}{2} A_c A_1 \cos 2\pi(f_c + f_1)t + \tfrac{1}{2} A_c A_1 \cos 2\pi(f_c - f_1)t \\ & + \tfrac{1}{2} A_c A_2 \cos 2\pi(f_c + f_2)t + \tfrac{1}{2} A_c A_2 \cos 2\pi(f_c - f_2)t + \cdots \end{aligned} \tag{5-4}$$

† It is not always essential that the carrier be sinusoidal. The carrier could be a pulse train in which the modulation process affects either the amplitude, the position, or the width of the pulses, or some combination of these.

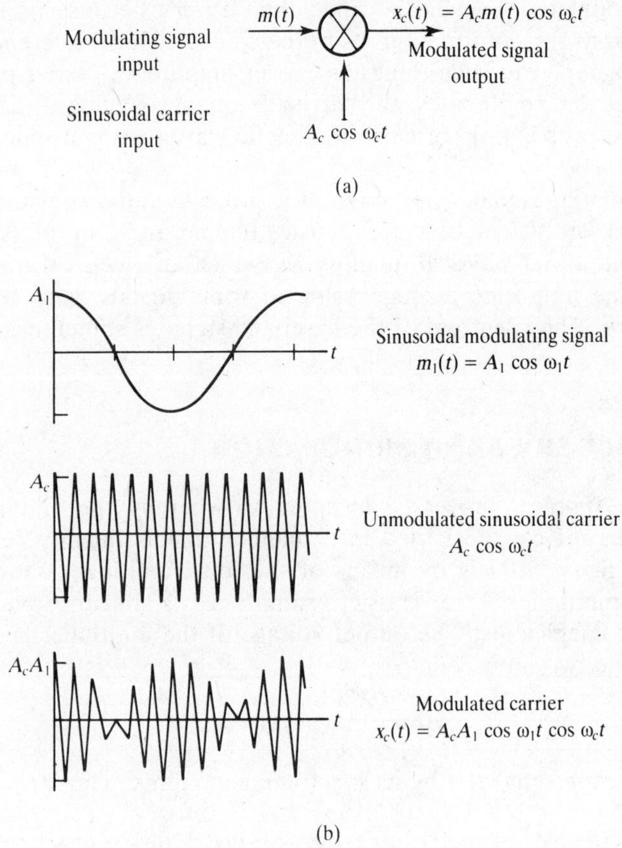

Fig. 5-1 (a) The ideal multiplier produces an output signal whose amplitude is linearly related to the modulating signal $m(t)$. (b) Some input and output waveforms.

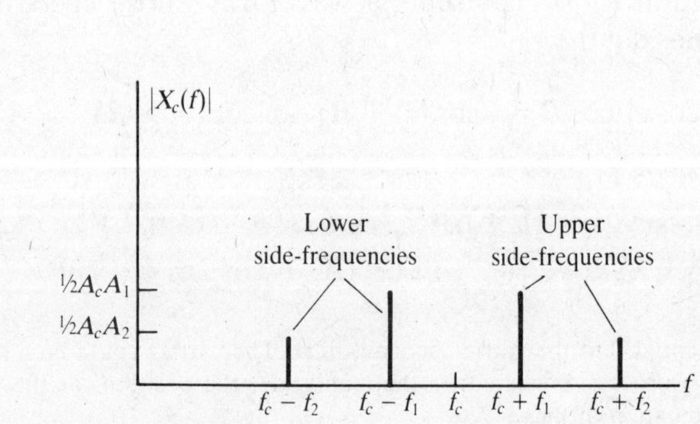

Fig. 5-2 Amplitude spectrum of $x_c(t)$ in Eq. (5-4) is $X_c(f)$.

This demonstrates that the muliplier has created a modulated signal having spectral components at various sum and difference frequencies as shown in Fig. 5-2. The new frequencies appear at either side of the carrier frequency f_c and are called *side frequencies.* The side frequencies differ from f_c by amounts equal to the modulating frequencies. It is also noticed from (5-4) that the amplitude of an upper side-frequency component is the same as the amplitude of the corresponding lower side-frequency component, and that *the output components at both of the input frequencies have been suppressed by the ideal multiplier.* The voltage $x_c(t)$ in (5-4) is periodic if the frequencies are commensurable.

In practice, the frequency composition of a modulating signal at the multiplier is determined not only by the spectrum of the original modulating signal itself but also by the frequency response of the circuitry that amplifies and filters the modulating signal before it arrives at the multiplier.

If $m(t)$ is represented by the sum of any number of baseband spectral components extending up to the component $A_W \cos 2\pi Wt$, where W is the highest frequency contained in the modulating signal, then each of the baseband spectral components produces a pair of equal-amplitude side-frequency components extending out to the ones located at $f_c \pm W$. This result is called a *double-sideband* (DSB) signal because its spectrum consists of a *band* of side frequencies above f_c, known as the *upper sideband* (USB) and a *band* of side frequencies below f_c, known as the *lower sideband* (LSB). The spectral width of the upper sideband is W Hz and the spectral width of the lower sideband is W Hz, making the total spectral width of the double-sideband signal equal to $2W$ Hz. Thus, if $M(f)$ represents the baseband spectrum in Fig. 5-3a, then the double-sideband spectrum is as shown in Fig. 5-3b. To prevent a doubling (or folding back) of components in the lower sideband when frequencies $f_c - W$ become $W - f_c$, it is required that $f_c > W$, that is, the carrier frequency should be larger than the highest frequency allowed in the baseband. Generally, in practice, the carrier frequency is considerably higher than the highest frequency allowed in the baseband.

It is important in the theory and operation of multipliers to note that if one of the spectral components of $m(t)$ is a constant having the value A then, from (5-1), the output of the multiplier contains a spectral component $AA_c \cos \omega_c t$. Consequently, whenever this constant value is present, the carrier component in the output of the multiplier is not suppressed.

5-2.1 Double-Sideband Demodulation

The process of recovering the modulating information at the receiver is known as *demodulation* or *detection.* The system used for that function is called a *demodulator* or *detector.* Depending on the nature of the transmission channel, it is possible that the components in one sideband may be

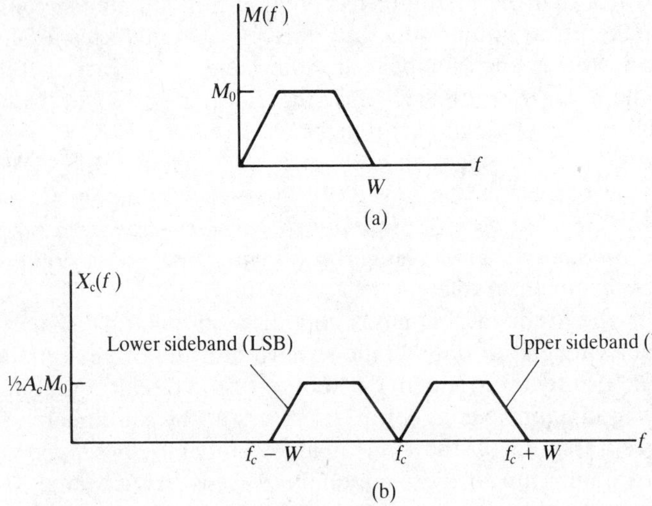

Fig. 5-3 (a) A baseband (i.e., spectral components of a modulating signal). (b) The spectrum of a double-sideband (DSB) signal produced by the $M(f)$ in (a).

propagated differently from those in the other sideband, causing a distortion known as *selective fading.*

When all components of a DSB signal are transmitted without distortion, the received signal $x_r(t)$ becomes an amplitude-scaled replica of the transmitted signal. Thus

$$x_r(t) = A_r m(t) \cos \omega_c t \tag{5-5}$$

The effect of a time delay caused by the finite propagation time between the transmitter and the receiver is eliminated in (5-5) by choosing a new time scale for the undistorted received signal that has been shifted an amount equal to the time delay. When the received DSB signal is applied to an ideal multiplier whose other input signal is the sinusoid from a local oscillator (LO)

$$A_{LO} \cos \phi_{LO} = A_{LO} \cos [\omega_c t + \theta(t)] \tag{5-6}$$

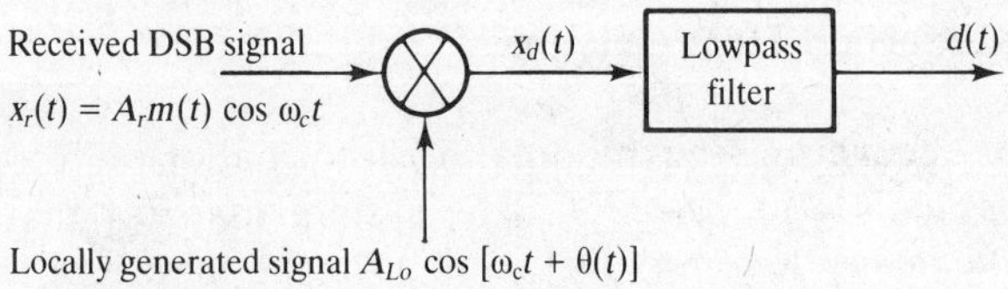

Fig. 5-4 Demodulation of a received DSB signal using an ideal multiplier.

as shown in Fig. 5-4, the output of the multiplier is

$$\begin{aligned} x_d(r) &= A_r A_{LO}\, m(t) \cos\,[\omega_c t + \theta(t)] \cos \omega_c t \\ &= \tfrac{1}{2} A_r A_{LO}\, m(t) \cos \theta(t) + \tfrac{1}{2} A_r A_{LO}\, m(t) \cos\,[2\omega_c t + \theta(t)] \end{aligned} \tag{5-7}$$

in which a familiar trigonometric identity has been used.

The spectrum of the second term in (5-7) consists of the upper and lower sidebands grouped around the vicinity of $2f_c$. These components are readily eliminated by passing $x_d(t)$ through a lowpass filter that cuts off slightly above W Hz. The design of this filter is not critical if $f_c \gg W$, so that a simple RC lowpass filter will usually be satisfactory.

The output of the lowpass filter is

$$d(t) = \tfrac{1}{2} A_r A_{LO}\, m(t) \cos \theta(t) \tag{5-8}$$

revealing that a replica of the original modulating signal $m(t)$ can be obtained provided that $\cos \theta(t)$ is a nonzero constant. It is preferable that $\cos \theta(t) = +1$ to obtain the greatest magnitude of $d(t)$; that is, the locally generated sinusoid should be *phase coherent* with the missing carrier of the DSB signal.

If $\theta(t)$ represents any possible random-phase noise of the locally generated signal caused by the oscillator or frequency synthesizer, the demodulated output $d(t)$ in (5-8) will also vary randomly, be noisy, and be distorted from the true waveform of $m(t)$ because of the effect of $\cos \theta(t)$.

If the frequency of the locally generated sinusoid is not the same as f_c, that is, it is not synchronous with, but differs from f_c by $\pm\Delta f$, then

$$\begin{aligned} A_{LO} \cos \theta_{LO} &= A_{LO} \cos 2\pi (f_c \pm \Delta f)t \\ &= A_{LO} \cos\,(\omega_c t \pm 2\pi\, \Delta f\, t) \end{aligned} \tag{5-9}$$

so that

$$\theta(t) = \pm 2\pi\, \Delta f\, t$$

and

$$\cos \theta(t) = \cos 2\pi\, \Delta f\, t \tag{5-10}$$

Therefore $d(t)$ in (5-8) will increase and decrease in a manner determined not only by $m(t)$ but also by how much the frequency of the locally generated sinusoid differs from the frequency f_c of the missing carrier of the received DSB signal. The resulting distortion can be eliminated only if $\Delta f = 0$.

This demodulation process requires that the locally generated signal be coherent or synchronous, in both phase and frequency, with the missing carrier of the received DSB signal, and so the process is known as *coherent detection* and as *synchronous detection*.

5-3 MULTIPLERS

An ideal multiplier produces an output $y(t)$ that is directly proportional to the product of two input signals $x_1(t)$ and $x_2(t)$, or

$$y(t) = Kx_1(t)x_2(t) \tag{5-11}$$

where K is the proportionality constant whose value depends on the particular device. The ideal multiplication relationship expressed by (5-11) shows that the output is linearly related to either of the input variables but that the other variable acts as a time-varying coefficient. The output of such a linear time-varying system contains new frequencies that were not present in the input, as discussed in Section 1-7. Linear time-varying systems share this feature with nonlinear systems.

There are numerous ways to implement a multiplication of two signal voltages, not all of which are suitable for the high frequencies associated with radio communications. For example, the output voltage from the movable contact of a variable resistor known as a potentiometer is determined by the signal voltage across the entire potentiometer and by the position of the movable contact. If a second signal voltage controls the physical position of the movable contact in a linear fashion, then the output voltage is proportional to the product of the two signal voltages.

The Hall effect produces another multiplying mechanism. The Hall voltage across opposite sides of a semiconductor is proportional to the product of the current through the material times the magnetic field strength perpendicular to the current direction.

In a bipolar junction transistor the signal component of collector current is given by $h_{fe}i_b$, where h_{fe} is a transistor parameter determined by the operating point and i_b is the signal component of base current determined by an input signal. If now a second input signal varies the operating point, causing h_{fe} to change linearly with that signal, then the collector current contains a signal component proportional to the product of the two input signals. The same kind of mechanism can occur in other devices when one input signal can linearly control how much output the other signal produces.

A very useful multiplying mechanism occurs in a difference amplifier. The output voltage between the amplifying devices depends on their current imbalance and on the total amount of their currents. One input signal controls the current imbalance and the other input signal controls the total amount of current. The device is called a 4-*quadrant* multiplier if *both* input voltages are allowed to take on both positive and negative polarities and the multiplier develops an output having the correct polarities. The multiplier required in a double-sideband system should be of this type. In a 2-*quadrant* multiplier only one of the input voltages is allowed to take on both polarities while the other input voltage accepts only one polarity (either positive or negative depending on the design), and the output has the correct polarities. In a 1-*quadrant* multiplier both input voltages accept signals of only one polarity each. Several integrated-circuit 4-quadrant multipliers have been developed and are available commercially. Some of them tend to suffer from limited large-signal-handling capabilities and rather noisy operation because of the numerous active and passive components required. Multipliers required in many radio communications applications are fabricated from discrete components to obtain the desired operating characteristics.

Nonlinear devices whose transfer characteristic is given by a power series that contains a squared term can be arranged to provide the multiplying action. That this is true can be seen by virtue of the fact that voltages v_1 and v_2 can be combined as

$$(v_1 + v_2)^2 - (v_1 - v_2)^2 = 4v_1v_2 \tag{5-12}$$

Figure 5-5 shows how (5-12) can be implemented with transformers and identical devices in a balanced configuration. Thus,

$$v_{b1} = v_1 + v_2 \tag{5-13a}$$

and

$$v_{b2} = v_1 - v_2 \tag{5-13b}$$

The device output currents are related to the input voltages by their power series. If the terms with exponents above 2 are omitted for simplicity, then

$$i_{c1} = Av_{b1} + Bv_{b1}^2 \tag{5-14a}$$

and

$$i_{c2} = Av_{b2} + Bv_{b2}^2 \tag{5-14b}$$

(see Problem 5-2 for analysis that includes the cubed term in the power series).

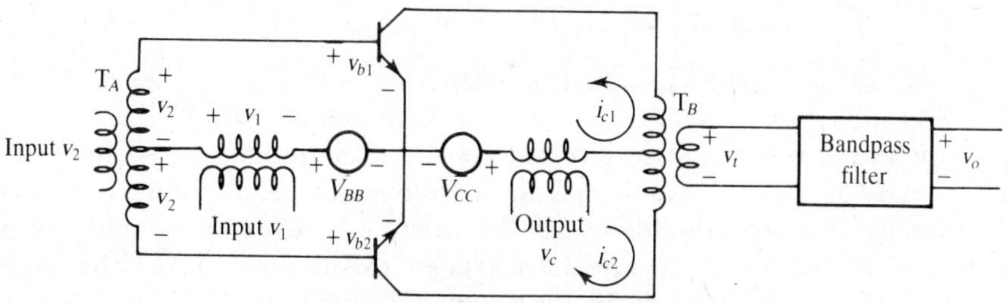

Fig. 5-5 A balanced modulator.

The output voltage v_t produced by the output transformer T_B is

$$v_t = K_1(i_{c1} - i_{c2}) = K_1A(v_{b1} - v_{b2}) + K_1B(v_{b1}^2 - v_{b2}^2) \tag{5-15a}$$

$$= K_1A(2v_2) + K_1B[(v_1 + v_2)^2 - (v_1 - v_2)^2] \tag{5-15b}$$

$$= K_1A(2v_2) + K_1B(4v_1v_2) \tag{5-15c}$$

There is no output here containing v_1 alone because of balanced condition.

If the double-sideband frequencies associated with the product v_1v_2 are sufficiently removed from the frequencies associated with v_2, then a simple bandpass filter (such as may be obtained by tuning the output transformer), will effectively eliminate the frequencies associated with v_2 in the output of the filter. In this case, the output of the filter is proportional to the product v_1v_2. This also indicates that v_2 should be the input signal having the lowest frequency content and that v_1 should have the highest frequency content for the filtering problem to be simplified as much as possible. Thus,

$$v_0 = Kv_1v_2 \tag{5-16}$$

where K is the proportionality constant for the multiplier. When v_2 represents the low-frequency modulating signal and v_1 represents the carrier sinusoid, the multiplier is known as a *balanced modulator*. A multiplier used as a demodulator in a radio receiver to combine an incoming modulated signal with a locally generated sinusoid is usually called a *product detector*. When used to convert the frequencies associated with an incoming signal, the multiplier is generally known as a *mixer*. There is no need for the filter if the linear terms in (5-14a) and (5-14b) are absent and the devices have square-law transfer characteristics.

When only v_1 in (5-16) changes sign, the output becomes $-K|v_1|v_2$ as expected. When only v_2 in (5-16) changes sign, the output becomes $-Kv_1|v_2|$ as expected. When both v_1 and v_2 change sign simultaneously, the output is $+K|v_1||v_2|$ as expected. Therefore, the balanced modulator circuit in Fig. 5-5 is a 4-quadrant multiplier.

If output is taken at v_c in Fig. 5-5, then

$$v_c = K_2(i_{c1} + i_{c2}) = K_2A(v_{b1} + v_{b2}) + K_2B(v_{b1}^2 + v_{b2}^2) \tag{5-17a}$$

$$= K_2A(2v_1) + K_2B[(v_1 + v_2)^2 + (v_1 - v_2)^2] \tag{5-17b}$$

$$= K_2A(2v_1) + K_2B(2v_1^2 + 2v_2^2) \tag{5-17c}$$

The squared terms v_1^2 and v_2^2 give rise to second harmonics, suggesting that the method is suitable for frequency doubling. Since v_2 alone has been canceled here, it is preferable to use the circuit for frequency doubling of v_2 with $v_1 = 0$. The resulting circuit has been called *push–push*; the input signal v_2 drives the devices in push–pull but their outputs are taken in parallel, and the transformer producing the output v_t is omitted.

5-4 SWITCHING-MODE PROCESSES

Also known as chopping mode, gating mode, and sampling mode, the switching-mode process is based on the abrupt changes that are possible between the conducting and nonconducting states of electronic devices. The switching action is most often implemented with hot-carrier or Schottky-barrier diodes† because of their fast response time, low-noise property, and low forward voltage. The switching action is similar to the alternate opening and closing of a switch, hence the name.

5-4.1 Singly Balanced

Consider the circuit shown in Fig. 5-6a. Whether or not all four diodes conduct simultaneously is determined primarily by the polarity of the voltage from C to D rather than by the voltage from A to B. Therefore the voltage v_2 controls the switching action of the diodes provided that the amplitude of v_2 is significantly larger than that of v_1. If the diodes are a matched set, that is, they have identical electrical characteristics, there will be no voltage from A to B when v_2 causes the diodes to conduct. This is because the conducting diodes form a balanced bridge. When the diodes conduct they act like the switch SW in Fig. 5-6b in its closed position, shorting v_1 and removing it from the output. When v_2 changes polarity, the diodes do not conduct, switch SW is now in its open position, and the other voltage v_1 is connected directly to the output transformer T_B. An important feature of this circuit is that none of v_2 itself is delivered to either of the

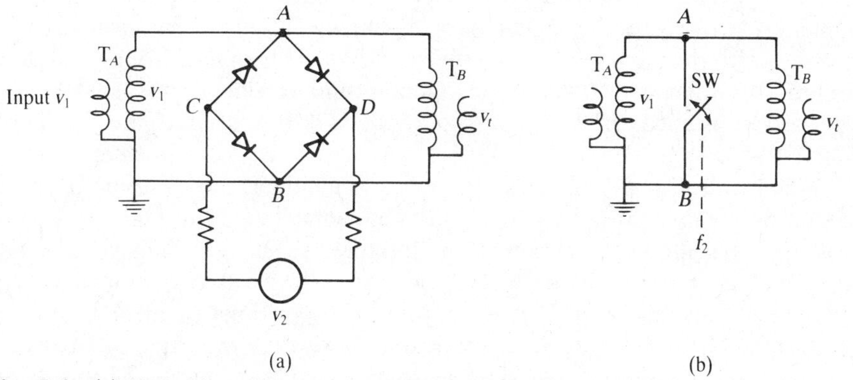

Fig. 5-6 (a) Switching diodes in parallel with the signal path from v_1 to v_t are turned on and off by the controlling voltage v_2 rather than by v_1. (b) The diodes act like a switch SW opening and closing at the rate set by f_2.

† Hot-carrier or Schottky-barrier diodes are fabricated to have a rectifying metal–semiconductor contact. The metal is usually platinum and the semiconductor is n-type silicon.

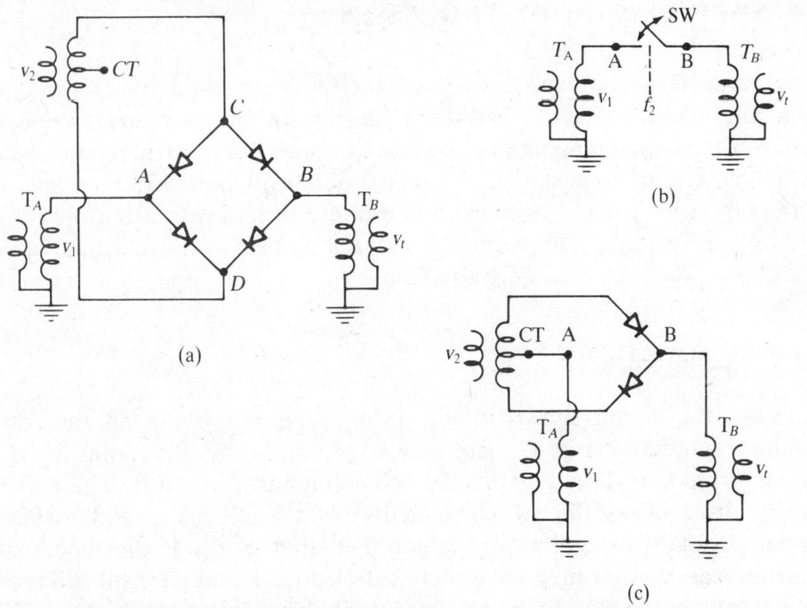

Fig. 5-7 (a) Switching diodes in series with the signal path from v_1 to v_t. (b) Mechanical-switch analog of the series diode-quad controlled by f_2. (c) Simplified two-diode equivalent of (a).

other ports because of the balanced condition. Also, only one of the three ports needs to be balanced with respect to ground, in this case v_2. Therefore, the term *singly balanced* is applied to the circuit.

A more useful singly-balanced switching-mode circuit places the diode quad in series with the signal path as shown in Fig. 5-7a, instead of in parallel with the signal path. The mechanical-switch analog is represented in Fig. 5-7b, where v_2 controls the opening and closing of the switch. The forward resistance of a conducting diode is considered to be negligibly small.

When v_2 in Fig. 5-7a causes the diodes to conduct, the resulting balanced condition means that points A, B, and the center tap CT are all at the same voltage. This allows the circuit to be simplified as shown in Fig. 5-7c. *Either* A *or* B can be connected to CT and the associated two diodes can be eliminated. Symmetry in Fig. 5-7a allows the positions of v_1 and v_2 to be interchanged. None of v_2 is delivered to either of the other ports because of the balanced condition, and only v_2 is balanced to ground; therefore, the circuit of Fig. 5-7c is singly balanced. The switch in Fig. 5-6b or in Fig. 5-7b is either open or closed. The operation of the diodes can be represented mathematically by describing their switching action by an appropriate switching function $s_1(t)$ as shown in Fig. 5-8. The waveform of v_2 itself does not need to be square-wave shape and usually it is sinusoidal, but v_2 must

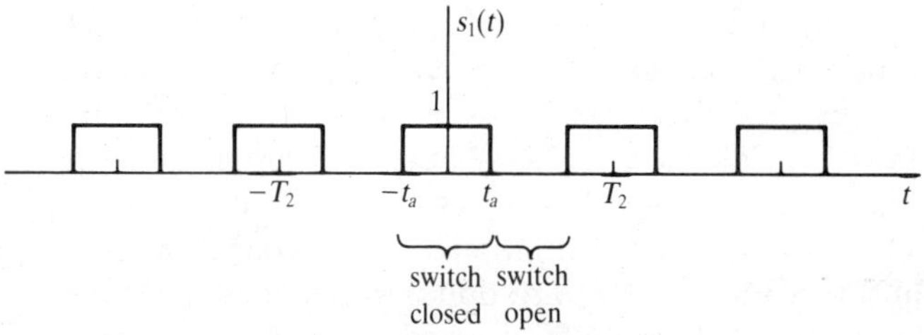

Fig. 5-8 The periodic switching function $s_1(t)$ for a singly-balanced switching-mode circuit.

have sufficient amplitude to cause the diodes to conduct heavily. Then the output voltage is given by the product

$$v_t = Kv_1 s_1(t) \tag{5-18}$$

The theory behind this use of a switching function multiplying v_1 is that when the switch is closed, $s_1(t) = 1$ and v_1 is delivered directly to the output, but when the switch is open, $s_1(t) = 0$ and no v_1 is delivered to the output, as expected.

Since $s_1(t)$ is periodic with period $T_2 = 1/f_2$, it has a Fourier series given by

$$s_1(t) = \frac{2t_a}{T_2} + \frac{4t_a}{T_2} \sum_{n=1}^{\infty} \operatorname{sinc} \frac{2nt_a}{T_2} \cos 2\pi n f_2 t \tag{5-19}$$

where

$$\operatorname{sinc} \frac{2nt_a}{T_2} = \frac{\sin (2\pi n t_a/T_2)}{2\pi n t_a/T_2} \tag{5-20}$$

and n represents the harmonic number. The function $\operatorname{sinc} z = (\sin \pi z)/(\pi z)$ is plotted in Fig. 5-9.

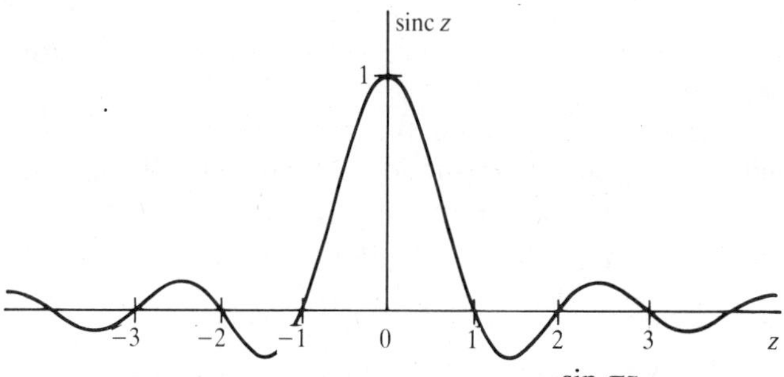

Fig. 5-9 The function $\operatorname{sinc} z = \dfrac{\sin \pi z}{\pi z}$.

The switching function $s_1(t)$ has an average value. When (5-19) is substituted into (5-18), that average value gives rise to output spectral components that are an amplitude-scaled replica of all the spectral components of v_1. These additional output components can cause undesirable spurious responses in some situations, and this effect is a major limitation of a single-balanced multiplier. The other terms that result when (5-19) is substituted into (5-18) are double-sidebands (i.e., they are created from the sum and difference frequencies) grouped around the harmonic locations of f_2, but there are no actual spectral components at any harmonics of f_2 in the output.

As the amplitude of v_1 in Fig. 5-7c is increased continually, the amplitude of an output component at, say, the difference frequency, increases linearly, at first. This is because the amplitude of the voltage v_2 that controls the diodes is still large compared to the amplitude of v_1. But a point is reached at which v_1 begins to influence the conduction of the diodes so that they are significantly different. This causes a circuit imbalance that becomes continually worse. An increasing proportion of v_1 appears across A and B, and the output amplitude levels off. This is *compression.*

5-4.2 Doubly Balanced

In order for the output voltage given by

$$v_t = K v_1 s_2 \tag{5-21}$$

to contain no spectral components of v_1, the switching function s_2 must have zero average. Thus, the switching function must be alternately positive and negative. This means that the switching diodes must operate like a reversing switch between input v_1 and output v_t. One way to do this is illustrated in Fig. 5-10a. Normally v_2 controls which of the diodes conduct, so its amplitude must be large compared to the amplitude of v_1. When v_2 has the polarity shown, diodes D1 and D2 conduct heavily, while diodes D3 and D4 cannot because of the large reverse voltage across them. Thus, v_1 is connected through transformer T_A directly to the output transformer T_B and v_t is in phase with v_1.

During alternate intervals when the polarity of v_2 reverses, diodes D3 and D4 conduct heavily, while diodes D1 and D2 do not conduct. This connects the bottom of T_A to the top of T_B, and the top of T_A to the bottom of T_B. Thus, v_t is opposite in polarity to v_1. The required switching function, shown in Fig. 5-10b, has the Fourier series

$$s_2(t) = \frac{8t_a}{T_2} \sum_{k=1}^{\infty} \operatorname{sinc} \frac{2nt_a}{T_2} \cos 2\pi(2k-1)t \tag{5-22}$$

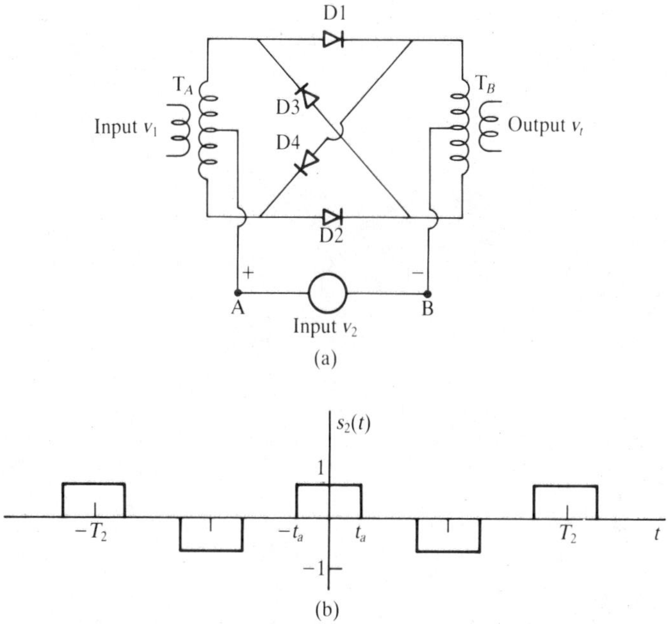

Fig. 5-10 (a) A doubly-balanced switching-mode circuit in which the polarity of v_2 normally controls which pairs of diodes conduct, and this determines whether v_1 passes to the output directly or reversed in polarity. (b) The periodic switching function $s_2(t)$ for (a).

where k is a positive-integer index and the quantity $(2k-1)$ represents odd harmonic numbers only. Substituting (5-22) into (5-21) reveals that there will be double-sidebands grouped around the odd harmonics of f_2, but that there are no actual spectral components at any harmonics of f_2 in the output.

The circuit in Fig. 5-10a is symmetrical, so the locations of v_1 and v_t can be interchanged. Further, either point A or point B can be grounded and then the other two ports are balanced with respect to ground. The voltage v_2 does not appear at either or the other ports and the voltage v_1 is not present in the output v_t. The circuit is doubly balanced.

The crossed-diode arrangement in Fig. 5-10a can be redrawn as in Fig. 5-11a. It is seen that the diodes form what is called a *ring*, with the diode symbols all pointing in the same direction around the ring. This contrasts with the way the diodes connect in Figs. 5-6 and 5-7 because of the different operating modes.

Another useful doubly balanced circuit is given in Fig. 5-11b. When v_2 causes D1 and D2 to conduct, their center point, also connected to the *bottom* side of T_B, is at zero volts (i.e., ground). When D3 and D4 conduct during alternate intervals, their center point, also connected to the *top* side

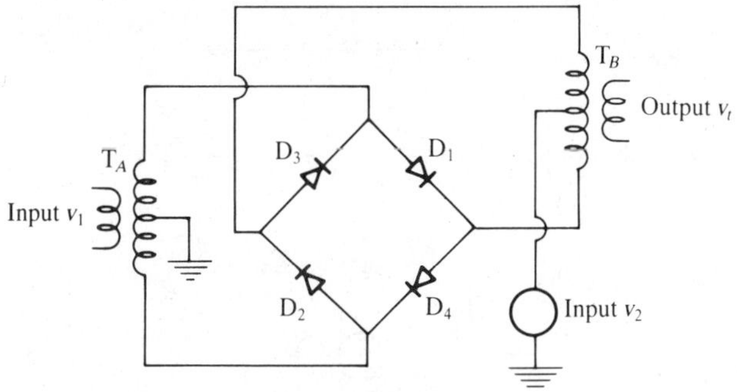

(a)

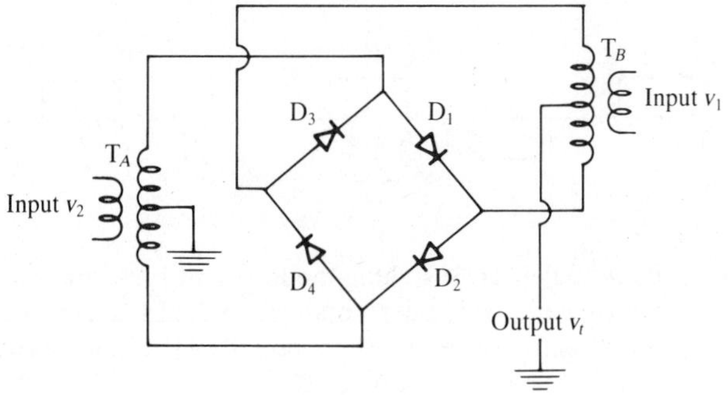

(b)

Fig. 5-11 (a) The crossed diodes in Fig. 5-10a redrawn as a diode ring (b) A doubly-balanced circuit in which the polarity of the output v_t (which is produced by v_1) is determined by v_2 and its polarity, not just by v_1 alone. Circuit symmetry allows v_1 and v_2 to be interchanged.

of T_B, is grounded. Thus, the polarity of voltage v_t (which is produced by v_1) is determined by v_2 and its polarity, not just by v_1 alone. The inherent symmetry of the circuit allows the positions of v_1 and v_2 to be interchanged.

The desired operating properties of these circuits are often difficult to obtain in practice. Diodes can be matched closely, but not perfectly. What you think is a center tap may be effectively offset somewhat by the proximity of other components. Stray capacitances of the wiring and the devices themselves can allow undesired feedthrough from port to port. Physical placement of parts, lead dress, and shielding may be critical.

5-5 AMPLITUDE MODULATION

In the early 1920s, when decisions were being made about what kind of system to use for the new phenomenon of radio broadcasting, the double-sideband system was not a serious contender for preference, because coherent demodulation was so complicated in that era. If a mass market for radio receivers was to become a reality, the receivers had to be as easy to design, manufacture, and use as possible. The people who used the receivers generally would not have any technical training and could not be expected to purchase and use receivers that required meticulous tuning and tender, loving care.

A major contributor to receiver simplicity was the *envelope detector*. This meant that the message to be transmitted should be contained in the envelope of the modulated signal. The envelope of the double-sideband system was not the same as the waveform of the modulating signal $m(t)$. Using an envelope detector to demodulate a DSB signal results in an unacceptably distorted demodulated output. On the other hand, *amplitude modulation* is said to occur when the amplitude of the modulated signal, which is the envelope, is linearly related to the modulating signal $m(t)$ in such a way that the envelope has the same waveform as that of $m(t)$.

Consider the system depicted in Fig. 5-12 where a *positive* constant A is added to $m(t)$ prior to the multiplier. The modulating signal $m(t)$ itself is assumed to have zero average. Thus, the output of the multiplier is the modulated carrier given by

$$x_c(t) = A_c'[A + m(t)] \cos \omega_c t \tag{5-23a}$$

$$= A_c\left[1 + \frac{m(t)}{A}\right] \cos \omega_c t \tag{5-23b}$$

where $A_c = AA_c'$ represents the amplitude of the unmodulated carrier.

The quantity $A_c'[A + m(t)]$ in (5-23a) or its equivalent $A_c[1 + m(t)/A]$ in (5-23b) represents what happens to the amplitude of the carrier and it is used to describe the envelopes. There is a positive envelope $+A_c'[A + m(t)]$ and a negative envelope $-A_c'[A + m(t)]$, because $\cos \omega_c t$ varies from +1 to −1, as shown in Fig. 5-13 for a specified $m(t)$. Both envelopes must remain

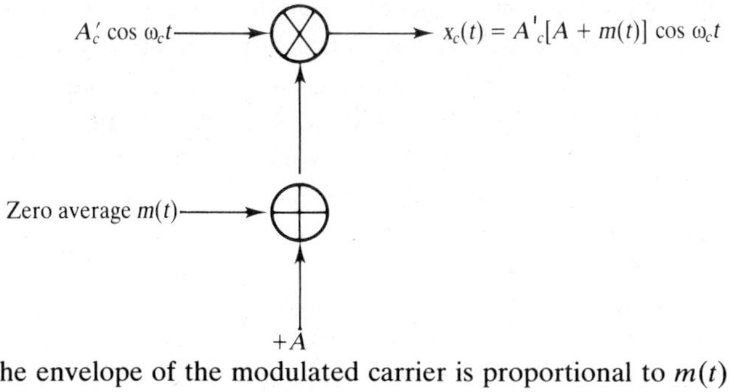

Fig. 5-12 The envelope of the modulated carrier is proportional to $m(t)$ if constant A is added to $m(t)$ prior to the multiplier.

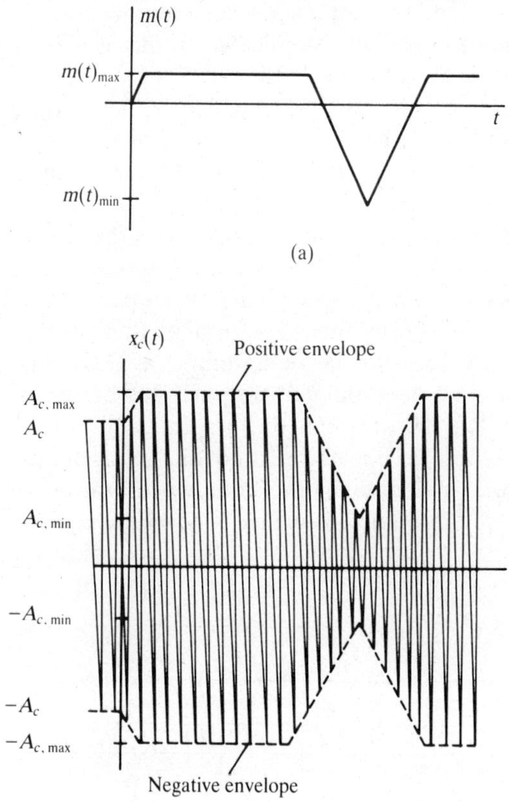

Fig. 5-13 (a) A periodic modulating signal having zero average. (b) A carrier amplitude modulated with the signal in (a).

separate to prevent a type of envelope distortion known as *over modulation.* This means that the positive envelope must always remain positive and the negative envelope must always be negative. Otherwise, when a 4-quadrant multiplier is used, what should be part of the positive envelope changes polarity and becomes part of the negative envelope and vice versa, as shown in Fig. 5-14a. If the modulated stage is a 2-quadrant multiplier that cuts off (i.e., if its output is zero) when the polarity of the sum $A + m(t)$ reverses, then both the positive and negative envelopes are zero for that interval as shown in Fig. 5-14b.

The worst possibility of over modulation occurring with $A > 0$ is when $m(t)$ is at its most negative value, given by $-|m(t)_{\min}|$. Thus, over modulation is prevented if

$$A - |m(t)_{\min}| \geqslant 0$$

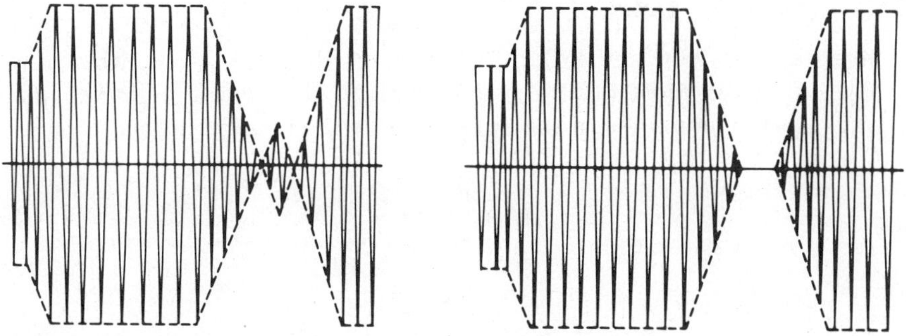

(a) (b)

Fig. 5-14. (a) Over modulation in a 4-quadrant multiplier whose output reverses polarity when $A + m(t)$ reverse polarity for a modulating $m(t)$ as in Fig. 5-13a. (b) Over modulation in a 2-quadrant multiplier that cuts off when $A + m(t)$ reverses polarity for $m(t)$ as in Fig. 5-13a.

or

$$\frac{|m(t)_{\min}|}{A} \leqslant 1 \tag{5-24}$$

where $A > 0$. It is necessary to keep $|m(t)_{\min}|$ from becoming too large at the transmitter by a combination of "riding" a manual volume control, using an automatic volume control system, and ultimately relying on a peak-limiter circuit.

In practice, the constant A is usually the DC voltage supplied to an electrode of the amplifying device in the modulated stage, and $m(t)$ is often the secondary voltage from a modulation transformer in series with that supply voltage. The name given to a particular amplitude-modulation process is taken from the name of the electrode to which $m(t)$ is applied. Thus, when the modulated stage is a vacuum tube, there are, for example, plate modulation, grid modulation, and cathode modulation. When the modulated stage is a bipolar junction transistor, there are collector modulation, base modulation, and emitter modulation.

When the modulated stage uses a device that requires a *negative* DC supply voltage for the electrode to which $m(t)$ is added in series, the constant A in (5-23a) and (5-23b) is negative, written as $-|A|$. In that situation, over modulation is prevented if

$$\frac{|m(t)_{\max}|}{|A|} \leqslant 1 \tag{5-25}$$

where $A = -|A|$.

5-5.1 Modulation Index

Reference to (5-23b) and Fig. 5-13b indicates that, for $A > 0$, when $m(t)$ changes from its most positive value $m(t)_{\max}$ to its most negative value $-|m(t)_{\min}|$, the amplitude of the modulated carrier changes from

$$A_{c,\max} = A_c\left[1 + \frac{m(t)_{\max}}{A}\right] \tag{5-26}$$

to

$$A_{c,\min} = A_c\left[1 - \frac{|m(t)_{\min}|}{A}\right] \tag{5-27}$$

where $A > 0$. Since $m(t)$ does not necessarily change in amplitude symmetrically above and below zero, it is possible for the amplitude of the modulated signal to increase above the unmodulated value A_c (i.e., in the upward direction) by an amount that differs from how much it decreases below A_c (i.e., in the downward direction). The fractional amount of each amplitude change is a measure of the degree of modulation in each direction. Thus, the degree of modulation in the upward direction is measured by an upward modulation index given by

$$m_{a+} = \frac{A_{c,\max} - A_c}{A_c} = \frac{m(t)_{\max}}{A} \tag{5-28}$$

where $A > 0$, and the degree of modulation in the downward direction is measured by a downward modulation index given by

$$m_{a-} = \frac{A_c - A_{c,\min}}{A_c} = \frac{|m(t)_{\min}|}{A} \tag{5-29}$$

where $A > 0$. Each modulation index is usually expressed in percent simply by multiplying (5-28) or (5-29) by 100. Further, m_{a+} can differ from m_{a-} if $m(t)_{\max}$ is not equal to $|m(t)_{\min}|$.

An average modulation index is defined as

$$m_a = \frac{m_{a+} + m_{a-}}{2} = \frac{A_{c,\max} - A_{c,\min}}{2A_c} \tag{5-30}$$

When $m(t)$ is symmetrical in extent about zero, then $m_{a+} = m_{a-}$. The modulation index for such a situation is found by equating (5-28) to (5-29) to obtain

$$2A_c = A_{c,\max} + A_{c,\min} \tag{5-31}$$

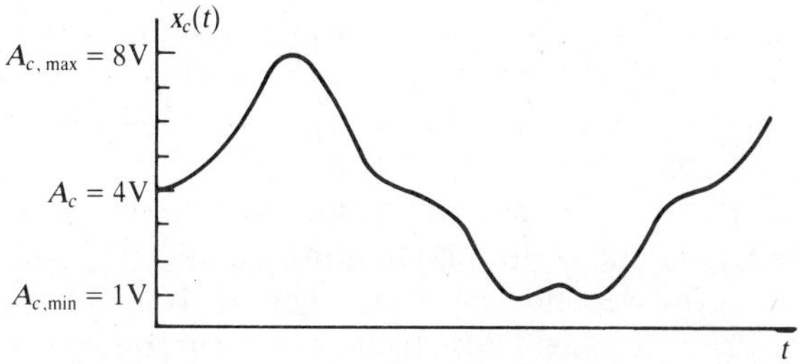

Fig. 5-15 Positive envelope for Example 5-1.

and then by substituting this result into (5-30). Thus,

$$m_a = \frac{A_{c,\max} - A_{c,\min}}{A_{c,\max} + A_{c,\min}} \tag{5-32}$$

and this yields another way to calculate the modulation index for a symmetrical $m(t)$.

Example 5-1

The periodic positive envelope of an AM signal with $A = +9$ V is shown in Fig. 5-15. The average value of $m(t)$ itself is zero. Calculate: (a) the upward, downward, and average modulation indexes in percent, (b) $m(t)_{\max}$ and $m(t)_{\min}$, (c) A_c'.

Solution

(a) $m_{a+} = (8-4)/4 = 1.00$ or 100 percent.
$m_{a-} = (4-1)/4 = 0.75$ or 75 percent.
$m_a = (1.00 + 0.75)/2$ or $(8-1)/(2 \times 4) = 0.875$ or 87.5 per cent.

(b) Since $m_{a+} = m(t)_{\max}/A$, then $m(t)_{\max} = Am_{a+} = 9$ V. Also, $m_{a-} = |m(t)_{\min}|/A$, so $m(t)_{\min} = -|m(t)_{\min}| = -m_{a-}A = 6.75$ V.

(c) Since $A_c = AA_c'$, then $A_c' = A_c/A = 4/9 = 0.44$ V.

5-5.2 The AM Spectrum

It was pointed out in connection with DSB that if one of the input signals to a multiplier has an average value, then the output of the multiplier contains a component at the frequency of the other input signal. That this occurs with an amplitude-modulated signal can be shown by expanding (5-23a) or (5-23b) to obtain

$$\begin{aligned} x_c(t) &= AA_c' \cos \omega_c t + A_c' m(t) \cos \omega_c t \\ &= A_c \cos \omega_c t + A_c \frac{m(t)}{A} \cos \omega_c t \end{aligned} \tag{5-33}$$

where $A_c = AA'_c$. It is seen that the AM signal contains a component at the carrier frequency and a term representing the double sideband. Therefore, the AM spectrum is composed of a carrier, an upper sideband, and a lower sideband. All AM transmitters share this feature.

It is also seen in (5-33) that the amplitude of the spectral carrier is constant and independent of the modulating signal $m(t)$ and that only the double-sideband term depends on $m(t)$. The AM signal occupies just as much of the spectrum as does DSB. If the highest frequency associated with $m(t)$ is W Hz, the total spectrum space occupied by the AM signal is $2W$ Hz.

The standard AM broadcast band in the United States extends from 535 kHz to 1605 kHz. Carrier frequencies are assigned to licensed AM stations by the Federal Communications Commission (FCC) every 10 kHz from 540 kHz to 1600 kHz. There are $(1605–535)/10 = 107$ different channels available, each 10 kHz wide. But this does not mean that there can only be 107 different stations, because stations assigned the same carrier frequency are hundreds of miles apart. Listening audiences are protected from interference from other stations on the same or adjacent carrier frequencies by a combination of low power, directive antennas, or daytime-only operation. There are, for example, 178 U.S. stations and 36 Canadian stations operating on 1450 kHz, mostly with 1 kW daytime power and 250 W night-time power. In all of North America there are 574 AM broadcasting stations having powers of 10 kW or more and 3768 stations radiating less than 10 kW. To prevent interference, no commerical broadcasting station in North America is allowed more than 50 kW of power radiated. Worldwide there are 11,239 AM broadcasting stations in the frequency range from 526.5 kHz to 1606.5 kHz, and their total power radiated is approximately 466,000 kW.†

If the spectral width of each AM station is restricted to 10 kHz, then $2W = 10$ kHz, so that $W = 5$ kHz represents the highest frequency that can be allowed in the modulating signal for the spectrum to be within the available 10-kHz channel. This severely limits the fidelity of musical reproductions of such stations. On the other hand, audio filters do not cut off exactly at 5 kHz, so some higher frequencies are actually present in the modulating signals but they will usually have small amplitudes. This means that each transmitter in the standard AM broadcast band occupies somewhat more spectrum space than 10 kHz. Fortunately, this does not ordinarily cause much interference between stations in adjacent channels because the FCC does not assign carrier frequencies in adjacent channels to the same locality.

It is convenient to express (5-33) in terms of the downward modulation index m_{a-} for $A > 0$. Thus, multiplying and dividing the second term by $|m(t)_{\min}|$ (because $|m(t)_{\min}|$ controls whether over modulation occurs) and

† *Source*: *World Radio TV Handbook,* 37th edition, Billboard A.G., 1983.

rearranging,

$$x_c(t) = A_c \cos \omega_c t + A_c \frac{|m(t)_{\min}|}{A} \frac{m(t)}{|m(t)_{\min}|} \cos \omega_c t$$

$$= A_c \cos \omega_c t + A_c m_{a-} m_n(t) \cos \omega_c t = A_c[1 + m_{a-} m_n(t)] \cos \omega_c t \qquad (5\text{-}34)$$

where $m_n(t) = m(t)/|m(t)_{\min}|$ represents the modulating signal $m(t)$ normalized to the magnitude of its most negative value. When $A < 0$, the normalizing quantity becomes $m(t)_{\max}$ and the upward modulation index m_{a+} controls whether over modulation occurs.

5-5.3 Power Relationships

If $x_c(t)$ in (5-34) represents the instantaneous current through a resistance R, the total long-time-average power delivered to R is

$$P = \lim_{T \to \infty} \frac{1}{2T} \int_{-T}^{T} x_c^2(t) R \, dt = R \langle x_c^2(t) \rangle \qquad (5\text{-}35)$$

where the symbol $\langle \ \rangle$ denotes the average value. If $m(t)$ is periodic with period T_m, the long-time average expressed by (5-35) is taken over T_m. Using (5-34) and a familiar trigonometric identity for $\cos^2 \phi$, (5-35) becomes

$$P = \frac{RA_c^2}{2} \langle [1 + m_{a-} m_n(t)]^2 \cos 2\omega_c t + 1 + 2m_{a-} m_n(t) + m_{a-}^2 m_n^2(t) \rangle \qquad (5\text{-}36)$$

The average value of the sum of the individual terms is the same as the sum of the individual averages. Now $\langle [1 + m_{a-} m_n(t)]^2 \cos 2\omega_c t \rangle$ is zero if $m_n(t)$ is slowly varying with respect to $\cos 2\omega_c t$ (which is the usual situation), and $\langle 2m_{a-} m_n(t) \rangle$ is zero because $\langle m_n(t) \rangle = 0$. Thus (5-36) becomes

$$P = \frac{RA_c^2}{2} [1 + m_{a-}^2 \langle m_n^2(t) \rangle] \qquad (5\text{-}37)$$

The long-time-average power in the spectral carrier component alone is that part of (5-37) given by

$$P_c = \tfrac{1}{2} R A_c^2 \cdot 1 \qquad (5\text{-}38)$$

The long-time-average power in both sidebands together is the other part of (5-37) given by

$$P_{sb} = \tfrac{1}{2} R A_c^2 m_{a-}^2 \langle m_n^2(t) \rangle \qquad (5\text{-}39)$$

with half of that power in one sideband and half in the other.

A measure of the amount of power contained in the information-bearing sidebands to the total power is the *modulation efficiency* η_{mod} defined as

$$\eta_{\text{mod}} = \frac{P_{sb}}{P} \times 100\% = \frac{m_{a-}^2\langle m_n^2(t)\rangle}{1 + m_a^2\langle m_n^2(t)\rangle} \times 100\% \tag{5-40}$$

The modulation efficiency varies from 0 percent when $m_{a-} = 0$ to 100 percent when $m_{a-} = 1$ and $\langle m_n^2(t)\rangle$ has a large value.

Example 5-2

Calculate the modulation efficiency when $A > 0$ and $m(t)$ is as shown in Fig. 5-16a.

Solution First check whether $\langle m(t)\rangle = 0$ (which it does) and then normalize $m(t)$ to $|m(t)_{\text{min}}| = 2$ since $A > 0$. Thus $m_n(t)$ is as shown in

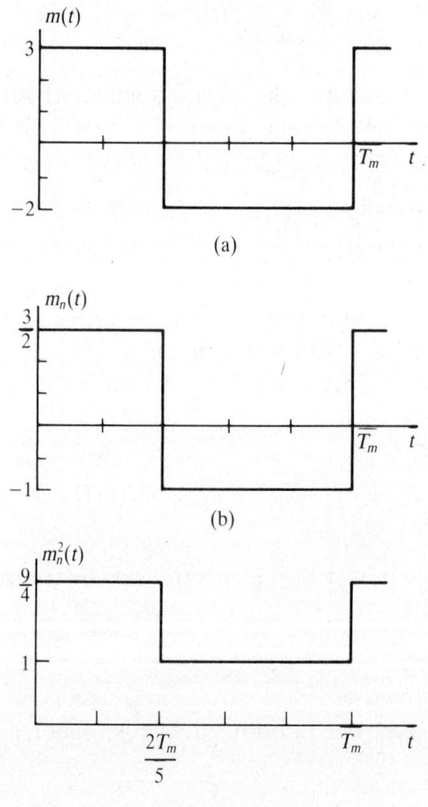

Fig. 5-16 (a) Periodic modulating signal for Example 5-2. (b) $m(t)$ normalized to $|m(t)_{\text{min}}| = 2$. (c) $m_n^2(t)$ obtained by squaring instantaneous values of (b).

Fig. 5-16b. Then, from Fig. 5-16c,

$$\langle m_n^2(t) \rangle = \frac{\text{area}}{\text{base}} = \frac{\frac{9}{4} \cdot \frac{2T_m}{5} + 1 \cdot \frac{3T_m}{5}}{T_m} = \frac{3}{2}$$

Therefore,

$$\eta_{\text{mod}} = \frac{m_{a-}^2 \cdot \frac{3}{2}}{1 + m_{a-}^2 \cdot \frac{3}{2}} \times 100\% = \frac{3m_{a-}^2}{2 + 3m_{a-}^2} \times 100\%$$

When $m_{a-} = 1$, the largest allowable value of modulation efficiency is

$$(\eta_{\text{mod}})_{\max} = \frac{300}{5}\% = 60\%$$

That is, 60% of the total average power is in the sidebands.

In practice, the ability of the modulated stage to provide the required undistorted output power in the vicinity of $m(t)_{\max}$ is determined by its saturation or limiting properties. If $m(t)_{\max}$ is too large, the output power may remain constant or even decrease during the saturated interval, as determined by the operating conditions of the modulated stage.

A measure of the ability to accommodate $m(t)_{\max}$ is the *peak envelope power* (PEP). Peak envelope power is the short-time-average power of the carrier sinusoid (taken over one period of the carrier) when its amplitude is the most that can occur without distortion when $m(t)_{\max}$ occurs, that is, when the envelope is at its peak value. Thus, from (5-26) and (5-28), for an amplitude-modulated current,

$$\text{PEP} = \tfrac{1}{2}RA_{c,\max}^2 = \tfrac{1}{2}RA_c^2\left[1 + \frac{m(t)_{\max}}{A}\right]^2 = \tfrac{1}{2}RA_c^2[1 + m_{a+}]^2 \tag{5-41}$$

5-6 ENVELOPE DETECTOR

The name *envelope* (*or peak*) *detector* indicates that the output of the detector is supposed to follow the envelope (or peaks) of the AM input signal. Since the envelope is proportional to the original modulating signal, the output of the envelope detector yields the waveform of the recovered modulating signal. A commonly used and economical envelope detector

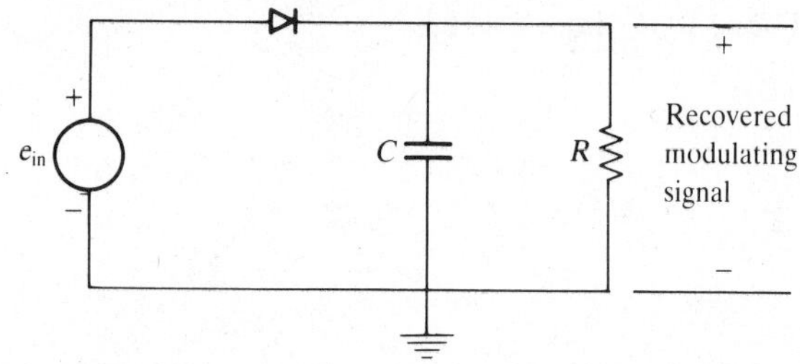

Fig. 5-17 An envelope detector using a diode as a half-wave rectifier.

employs a diode as a half-wave rectifier as shown in Fig. 5-17, but full-wave and voltage-doubling diode circuits are occasionally encountered.†

If e_{in} is a constant-amplitude unmodulated carrier whose amplitude is greater than the threshold voltage of the diode, and, further, if the RC time constant is long compared to the period T_c of the carrier frequency (i.e., $RC \gg T_c$), then the output voltage is nearly constant and has a small ripple component. If e_{in} is an AM carrier whose amplitude is always greater than the threshold voltage of the diode, and if the RC time constant is short compared to the period $T_{m,\min}$ of the highest modulating frequency (i.e., $RC \ll T_{m,\min}$) in addition to $RC \gg T_c$, then the output voltage is able to follow the envelope variations with a small ripple component. Since $T_c = 1/f_c$ and $T_{m,\min} = 1/f_{m,\max}$, the condition $T_c \ll RC \ll T_{m,\min}$ requires that f_c is very much larger than $f_{m,\max}$. This latter requirement is usually much more than satisfied in practice, especially at the higher carrier frequencies, so that many different values of the RC product are satisfactory. The choice of the RC product is more critical when f_c is only about 10 or 20 times higher than $f_{m,\max}$, as is often the case in figures that illustrate the details of envelope detection.

5-6.1 Diagonal Clipping

Diagonal clipping is said to occur when the voltage from an envelope detector does not decrease rapidly enough to follow the actual envelope of an amplitude-modulated signal when it is decreasing. Whether diagonal clipping occurs is determined by the RC time constant of the output circuit of the envelope detector.

Consider the exponential discharge voltage of a resistor–capacitor parallel combination given by

$$e_c(t) = Ee^{-t/RC}$$

† If the capacitor C is omitted, the circuit is known as an *average* detector because the average of the rectified AM signal follows the waveform of the modulating signal.

The slope of the voltage curve is

$$\frac{de_c(t)}{dt} = -\frac{E}{RC}e^{-(t/RC)} = -\frac{e_c(t)}{RC}$$

The initial slope at $t = 0$ is $-E/RC$. A straight line having this slope drawn through E will intersect the time axis at $t = RC$, at which instant the actual voltage from the exponential is $0.368E$. If E takes on different values, as it does in an AM diode envelope detector, the slope of this line varies, being steeper when E is higher and flatter when E is lower. In spite of the relative simplicity of this action, figures in the pertinent literature on envelope detectors, which purport to represent the capacitor voltage as it discharges, almost universally fail properly to relate the slopes of the various discharge curves to the initial amplitude of the voltage at the onset of each discharge. Drawings that portray a typical lack of comprehension of this simple action are shown in Fig. 5-18.

In Fig. 5-18a the erroneous impression is given that, since the RC time constant is fixed, the capacitor discharges the same way between adjacent voltage peaks regardless of their amplitudes. Figure 5-18b depicts the

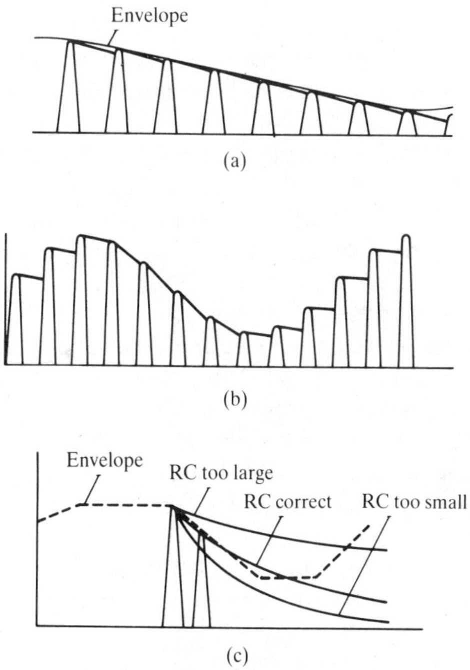

Fig. 5-18 Erroneous portrayals of the output voltage of an envelope detector. (a) Identical capacitor discharges between adjacent voltage peaks. (b) Capacitor discharges opposite to the actual situation. (c) Details are vague.

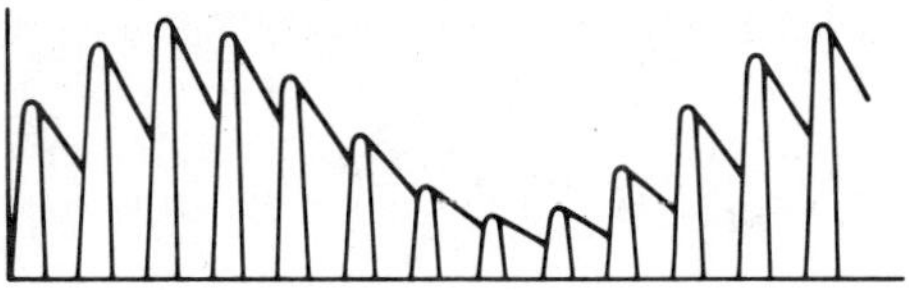

(a)

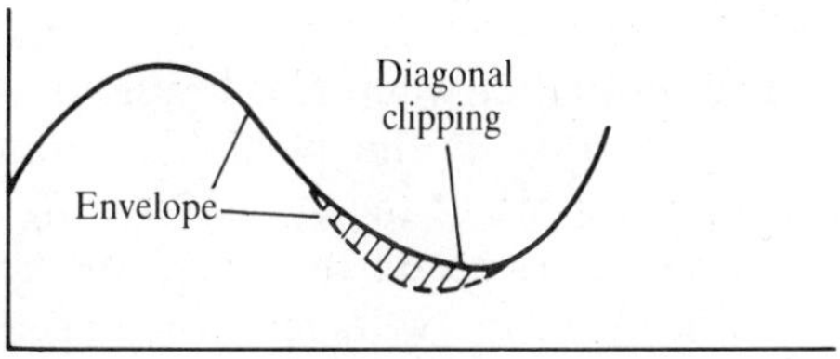

(b)

Fig. 5-19 (a) A correct representation of the capacitor discharges in a diode envelope detector. (b) Diagonal clipping occurs when the capacitor voltage cannot decrease as rapidly as the envelope.

capacitor discharges in a manner almost opposite to the way in which they actually occur, and there is the additional error that there are different discharge rates even after the same initial voltages. Figure 5-18c is sufficiently vague about the details to have little value, and there is a strong suspicion that the curve marked "*RC* too small" actually has *RC* too large for satisfactory operation even with the envelope as marked.

The correct way to approximate the various capacitor discharges is shown in Fig. 5-19a. There, the *RC* time constant is chosen equal to twice the period T_c of the carrier, or $2T_c$. Each line that represents a capacitor discharge is drawn from each maximum voltage point to the point $RC = 2T_c$ time units later on the time axis.

Capacitor discharge occurs between cutout and cutin of the diode. Cutout occurs at an instant just past, but near, the peak of a half-cycle of the modulated input sinusoid when the diode ceases to conduct. Cutin occurs at an instant before the peak of a following half cycle of the modulated input sinusoid when the diode begins to conduct again. The action is like that of any half-wave rectifier having a parallel *RC* combination as its load.

A conservative estimate of the conditions for preventing diagonal clipping can be made after formulating some simplifying assumptions:

1. The diode is ideal.
2. The envelope follows the peaks of the modulated input sinusoid.
3. Cutout and cutin occur at points on the envelope.

Diagonal clipping is illustrated in Fig. 5-19b.

For no diagonal clipping to occur, the voltage from the exponential discharge of the capacitor at an interval T_c after any instant t' must be equal to or less than the voltage given by the envelope after the same time interval T_c from any instant t'. Thus, at t', both the envelope and the exponential have the value

$$A_c(1 + m_{a-} \cos \omega_m t') \tag{5-42}$$

At the instant of the next peak of the carrier, the exponential has the value

$$A_c(1 + m_{a-} \cos \omega_m t')e^{-T_c/RC} \tag{5-43}$$

and the envelope has the value

$$A_c[1 + m_{a-} \cos \omega_m (t' + T_c)] \tag{5-44}$$

which can be written as

$$A_c(1 + m_{a-} \cos \omega_m t' \cos \omega_m T_c - m_{a-} \sin \omega_m t' \sin \omega_m T_c) \tag{5-45}$$

Thus, for no diagonal clipping, the voltage given by (5-43) must be equal to or less than the voltage given by (5-45), or

$$A_c(1 + m_{a-} \cos \omega_m t')e^{-T_c/RC} \leqslant A_c(1 + m_{a-} \cos \omega_m t' \cos \omega_m T_c - m_{a-} \sin \omega_m t' \sin \omega_m T_c)$$

This can be rearranged as

$$(\sin \omega_m T_c) \sin \omega_m t' + (e^{-T_c/RC} - \cos \omega_m T_c) \cos \omega_m t' \leqslant \frac{1 - e^{-T_c/RC}}{m_{a-}} \tag{5-46}$$

But the left-hand side of (5-46) is like

$$A \sin \omega_m t' + B \cos \omega_m t'$$

where $A = \sin \omega_m T_c$ and $B = e^{-T_c/RC} - \cos \omega_m T_c$. Therefore, (5-46) becomes

$$\sqrt{A^2 + B^2} \sin (\omega_m t' + \theta) \leqslant \frac{1 - e^{-T_c/RC}}{m_{a-}} \tag{5-47}$$

where $\theta = \tan^{-1} (B/A)$. Then, from (5-47),

$$\sin (\omega_m t' + \theta) \leqslant \frac{1 - e^{-T_c/RC}}{m_{a-}\sqrt{A^2 + B^2}} \tag{5-48}$$

Since $-1 \leqslant \sin(\omega_m t' + \theta) \leqslant 1$, it is assured that there will be no diagonal clipping if

$$1 - e^{-T_c/RC} \geqslant m_{a-}\sqrt{A^2 + B^2} \tag{5-49}$$

because then (5-48) is satisfied for all t'. Whether diagonal clipping occurs depends on the values of T_c, RC, f_m, and m_{a-}. Thus, for certain values of T_c, RC, and f_m, (5-49) yields, after substituting for A and B and simplifying,

$$m_{a-} \leqslant \frac{1 - e^{-T_c/RC}}{\sqrt{1 + e^{-2T_c/RC} - 2e^{-T_c/RC}\cos \omega_m T_c}} \tag{5-50}$$

The numerator of the right-hand side of (5-50) is always less than the denominator, therefore the maximum value of m_{a-} is always less than unity. Further, the right-hand side of this inequality is least when the highest modulating frequency and the longest RC time constant encountered in a particular application are used. See Problem 5-10 for a simplification of (5-50).

Example 5-3

The carrier frequency of an AM signal is 100 times the frequency of a sinusoidal modulating signal. The AM signal is applied to a diode envelope detector. For $RC = 10T_c$: (a) Calculate $m_{a-,\max}$ for no diagonal clipping; (b) calculate $\omega_m t'$ if $m_{a-} = 1$. (c) Repeat the calculations of (a) and (b) for $RC = 50T_c = T_m/2$.

Solution Since $RC = 10T_c = T_m/10$, then $T_c/RC = 0.1$ and $\omega_m T_c = 2\pi f_m/f_c = 2\pi/100$ radians or 3.6°.

(a) Substituting values into (5-50) and calculating to four decimal places because of subtracting nearly equal numbers, it is found that there is no diagonal clipping if $m_{a-} \leqslant 0.8472$ or 84.72 percent.

(b)

$$\theta = \tan^{-1}\frac{B}{A} = \tan^{-1}\frac{e^{-T_c/RC} - \cos \omega_m T_c}{\sin \omega_m T_c}$$

$$= \tan^{-1}\frac{0.9048 - 0.9980}{0.0628} = -56.03°$$

From (5-48), with $m_{a-} = 1$, $\sin(\omega_m t' + \theta) \leqslant 0.8472$, therefore $\omega_m t' \leqslant 57.91° - (-56.03°) = 113.94°$, showing that diagonal clipping starts at about 113.94° into the cycle of the modulating signal.

(c) Since now $T_c/RC = 0.02$, (5-50) yields $m_{a-,\max} = 0.3034$ for no diagonal clipping. Also, $\theta = -15.85°$, so, for $m_{a-} = 1$, $\sin(\omega_m t' + \theta) \leqslant 0.3034$. Therefore, $\omega_m t' \leqslant 33.55°$ for no diagonal clipping. The RC time constant is now so large that diagonal clipping starts very early into the cycle of the modulating signal.

5-6.2 Bottom Clipping

Another deficiency of the envelope detector in Fig. 5-17 can occur when the modulation percentage is high but not when the modulation percentage is low. When it occurs, the bottom part (i.e., the low-amplitude section) of the output voltage waveform is clipped off, hence the name *bottom clipping*. The effect occurs when the load resistance of the signal circuit is less than the DC load resistance.

The recovered modulating signal from the circuit in Fig. 5-17 is delivered to the input resistance R_{in} of the following amplifier through a coupling capacitor. The DC load resistance of the diode in Fig. 5-17 is simply R, but the load resistance of the signal circuit is the parallel combination of R and R_{in}. How this difference affects the output waveform is readily shown with the aid of rectification characteristics for the diode.

Rectification characteristics can be calculated (see Problem 5-11) or measured using the circuit in Fig. 5-20. The frequency can be in the audio range. The procedure is to set the rms amplitude A of the input sinusoid to a certain value, then to vary the DC voltage V_{DC} in small steps, recording the average DC current I at each step. Repeat using several other values of rms amplitude, then plot the points and draw the curves as in Fig. 5-21. Note that values of V_{DC} increase toward the left to make load lines have their usual negative slopes.

Suppose that $R = 2\,k\Omega$, and that the output is capacitively coupled to an input resistance $R_{in} = 2\,\text{k}\Omega$ for the following amplifier. The DC load line for $2\,\text{k}\Omega$ is drawn as shown in Fig. 5-21 and a Q point selected, say for $A = 3.5$ V rms. The current at the Q point is approximately 2 mA and the DC voltage across R is about 4 V. Next draw the AC load line for 1 kΩ through the Q point as in Fig. 5-21. If now the input signal is amplitude-modulated, the path of operation is up and down along the AC load line as the input amplitude varies. The output voltage at each instant during the modulation cycle is found by projecting downward to the V_{DC} axis, and the output voltage varies about its unmodulated value of 4 V with little distortion. Another purpose of the coupling capacitor is to block this 4 V from interfering with the bias of the following stage.

When the depth of modulation makes the input voltage drop to where

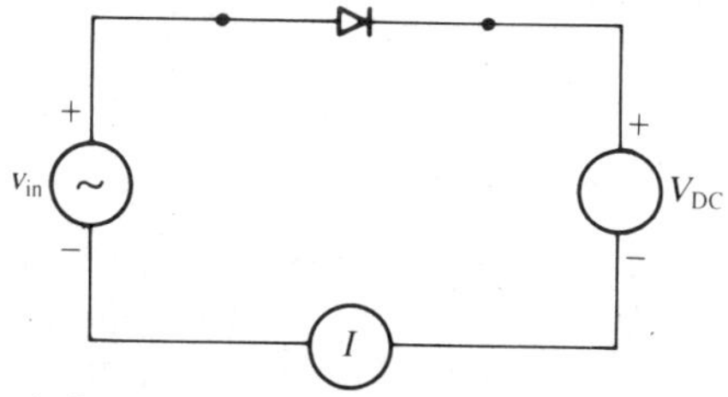

Fig. 5-20 Circuit for measuring rectification characteristics of a diode.

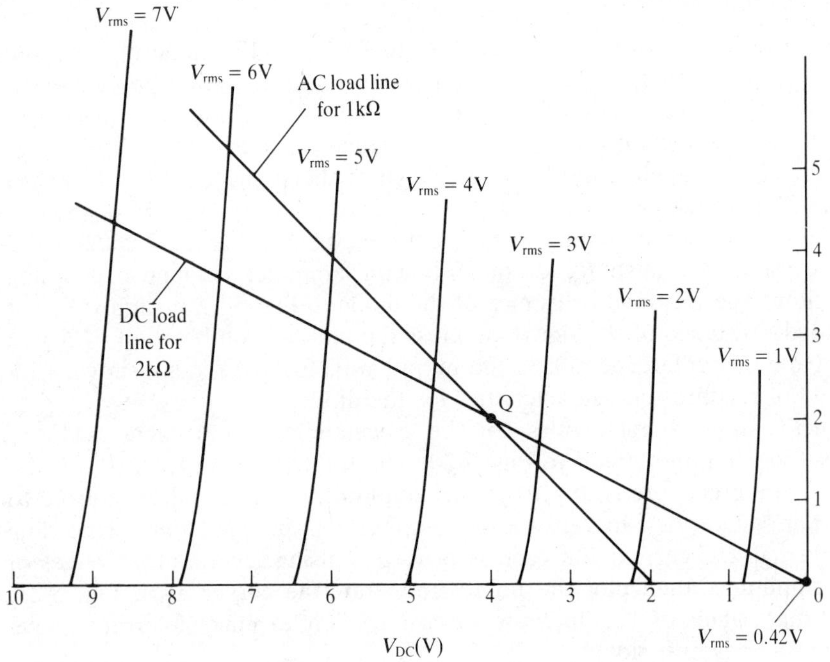

Fig. 5-21 DC and AC load lines drawn on the rectification characteristics.

the AC load line intersects the V_{DC} axis, then the output voltage drops to 2 V; it can drop no farther. The input voltage can drop to approximately 1.86 V. This is a decrease of 1.64 V below 3.5 V. Therefore, the maximum percentage of modulation without bottom clipping is, in this example, $m_{a-} = 1.64/3.5 \times 100\% = 47\%$. For symmetrical modulation, an increase of 1.64 V above 3.5 V will make $A = 5.14$ V and this will cause $V_{DC} = 6.12$ V, approximately. The peak-to-peak swing in output voltage is $6.12 - 2 = 4.12$ V.

Even if the AC load line falls along the DC load line, the maximum modulation percentage would be $(3.5 - 0.42)/3.5 \times 100\% = 88\%$. This is limited by the approximately 0.6 V cutin voltage of the diode. This problem can be minimized by using a diode with smaller cutin voltage or by forward-biasing this diode to 0.6 V.

5-7 SINGLE-SIDEBAND PROCESSES

Recall that the typical spectrum of a standard amplitude-modulated signal consists of the carrier and the two sidebands. The spectral carrier itself contains no information about the modulating signal, thus the power contained in the spectral carrier is, in effect, wasted, even though it is useful in demodulation with diode detectors. Further, the amplitudes of the components in one of the sidebands are duplicated in the other sideband. Therefore, the power supplied to one of the sidebands is also, in effect, wasted. It appears, then, that the transmitter should be designed to concentrate all of its output power in the one desired sideband, rather than having to share its power capabilities with the other sideband and the carrier too. Also, the spectrum space occupied by a single-sideband signal is one-half of that for a standard AM signal.

There are at least two ways to generate a single-sideband signal: (1) a sharp-cutoff bandpass filter can eliminate one sideband and pass the other, and (2) two voltages equal in amplitude but 180° out of phase at all their frequencies will cancel when added together, giving rise to a phasing method for eliminating one sideband.

5-7.1 Sideband Filtering

Quartz crystals can be arranged in a bandpass filter circuit to give good sideband elimination up to about 10 MHz, and bandpass mechanical filters are best suited for use at frequencies up to about 1 MHz. Both types are readily available commercially, but good home-built crystal filters can be fabricated at 455 kHz by those so inclined. The power-handling capabilities of these filters are severely limited and therefore sideband elimination is accomplished at very low power levels. This means that the newly created low-power single-sideband signal must be amplified to the desired power level by using linear amplifiers. Such amplifiers are operated Class AB or Class B to take advantage of the fact that their power efficiency is higher than for Class A. Further, the desired output frequency must be obtained by heterodyning another sinusoidal carrier frequency one or more times with the single-sideband signals.

The filter is not expected to eliminate the carrier as well as one sideband, because the carrier is suppressed in the balanced modulator that produces only the double sidebands. The block diagram in Fig. 5-22 illustrates the basic system.

The audio input signal, usually speech, is amplified as needed, then processed by a combination of compression, clipping, and filtering. The purpose of speech processing is to increase the average-to-peak ratio of the speech signal. This processing increases the average power level of the single-sideband signal at the output of the transmitter, and also controls the bandwidth of the output signal.

Suppose the filter in Fig. 5-22 has a passband extending from 453.5 kHz to 456.5 kHz and that oscillator #1 operates at 543.2 kHz. The balanced modulator (mixer #1) produces the sum frequencies of an initial upper sideband extending from $(453.2 + 0.3)\,\text{kHz} = 453.5\,\text{kHz}$ to $(453.2 + 3.3)\,\text{kHz} = 456.5\,\text{kHz}$, and also the difference frequencies of an initial lower

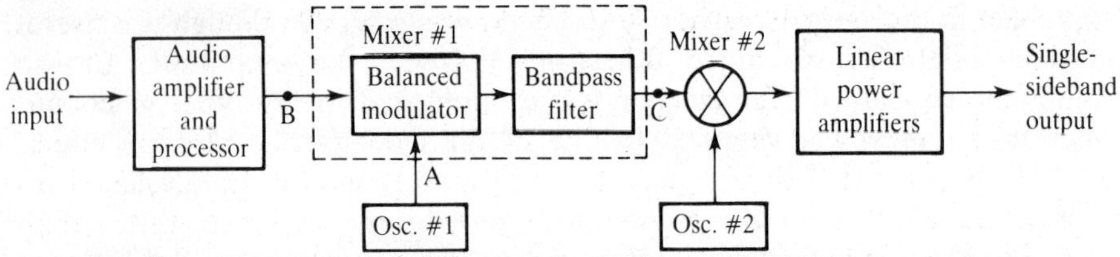

Fig. 5-22 Basic system using a bandpass filter to eliminate one sideband.

sideband extending from (453.2 − 0.3) kHz = 452.9 kHz to (453.2 − 3.3) kHz = 449.9 kHz. The original carrier frequency 453.2 kHz has been suppressed by the balanced modulator and only the initial upper sideband passes through the filter. Suppose further that the transmitter output to the antenna is to be upper sideband with a missing carrier at 3950 kHz. The tuned circuits of the power amplifiers are adjusted to pass frequencies above 3950 kHz in the upper sideband from 3950.3 kHz to 3953.3 kHz. The frequency f_2 of oscillator #2 must be used in mixer #2 to convert 453.5 kHz into 3950 kHz and to convert 456.5 kHz into 3953.3 kHz. This requires that the sum frequency of mixer #2 be $f_2 + 453.5 = 3950.3$ and $f_2 + 456.5 = 3953.3$. Therefore,

$$f_2 = (3950.3 - 453.5)\text{ kHz} = 3496.8\text{ kHz}$$

$$= (3953.3 - 456.5)\text{ kHz} = 3496.8\text{ kHz}$$

This mixer also produces the difference frequencies between (3496.8 − 453.5) kHz = 3043.3 kHz and (3496.8 − 456.5) kHz = 3040.3 kHz. These are the frequency limits for a lower sideband. The initial sideband that entered mixer #2 has been converted into a lower sideband, that is, sideband inversion has occurred. However, continuing this example, these latter frequencies are rejected by the tuned circuits of the power amplifiers, and are not delivered to the antenna; only the upper sideband is transmitted. The carrier that is supposed to be at 3950 kHz is missing, and that fact gives rise to the terminology *single sideband, suppressed carrier*. That condition is what is usually meant when *single sideband* is used, which, in turn, is often shortened to just *sideband*.

5-7.2 The Phasing Method

The phasing-method concepts can be developed readily with the aid of the two basic trigonometric functions

$$2 \cos A \cos B = \cos (A - B) + \cos (A + B) \qquad (5\text{-}51)$$

$$2 \sin A \sin B = \cos (A - B) - \cos (A + B) \qquad (5\text{-}52)$$

Think of A as being $\omega_c t$ of a fixed sinusoidal carrier, and of B as being an $\omega_m t$ for each component in a modulating signal. Then $\cos A$ and $\cos B$ are the input signals to a multiplier (or mixer, balanced modulator, or product detector). Then, at the output of the multiplier, $\cos(A - B)$ represents a component in the lower sideband, and $\cos(A + B)$ represents one component in the upper sideband. The only reason that the phasing method can eliminate a sideband is indicated by the central minus sign on the right-hand side of (5-52), a sign that is not also present on the right-hand side of (5-51). Thus, if (5-51) and (5-52) are added, the result does not contain a $\cos(A + B)$ term. The upper-sideband component has been phased out and eliminated. Consequently, any component in the lower sideband can be represented as

$$\text{For the LSB:} \quad \cos(A - B) = \cos A \cos B + \sin A \sin B \tag{5-53}$$

Similarly, the result found by subtracting (5-52) from (5-51), performed by changing the signs in (5-52) and then adding, does not contain the term $\cos(A - B)$. Hence, the lower-sideband component has been phased out and eliminated. Therefore, any component in the upper sideband can be represented as

$$\text{For the USB:} \quad \cos(A + B) = \cos A \cos B - \sin A \sin B \tag{5-54}$$

Of course, (5-53) and (5-54) are basic trigonometric relations in their own right, and they could have been used as the starting points of this development. In any event, (5-53) and (5-54) indicate what hardware is required in the phasing method of sideband suppression : (1) two multipliers to produce the product terms; (2) an adder, with perhaps a sign changer or inverter, to add the product terms; (3) a $\mp 90°$ phase shifter to change the single-frequency carrier sinusoid $\cos A$ into $\pm\sin A$; and (4) a wideband $\mp 90°$ phase shifter to change each $\cos B$ component of the modulating signal into $\pm\sin B$. A block diagram of the phasing method is shown in Fig. 5-23.

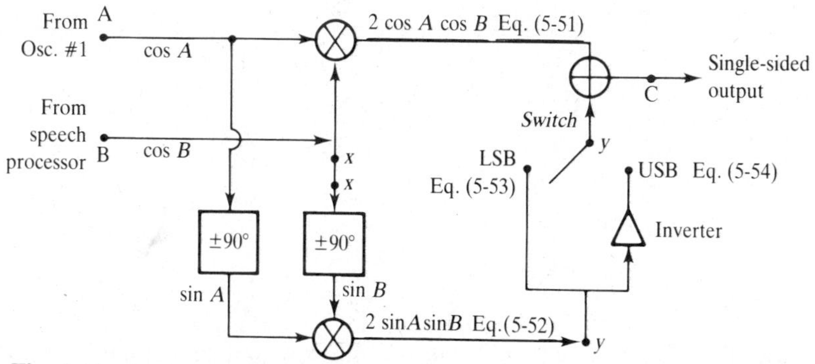

Fig. 5-23 Basic system using the phasing method to eliminate a sideband.

Everything shown in Fig. 5-23 can replace the contents of the dashed-line box in Fig. 5-22 between terminals A, B, and C.

5-7.3 Sideband Suppression

In practice, a balance adjustment should be provided to assure that the amplitudes of the sideband components produced by the multipliers can be made as nearly equal as possible. The ±90° phase-shift network for the carrier is relatively easy to obtain because only the single carrier frequency is involved. An *RC* circuit that can be used is shown in Fig. 5-24. Straightforward circuit analysis reveals that

$$V_o = \frac{V}{2} \underline{/180° - 2\tan^{-1}\omega_c RC} \tag{5-55}$$

Therefore, if $R = 1/(\omega_c C)$, then $V_o = \frac{V}{2}\underline{/90°}$, as desired. Of course, if the carrier frequency is changed, provision must be made to adjust R or C to a new value. Sometimes, however, the 90° angle between the two carrier sinusoids is obtained by using a series *RL* circuit to provide +45° phase shift for one sinusoid, and a series *RC* circuit to provide −45° phase shift for the other sinusoid. But again, if the carrier frequency is changed, each circuit must be readjusted to maintain the desired total phase shift.

To determine what effect phase shifts differing from 90° have on sideband suppression, but with proper balanced amplitudes, let (5-52) become

$$2\sin(A + \alpha)\sin(B + \beta) \tag{5-56}$$

where α and β are amounts of phase shift above or below 90°. Using

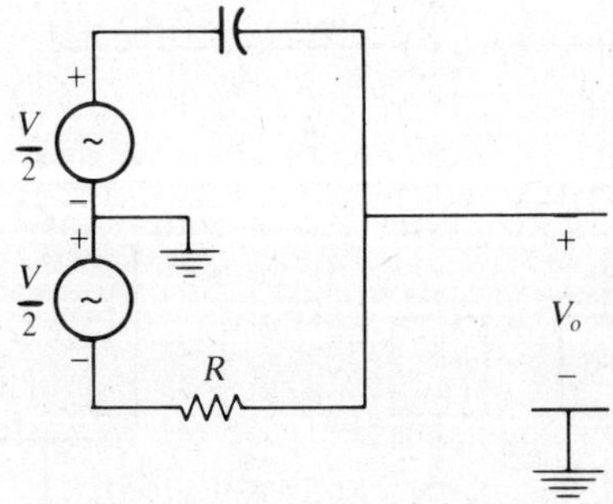

Fig. 5-24 The output voltage V_o is 90° out of phase with the voltage $V/2$ when $R = 1/(\omega_c C)$.

familiar trigonometric identities, this expands into

$$\begin{aligned}&\cos\alpha\cos\beta[\cos(A-B)-\cos(A+B)]\\&\quad+\cos\alpha\sin\beta[\sin(A+B)+\sin(A-B)]\\&\quad+\sin\alpha\cos\beta[\sin(A+B)-\sin(A-B)]\\&\quad+\sin\alpha\sin\beta[\cos(A+B)+\cos(A-B)]\end{aligned}\tag{5-57}$$

Now add (5-57) and (5-51) and collect terms to obtain

$$\begin{aligned}&[\cos\alpha\cos\beta+\sin\alpha\sin\beta+1]\cos(A-B)\\&\quad+[\cos\alpha\sin\beta-\sin\alpha\cos\beta]\sin(A-B)\\&\quad+[\sin\alpha\sin\beta-\cos\alpha\cos\beta+1]\cos(A+B)\\&\quad+[\cos\alpha\sin\beta+\sin\alpha\cos\beta]\sin(A+B)\end{aligned}\tag{5-58}$$

Only a lower-sideband component is desired by this process, but (5-58) shows that an upper sideband component is also present.

The amplitude of the desired lower-sideband component is

$$\sqrt{[(\cos\alpha\cos\beta+\sin\alpha\sin\beta+1)^2+(\cos\alpha\sin\beta-\sin\alpha\cos\beta)^2]}$$
$$=2\cos\left(\frac{\alpha-\beta}{2}\right)\tag{5-59}$$

The amplitude of the undesired upper sideband component is

$$\sqrt{[(\sin\alpha\sin\beta-\cos\alpha\cos\beta+1)^2+(\cos\alpha\sin\beta+\sin\alpha\cos\beta)^2]}$$
$$=2\sin\left(\frac{\alpha+\beta}{2}\right)\tag{5-60}$$

The ratio of the stronger (lower sideband) component to the weaker (upper sideband), expressed in dB, is a measure of the amount of suppression or attenuation A_u of the undesired weaker component relative to the desired stronger component. Hence,

$$A_u=20\log\left|\frac{\cos\left(\dfrac{\alpha-\beta}{2}\right)}{\sin\left(\dfrac{\alpha+\beta}{2}\right)}\right|\quad\text{dB}\tag{5-61}$$

The upper-sideband components for which $\alpha=-\beta$ as well as $\alpha=\beta=0$ are completely suppressed.

If it is assumed that $\alpha = 0$ is not too difficult to obtain, then

$$A_u = 20 \log \left| \cot \frac{\beta}{2} \right| \quad \text{dB} \qquad (\text{for } \alpha = 0) \tag{5-62}$$

Therefore, if $\beta = \pm 0.25°$, then $A_u = 53.22$ dB; if $\beta = \pm 2.5°$, then $A_u = 33.22$ dB; and if $\beta = \pm 10°$, then $A_u = 21.16$ dB. Only a few degrees of departure from the desired 90° phase shift can be tolerated for reasonably good sideband suppression.

5-7.4 Wideband 90° Phase Shift

For best sideband suppression, all components of the modulating signal, whatever their frequencies, must be shifted 90° without amplitude distortion. The process is sometimes called taking the Hilbert transform. Unfortunately, there is no known circuit that can do this over a very wide frequency range, such as for a video signal, for example. There are circuits, however, that provide almost the ideal 90° phase shift with no amplitude distortion, over a few kilohertz. These circuits would be suitable for intelligible speech, but not for high-fidelity music. One such circuit is shown in Fig. 5-25. The voltage sources are assumed to have negligible output resistances. Straightforward nodal analysis reveals that

$$V_{o1} = V \frac{Z_3(Z_2 - Z_1)}{Z_1 Z_2 + Z_1 Z_3 + Z_2 Z_3} \tag{5-63a}$$

and

$$V_{o2} = V \frac{Z_6(Z_5 - Z_4)}{Z_4 Z_5 + Z_4 Z_6 + Z_5 Z_6} \tag{5-63b}$$

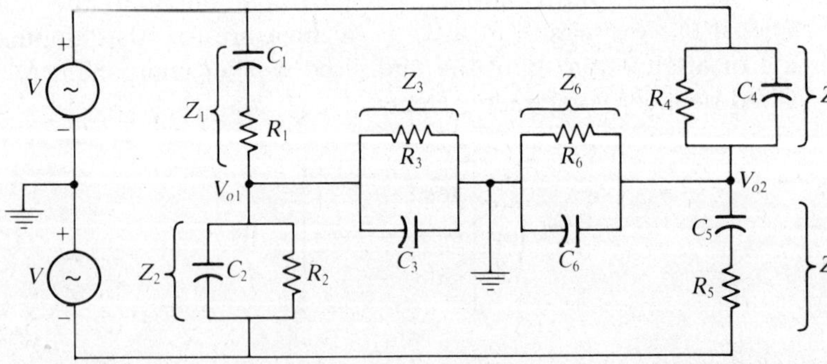

Fig. 5-25 When the components are properly selected, V_{o1} and V_{o2} are almost 90° out of phase over several kilohertz with no amplitude distortion.

To obtain some simplicity, assume that

$$R_1C_1 = R_2C_2 = R_3C_3 = 1/(2\pi f_1) \tag{5-64a}$$

$$R_4C_4 = R_5C_5 = R_6C_6 = 1/(2\pi f_2) \tag{5-64b}$$

$$R_2 = aR_3, \qquad R_4 = aR_6 \tag{5-64c}$$

$$R_3 = bR_1, \qquad R_6 = bR_5 \tag{5-64d}$$

Consequently,

$$Z_1 = R_1 \frac{1 + j\dfrac{f}{f_1}}{j\dfrac{f}{f_1}}; \qquad Z_2 = \frac{abR_1}{1 + j\dfrac{f}{f_1}}; \qquad Z_3 = \frac{bR_1}{1 + j\dfrac{f}{f_1}}$$

$$Z_4 = \frac{abR_5}{1 + j\dfrac{f}{f_2}}; \qquad Z_5 = R_5 \frac{1 + j\dfrac{f}{f_2}}{j\dfrac{f}{f_2}}; \qquad Z_6 = \frac{bR_5}{1 + j\dfrac{f}{f_2}} \tag{5-65}$$

When these are substituted into (5-63a) and (5-63b), it is found that

$$V_{o1} = \frac{V}{a+1} \frac{\left(\dfrac{f}{f_1}\right)^2 - 1 - j\dfrac{f}{f_1}(2 - ab)}{1 - \left(\dfrac{f}{f_1}\right)^2 + j\dfrac{f}{f_1}\left(2 + \dfrac{ab}{a+1}\right)} \tag{5-66}$$

and

$$V_{o2} = \frac{V}{a+1} \frac{1 - \left(\dfrac{f}{f_2}\right)^2 + j\dfrac{f}{f_2}(2 - ab)}{1 - \left(\dfrac{f}{f_2}\right)^2 + j\dfrac{f}{f_2}\left(2 + \dfrac{ab}{a+1}\right)} \tag{5-67}$$

Next, examine the magnitudes

$$|V_{o1}|^2 = \frac{|V|^2}{(a+1)^2} \frac{\left(\dfrac{f}{f_1}\right)^4 + \left(\dfrac{f}{f_1}\right)^2 (2 - 4ab + a^2b^2) + 1}{\left(\dfrac{f}{f_1}\right)^4 + \left(\dfrac{f}{f_1}\right)^2 \left[2 + \dfrac{4ab}{a+1} + \left(\dfrac{ab}{a+1}\right)^2\right] + 1} \tag{5-68}$$

$$|V_{o2}|^2 = \frac{|V|^2}{(a+1)^2} \frac{\left(\dfrac{f}{f_2}\right)^4 + \left(\dfrac{f}{f_2}\right)^2 (2 - 4ab + a^2b^2) + 1}{\left(\dfrac{f}{f_2}\right)^4 + \left(\dfrac{f}{f_2}\right)^2 \left[2 + \dfrac{4ab}{a+1} + \left(\dfrac{ab}{a+1}\right)^2\right] + 1} \tag{5-69}$$

To have zero amplitude distortion, the numerators must track the denominators in both (5-68) and (5-69). This is accomplished when

$$2 - 4ab + a^2b^2 = 2 + \frac{4ab}{a+1} + \left(\frac{ab}{a+1}\right)^2 \tag{5-70}$$

from which is obtained

$$b = \frac{4(a+1)}{a^2} \tag{5-71}$$

This means that certain resistance ratios and capacitance ratios must be selected to have no amplitude distortion. If $a = 2$ is chosen, then $b = 3$.

In the circuit of Fig. 5-25, it is desirable to choose C_3 and C_6 large to "swamp" the input capacitances of the system to which V_{o1} and V_{o2} are applied, and choose R_3 and R_6 small to "swamp" the input resistances of the systems. Otherwise, the input capacitances and input resistances must be accounted for in the values of C_3, C_6, R_3, and R_6.

Now, the phase-shift properties need to be determined. Substitute (5-71) into (5-66) and (5-67), rationalize the denominators, and simplify using $F_1 = f/f_1$ and $F_2 = f/f_2$. Thus,

$$\mathrm{V_{o1}} = \frac{\mathrm{V}}{\mathrm{a}+1}\,\frac{\left\{-\mathrm{F}_1^4 + \mathrm{F}_1^2\left[6 + \frac{16(\mathrm{a}+1)}{\mathrm{a}^2}\right] - 1\right\} + \mathrm{j}\,4\mathrm{F}_1(1-\mathrm{F}_1^2)\left(\frac{\mathrm{a}+2}{\mathrm{a}}\right)}{(1-\mathrm{F}_1^2)^2 + \mathrm{F}_1^2\left(\frac{4+2\mathrm{a}}{\mathrm{a}}\right)^2} \tag{5-72}$$

$$\mathrm{V_{o2}} = \frac{\mathrm{V}}{\mathrm{a}+1}\,\frac{\left\{\mathrm{F}_2^4 + \mathrm{F}_2^2\left[-6 - \frac{16(\mathrm{a}+1)}{\mathrm{a}^2}\right] + 1\right\} - \mathrm{j}\,4\mathrm{F}_2(1-\mathrm{F}_2^2)\left(\frac{\mathrm{a}+2}{\mathrm{a}}\right)}{(1-\mathrm{F}_2^2)^2 + \mathrm{F}_2^2\left(\frac{4+2\mathrm{a}}{\mathrm{a}}\right)^2} \tag{5-73}$$

It has already been shown that $V_{o1} = V_{o2} = V/(a+1)$ when b is chosen according to (5-71). Therefore, to keep V_{o1} and V_{o2} from being quite small, the value of a should not be a large number, and $a = 2$ is a good compromise.

Let ϕ_1 and ϕ_2 represent the phase angles of V_{o1} and V_{o2}, respectively. They are determined from the inverse tangents of the imaginary parts divided by the real parts in the numerators of (5-72) and (5-73). Hence, using $a = 2$,

$$\phi_1 = \tan^{-1}\frac{8F_1(1-F_1^2)}{-F_1^4 + 18F_1^2 - 1} \tag{5-74}$$

and

$$\phi_2 = \tan^{-1}\frac{8F_2(F_2^2 - 1)}{F_2^4 - 18F_2^2 + 1} \tag{5-75}$$

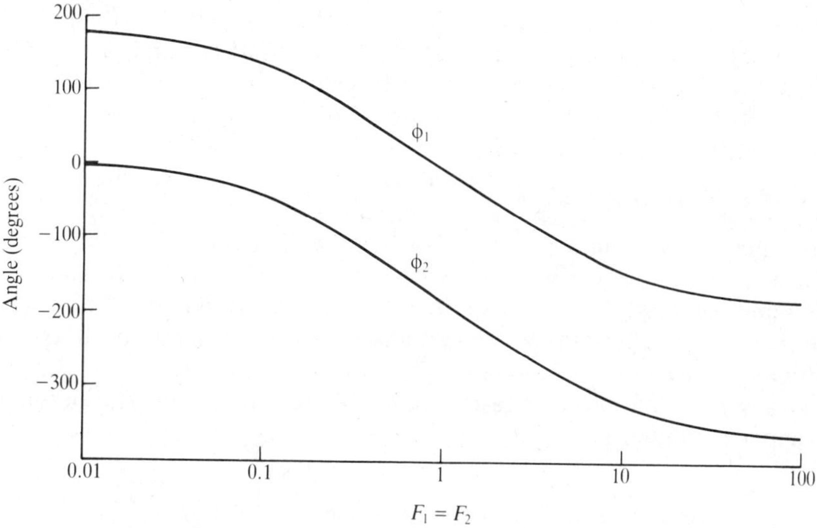

Fig. 5-26 Phase angles of V_{o1} and V_{o2} vs. f when $f_1 = f_2$.

These are plotted in Fig. 5-26 when $F_1 = F_2$, or $f_2 = f_2$, meaning that the individual time constants identified in (5-64a) and (5-64b) are all equal. The curves in Fig. 5-26 show that $\phi_2 - \phi_1 = -180°$ over the entire frequency range. But what we want is $\phi_2 - \phi_1 = -90°$, so f_1 and f_2 must be different. In effect, the curve for ϕ_2 must be moved to the right, to make it closer to the curve for ϕ_1. Consequently, f_2 must be adjusted to make $\phi_2 - \phi_1$ as nearly equal to $-90°$ as possible over part of the frequency range. Then at very low

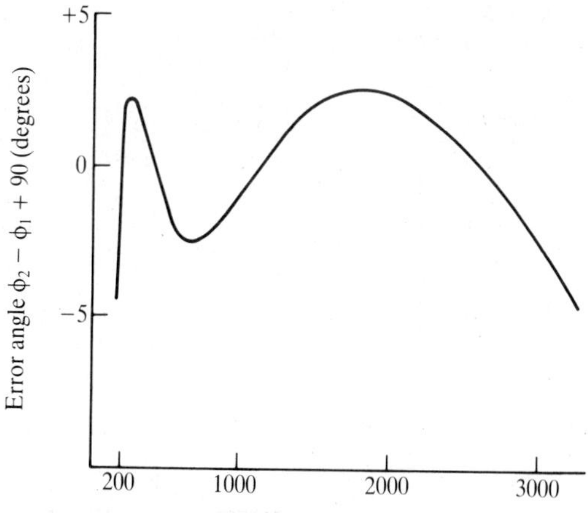

Fig. 5-27 Error angle vs. frequency for the voltages V_{o1} and V_{o2} in Fig. 5-25.

and at very high frequencies, $\phi_2 - \phi_1$ reverts back to $-180°$. After trying a few values, it is quickly determined that $f_1 = 318.31$ Hz and $f_2 = 1385$ Hz causes an error angle $\phi_2 - \phi_1 + 90°$ to be less than 3° over the frequency range 200–3000 Hz (see Fig. 5-27). From (5-62), the upper sideband will be suppressed by at least 31.6 dB provided that the amplitudes are properly balanced. Additional details about this circuit are addressed in the Problems of this chapter.

It is possible to obtain two sinusoids that are 90° out of phase over a wide frequency range with the arrangement shown in Fig. 5-28. The frequency of one input sinusoid $\cos 2\pi ft$ can be continuously varied over the range $f_1 \leqslant f \leqslant f_2$. That sinusoid is applied directly to one input each of two multipliers. The other sinusoidal inputs are in quadrature at the constant frequency f_0, so an uncomplicated phase shifter is required. The output of the upper multiplier is

$$2\cos 2\pi f_0 t \cos 2\pi ft = \cos 2\pi(f_0 + f)t + \cos 2\pi(f_0 - f)t$$

and the output of the lower multiplier is

$$2\sin 2\pi f_0 t \cos 2\pi ft = \sin 2\pi(f_0 + f)t + \sin 2\pi(f_0 - f)t$$

If these two signals are delivered separately to low pass filters, the outputs are $\cos 2\pi(f_0 - f)t$ and $\sin 2\pi(f_0 - f)t$. Their frequencies are continuously variable from $f_0 - f_2$ to $f_0 - f_1$, provided that $f_0 > f_2$. If a highpass filter is used instead, the outputs are $\cos 2\pi(f_0 + f)t$ and $\sin 2\pi(f_0 + f)t$, and their frequencies are variable from $f_0 + f_1$ to $f_0 + f_2$.

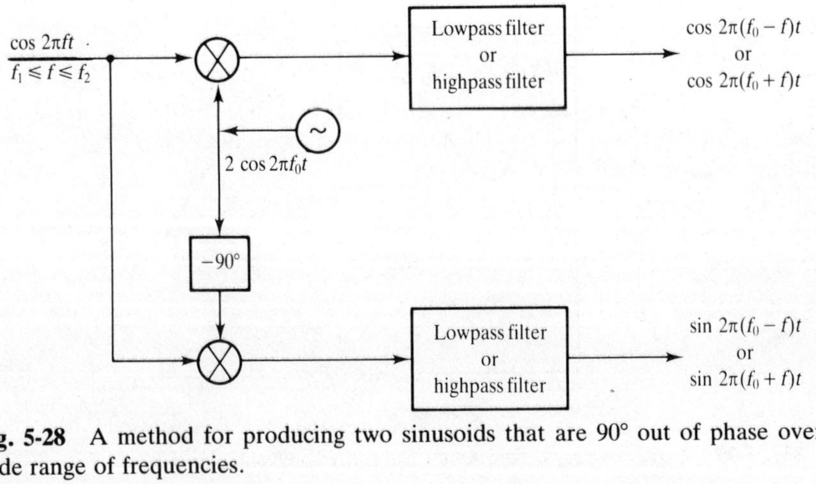

Fig. 5-28 A method for producing two sinusoids that are 90° out of phase over a wide range of frequencies.

5-8 SINGLE-SIDEBAND DEMODULATION

Suppose that the original modulating signal is the voltage $3K_0 \cos 2\pi(1000)t + 2K_0 \cos 2\pi(1500)t$, where K_0 is a constant that sets the voltage level, and that only the upper side-frequencies above the (missing) sinusoidal carrier at 3800 kHz are transmitted. Thus, the SSB output is composed of the sum of the two sinusoids $3K_1 \cos 2\pi(3801 \times 10^3)t + 2K_1 \cos 2\pi(3801.5 \times 10^3)t$, where K_1 is a constant that sets the voltage level that corresponds to the output power. The signal picked up by the receiving system and delivered to the demodulator has only the sum of the two components $3K_2 \cos 2\pi(3801 \times 10^3)t + 2K_2 \cos 2\pi(3801.5 \times 10^3)t$, where K_2 sets the voltage level, to work with. It is the function of the demodulator to create from its input signal, two new frequencies of 1 kHz and 1.5 kHz in the correct amplitude ratio to recover something directly proportional to the original modulating signal.

We know that when two sinusoids at different frequencies are applied, one each of the two inputs of a multiplier (or product detector), the output contains components at the sum and difference frequencies. If the desired output frequency is less than either of the input frequencies, then the difference frequency must be chosen and the undesired sum frequency filtered out. Consider Fig. 5-29. The output of the multiplier (product detector) is

$$\begin{aligned} y_1(t) &= 6K_2 \cos 2\pi(3801 \times 10^3)t \cos 2\pi(3800 \times 10^3)t \\ &\quad + 4K_2 \cos 2\pi(3801.5 \times 10^3)t \cos 2\pi(3800 \times 10^3)t \\ &= 3K_2 \cos 2\pi(7601 \times 10^3)t + 3K_2 \cos 2\pi(1000)t \\ &\quad + 2K_2 \cos 2\pi(7601.5 \times 10^3)t + 2K_2 \cos 2\pi(1500)t \end{aligned}$$

It is relatively easy for the filter to reject the sum-frequency components because they are so high in frequency compared to the difference-frequency components. Hence, the filter output is

$$y_D(t) = 3K_2 \cos 2\pi(1000)t + 2K_2 \cos 2\pi(1500)t$$

The frequencies and amplitudes of the two components are correct, so the demodulator has done its job.

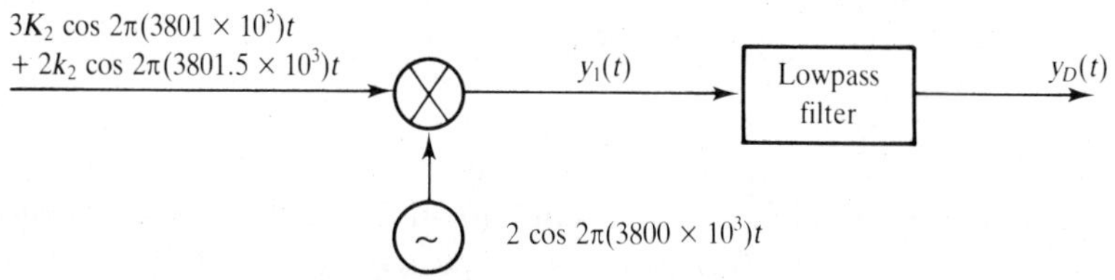

Fig. 5-29 A demodulator for SSB.

However, suppose the oscillator output in Fig. 5-29 is

$$2 \cos 2\pi(3802.5 \times 10^3)t$$

This could occur if the receiver operator thought that the incoming signal was the lower sideband. Then,

$$\begin{aligned} y_1(t) &= 6K_2 \cos 2\pi(3802.5 \times 10^3)t \cos 2\pi(3801 \times 10^3)t \\ &\quad + 4K_2 \cos 2\pi(3802.5 \times 10^3)t \cos 2\pi(3801.5 \times 10^3)t \\ &= 3K_2 \cos 2\pi(7603.5 \times 10^3)t + 3K_2 \cos 2\pi(1500)t \\ &\quad + 2K_2 \cos 2\pi(7604 \times 10^3)t + 2K_2 \cos 2\pi(1000)t \end{aligned}$$

The correct frequencies are produced in this example, but their amplitude ratio is not correct. If the original modulating signal had been speech, it would sound distorted and garbled, a sound that has been called monkey chatter. Upon hearing this useless result, the receiver operator would know to retune the oscillator to recover intelligible speech.

5-9 SINGLE-SIDEBAND TESTING

The carrier is supposed to be suppressed by the balanced modulator, if it is in fact balanced. The presence of the carrier in the output of the transmitter can be detected by a nearby radio receiver tuned to the carrier frequency. It is convenient if the receiver has a signal-strength meter, because then the carrier-balance controls can be adjusted and readjusted several times, as needed, until the least amount is indicated on the meter.

The presence of the carrier can also be detected by observing the output on an oscilloscope screen when a single sinusoid (or tone) at about 500 Hz is used as a modulating signal. If the carrier and the unwanted side frequency are adequately suppressed, a constant amplitude of the desired side frequency is obtained. If some carrier is also present, the observed waveform is like that of AM with a small modulation index, and is produced by the linear mixing of the small-amplitude carrier and the larger-amplitude side frequency. If the carrier is nulled out, but the unwanted side frequency has an appreciable amplitude, the resulting waveform again is like that of AM with a small modulation index, but the envelope has twice the frequency of the previous case because now the upper and lower side frequencies are mixing rather than one side frequency and the carrier. When the carrier and unwanted side frequency are both present, the envelope is like a combination of a fundamental and its second harmonic. The envelope is stationary on the oscilloscope screen if the horizonal sweep is synchronized with the frequency of the single tone.

Probably the best way to determine carrier and side-frequency suppression and nonlinearity of the power amplifiers is to look at the output with a spectrum analyzer. The input speech signal is simulated by two equal-amplitude sinusoids having frequencies of about 800 Hz and 1800 Hz, hence the name *two-tone test.* When the transmitter is properly adjusted, the spectrum analyzer will reveal only the desired two equal-amplitude output side frequencies. In this case, an oscilloscope screen will show the same kind of waveform as discussed in Chapter 1 on the linear mixing of equal-amplitude sinusoids.

If there are some carrier as well as undesired side frequencies present, they too will show up in the spectrum-analyzer display. If nonlinearity is present in the power amplifiers following the SSB modulator, odd-order intermodulation products will also be observed. When distortion is caused by overdriving one or more of the power amplifiers, the waveform on an oscilloscope screen will become flatter on top and bottom and any distortion products will make the valleys of the pattern more rounded. Severe "flat-topping" is prevented by automatic level-control circuitry, otherwise the sideband will splatter over a considerable width of spectrum, possibly interfere with other stations, and sound distorted upon demodulation.

PROBLEMS

5-1 The local oscillator in a DSB receiver is offset from its proper value by $\Delta f = 10$ Hz. If the modulating signal is a sinusoid whose frequency is 1000 Hz, sketch the waveform for the output from the demodulator/filter combination.

5-2 Determine the output voltages $v_t(t)$ and $v_c(t)$ of the balanced modulator in Fig. 5-5 when the power series for each device includes the cubed term. Let $v_1 = V_c \cos \pi f_c t$ and $v_2 = V_2 \cos 2\pi f_2 t$ represent the carrier and the modulating signal, respectively.

5-3 One signal applied to the input of an ideal multiplier is

$$x_1(t) = 3 \cos 1500\pi t + 2 \cos 2000\pi t$$

The other input signal is

$$x_2(t) = 4 \cos 18{,}000\pi t + 5 \cos 24{,}000\pi t$$

The output from the multiplier is applied to an ideal filter that passes all frequencies between 9000 Hz and 12,000 Hz. List all frequencies that are present in the output of the *filter*.

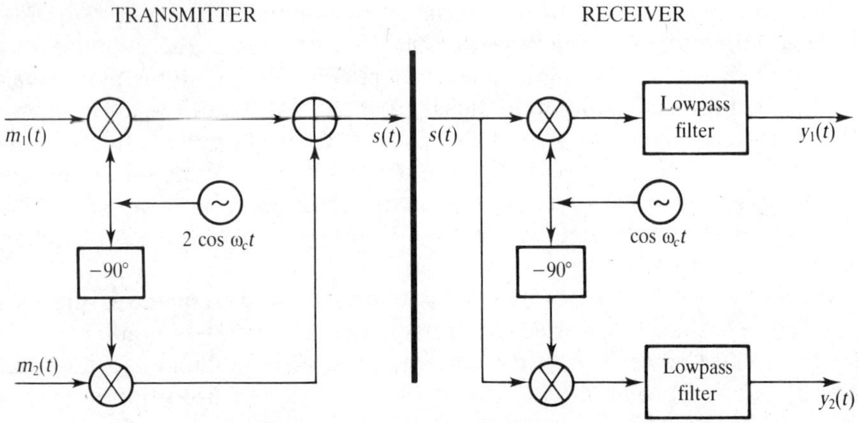

Fig. P5-4 System for Problem 5-4.

5-4 The quadrature-carrier multiplexing scheme in Fig. P5-4 allows two DSB signals to occupy the same transmission bandwidth when produced by independent messages $m_1(t)$ and $m_2(t)$. (a) Determine the transmitted signal $s(t)$. (b) A quadrature-carrier multiplexed signal $s(t)$ is delivered to the receiver and applied to coherent detectors as shown in Fig. P5-4. Evaluate the outputs $y_1(t)$ and $y_2(t)$.

5-5 If the carrier frequency f_c in a balanced modulator is too low relative to the highest frequency W of a modulating signal $m(t)$, the spectrum of the DSB signal overlaps that of $m(t)$ and exact recovery of $m(t)$ in the receiver is impossible. Using a spectral diagram, determine an inequality for the carrier frequency f_c such that spectral overlap does not occur.

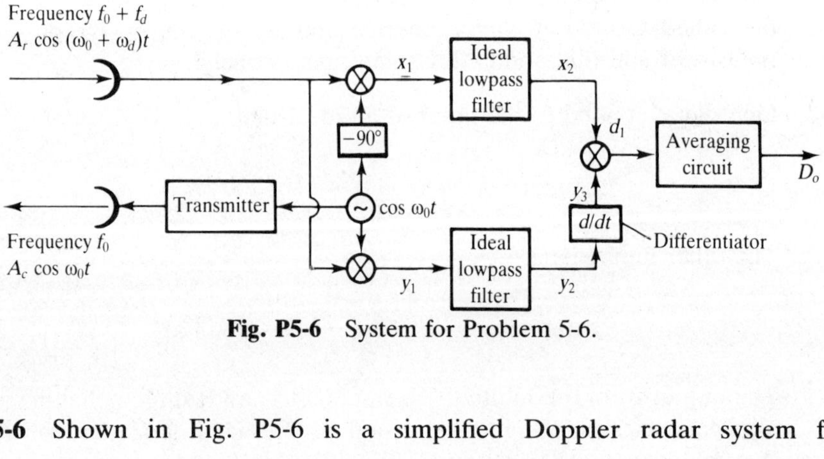

Fig. P5-6 System for Problem 5-6.

5-6 Shown in Fig. P5-6 is a simplified Doppler radar system for measuring the speed v of a moving automobile. When the transmitter is pulsed ON, it radiates a signal at frequency f_0. The moving target reflects a return signal at frequency $f_0 + f_d$, where the doppler shift is $f_d = Kv$ and K is a constant. Evaluate the signals at each stage of the diagram and demonstrate that the DC output D_0 is proportional to v.

5-7 By evaluating the signals at each stage, describe the output $y(t)$ from Fig. P5-7.

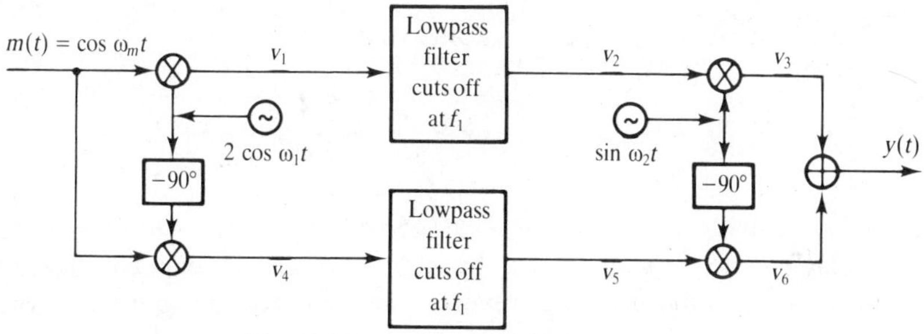

Fig. P5-7 System for Problem 5-7.

5-8 The periodic modulating signal shown in Fig. P5-8 has period T_m and zero average. Calculate $|V_{\min}|$.

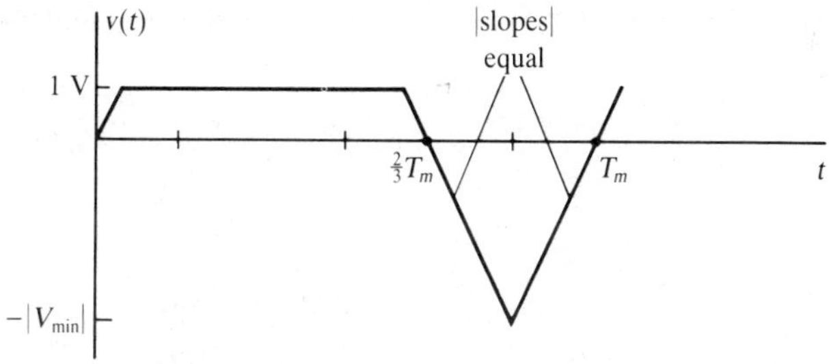

Fig. P5-8 Signal for Problem 5-8.

5-9 The carrier frequency of an AM signal is 16 times the frequency of the sinusoidal modulating signal. The AM signal is applied to an envelope detector using one diode and an RC filter. Sketch the half-wave-rectified carrier waveform and the DC output voltage when: (a) $RC = 10T_c$, and (b) $RC = 3T_c$ (where T_c = carrier period). Identify where diagonal clipping occurs if it does.

5-10 Calculations using Eq. (5-50) involve the differences between nearly equal numbers. That inequality can be simplified by recognizing that usually $f_c \gg f_m$ (which makes $\omega_m T_c$ very small) and that T_c/RC is often very small. Let $x = T_c/RC$ and $y = \omega_m T_c$. Then, using $e^{-x} = 1 - x + x^2/2$ and $\cos y = 1 - y^2/2$, and neglecting higher orders above 2 in e^{-2x}, demonstrate that (5-50) can be approximated by

$$m_{a-} \leqslant \frac{x(1 - x/2)}{\sqrt{x^2 + y^2(1 - x + x^2/2)}}$$

which, for small x, is further approximated by

$$m_{a-} \leqslant \frac{x}{\sqrt{x^2 + y^2}} = \frac{1}{\sqrt{1 + (2\pi RC/T_m)^2}}$$

5-11 The average current I in Fig. 5-20 can be calculated approximately by replacing the actual diode by its piecewise linear circuit representation composed of a reverse bias voltage V_γ in series with a constant resistance r, as in Fig. P5-11. Then,

$$i = \begin{cases} \dfrac{\sqrt{2}\,A \cos \omega_0 t - V_\gamma - V_{\text{DC}}}{r}, & \text{for } \sqrt{2}\,A > V_\gamma + V_{\text{DC}} \\ 0, & \text{otherwise} \end{cases}$$

where A is the rms input voltage. By integrating i over the conduction interval of the diode, show that the average current is

$$I = \frac{V_\gamma + V_{\text{DC}}}{\pi r} \left\{ \sqrt{\left[\left(\frac{\sqrt{2}\,A}{V + V_{\text{DC}}} \right)^2 - 1 \right]} - \cos^{-1} \left(\frac{V_\gamma + V_{\text{DC}}}{\sqrt{2}\,A} \right) \right\}$$

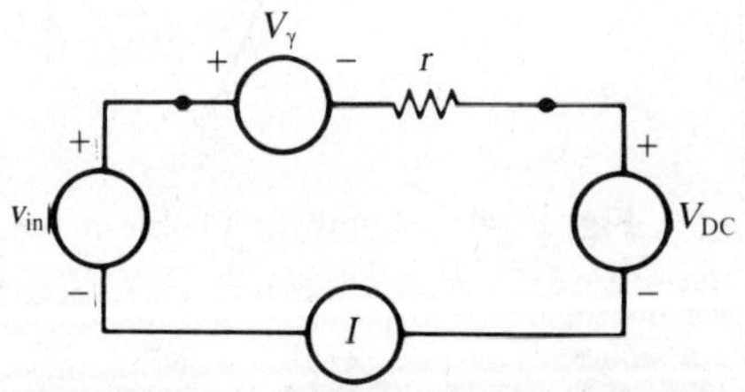

Fig. P5-11 Circuit for Problem 5-11.

5-12 (a) Using the result of Problem 5-11, plot a set of rectification characteristics as in Fig. 5-21. Assume that $V_\gamma = 0.5$ V, and $r = 100\ \Omega$.

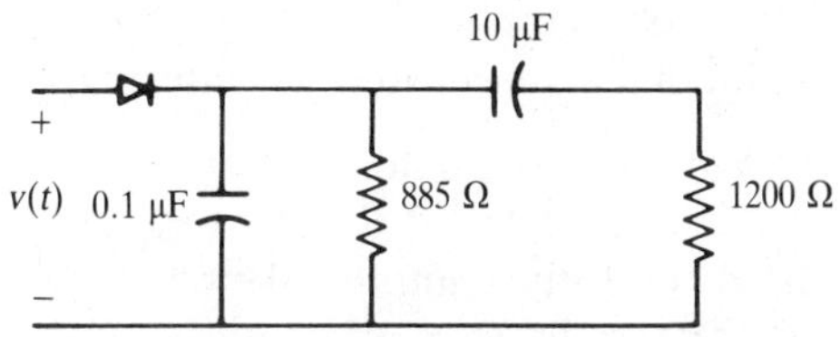

Fig. P5-12 Circuit for Problem 5-12.

(b) The diode is used in the envelope detector of Fig. P5-12. The input AM signal is $x_c(t) = 3\sqrt{2}\,(1 + m_a \cos 2\pi 1000t) \cos 2\pi 10^6 t$ V. Draw the DC load line and locate the Q point.

(c) Draw the AC load line and calculate the maximum allowable downward modulation percentage to prevent bottom clipping.

5-13 The AM signal $x_c(t) = 2(1 + 0.9 \cos 2\pi 5000t) \cos 2\pi 10^6 t$ V passes through a tuned amplifier circuit whose voltage amplification is its maximum of 10 at 1000 kHz but only 8 at both 995 kHz and 10005 kHz. Calculate the modulation index in percent of (a) the input signal (b) the output signal.

5-14 The symmetrical waveform for one period of a periodic modulating voltage $m(t)$ is shown in Fig. P5-14. A carrier voltage is represented by $10 \cos 2\pi 10^6 t$. These two signals are applied to the two inputs of a multiplier. The output from the multiplier is delivered to an ideal envelope detector. Sketch to scale and label the output waveform from the envelope detector for two periods of $m(t)$.

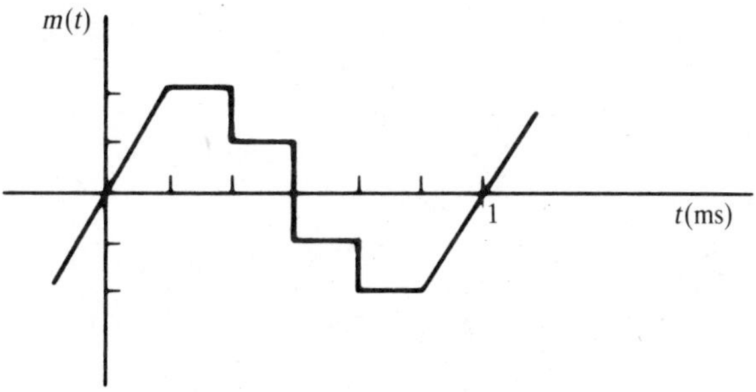

Fig. P5-14 Waveform for Problem 5-14.

5-15 One input to the multiplier circuit of a DSB demodulator is a received signal represented by $A_r m(t) \cos \omega_c t$. The other input to the multiplier is a locally generated sinusoid given by $2 \cos (\omega_c + \Delta\omega_c)t$, where $\Delta\omega_c$ represents the amount of off-tuning in the radian frequency of the local oscillator ($\Delta\omega_c = 2\pi\,\Delta f_c$). Determine the output of the demodulator after lowpass filtering.

5-16 The carrier voltage from oscillator #1 in Fig. 5-23 is $V_1 \cos A$ and from the speech processor $V_2 \cos B$. The output of the carrier phase shifter is $V_1 \sin A$ and the output of the other phase shifter is $(V_2 + \Delta V) \sin B$. Determine the attenuation A_u in dB of the undesired weaker component relative to the desired stronger component in the output. Express the result in terms of $\Delta V/V_2$. Then calculate A_u in dB as $\Delta V/V_2$ varies from 0.05 to 0.25 in steps of 0.05.

5-17 The circuit of Fig. 5-25 has $f_1 = 318.31$ Hz and $f_2 = 1385$ Hz with $a = 2$ and $C_3 = C_6 = 0.03\ \mu$F. Calculate all other capacitances and resistances.

5–18 Repeat Problem 5-17 starting with $R_3 = R_6 = 5$ kΩ instead of with the capacitances $C_3 = C_6$.

6

ANGLE MODULATION

6-1 ANGLE MODULATION

Again consider the carrier voltage $x_c(t) = V(t) \cos \phi(t)$. Angle modulation is said to occur when the angle $\phi(t)$ is caused to vary by an amount determined by the instantaneous amplitudes of the modulating signal $m(t)$, and the amplitude $V(t)$ of the carrier remains constant. Angle modulation is not linear modulation, and for that reason it is a more complicated process mathematically. The commonly encountered forms of angle modulation are phase modulation and frequency modulation.

Phase modulation (PM) is said to occur when the deviation of the angle $\phi(t)$ from its unmodulated value is caused to vary by an amount that is determined (usually linearly) by the instantaneous amplitudes of the modulating signal $m(t)$, and the amplitude $V(t)$ of the carrier remains constant.

Frequency modulation (FM) is said to occur when the deviation of the carrier frequency from its unmodulated value f_c is caused to vary by an amount that is determined (usually linearly) by the instantaneous amplitudes of the modulating signal $m(t)$, and the amplitude of the carrier $V(t)$ remains constant.

6-2 PHASE MODULATION

Let us consider phase modulation (PM) first, and then duplicate everything for frequency modulation (FM). Represent the carrier voltage by

$$x_c(t) = A_c \cos \phi(t) = A_c \cos [\omega_c t + \theta(t)] \tag{6-1}$$

where $\theta(t)$ = instantaneous phase in radians:

$$\phi(t) = \omega_c t + \theta(t) = \text{instantaneous angle (radians)}$$

In PM, the instantaneous phase deviation of the carrier, from its unmodulated phase $\theta(0)$, is directly proportional to the instantaneous amplitudes of the modulating signal $m(t)$. Thus,

$$\theta(t) = \theta(0) + k_p m(t) \tag{6-2}$$

where k_p is a phase deviation constant in radians per unit amplitude of $m(t)$ and its value is determined by the design of the phase modulator used. If $m(t)$ is represented by its various spectral components, then

$$m(t) = A_1 m_{n1}(t) + A_2 m_{n2}(t) + \cdots = \sum_k A_k m_{nk}(t) \tag{6-3}$$

where the values of A_k are those amplitudes required to normalize each spectral component, and the $m_{nk}(t)$ are the normalized spectral components. Then,

$$\begin{aligned} \theta(t) &= \theta(0) + k_p A_1 m_{n1}(t) + k_p A_2 m_{n2}(t) + \cdots \\ &= \theta(0) + \Delta\theta m_{n1}(t) + \Delta\theta_2 m_{n2}(t) + \cdots \end{aligned} \tag{6-4}$$

where $\Delta\theta_1 = k_p A_1$ and $\Delta\theta_2 = k_p A_2$, etc., are the individual maximum phase deviations in radians produced in the PM modulator by the components of $m(t)$. It is usual to select a time scale such that $\theta(0) = 0$.

Assume that the normalized components of $m(t)$ are given by

$$\begin{aligned} m_{n1}(t) &= \sin \omega_1 t \\ m_{n2}(t) &= \sin \omega_2 t \\ &\vdots \end{aligned}$$

Then, with $\theta(0) = 0$,

$$\phi(t) = \omega_c t + \Delta\theta_1 \sin \omega_1 t + \Delta\theta_2 \sin \omega_2 t + \cdots \tag{6-5}$$

When this is substituted into (6-1) and expanded, the spectrum of the phase-modulated signal is obtained. We will demonstrate the procedure after considering FM.

Even though this is phase modulation, there is an accompanying change in the carrier frequency that is obtained from

$$\omega(t) = \frac{d\phi(t)}{dt} = \omega_c + \Delta\theta_1 \frac{dm_{n1}(t)}{dt} + \Delta\theta_2 \frac{dm_{n2}(t)}{dt} + \cdots \tag{6-6}$$

which becomes, using $m_{n1}(t) = \sin \omega_1 t$ and $m_{n2}(t) = \sin \omega_2 t_1$

$$\omega(t) = \omega_c + \omega_1 \, \Delta\theta_1 \cos \omega_1 t + \omega_2 \, \Delta\theta_2 \cos \omega_2 t + \cdots$$

or, dividing by 2π,

$$\begin{aligned} f(t) &= f_c + f_1 \, \Delta\theta_1 \cos \omega_1 t + f_2 \, \Delta\theta_2 \cos \omega_2 t + \cdots \\ &= f_c + \Delta f_1 \cos \omega_1 t + \Delta f_2 \cos \omega_2 t + \cdots \end{aligned} \tag{6-7}$$

where $\Delta f_1 = f_1 \, \Delta\theta_1$, $\Delta f_2 = f_2 \, \Delta\theta_2$, etc., represent the individual maximum frequency deviations in hertz produced in the PM modulator by the individual components of $m(t)$. Notice that each frequency deviation varies directly with both the frequency and the amplitude of a modulating component.

Example 6-1

The sinusoidal output from a 100-kHz crystal oscillator is phase-modulated. The modulating signal is $m(t) = 2 \sin 1000t + \sin 5000t$ V. The phase deviation constant of the phase modulator is $k_p = 0.05$ radian per volt of $m(t)$. Calculate (a) the individual maximum phase deviations in degrees produced by the phase modulator; (b) the total maximum phase deviation $\theta(t)_{\max}$ produced by both modulating components together; (c) the individual maximum frequency deviations in hertz produced by the components of $m(t)$; (d) the total maximum frequency deviation Δf produced by both modulating components together.

Solution

(a) $\Delta\theta_1 = k_p A_1 = 0.05(2) = 0.1$ radian or 5.73°.
$\Delta\theta_2 = k_p A_2 = 0.05(1) = 0.05$ radian or 2.865°.

(b) $m(t)$ is periodic with $f_2/f_1 = 5/1$. Both (1-32) and (1-33) are satisfied; therefore, $m(t)_{\max} = 3$ V and $m(t)_{\min} = -3$ V. Thus, $\theta(t)_{\max} = 5.73 + 2.865 = 8.59°$, or $\theta(t)_{\max} = k_p m(t)_{\max} = 0.15$ radian or 8.59°. Also, because of the symmetry, $\theta(t)_{\min} = -0.15$ radian or −8.59°. The phase of the signal in the output of the phase modulator is alternately advanced and retarded by a maximum of 8.59° with this $m(t)$.

(c) $\Delta f_1 = f_1 \, \Delta\theta_1 = (1000/2)(0.1) = 15.92$ Hz.
$\Delta f_2 = f_2 \, \Delta\theta_2 = (5000/2)(0.05) = 39.79$ Hz.

(d) The sum of the two cosine terms in (6-7) is periodic and symmetrical, thus $\Delta f = 15.92 + 39.79 = 55.71$ Hz. The frequency of the signal at the output of the phase modulator is alternately increased and decreased by a maximum of 55.71 Hz with this $m(t)$.

6-2.1 Phase Modulators

The simplified circuit of a type of phase modulator is shown in Fig. 6-1. Details of the biasing circuit have been omitted to concentrate on the basic

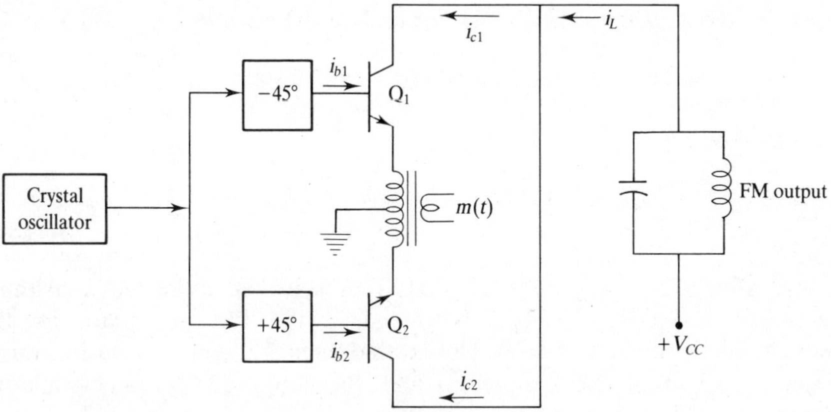

Fig. 6-1 A phase modulator (circuit details have been omitted).

concepts involved. The crystal oscillator establishes a stable carrier frequency, and its output is split into two sinusoidal base currents that are 90° out of phase. Therefore, the two collector currents at the carrier frequency are also 90° out of phase, and they combine vectorially to produce the load current i_L, as illustrated in Fig. 6-2.

The modulating signal $m(t)$ is applied to the two emitters through a center-tapped transformer secondary. When the polarity of $m(t)$ is such that the emitter of Q1 is driven more positive, then i_{b1} at the carrier frequency decreases, and this makes i_{c1} at the carrier frequency also decrease. But at the same time the emitter of Q_2 is driven more negative because of the transformer's action, so that i_{b2} at the carrier frequency increases, and this

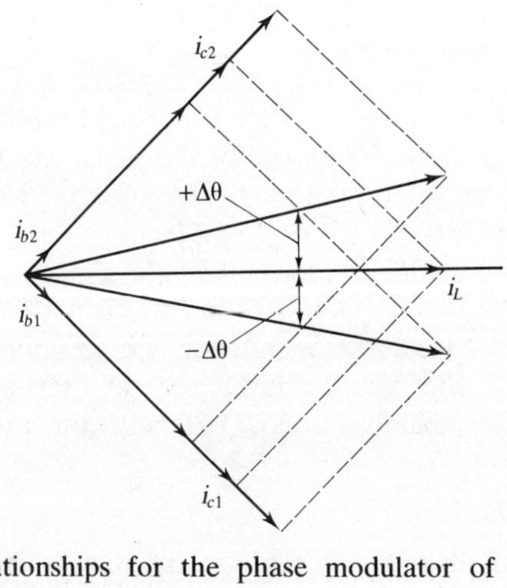

Fig. 6-2 Phase relationships for the phase modulator of Fig. 6-1 in which i_{c1} decreases when i_{c2} increases and vice versa.

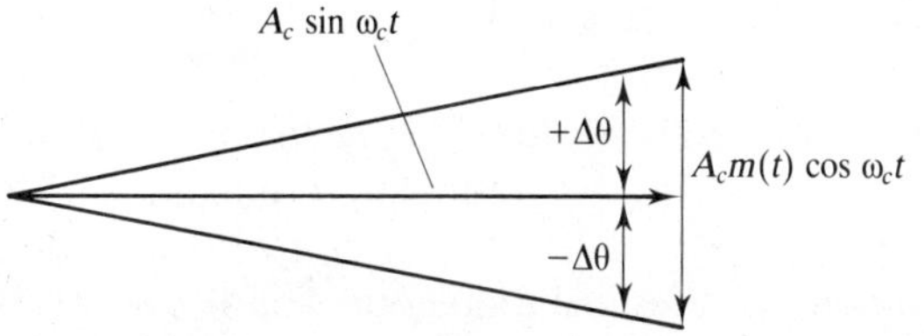

Fig. 6-3 Phase-shift modulation by adding a carrier and modulation product in quadrature.

causes i_{c2} at the carrier frequency to increase. When the smaller i_{c1} combines vectorially with the larger i_{c2}, the phase of the resultant i_L is advanced by $+\Delta\theta$. If $\Delta\theta$ is restricted to only a few degrees, then the amplitude of the phase-shifted resultant is essentially the same as the unmodulated value, and this keeps spurious AM to a minimum.

When $m(t)$ assumes the opposite polarity, then i_{c1} increases and i_{c2} decreases, so that i_L is shifted in phase through $-\Delta\theta$. Therefore, this circuit causes the phase of the load current to be alternately advanced and retarded by an amount $\Delta\theta$ that depends on the amplitude of the modulating signal $m(t)$.

Another method, originally proposed by Armstrong, is useful as a phase modulator. As described in Section 5-2, the output of an ideal double-sideband modulator is $x_c(t) = A_c m(t) \cos \omega_c t$. Normal AM can be obtained by adding $A_c \cos \omega_c t$, which is in phase with the above term. But if $A_c \sin \omega_c t$ is added instead, as shown in Fig. 6-3, then the phase of the resultant phasor is alternately advanced and retarded by an amount $\Delta\theta$ that depends on the amplitude of $m(t)$. A block diagram of this system is given in Fig. 6-4.

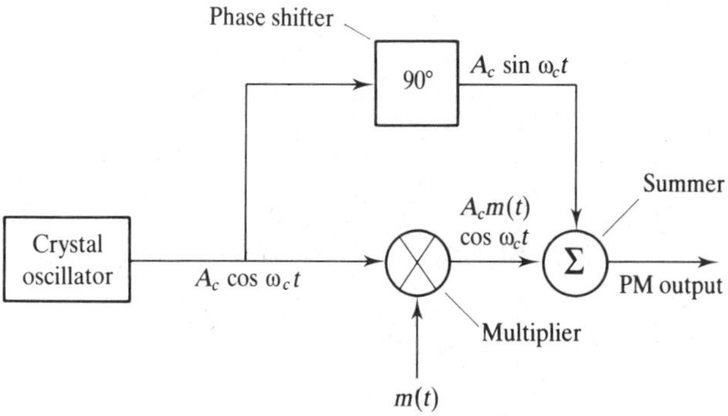

Fig. 6-4 Block diagram of a phase modulator.

6-3 FREQUENCY MODULATION

Before continuing with details of the PM spectrum, we now focus attention on the frequency modulator. In frequency modulation (FM), the instantaneous frequency deviation of the carrier from its unmodulated value f_c is directly proportional to the instantaneous amplitude of the modulating signal $m(t)$. Thus,

$$f(t) = f_c + k_f m(t) \tag{6-8}$$

where k_f is a frequency deviation constant in hertz per unit amplitude of $m(t)$ and its value is determined by the design of the frequency modulator used. If $m(t)$ is represented by its spectral components as given in (6-3), then

$$f(t) = f_c + k_f A_1 m_{n1}(t) + k_f A_2 m_{n2}(t) + \cdots \tag{6-9}$$

or, multiplying by 2π,

$$\omega(t) = \omega_c + 2\pi\, \Delta f_1\, m_{n1}(t) + 2\pi\, \Delta f_2\, m_{n2}(t) + \cdots \tag{6-10}$$

where $\Delta f_1 = k_f A_1$, $\Delta f_2 = k_f A_2$, etc., represent the maximum frequency deviations produced in the FM modulator by the individual components of $m(t)$.

Assume that the normalized components of $m(t)$ for FM can be given by

$$m_{n1}(t) = \cos \omega_1 t$$
$$m_{n2}(t) = \cos \omega_2 t$$
$$\vdots$$

Then,

$$f(t) = f_c + \Delta f_1 \cos \omega_1 t + \Delta f_2 \cos \omega_2 t + \cdots \tag{6-11}$$

or

$$\omega(t) = \omega_c + \Delta\omega_1 \cos \omega_1 t + \Delta\omega_2 \cos \omega_2 t + \cdots \tag{6-12}$$

Even though this is frequency modulation, there is an accompanying change in the carrier phase that is obtained from

$$\phi(t) = \int \omega(t)\, dt \tag{6-13}$$

where the constant of integration has been chosen equal to zero for simplicity. Thus, using (6-10) in (6-13),

$$\phi(t) = \int [2\pi f_c + 2\pi\,\Delta f_1\, m_{n1}(t) + 2\pi\,\Delta f_2\, m_{n2}(t) + \cdots]\, dt$$

$$= 2\pi f_c t + 2\pi\,\Delta f_1 \int m_{n1}(t)\, dt + 2\pi\,\Delta f_2 \int m_{n2}(t)\, dt + \cdots \quad (6\text{-}14)$$

which, for $m_{n1}(t) = \cos\omega_1 t$ and $m_{n2}(t) = \cos\omega_2 t$, becomes

$$\phi(t) = 2\pi f_c t + \frac{2\pi\,\Delta f_1}{\omega_1}\sin\omega_1 t + \frac{2\pi\,\Delta f_2}{\omega_2}\sin\omega_2 t + \cdots$$

$$= \omega_c t + \frac{\Delta f_1}{f_1}\sin\omega_1 t + \frac{\Delta f_2}{f_2}\sin\omega_2 t + \cdots \quad (6\text{-}15)$$

$$= \omega_c t + \Delta\theta_1 \sin\omega_1 t + \Delta\theta_2 \sin\omega_2 t + \cdots \quad (6\text{-}16)$$

where $\Delta\theta_1 = \Delta f_1/f_1$, $\Delta\theta_2 = \Delta f_2/f_2$, etc., represent the maximum phase deviations produced in the FM modulator by the individual components of $m(t)$.

Example 6-2

The sinusoidal output of an 88.5 MHz oscillator is frequency-modulated. The modulating voltage is $m(t) = 2\cos 1000t + \cos 5000t$ V. The frequency deviation constant of the frequency modulator is 25 kHz per volt of $m(t)$. Calculate

(a) the individual maximum frequency deviations in kHz produced by the frequency modulator
(b) The total maximum frequency deviation Δf_{max} in kHz produced by both modulating components together
(c) The individual maximum phase deviations produced by the components of $m(t)$
(d) The total maximum phase deviation $\Delta\theta$ produced by both modulating components together

Solution

(a) $\Delta f_1 = k_f \theta_1 = 25(2)$ kHz $= 50$ kHz.
$\Delta f_2 = k_f A_2 = 25(1)$ kHz $= 25$ kHz.
(b) $m(t)$ is periodic with $f_2/f_1 = 5/1$, and symmetrical with $m(t)_{max} = 3$ V and $m(t)_{min} = -3$ V. Thus, $\Delta f_{max} = k_f m(t)_{max} = 25(3)$ kHz $= 75$ kHz. Or, $\Delta f_{max} = \Delta f_1 + \Delta f_2 = 75$ kHz. The frequency of the output signal from the frequency modulator is alternately increased and decreased by a maximum of 75 kHz with this $m(t)$.

(c) $\Delta\theta_1 = \Delta f_1/f_1 = 50\ \text{kHz}/(1000/2\pi) = 314.16$ radians or 18,001°.
$\Delta\theta_2 = \Delta f_2/f_2 = 25\ \text{kHz}/(5000/2\pi) = 31.42$ radians or 1800°.

(d) The sum of the two sine terms in (6-16) is periodic and symmetrical in this example. Thus, $\Delta\theta = 314.16 + 31.42$ radians $= 345.58$ radians or 19,801°. The phase of the output signal from the frequency modulator is alternately advanced and retarded by a maximum of 19,801° with this $m(t)$.

6-3.1 Frequency Modulators

One of the simplest schemes conceptually for a frequency modulator is to use a capacitor microphone as part of the tuning capacitance of an *LC* oscillator. Then, when sound waves impinge on the microphone diaphragm, its capacitance changes slightly, and the frequency of oscillation changes accordingly. The louder the sound, the greater the frequency deviation.

A more practical method in some situations is to use the voltage-dependent capacitance of a reverse-biased semiconductor diode (or varactor) in the frequency-determining *LC* circuit of the oscillator. The frequency of oscillation can be changed by changing the reverse voltage applied to the diode, and the general arrangement is known as a *voltage-controlled oscillator,* or VCO. For an FM modulator, the controlling voltage is the sum of a DC reverse voltage and a smaller-amplitude modulating voltage.

A classical method for producing inductive or capacitive reactance that can be varied by a modulating or controlling voltage is the reactance transistor, one form of which is shown in Fig. 6-5a. Straightforward analysis of the simplified signal circuit in Fig. 6-5b shows that its input impedance is

$$Z_{\text{in}} = \frac{Z + h_{ie}}{h_{fe} + 1}$$

For Z_{in} to have a reactive component, Z must have a reactive component. For Z_{in} to be as reactive as possible, Z must have either inductive or capacitive reactance only whose magnitude is much larger than h_{ie}. Consequently,

$$Z_{\text{in}} = \pm j\frac{X}{h_{fe} + 1}$$

This reactance is injected into the frequency-determining circuit of the *LC* oscillator. But h_{fe} depends on the operating point of the transistor. Therefore, changing the operating point with a controlling or modulating voltage will cause the frequency of the oscillator to change. This process is a manifestation of the classical Miller effect in which a capacitance located between the input and output terminals of an inverting amplifier is effectively increased and controlled by the voltage gain of the amplifier.

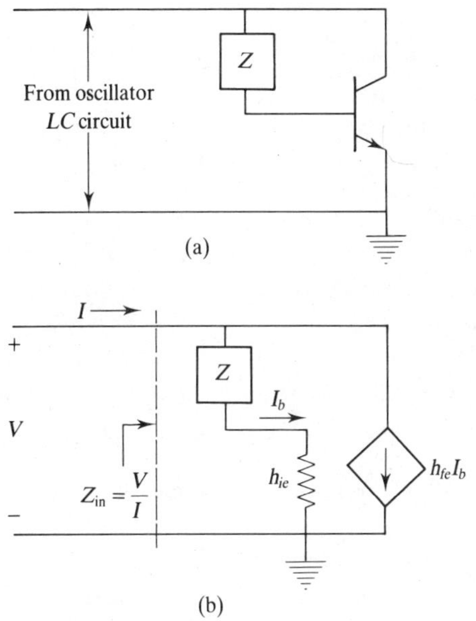

Fig. 6-5 (a) A reactance transistor. (b) Simplified signal circuit of (a).

At low frequencies where *RC* phase-shift oscillators are possible, the frequency can be varied by changing one or more of the resistors in the phase-shifting *RC* cascade. The dynamic resistance of another device can be used as one of the phase-shift-circuit resistors, and its value can be varied by the controlling voltage. Also, at low frequencies, an *RC* relaxation oscillator, such as the astable multivibrator, can have its frequency controlled by changing the triggering point during the operating cycle. This can be done simply by changing the DC bias voltage. The frequency-modulated square-wave output is then delivered to a bandpass filter to eliminate the harmonics and their associated side frequencies.

6-4 PM AND FM SPECTRA

The spectra of PM and FM waves are obtained by expanding $x_c(t) = A_c \cos \phi(t)$ for each type of modulation. Thus, for PM, from (6-5), and for FM, from (6-16),

$$\phi(t) = \omega_c t + \Delta\theta_1 \sin \omega_c t + \Delta\theta_2 \sin \omega_2 t + \cdots$$

Both PM and FM have the same expressions for $\phi(t)$ and therefore they can be combined into a single form that includes both. Thus,

$$x_c(t) = A_c \cos(\omega_c t + \beta_1 \sin \omega_1 t + \beta_2 \sin \omega_2 t + \cdots) \tag{6-17}$$

where

$$\beta_1 = \begin{cases} \Delta\theta_1 & \text{for PM,} \\ \dfrac{\Delta f_1}{f_1} & \text{for FM,} \end{cases} \qquad \beta_2 = \begin{cases} \Delta\theta_2 & \text{for PM,} \\ \dfrac{\Delta f_2}{f_2} & \text{for FM,} \end{cases} \qquad \text{etc.}$$

When (6-17) is expanded, the spectrum for either PM or FM is obtained simply by inserting the appropriate values of β. Each β is a measure of the "degree" of modulation produced by each component of $m(t)$; hence, each β becomes a modulation index.

A mathematical procedure will now be illustrated to show the expansion of (6-17) when $m(t)$ is composed of only two sinusoidal components. Thus, (6-17) becomes

$$x_c(t) = A_c \cos(\omega_c t + \beta_1 \sin \omega_1 t + \beta_2 \sin \omega_2 t) \tag{6-18}$$

The trigonometric identity

$$\begin{aligned}\cos(A + B + C) = {} & \cos A \cos B \cos C - \sin A \sin B \sin C \\ & - \sin A \cos B \sin C - \cos A \sin B \sin C\end{aligned}$$

allows (6-18) to be written as

$$\begin{aligned} x_c(t) = A_c[& \cos(\beta_1 \sin \omega_1 t) \cos(\beta_2 \sin \omega_2 t) \cos \omega_c t \\ & - \sin(\beta_1 \sin \omega_1 t) \cos(\beta_2 \sin \omega_2 t) \sin \omega_c t \\ & - \cos(\beta_1 \sin \omega_1 t) \sin(\beta_2 \sin \omega_2 t) \sin \omega_c t \\ & - \sin(\beta_1 \sin \omega_1 t) \sin(\beta_2 \sin \omega_2 t) \cos \omega_c t] \end{aligned} \tag{6-19}$$

As shown in Chapter 4,

$$\cos(\beta \sin \omega t) = J_0(\beta) + \sum_{n=2}^{\infty} 2J_n(\beta) \cos n\omega t \qquad \text{for even } n$$

and

$$\sin(\beta \sin \omega t) = \sum_{k=1}^{\infty} 2J_k(\beta) \sin k\omega t \qquad \text{for odd } k$$

where $J_0(\beta)$, $J_n(\beta)$, and $J_k(\beta)$ are Bessel functions of the first kind. See Chapter 4 for a review of Bessel functions and Table 4-3 for some tabulated values.

Using the above in (6-19), with appropriate subscripts, gives rise to both single and double summations and factors involving the products of three sinusoids at different frequencies. Then, using the trigonometric identities

$$4 \cos A \cos B \cos C = \cos (A + B - C) + \cos (B + C - A) + \cos (C + A - B) + \cos (A + B + C)$$

and

$$4 \sin A \sin B \cos C = -\cos (A + B - C) + \cos (B + C - A) + \cos (C + A - B) - \cos (A + B + C)$$

simplifying, and combining summations over even and odd indexes, we find, after tedious manipulations,

$$x_c(t) = A_c \sum_{n=-\infty}^{\infty} \sum_{k=-\infty}^{\infty} J_n(\beta_1) J_k(\beta_2) \cos 2\pi[f_c + (nf_1 + kf_2)]t \qquad (6\text{-}20)$$

in which each index now ranges over all integers, positive, negative, even, and odd, and $J_{-n}(\beta) = (-1)^n J_n(\beta)$.

Several important observations can be obtained from inspection of (6-20). There is a carrier-frequency component located at f_c for $n = k = 0$. There are an infinite number of side-frequency components located at $f_c + nf_1$ for $k = 0$ and all n, both positive and negative; at $f_c + kf_2$ for $n = 0$ and all k, both positive and negative; and at $f_c + nf_1 + kf_2$ for all nonzero n and all nonzero k, both positive and negative. The amplitudes of the various components are given by $A_c J_n(\beta_1) J_k(\beta_2)$. The absolute values of the indexes n and k are the harmonic numbers associated with each modulating frequency. As the harmonic numbers increase from zero, the values of $J_n(\beta_1)$ and $J_k(\beta_2)$ will oscillate, ultimately reach a maximum positive value, and then decrease monotonically toward zero as demonstrated in Fig. 6-6 for $J_n(20)$, $0 \leq n < 30$. This means that the amplitudes of side-frequency components with high harmonic numbers will become negligibly small. Consequently, the bandwidth of a circuit that is to pass an angle-modulated signal can be finite instead of infinite. But the bandwidth should be wide enough to pass all side-frequency components that have significant amplitudes. The criterion used in this book for what constitutes a significant amplitude is that any Bessel functions whose ultimate values are equal to or greater than 0.01 are taken to be significant. In practice, even if the bandwidth is more than adequate in a certain situation, side-frequency components having very high harmonic numbers will also have such very small amplitudes that they will be buried in the noise level and rendered undetectable.

The form given in (6-20) is for $m(t)$ composed of only two sinusoidal components. It can be readily extended to include any number of components, because each additional component will introduce a summation over its index from minus infinity to plus infinity, insert the Bessel

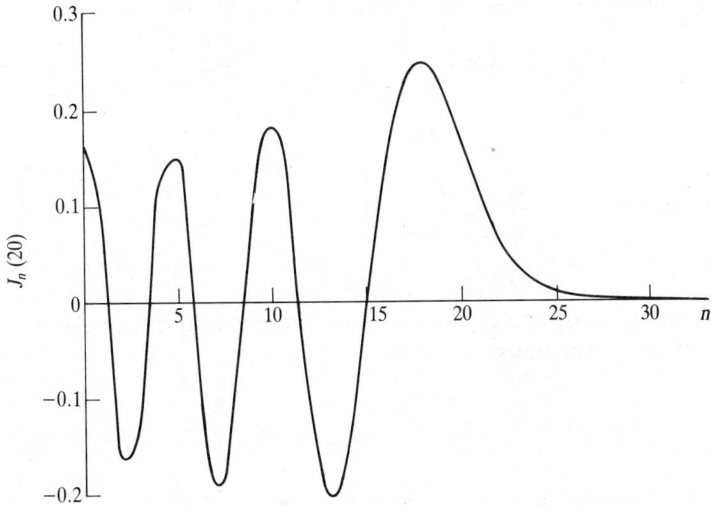

Fig. 6-6 This graph for $J_n(20)$ shows the typical oscillation initially of Bessel functions, then they ultimately attain a maximum positive value, after which they decrease monotonically toward zero, as the order n increases.

function of its modulation index as a factor, and add its harmonics to the others present. This result is traced back to the facts that an identity for the sine or cosine of the sum of n angles requires 2^{n-1} different terms, each comprising the product of n sinusoids involving the n angles, and that an identity for the product of n different sines or cosines requires 2^{n-1} different terms, each comprising the sine or cosine of the sum or difference between certain of the n angles. Unfortunately, a prodigious amount of time, effort, and book-keeping is required to determine and keep track of the numerous amplitudes and frequencies that are produced, even when $m(t)$ has only two sinusoidal components. For this reason, a BASIC computer program, described in Appendix E, was written to do the work.

Example 6-3

A certain modulating signal $m(t)$ is composed of two components having frequencies of 5 kHz and 1 kHz that produce modulation indexes of 1 and 2 respectively in an FM modulator. Using the BASIC program in Appendix E, determine: (a) the highest frequency of a significant component in the output spectrum whose amplitude is $\geqslant 0.01$; (b) the bandwidth required to pass all significant side frequencies; and (c) the maximum frequency deviation.

Solution After the information is entered, the program calculates four Bessel functions for $J_n(1)$ and five Bessel functions for $J_n(2)$ and displays or makes a hard copy of the values like:

```
n    Jn(1)
-    -----
0    0.7652
1    0.4401
2    0.1149
3    0.0196
* *  * * * * *
n    Jn(2)
-    -----
0    0.2239
1    0.5767
2    0.3528
3    0.1289
4    0.034
```

Instead of using the program, these values could be obtained from Table 4-3 and all of the calculations done by hand. The program then gives the double-sided spectral components for each of the two modulating components, with appropriate sign:

$$-0.0196 \quad 0.1149 \quad -0.4401 \quad 0.7652 \quad 0.4401 \quad 0.1149 \quad 0.0196$$

and

$$0.034 \quad -0.1289 \quad 0.3528 \quad -0.5767 \quad 0.2239$$

$$0.5767 \quad 0.3528 \quad 0.1289 \quad 0.034$$

The program uses the index $1 \leqslant X \leqslant 7$ to identify the seven values of the first set and the index $1 \leqslant Y \leqslant 9$ to identify the nine values of the second set. Each value in the first set must be multiplied by every value of the second set, so there will be 63 products. These results can be shown in 63 rows of four columns each giving the particular values of X and Y, the side frequencies above and below the unmodulated center frequency, and the value of each product to three decimal places. The results are shown in Table 6-1.

The last entry in the left-hand columns of Table 6-1 is for the amplitude 0.171 of the center frequency. It comes from the product of the $X = 4$ Bessel function 0.7652 and the $Y = 5$ Bessel function 0.2239 rounded to three decimal places. According to (6-20), all of the values listed should be multiplied by A_c, so the significant amplitudes will be $\geqslant 0.01A_c$. This will not be done here because it will not change the relative amplitudes.

TABLE 6-1 Calculations for Example 6-3

X	Y	Frequency	Value	X	Y	Frequency	Value
1	1	−19.00	−0.001	4	6	1.00	0.441
1	2	−18.00	0.003	4	7	2.00	0.270
1	3	−17.00	−0.007	4	8	3.00	0.099
1	4	−16.00	0.011	4	9	4.00	0.026
1	5	−15.00	−0.004	5	1	1.00	0.015
1	6	−14.00	−0.011	5	2	2.00	−0.057
1	7	−13.00	−0.007	5	3	3.00	0.155
1	8	−12.00	−0.003	5	4	4.00	−0.254
1	9	−11.00	−0.001	5	5	5.00	0.099
2	1	−14.00	0.004	5	6	6.00	0.254
2	2	−13.00	−0.015	5	7	7.00	0.155
2	3	−12.00	0.041	5	8	8.00	0.057
2	4	−11.00	−0.066	5	9	9.00	0.015
2	5	−10.00	0.026	6	1	6.00	0.004
2	6	−9.00	0.066	6	2	7.00	−0.015
2	7	−8.00	0.041	6	3	8.00	0.041
2	8	−7.00	0.015	6	4	9.00	−0.066
2	9	−6.00	0.004	6	5	10.00	0.026
3	1	−9.00	−0.015	6	6	11.00	0.066
3	2	−8.00	0.057	6	7	12.00	0.041
3	3	−7.00	−0.155	6	8	13.00	0.015
3	4	−6.00	0.254	6	9	14.00	0.004
3	5	−5.00	−0.099	7	1	11.00	0.001
3	6	−4.00	−0.254	7	2	12.00	−0.003
3	7	−3.00	−0.155	7	3	13.00	0.007
3	8	−2.00	−0.057	7	4	14.00	−0.011
3	9	−1.00	−0.015	7	5	15.00	0.004
4	1	−4.00	0.026	7	6	16.00	0.011
4	2	−3.00	−0.099	7	7	17.00	0.007
4	3	−2.00	0.270	7	8	18.00	0.003
4	4	−1.00	−0.441	7	9	19.00	0.001
4	5	0.00	0.171				

(a) By scanning the Frequency and Value columns, it is found that 16 kHz is the highest frequency of a significant component in the output spectrum. The program is designed to do this searching and stores the result in a variable FMAX.

(b) The side-frequency components are disposed either side of f_c as shown in Fig. 6-7. The various amplitudes are found from Table 6-1. The total bandwidth required for this modulated signal is

$$\text{BW} = 2(16)\ \text{kHz} = 32\ \text{kHz}$$

The program calculates the bandwith using BW = 2 FMAX kHz.

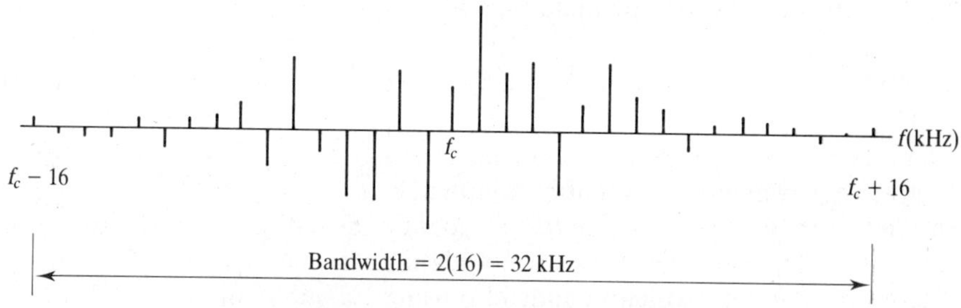

Fig. 6-7 Output spectrum of Example 6-3.

(c) Assuming that $m(t)$ is symmetrical and that its positive and negative extents are equal, the maximum frequency deviation is

$$\beta_1 \cdot f_1 + \beta_2 \cdot f_2 = 1(5) + 2(1) = 7 \text{ kHz}$$

Notice that several different pairs of harmonic numbers in Table 6-1 yield the same side frequencies; consequenctly, their paired values are added to find the total amplitude for that particular side frequency. The program includes a (slow) sequential search for those pairs whose total amplitudes ≥ 0.01 and whose frequencies are >FMAX. If there are any, use the greatest, instead of FMAX, to calculate the bandwidth; the program itself does not revise the bandwidth to take this effect into account.

If only the 5 kHz modulating signal had been present, the bandwidth could have been found directly from the three pairs of double-sided spectral components to be 2(3)(5) kHz = 30 kHz. Similarly, if only the 1 kHz modulating signal had been present, the bandwidth could have been found directly from the four pairs of double-sided spectral components to be 2(4)(1) kHz = 8 kHz. Having both modulating signals present simultaneously caused the bandwidth to increase.

6-4.1 Carson's Rule for Bandwidth

The primary difficulty in determining the bandwidth required in Example 6-3 is obtaining the products of the Bessel functions whose amplitudes are ≥ 0.01. This is not too difficult with only two modulating components, but most modulating signals are not so simple. The computational problems increase greatly as the number of modulating components increases to the more typical value provided by the harmonics and overtones of voice and music.

In an effort to determine the required bandwidth without undue expenditure of time and effort. John Carson, one of the ablest mathematical physicists associated with the communications industry, proposed that the bandwidth could be approximated by

$$BW = 2(\Delta f_{max} + f_{max}) \tag{6-21}$$

where Δf_{max} = maximum frequency deviation
f_{max} = highest modulating frequency
In Example 6-3 the maximum frequency deviation was found to be $\Delta f_{max} = 7$ kHz and the highest modulating frequency was $f_{max} = 5$ kHz. Therefore, the approximate bandwidth using Carson's rule is

$$BW = 2(7 + 5)\text{ kHz} = 24\text{ kHz}$$

whereas the bandwidth required to pass all components with amplitudes $\geqslant 0.01$ was found to be 32 kHz.

Equation (6-21) may be written as

$$BW = 2f_{max}\left(\frac{\Delta f_{max}}{f_{max}} + 1\right) \tag{6-22a}$$

in which $\Delta f_{max}/f_{max}$ is called the *deviation ratio*. For a transmitter operating in the standard FM broadcast band, the maximum allowable frequency deviation is 75 kHz and the highest modulating frequency is 15 kHz, nominally. By (6-22a), the bandwidth required is $2(15)(75/15 + 1) =$ 180 kHz.

When multistone modulation is considered, a better approximation to the bandwidth than Carson's rule is rule #2 given by

$$BW = 2(\Delta f_{max} + 2f_{max}) \tag{6-22b}$$

Thus, in Example 6-3 where $\Delta f_{max} = 7$ kHz and $f_{max} = 5$ kHz, rule #2 yields

$$BW = 2[7 + 2(5)]\text{ kHz} = 34\text{ kHz}$$

which is a better approximation to 32 kHz than given by Carson's rule. Also, if $\Delta f_{max} = 75$ kHz and $f_{max} = 15$ kHz, then rule #2 yields

$$BW = 2[75 + 2(15)] = 210\text{ kHz}$$

instead of 180 kHz by Carson's rule.

The standard FM broadcast consists of 200 channels that extend from 88 MHz to 108 MHz. Carrier frequencies are assigned to stations every 200 kHz from 88.1 MHz to 107.9 MHz. This allows a small guard band and transition region between adjacent 200-kHz segments. Adjacent channels are not normally assigned to stations in the same geographical area.

6-5 POWER RELATIONSHIPS

The amplitude of an angle-modulated signal is constant if an infinite number of components at the side frequencies are combined with that at the carrier frequency. If too many of the side-frequency components having significant amplitudes are removed because some circuit has inadequate bandwidth, then the amplitude of the angle-modulated signal will no longer be constant but will vary. This is a spurious AM, and if the demodulator output responds to AM in addition to PM or FM, then a distortion will occur that is caused by the spurious AM. This is the reason why the amplifiers just preceding the demodulator of many FM receivers are designed to operate in their saturation mode. Then, even though the amplitude of the input FM signal changes owing to spurious AM or to any other rapid changes or flutter in the amplitude, the output of the saturated amplifiers does not change. The signal delivered to the demodulator has a constant amplitude and therefore the effect of the spurious AM or flutter is eliminated.

If the highest frequency associated with the modulating signal is low compared to the carrier frequency, then the average power P_o of an angle modulated signal with a constant amplitude A_c volts, across a load resistance R, is

$$P_o = \frac{1}{2}\frac{A_c^2}{R} \tag{6-23}$$

Assuming that the load resistance is the same at each side frequency, the power is also equal to the sum of the powers delivered to R by each of the infinite number of spectral components. But if only a finite number of components having significant amplitudes are kept, then the power P_f delivered to R will be less than P_o, and their ratio $P_r = P_f/P_o$ will be less than unity. Thus,

$$P_f = \frac{A_c^2}{2R}[A_0^2 + 2(A_1^2 + A_2^2 + A_3^2 + \cdots + A_k^2)] \tag{6-24}$$

where A_0 = carrier value from the product of Bessel functions
A_1 = 1st side-frequency value from Bessel product
⋮ ⋮
A_k = kth side-frequency value from Bessel product

Then, setting up the ratio of powers and dividing out common factors yields

$$P_r = A_0^2 + 2[A_1^2 + A_2^2 + \cdots + A_k^2) \tag{6-25}$$

In case there is only one component in the modulating signal, this becomes

$$P_r = J_0^2(\beta) + 2[J_1^2(\beta) + \cdots + J_k^2(\beta)] \tag{6-26}$$

TABLE 6-2 Amplitude Values for Example 6-4

k	A_k	k	A_k	k	A_k
0	0.171	6	0.258	12	0.038
1	0.456	7	0.14	13	0.022
2	0.213	8	0.098	14	−0.007
3	0.254	9	−0.051	15	0.004
4	−0.228	10	0.026	16	0.011
5	0.099	11	0.067		

Example 6-4

Find the bandwidth required in Example 6-3 for a power ratio of at least (a) 0.98, (b) 0.99.

Solution The amplitude values from Table 6-1 are combined as shown in Table 6-2. Starting with the carrier value, substitute values from Table 6-2 one at a time into (6-25) until the result is $\geqslant 0.98$ for part (a) or 0.99 for part (b). Thus, for (a), $0.171^2 + 2(0.456^2 + 0.213^2 + 0.254^2 + 0.228^2 + 0.99^2 + 0.258^2 + 0.14^2 + 0.098^2) = 0.980$. Consequently, eight side-frequency pairs are required to obtain at least 98.0 percent of the total power possible. The bandwidth is $BW = 2(8)1\text{ kHz} = 16\text{ kHz}$. For (b), $0.171^2 + 2(0.456^2 + 0.213^2 + 0.254^2 + 0.228^2 + 0.099^2 + 0.258^2 + 0.14^2 + 0.098^2 + 0.051^2 + 0.026^2 + 0.067^2) = 0.996$. Consequently, twelve side-frequency pairs are required to obtain at least 99 percent of the total power possible. The bandwidth is $BW = 2(12)1\text{ kHz} = 24\text{ kHz}$.

6-6 ANALYSIS USING THE EXPONENTIAL FOURIER SERIES

Consider the angle-modulated waveform given by (6-1),

$$x_c(t) = A_c \cos\left[\omega_c t + \theta(t)\right] \tag{6-27}$$

Using a familiar trigonometric identity, expand this into

$$x_c(t) = A_c[\cos \omega_c t \cos \theta(t) - \sin \omega_c t \sin \theta(t)]$$

If the modulating signal is periodic, then $\cos \theta(t)$ and $\sin \theta(t)$ are also periodic and may be expanded into their respective exponential Fourier series. Thus,

$$\cos \theta(t) = \sum_{n=-\infty}^{\infty} a_n e^{jn\omega_m t} \tag{6-28}$$

where

$$a_n = \frac{1}{T}\int_T \cos\theta(t) e^{-jn\omega_m t}\, dt$$

$$T = \frac{1}{f_m}$$

and

$$\sin\theta(t) = \sum_{n=-\infty}^{\infty} b_n e^{jn\omega_n t} \tag{6-29}$$

where

$$b_n = \frac{1}{T}\int_T \sin\theta(t) e^{-jn\omega_m t}\, dt$$

The a_n coefficients will be real because $\cos\theta(t)$ is an even function of time and the b_n coefficients will be imaginary because $\sin\theta(t)$ is an odd function of time. The integrals are taken over a fundamental period T of the modulating signal.

Therefore, (6-27) becomes

$$x_c(t) = A_c \cos\omega_c t \sum_{n=-\infty}^{\infty} a_n e^{jn\omega_m t} - A_c \sin\omega_c t \sum_{n=-\infty}^{\infty} b_n e^{jn\omega_m t}$$

But $\text{Re}\, e^{j\omega_c t} = \cos\omega_c t$ and $\text{Re}\, je^{j\omega_c t} = -\sin\omega_c t$, where Re stands for "the real part of."

$$x_c(t) = \text{Re}\, A_c \sum_{n=-\infty}^{\infty} c_n e^{j\omega_c t} e^{jn\omega_m t} \tag{6-30}$$

where

$$c_n = a_n + jb_n = T\int_T [\cos\theta(t) + j\sin\theta(t)] e^{-jn\omega_m t}\, dt$$

$$= T\int_T e^{j\theta(t)} e^{-jn\omega_m t}\, dt \tag{6-31}$$

The c_n coefficients are real because both a_n and jb_n are real.

The integral in (6-31) can be evaluated in terms of elementary functions when $\theta(t)$ contains constant and linearly varying angles, such as square waves, pulses, and sawtooth shapes. When $\theta(t)$ varies sinusoidally, (6-31) leads to integral forms of the Bessel functions $J_n(\beta)$. When $\theta(t)$ varies as t^2, (6-31) leads to the Fresnel integrals which look like

$$C_1(x) = \sqrt{\frac{2}{\pi}}\int_0^x \cos t^2\, dt \qquad \text{and} \qquad C_2(x) = \frac{1}{\sqrt{2\pi}}\int_0^x \frac{\cos t}{\sqrt{t}}\, dt$$

and their counterparts that contain sines instead of cosines. Values for $C_1(x)$ and $C_2(x)$ are tabulated in the literature, but they will not be considered further here.

A major disadvantage of this analysis procedure for angle modulation is that the integrals become intractable for many relatively simple modulating waveforms. Examples will illustrate the procedures.

Example 6-5

A phase modulator, modulated by a square-wave signal whose fundamental frequency is f_m, produces the phase variation shown in Fig. 6-8 (note that $\theta(0) \equiv 0$). Determine the spectrum.

Solution By inspection, the angle variation is found to be

$$\theta(t) = \begin{cases} \Delta\theta, & 0 \leq t \leq T/2 \\ -\Delta\theta, & T/2 \leq t \leq T \end{cases} \tag{6-32}$$

The coefficients are found from (6-31). Thus,

$$c_n = \frac{1}{T}\left[\int_0^{T/2} e^{j\Delta\theta} e^{-jn\omega_m t}\,dt + \int_{T/2}^{T} e^{-j\Delta\theta} e^{-jn\omega_m t}\,dt\right] \tag{6-33}$$

which becomes

$$c_n = \begin{cases} \dfrac{2\sin\Delta\theta}{n\pi}, & n \text{ odd} \\ 0, & n \text{ even and } n \neq 0 \end{cases} \tag{6-34}$$

For $n = 0$, return to (6-33), set $n = 0$, and evaluate to find

$$c_0 = \cos\Delta\theta \tag{6-35}$$

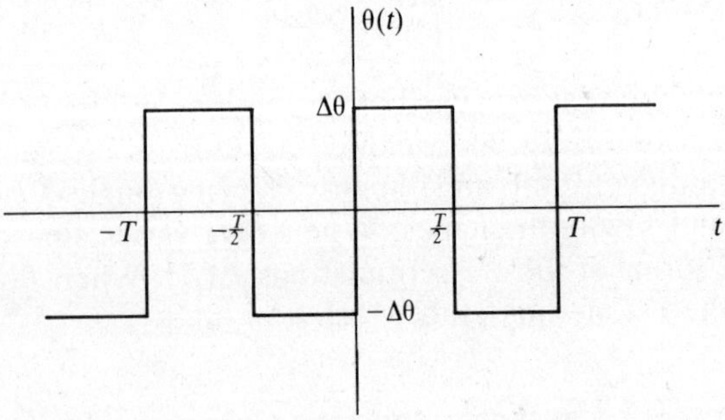

Fig. 6-8 Phase variation produced by the phase modulator of Example 6-5.

Finally, using (6-30),

$$x_c(t) = A_c \cos \Delta\theta \cos 2\pi f_c t + A_c \sum_{n=-\infty}^{\infty} \frac{2}{n\pi} \sin \Delta\theta \cos 2\pi(f_c + nf_m)t, \qquad n \text{ odd} \quad (6\text{-}36)$$

Equation (6-36) reveals that there will be no side frequencies for $\sin \Delta\theta = 0$. This occurs when $\Delta\theta$ is equal to an integral multiple of π radians. This is an expected result, however, because then the periodic square-wave modulation of $\Delta\theta$ changes instantaneously by an integral multiple of 2π radians at each edge of the square wave, and that puts it right back where it was. In effect, no modulation has occurred, no side frequencies are created, and the carrier has its maximum unmodulated amplitude. When present, the amplitudes of the side-frequency components vary inversely with the harmonic number n.

Example 6-6

A frequency modulator, modulated with a square-wave signal whose fundamental frequency is f_m, produces the frequency variation shown in Fig. 6-9. Determine the spectrum.

Solution By inspection, the instantaneous frequencies are

$$f(t) = \begin{cases} f_c - \Delta f, & -T/2 \leq t \leq -T/4 \\ f_c + \Delta f, & -T/4 \leq t \leq T/4 \\ f_c - \Delta f, & T/4 \leq t \leq T/2 \end{cases} \qquad (6\text{-}37)$$

The instantaneous phase is found from $\int 2\pi[f(t) - f_c]\, dt$ subject to the

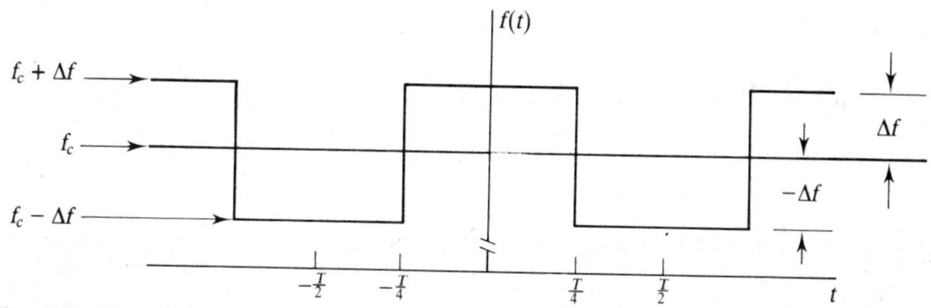

Fig. 6-9 Square-wave frequency variation produced by the frequency modulator of Example 6-6.

conditions that $\theta(0) = 0$ and that this phase function is continuous. Thus,

$$\theta(t) = \begin{cases} -\dfrac{\Delta\omega}{\omega_m}(\pi + \omega_m t), & -\dfrac{T}{2} \leqslant t \leqslant -\dfrac{T}{4} \\ \Delta\omega t, & -\dfrac{T}{4} \leqslant t \leqslant \dfrac{T}{4} \\ \dfrac{\Delta\omega}{\omega_m}(\pi - \omega_m t), & \dfrac{T}{4} \leqslant t \leqslant \dfrac{T}{2} \end{cases} \tag{6-38}$$

Consequently, the coefficients are found from

$$c_n = \frac{1}{T}\left\{\int_{-T/2}^{-T/4} \exp\left[-j\frac{\Delta\omega}{\omega_m}(\pi + \omega_m t - jn\omega_m t)\right] dt\right.$$

$$+ \int_{-T/4}^{T/4} \exp\left[j\Delta\omega t - jn\omega_m t\right] dt$$

$$\left. + \int_{T/4}^{T/2} \exp\left[j\frac{\Delta\omega}{\omega_m}(\pi - \omega_m t - jn\omega_m t\right] dt\right\} \tag{6-39}$$

which becomes, for $n \neq \pm\beta$,

$$c_n = \frac{2\beta \sin{(\beta - n)}\dfrac{\pi}{2}}{\pi(\beta^2 - n^2)} \tag{6-40}$$

where $\beta = \Delta\omega/\omega_m = \Delta f/f_m =$ a positive number, sometimes an integer. Equation (6-40) can be expressed as

$$c_n = \frac{\beta}{\beta + n} \operatorname{sinc}\left(\frac{\beta - n}{2}\right) \tag{6-41}$$

When $n = \beta$, (6-40) or (6-41) reduces to

$$c_{n=\beta} = \tfrac{1}{2} \tag{6-42}$$

To calculate c_n for $n = -\beta$ when β is a positive integer, return to (6-40), set $n = -\beta$, integrate, and simplify to obtain

$$c_{n=-\beta} = \tfrac{1}{2}(\cos \beta\pi - \operatorname{sinc} \beta)$$

But here sinc $\beta = 0$ and cos $\beta\pi = (-1)^\beta$, so that

$$c_{n=-\beta} = \frac{(-1)^\beta}{2}, \qquad \text{for } \beta = \text{a positive integer} \tag{6-43}$$

Finally, using (6-41) in (6-30),

$$x_c(t) = A_c \sum_{n=-\infty}^{\infty} \frac{\beta}{\beta + n} \operatorname{sinc}\left(\frac{\beta - n}{2}\right) \cos 2\pi(f_c + nf_m)t \tag{6-44}$$

If β is a small value, then sinc$[(\beta - n)/2]$ has its largest values for small values of n, and the side frequencies are clustered near f_c. Calculations show that when $\beta = 1$, then eight pairs of side frequencies are significant. If $\beta = 20$, then sinc$[(\beta - n)/2]$ has its largest values for large values of n, and the side frequencies are clustered near $f_c \pm \Delta f$. Calculations show that when $\beta = 20$, then 41 pairs of side frequencies are significant.

6-7 PHASE DETECTORS

Most circuits that are suitable as multipliers, balanced modulators, product detectors, or mixers, are also useful, under certain conditions, as phase detectors (also known as phase comparators and as phase discriminators). For example, the two-diode switching-mode mixer of Fig. 5-7c and the doubly balanced mixer of Fig. 5-11 will perform as phase detectors.

A classic phase detector is represented in Fig. 6-10. The carrier frequency f_c of the incoming signal voltage V_i whose phase is varying and the reference frequency of the locally generated voltage V_r should be equal. Because of the center-tapped secondary, the total signal voltages applied to diodes D1 and D2 are $V_{D1} = V_r + V_2$ and $V_{D2} = V_r - V_2$, respectively. Both diode

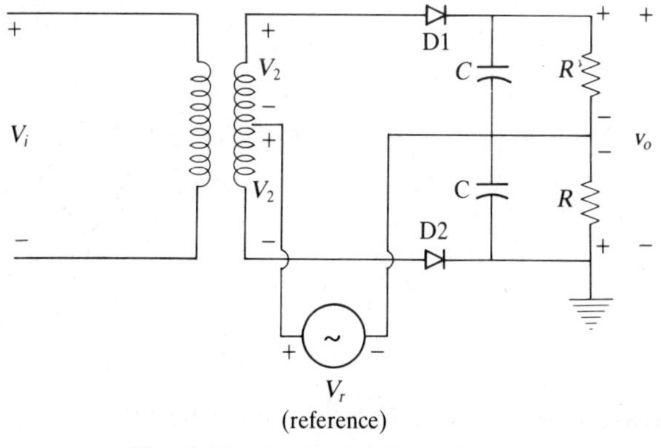

Fig. 6-10 A classic phase detector.

circuits are peak detectors and produce DC voltages across their respective RC loads with opposing polarities as shown in Fig. 6-10. Each RC time constant should be long compared to the period of the frequency f_c and short compared to the period of the lowest modulating frequency f_m. Assume that V_i and V_2 are in phase.

When V_2 is in phase with V_r as in Fig. 6-11a, then $V_{D1} = V_r + V_2$ will be maximum and $V_{D2} = V_r - V_2$ will be minimum. The DC output voltage from D1 will be greater than that from D2. Thus, v_o is the greatest with the polarity shown in Fig. 6-10.

If the phase of V_2 advances or retards from the in-phase conditions, as in Fig. 6-11b, c, V_{D1} decreases and V_{D2} increases for both conditions, but v_o does not change polarity. This is not the desired operating condition for a phase detector.

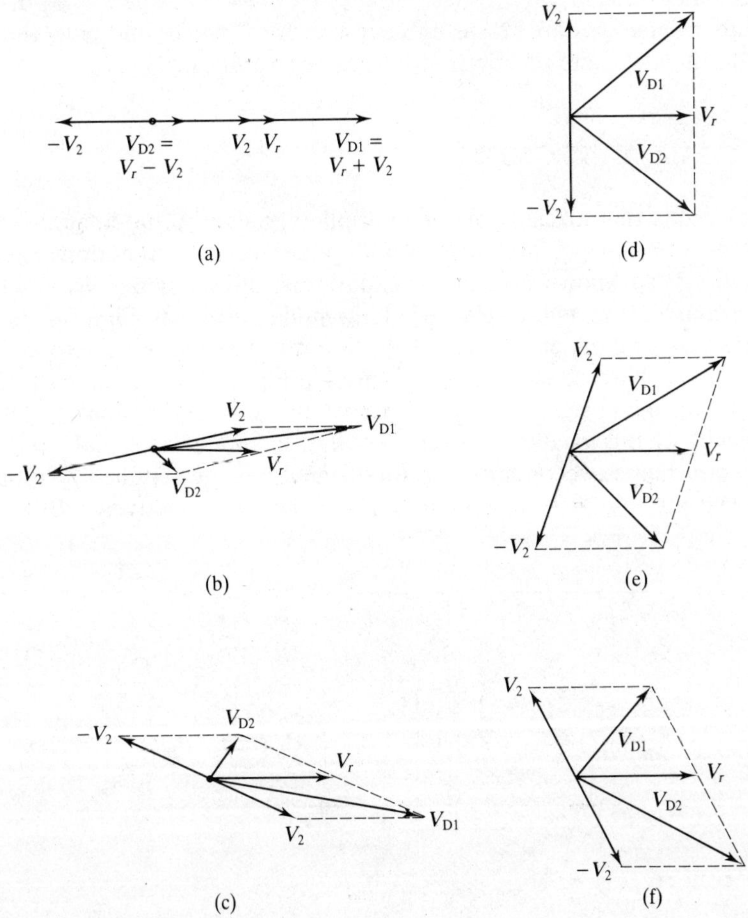

Fig. 6-11 Phase relationships for the classic phase detector: (a), (b), (c), undesired; (d), (e), (f), desired.

Now consider Fig. 6-11d. The reference voltage V_e has been adjusted to be 90° out of phase with V_2. Therefore, $V_{D1} = V_{D2}$ and $v_o = 0$. If now V_2 is retarded in phase from this zero-output condition as in Fig. 6-11e, then $V_{D1} > V_{D2}$ and $v_o > 0$. However, if V_2 is advanced in phase from the zero-output condition as in Fig. 6-11f, then $V_{D2} > V_{D1}$ and $v_o < 0$. This, then, is the desired operating condition for the phase detector: the amplitude of v_o depends on the amount of phase difference between V_i and V_r relative to the zero-output phase, and the polarity of v_o depends on whether the phase of V_i leads or lags that of V_r around the zero-output phase condition.

The action can be analyzed mathematically by using

$$v_2 = V_2 \sin\left(\omega t + \frac{\phi}{2}\right) \tag{6-45a}$$

$$v_r = V_r \cos\left(\omega t - \frac{\phi}{2}\right) \tag{6-45b}$$

as the two sinusoids that are originally 90° out of phase. The angle ϕ is the total additional phase angle between v_2 and v_r. A positive value for ϕ means that v_2 lags v_r, and the phase angle has been divided equally between the two sinusoids for simplicity. Let $V_2 = V_r$ and define a detection coefficient to be

$$\eta = \frac{\text{DC output voltage}}{\text{peak AC input voltage}}$$

Then,

$$v_o = \eta(|v_r + v_2| - |v_r - v_2|) \tag{6-46}$$

Now, substituting (6-45) into (6-46) and simplifying, using familiar trigonometric identities, results in

$$v_o = 2\eta V_r \sin\frac{\phi}{2} \tag{6-47}$$

If ϕ is small, so that v_2 and v_r are almost 90° out of phase, then (6-47) becomes

$$v_o \cong \eta V_r \phi \tag{6-48}$$

A typical input–output characteristic for the classic phase detector is shown in Fig. 6-12.

The phase detector is often represented by the blocks shown in Fig. 6-13.

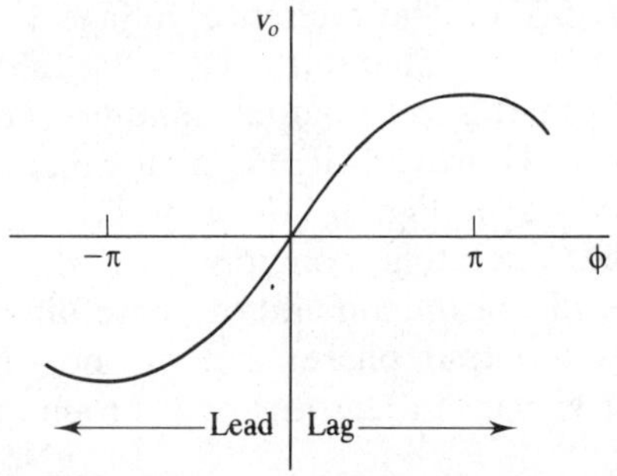

Fig. 6-12 Phase detector input–output characteristic.

The input sinusoids have the same frequency, otherwise the phase between them is meaningless, and they are 90° out of phase when $\phi(t) = 0$. The output from the multiplier is

$$d(t) = 2A_cA_r \sin[\omega_c t + \phi(t)] \cos \omega_c t = A_cA_r \sin \phi(t) + A_cA_r \sin[2\omega_c t + \phi(t)] \tag{6-49}$$

The lowpass filter eliminates the second-harmonic term. (The RC circuits in Fig. 6-10 comprise the filter in that classic phase detector.) Therefore, the phase detector output is

$$y(t) = A_cA_r \sin \phi(t) \tag{6-50}$$

which, for small angles becomes,

$$y(t) = A_cA_r\phi(t) \tag{6-51}$$

Notice that the amplitudes of the outputs from both the actual phase detector and the one modeled with a multiplier depend on the amplitudes of the input signals. This means that the phase-detector output will be corrupted when the amplitude of either one of the input signals changes unless such amplitude changes are eliminated.

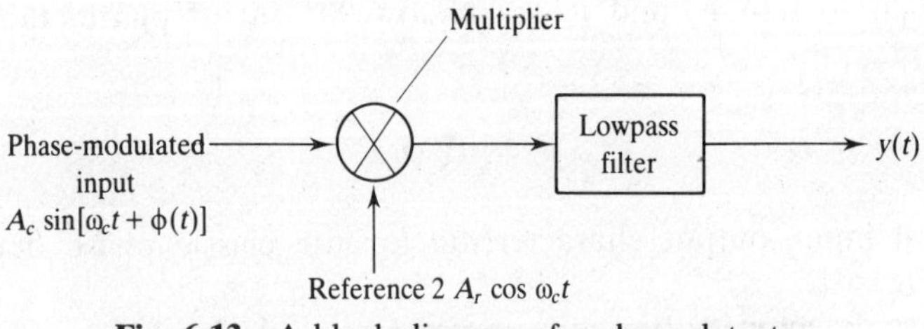

Fig. 6-13 A block diagram of a phase detector.

6-8 FREQUENCY DETECTORS

The circuit that demodulates an FM signal should provide an output voltage whose amplitude varies linearly with the instantaneous frequency deviation, whose polarity is determined by whether the deviation is above or below the unmodulated carrier frequency, and whose frequency is the same as the rate at which the frequency of the FM signal is modulated. Such demodulators are known as frequency detectors and discriminators. Any circuit or device whose transfer function is frequency-dependent becomes a candidate for a discriminator.

The classic Foster–Seeley discriminator is shown in Fig. 6-14. Its basic operation is similar to that of the classic phase detector. However, the reference frequency must automatically track the changing frequency of the input FM signal. This is accomplished by using RF bypass capacitors in addition to the capacitors of the two RC loads. This allows the input signal voltage V_i to also appear across RFC (radiofrequency choke—an inductor whose reactance is relatively high at the frequencies involved.) Therefore, the reference voltage is $V_r = V_i$. The primary and secondary circuits are both tuned to the unmodulated frequency f_c.

The phasor diagrams in Fig. 6-15 help understand how the circuit operates. In Fig. 6-15a, the incoming frequency is the same as the series resonant frequency of the secondary. Voltage V_i causes current I_{L1} in the primary inductance that lags V_i by essentially 90°. This current produces an in-phase magnetic field ψ_1. A voltage E_2 is induced into the entire secondary whenever this magnetic field changes. Half of the total induced voltage appears above the center tap and half below. The induced voltage depends on $d\psi_1(t)/dt$, so E_2 leads ψ_1 by 90°. Since the secondary is assumed to be series resonant at f_c so far as E_2 is concerned, the inductive and capacitive reactances cancel and current I_2 circulates through that series resonant circuit in phase with E_2. This produces a voltage $2V_2$ across the entire secondary inductance that leads I_2 by essentially 90°. With respect to

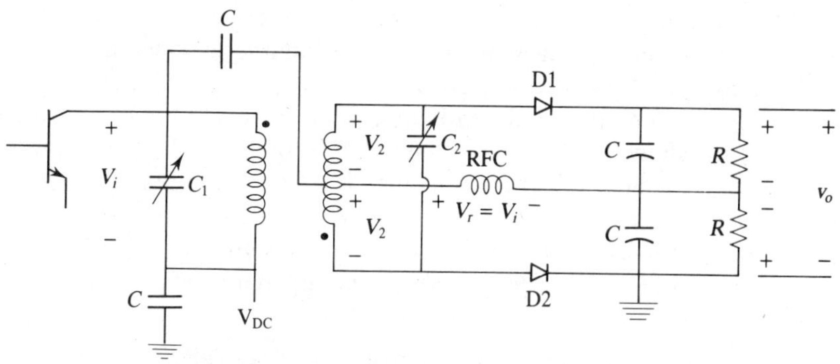

Fig. 6-14 The classic Foster–Seeley discriminator.

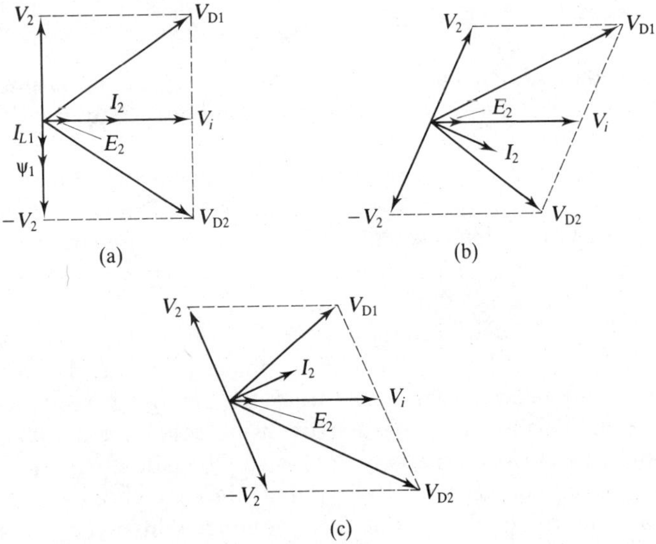

Fig. 6-15 Phase relationships for the Foster–Seeley discriminator: (a) $f = f_c$; (b) $f > f_c$; (c) $f < f_c$.

the center tap, the voltage of self-induction on one side of the center tap is V_2 and on the other side is $-V_2$. Similarly, E_2 becomes $E_2/2$ and $-E_2/2$ with respect to the center tap.

Then the total voltage applied to diode D1 is $V_{D1} = V_i + E_2/2 + V_2$ and the voltage applied to diode D2 is $V_{D2} + V_i - E_2/2 - V_2$. But the magnitude of V_2 is usually much larger than that of $E_2/2$ because of the resonant rise in voltage across the inductor of a series resonant circuit as determined by the circuit Q. Consequently, in Fig. 6-15a, $V_{D1} = V_i + V_2$ and $V_{D2} = V_i - V_2$, so that $|V_{D1}| = |V_{D2}|$ and $v_o = 0$.

When the incoming frequency is higher than f_c, the inductive reactance in the secondary series circuit is greater than the capacitive reactance, so I_2 lags E_2 as shown in Fig. 6-15b. This makes $|V_{D1}| > |V_{D2}|$ so that $v_o > 0$.

When the incoming frequency is below f_c, the capacitive reactance is greater than the inductive reactance of the secondary, so I_2 leads E_2 as in Fig. 6-15c. This makes $|V_{D1}| < |V_{D2}|$ so that $v_o < 0$.

The action can be analyzed mathematically by examining the series secondary circuit shown in Fig. 6-16. There, E_2 represents the total voltage induced into the secondary whose inductance is L, C is the secondary tuning capacitance, and R represents the loading effects of the diodes and their RC circuits. The secondary is series resonant at f_c, therefore $\omega_c^2 LC = 1$. Also, the circuit $Q = \omega_c L/R = 1/(\omega_c RC)$. The series impedance is

$$Z = R + j\left(\omega L - \frac{1}{\omega C}\right) = R\left[1 + jQ\left(\frac{f}{f_c} - \frac{f_c}{f}\right)\right]$$

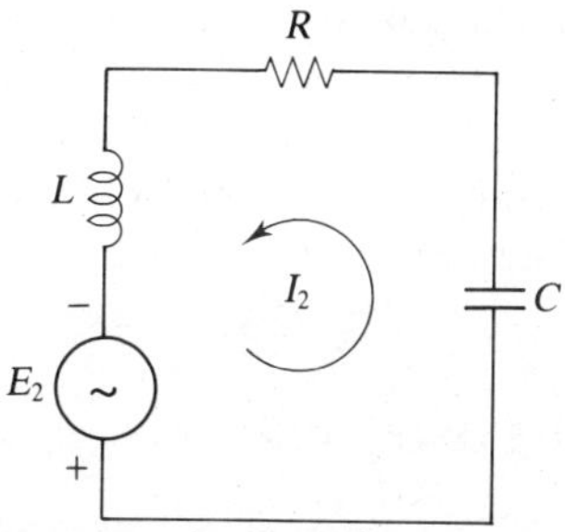

Fig. 6-16 Simplified secondary circuit of the Foster–Seeley discriminator.

Now let $f = f_c + \delta f$ with $\delta f \ll f_c$. Then

$$Z = R\left(1 + j\frac{2Q\delta f}{f_c}\right) \tag{6-52}$$

The current I_2 lags the induced voltage E_2 by

$$\phi = \tan^{-1}\frac{2Q\delta f}{f_c} \tag{6-53}$$

Further, if $2Q\delta f/f_c < 0.1$, then

$$\phi = \frac{2Q\delta f}{f_c} \tag{6-54}$$

It is seen from the phasor diagrams of Fig. 6-15 that the angle by which I_2 lags or leads E_2 is the same as the angle ϕ by which V_2 lags or leads V_2 for the zero-output condition. Also, from (6-54), it is possible for the phase difference to be proportional to the instantaneous frequency deviation.

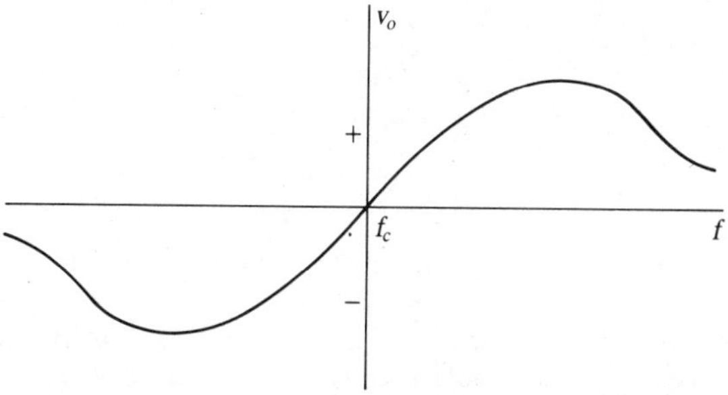

Fig. 6-17 Typical Foster–Seeley discriminator characteristic.

And, from (6-48), the output from the diode phase detector can be proportional to the phase difference. Therefore, it is possible for the output voltage v_o from the discriminator circuit of Fig. 6-14 to be proportional to the instantaneous frequency deviation, as desired. Of course, if the frequency deviation is increased too far, the linearity is destroyed and the output voltage begins to level off and to decrease as shown in Fig. 6-17.

6-9 PHASE-LOCKED LOOP (PLL)

Some of the most important applications of electronics in communications systems are:

(a) to automatically track the carrier from the transmitter of an orbiting satellite or other space vehicle when Doppler shift causes the frequency to change;

(b) to provide synchronization in television and radar systems;

(c) to synthesize signals having controlled frequencies that can be selected either manually or electrically and that have the frequency stability of a quartz crystal oscillator;

(d) to demodulate signals whose frequency is modulated, as in, for example, FM broadcasting, telemetry, frequency-shift keying of radioteletype, and weather-satellite transmissions.

All of these applications are satisfied by using a phase-locked loop† (PLL) as part of the electronics system. The concept of the PLL is so useful that the system has been fabricated into integrated-circuit forms by several manufacturers.

6-9.1 The Basic PLL Configuration

The basic PLL consists of a phase detector, a lowpass filter, and a voltage-controlled oscillator (VCO), as shown in Fig. 6-18. The voltage v_d depends on the phase difference between the input voltage v_i and the

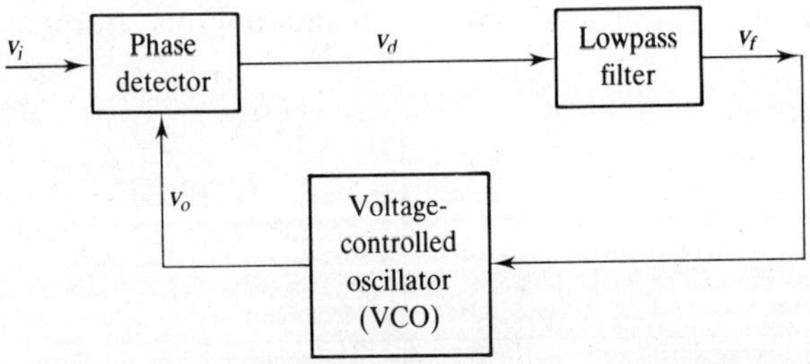

Fig. 6-18 The basic phase-locked loop (PLL).

† Also variously referred to as phase lock loop, phase locked loop, phaselock loop, phase-lock loop, and phase-locked-loop. Can there be any wonder why PLL is most generally used instead?

oscillator output voltage v_o. If v_i and v_o have different frequencies f_i and f_o, respectively, then their phase difference is continually increasing, and this causes the output voltage v_d from the phase detector to fluctuate at the difference or beat frequency $|f_i - f_o|$. Initially, the loop is not in lock, but after v_d is filtered, the filter output voltage v_f causes the frequency f_o of the VCO to move toward f_i and decrease the beat frequency, provided that f_i is not too far removed from f_o. The loop is said to be locked when $f_o = f_i$, and the phase error between v_i and v_o is such that there is enough voltage v_f to force f_o to be the same as f_i. The maximum allowable phase error depends on the type of phase detector used. The output from the classic phase detector of Fig. 6-10 is given by (6-47) and shown in Fig. 6-12. Its maximum allowable phase error, before the output repeats, is $\pm 180°$. The operating phase error can be decreased by including some amplification within the loop.

6-9.2 The PLL as an FM Demodulator

If the frequency f_i of the incoming modulated signal continually changes, as in FM, then the voltage v_f also changes by a proportional amount to force the loop to remain locked. Consequently, the demodulated FM signal is represented by the voltage v_f.

6-9.3 The PLL as a Frequency Synthesizer

Consider the system shown in Fig. 6-19. The output of a crystal oscillator at the frequency f_{in} enters a programmable counter of count M, which, in turn, provides one input to a phase detector at the reference frequency $f_r = f_{in}/M$. The VCO frequency f_0 is divided by the count N of another programmable counter. The other input to the phase detector is at frequency f_0/N. The loop is locked when the frequencies of the two signals applied to the phase detector are equal. Thus, $f_0/N = f_{in}/M$, from which is obtained the output frequency

$$f_0 = f_{in} \frac{N}{M} \tag{6-55}$$

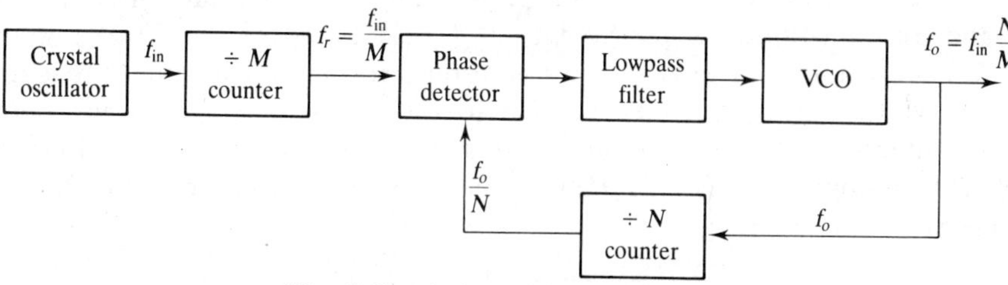

Fig. 6-19 A frequency synthesizer.

Suppose $f_{in} = 100$ kHz and you want the VCO to generate frequencies from 1000 kHz to 2000 kHz in 5 kHz steps for a total of 200 steps. Perhaps these are to be the local oscillator frequencies for a wideband receiver that is to tune over channels that are 5 kHz wide. If $M = 20$, then the reference frequency f_r for the phase detector is $f_{in}/M = 5$ kHz. Then, from (6-55),

$$1000 = \frac{100N_1}{20}, \qquad \text{or } N_1 = 200$$

and

$$2000 = \frac{100N_2}{20}, \qquad \text{or } N_2 = 400$$

As the N counter increments by 1, the VCO frequency increments by 5 kHz. When $N = 302$, $f_0 = 1510$ kHz, and for $N = 303$, $f_0 = 1515$ kHz. The incrementing can be made automatic, with the process repeating over and over and stopping only when a signal of sufficient strength enters the receiver. This is the basic principle involved in a system called a scanner.

Further applications of the PLL in radio receiving systems are described in Chapter 8.

6-9.4 A Microwave Frequency Synthesizer

Programmable counters have a maximum speed at which they can operate satisfactorily. This becomes a disadvantage when they are to be used to help generate stable frequencies above about 30 MHz. This is where a fixed counter with a fairly small count can be used effectively because it can operate at such high frequencies. Called a prescaler, it scales (or divides) the output frequency down to a range that the programmable counter can handle.

Consider Fig. 6-20. The two-modulus prescaler counts by $P + 1$ initially, and when it receives a control pulse, changes its count to P. Typically, P is a small integer such as 3 or 8 or 20. When counting begins, the counters are reset to zero. Initially, the prescaler divides the VCO output frequency by $P + 1$ and then both programmable counters increment. When the count reaches M, the M counter's output pulse changes the prescaler modulus to P and the M counter stops. So far, $M(P + 1)$ cycles of the VCO output have been counted. The N counter continues to increment until it reaches N, at which time it delivers a pulse to the phase detector, resets both programmable counters, and the counting cycle repeats. During the last part of the counting cycle, $(N - M)P$ cycles of the VCO output are counted. The total count is

$$M(P + 1) + (N - M)P = M + NP$$

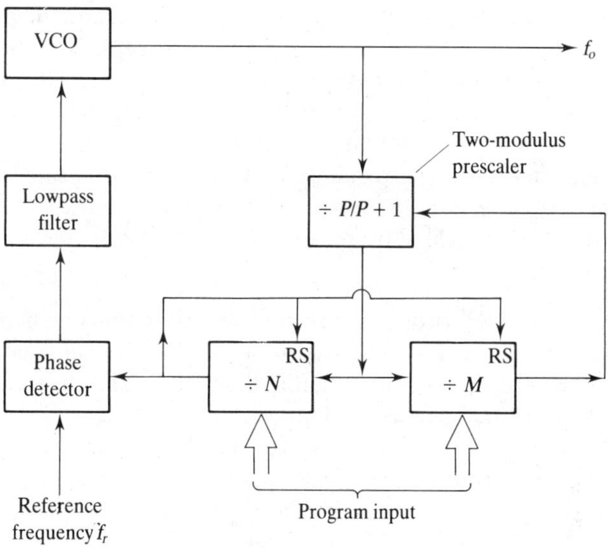

Fig. 6-20 A microwave frequency synthesizer.

Consequently, assuming that the loop is locked,

$$f_o = (M + NP)f_r \tag{6-56}$$

The output frequency can be made to hop about or slide around as determined by the way the counters are programmed to count.

PROBLEMS

6-1 The sinusoidal output from a 1-MHz crystal oscillator is phase-modulated. The modulating voltage is $m(t) = 1 \sin 500\,t + 2 \sin 2500\,t$ V. The phase deviation constant of the phase modulator is $k_p = 0.1$ radian per volt. Calculate (a) the individual maximum phase deviations in degrees produced by the phase modulator; (b) the total maximum phase deviation in degrees produced by both modulating components together; (c) the individual maximum frequency deviations in hertz produced by the components of $m(t)$; the (d) total maximum frequency deviation produced by both modulating components together.

6-2 Using the phase relationships shown in Fig. 6-2, calculate $\Delta\theta$ and the percent increase in i_L when i_{C1} decreases by 10 percent and i_{C2} increases by 10 percent from their unmodulated values.

6-3 In Fig. 6-3, calculate the peak amplitude of $m(t)$ to restrict $\Delta\theta$ to 0.1 radian.

6-4 The sinusoidal output of a 91.3 MHz source is frequency modulated. The modulating voltage is $3 \cos 500t + \cos 2500\,t$ V. The frequency deviation constant of the frequency modulator is 20 kHz per volt. Calculate (a) the individual maximum frequency deviations produced by the frequency modulator; (b) the total maximum frequency deviation in kilohertz produced by both modulating components together; (c) the individual maximum phase deviations; (d) the total maximum phase deviations produced by both modulating components together.

6-5 The SPECTRUM program for angle modulation given in Appendix E calculates, for a given β, the number N of Bessel functions whose amplitudes are equal to or greater than 0.01. The empirical relationship used for N is the integer part of $(3.24\beta + 1.1)^{0.76}$. For $\beta =$ 1, 5, 10, 15, and 20: (a) calculate N; (b) using a table of Bessel functions, determine how many pairs have amplitudes equal to or greater than 0.01; (c) compare the results of (a) and (b).

6-6 Verify several of the calculated values listed in Table 6-1 for Example 6-3.

6-7 (a) An FM transmitter operates on 89.7 MHz. At a certain instant, $\Delta f = 50$ kHz when the single sinusoidal modulating voltage has a frequency of 5 kHz and an rms amplitude of 0.5 V. (i) Use Eq. (6-26) and a table of Bessel functions to determine the number of pairs of side frequencies for a power ratio $\geqslant 0.6$. (ii) Calculate the minimum bandwidth required for this power ratio.

(b) The rms amplitude of the modulating voltage in part (a) is now decreased to 0.25 V and the frequency increased to 12.5 kHz. (i) Determine the number of pairs of side frequencies for a power ratio $\geqslant 0.8$. (ii) Calculate the minimum bandwidth required for this power ratio.

6-8 (a) A PM transmitter operates on 92.1 MHz. At a certain instant, $\Delta\theta = 2$ radians when the single sinusoidal modulating voltage has a frequency of 5 kHz and an rms amplitude of 0.2 V. (i) Use Eq. (6-26) and a table of Bessel functions to determine the number of pairs of side frequencies for a power ratio $\geqslant 0.7$. (ii) Calculate the minimum bandwidth required for this power ratio.

(b) The rms amplitude of the modulating voltage in part (a) is now increased to 0.8 V and the frequency reduced to 3 kHz. (i) Determine the number of pairs of side frequencies for a power ratio $\geqslant 0.98$, (ii) Calculate the minimum bandwidth required for this power ratio.

6-9 A sinusoidal modulating signal has a frequency f_m and period $T = 1/f_m$. It is applied to a frequency modulator to cause the instantaneous frequency to be $f = f_c + \Delta f \cos 2\pi f_m t$ Hz. Show that the coefficients C_n in Eq. (6-31) are equal to $J_n(\beta)$. Hint: Integral forms of $J_n(\beta)$ are:

$$J_n(\beta) = \frac{1}{\pi}\int_0^{\pi} \cos(\beta \sin x)\cos nx\, dx \qquad n = 0, 2, 4, \ldots$$

$$J_n(\beta) = \frac{1}{\pi}\int_0^{\pi} \sin(\beta \sin x)\sin nx\, dx \qquad n = 1, 3, 5, \ldots$$

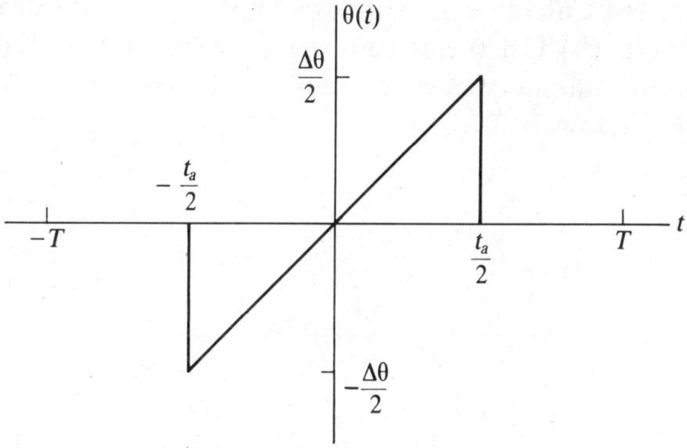

Fig. P6-10 Waveform for Problem 6-10.

6-10 The periodic modulating signal applied to a phase modulator

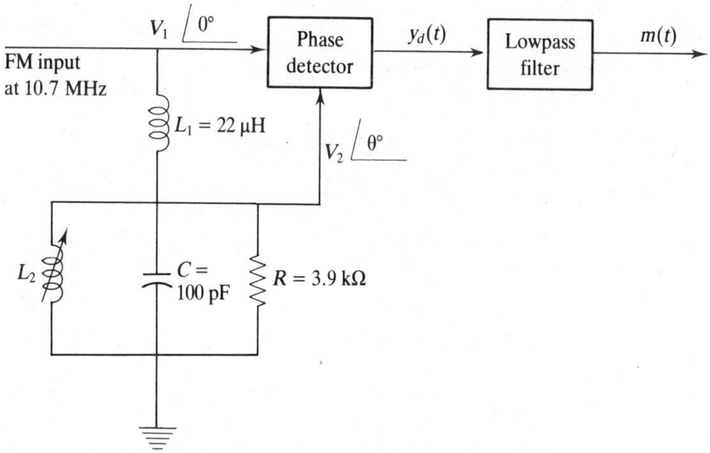

Fig. P6-12 FM demodulator for Problem 6-12.

produces the phase variation shown in Fig. P6-10. Show that

$$C_n = \frac{t_s}{T} \operatorname{sinc}\left(\frac{\Delta\theta\, T}{2\pi t_a} - n\right)\frac{t_a}{T}$$

6-11 In Example 6-6, verify that (a) when $\beta = 1$, then eight pairs of side frequencies are significant, (b) when $\beta = 20$, then 41 pairs of side frequencies are significant.

6-12 The block diagram of an FM demodulator is shown in Fig. P6-12. The phase detector is of such a design that its output is zero when $\theta = -90°$. (a) Calculate L_2 such that $y_d(t) = 0$ for an input frequency of 10.7 MHz. (b) Using this value of L_2, plot angle θ in degrees when the input frequency varies in steps of 10 kHz for a total of 80 kHz above and below 10.7 MHz.

7

NOISE II

It is desirable to develop methods for characterizing the noisiness of a complete operating system. Further, those methods should not require detailed mathematical analysis of the noise contribution of each individual component in the system, as briefly outlined in Chapter 3. Instead, the primary concern now is the total output noise present in the output of the noisy system as compared to what an ideal noiseless system would have and to what the level of the desired output signal is.

The analysis of the output noise produced by each internal component in a complete system, such as a radio receiver, is generally too complicated to undertake without requiring various simplifying assumptions. Those assumptions, in turn, lead to inaccuracies in calculations for total output noise. Consequently, noise measurements must be relied on, and those measurements will account for such things as system nonlinearities, unknown correlation coefficients, component modeling deficiencies, variations in distinct bandwidths perceived by each noisy component, and the effects of any incompletely understood physical process that generates noise.

7-1 NOISE FACTOR F

The extent to which an electronic system (having a single input and a single output) contributes noise can be described by a *noise factor* F that compares the output noise power of the actual noisy system to the output noise power that would be delivered by the system if the *only* source of noise were the thermal noise in the source resistance at the standard temperature of 290 K.

Thus, for a specified frequency,

$$F = \frac{\text{output noise power of the actual noisy system}}{\text{output noise power caused by thermal noise in the source resistance at 290 K}} \tag{7-1}$$

The noise factor F of an ideal noiseless system is unity because the system does not add any noise of its own. Practical systems always have $F > 1$ because they are always noisy to some extent. For the transistor amplifier of Fig. 3-9c, the numerator of (7-1) is found from Eq. (3-47) to be V_o^2/R_L and the denominator of (7-1) is found from Eq. (3-44a) to be V_{o1}^2/R_L, where $V_{o1} = W_{o1}B$. Thus (7-1) becomes

$$F = \frac{V_o^2}{V_{o1}^2} \tag{7-2}$$

Because of the ratio, it is seen that F is independent of R_L. Substituting Eqs. (3-47) and (3-44a) into (7-2) and simplifying yields

$$F = \left[\frac{(1 + g_m r_\pi)}{2g_m r_\pi^2}\right] R_s + \left[\frac{(r_b + r_\pi)^2 + r_b(1 + 2r_\pi)g_m r_\pi}{2g_m r_\pi^2}\right]\frac{1}{R_s} + \left[\frac{(r_b + r_\pi)(1 + g_m r_\pi)}{g_m r_\pi^2}\right] \tag{7-3}$$

Equation (7-3) has the form

$$F = AR_s + \frac{B}{R_s} + C \tag{7-4}$$

where A, B, and C are obvious from (7-3). Clearly, F depends on the value of R_s as well as on the various circuit constants. When a noise factor is given for an amplifier, the operating point, frequency, and temperature, as well as R_s, should all be specified. It is seen that $R_s = 0$ does not minimize F as it did for the output noise in Chapter 3. The optimum value of R_s for minimum F is found from $dF/dR_s = 0$ to be the value that makes the term AR_s equal to the term B/R_s. Hence,

$$R_{s,\text{opt}} = \sqrt{\left(\frac{B}{A}\right)} \tag{7-5}$$

Therefore,

$$F_{\min} = 2AR_{s,\text{opt}} + C \tag{7-6}$$

Using the previous numerical values for the transistor amplifier of Fig. 3-9c, $R_{s,\text{opt}} = 755\ \Omega$ and $F_{\min} = 1.36$. ($A = 2.02 \times 10^{-4}$, $B = 113.54$, $C = 1.05$).

Equation (7-2) gives a clue about how to measure F. Since F is independent of R_L because of the ratio, any R_L is satisfactory so long as the apparatus used to measure noise voltages is sensitive enough. Voltage V_{o1} is not measured directly because it is an inseparable part of the total noise voltage V_o, but it is calculated from $V_{o1} = |A_v|\, V_n$, where A_v is the overall voltage gain from the signal source voltage E_s to the output signal voltage across R_L. Then

$$F = \frac{V_o^2}{4kTR_sB\,|A_v|^2} \tag{7-7}$$

A disadvantage of this method is that the noise bandwidth B is needed. The circuit of Fig. 3-9c is resistive and there is nothing to limit the bandwidth. The bandwidth-limiting property must be provided by the instrument used to measure V_o and that is the bandwidth to be used for B in (7-7). Of course, other electronic systems have their own selectivity characteristic that determines the noise bandwidth B. Signal generators are calibrated either in terms of the terminal voltage across a matched load resistance ("soft" microvolts) or the source voltage E_s ("hard" microvolts). It is necessary to know how the calibration is made for a particular signal generator used to measure voltage gain A_v, because E_s is twice the terminal voltage across a matched load resistance.

A distinction is made between the "spot noise factor," which is measured for a narrow band centered at a specified frequency, and the "average noise factor," which is measured over the entire frequency range of the system. The spot noise factor is the only one we will be concerned with. After measurements are made, consider dividing both numerator and denominator of (7-7) by the narrow bandwidth B to obtain

$$F = \frac{V_o^2/B}{4kTR_s\,|A_v|^2} \tag{7-8}$$

Equation (7-8) represents the spot noise factor at a specified frequency over a 1-Hz bandwidth. If the spot noise factor is constant throughout the frequency range of interest for a certain system, the average and spot noise factors are equal.

7-1.1 Available Power Gain

In small-signal high-frequency amplifiers, lossless impedance-matching networks are usually included between the amplifier and its ultimate input and output terminations. The matching-network elements are chosen, or tuning adjustments are made, so as to maximize the output signal power. If the amplifier is inherently stable, a conjugate match is assured separately for both the input and output circuits when maximum output signal power is obtained. When a load is conjugately matched to a source, the power delivered to that load is called the *available power* from the source. When an amplifier is conjugately matched simultaneously at both its input and output ports, the power gain is called the *available power gain* G.

The noise power N_1 delivered by the rms thermal-noise voltage V_n of the internal resistance R_s of a source to a matched load $R_L = R_s$ produced by a lossless matching network having narrow noise bandwidth B at a specified frequency is

$$N_i = \left(\frac{V_n}{2}\right)^2 \frac{1}{R_L} = \frac{4kTR_sB}{4R_s} = kTB \tag{7-9}$$

The important and startling aspect of this result is that the thermal-noise power is constant at kTB watts regardless of the actual value of R_s provided only that there is a conjugate match between the source and its load. Thus $N_i = kTB$ watts is the available thermal-noise power from *any* source having positive internal resistance of any value whatever. When this N_i is multiplied by the available power gain G and the temperature is 290 K, the denominator of (7-1) is obtained. Hence, when the noise factor F is determined for the simultaneous conjugate matched conditions at both ports, it is not necessary to also specify R_s or R_L (or Z_s or Z_L perhaps). The reason for this remarkable conclusion is that an inherently stable system possesses unique input and output terminations that provide for maximum output power without sustained oscillation. Then, regardless of Z_s and Z_L, the matching-network adjustments provide the required conjugate matching for the active device as observed simply by the maximum output power.

The standard temperature of the input termination is taken to be $T_0 = 290\text{ K}$ ($= 17°\text{C} = 63°\text{F}$) so that the input available noise power is $N_i = k(290)B = 4 \times 10^{-21}B$ watts.

7-2 NOISE FIGURE *NF*

Officially, the terms "noise factor" and "noise figure" are synonymous and they are often casually used interchangeably, the choice being made by individual preference. Here, however, yielding to popular common usage during the past decades of intensive noisemanship by many investigators, we adopt the term *noise figure, NF,* for the result obtained by expressing noise factor F in decibels. Thus,

$$NF = 10 \log F \quad \text{dB} \tag{7-10}$$

This use of noise factor and noise figure is justified on the basis of expediency in much the same way that popular jargon, established by pundits and lingoists of various ilks and generally deplored by purists, refers to such things as:

- Current flow (actually current exists when charge flows)
- AC voltage, DC current, RF frequency (the first letters of certain well-known words are used together as adjectives)
- Cycles (for frequency; this malapropism finally led to the hertz unit)
- K (for kΩ, k bucks, 1024 bits, etc.)
- Drop (when used for the voltage across something; "rise" never caught on)
- LCD display (the D in LCD means display)
- The spoken use of the letter *oh* instead of the number zero, as in: *the time is* 1 "*oh*" 5 (meaning 5 minutes after 1); again—*the meeting is in room* 2 "*oh*" 3 (meaning 2 hundred and 3) or *the football game score was* 7 *to* "*oh*" meaning 7 to zero).

and other useful and informative, often redundent, colloquialisms, some of doubtful validity. Their pervasiveness, now firmly entrenched, contributes to the degradation of spoken and written language. Further, this rather imprecise manner of contemporary thinking often becomes a malevolent enticement that cramps learning and understanding because somehow the lingo must be assimilated before the essence. This leads to interpretation noise, comprehension noise, and application noise. This antiorthopraxy can be succinctly summarized with the cogent observation: "Us engineers don't need to know no English nohow."

7-3 SIGNAL-TO-NOISE RATIOS

Electronic systems, such as measuring instruments, radio receivers, and signal processors, are designed to modify input signals (accompanied by the inevitable noise) in some way as they are coupled from stage to stage, amplified, filtered, sampled, mixed, etc., and thereby make their way in an orderly manner from the input sides of the systems to their final output destinations. Each subsystem, device, and component performs its assigned operation on the signal and, unfortunately, also contributes additional noise that was not present with the original signal and its accompanying noise.

The extent to which noise accompanies the signal can be described by a signal-to-noise ratio, S/N or SNR. This ratio may be a voltage, current, or power ratio or its decibel equivalent. Signal levels may be measured and expressed as rms or peak voltage or current, or as average power, or as decibles relative to a reference. Noise levels are readily expressed as mean-square values of voltage or current or as decibels relative to a reference.

As used here, the signal-to-noise ratio SNR at some location in a system is defined to be the ratio of (1) the average signal power S delivered to that location from the input side only, to (2) the average noise power N delivered to that same location from the input side only. Thus SNR $= S/N$. Sometimes the decibel equivalent of SNR will be used and expressed as $\text{SNR}_{\text{dB}} = 10 \log \text{SNR}$.

The standard noise factor defined by (7-1) can be expressed as

$$F = \frac{N_o}{N_i G} \tag{7-11}$$

where N_o = output available noise power of the actual noisy system
N_i = available thermal-noise power from the source ($= kT_0B$ with $T_0 = 290$ K)
G = available signal-power gain

Let S_o = available output signal power and S_i = available input signal power. Then (7-11) becomes

$$F = \frac{S_i/N_i}{S_o/N_o} = \frac{\text{SNR}_i}{\text{SNR}_o} \tag{7-12}$$

It is tacitly assumed that the noise bandwidths associated with the two signal-to-noise ratios are the same, otherwise values of F apparently less than unity can be obtained.

Figure 7-1 depicts a sinusoidal signal source (e.g., a CW signal generator) connected to an electronic system whose input resistance is R_{in}. The output resistance R_s of the signal source has its own unavoidable thermal noise represented by the rms voltage source V_n. The signal power S_i delivered to R_{in} from the input side is

$$S_i = \left(\frac{E_s}{R_s + R_{\text{in}}}\right)^2 R_{\text{in}} \tag{7-13}$$

where E_s is the rms signal (or carrier) voltage. The thermal-noise power N_i

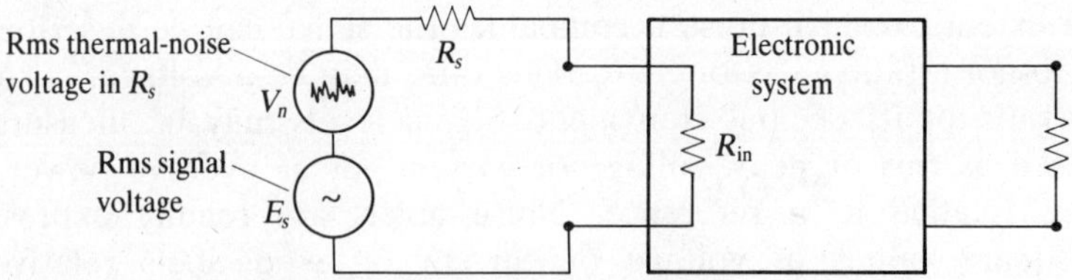

Fig. 7-1 A circuit with both a signal and thermal noise present.

delivered to R_{in} in a noise bandwidth B from the input side is

$$N_i = \left(\frac{V_n}{R_s + R_{\text{in}}}\right)^2 R_{\text{in}} = \frac{4kTR_sBR_{\text{in}}}{(R_s + R_{\text{in}})^2} \tag{7-14}$$

Note that thermal and other noises that originate within the system are not included in (7-14). The input signal-to-noise ratio becomes

$$\text{SNR}_i = \frac{S_i}{N_i} = \frac{\left(\dfrac{E_s}{R_s + R_{\text{in}}}\right)^2 R_{\text{in}}}{\dfrac{4kTR_sBR_{\text{in}}}{(R_s + R_{\text{in}})^2}} = \frac{E_s^2}{4kTR_sB} = \frac{\left(\dfrac{E_s^2}{4R_s}\right)}{kTB} \tag{7-15}$$

It is seen that SNR_i is independent of R_{in} because of the ratio and also that SNR_i is the ratio of the mean-square signal voltage to the mean-square thermal-noise voltage or the ratio of available input-signal power to the available input-noise power.

The largest possible SNR_i at any signal-voltage level is determined by the inherent and unavoidable thermal noise in R_s. In practice, any signal generator can also have other internal noise sources but one of the design objectives of a high-quality signal generator is to keep these additional noise sources at the lowest level achievable. The minimum amount of noise that can accompany any signal is the thermal noise of the source resistance.

SNR_i increases (i.e., improves) as R_s and B are decreased. The smallest value of R_s is limited by practical lower limits on the characteristic impedance of coaxial cables or of microstrip lines. The noise bandwidth B must be the same as the noise bandwidth associated with SNR_o, that is, the overall noise bandwidth of the system, not the actual noise bandwidth of the input circuit. This, plus the fact that the system delivers noise to the input circuit, means that SNR_i is not the actual signal-to-noise ratio but is higher than the actual value that measurements would indicate.

Example 7-1

A certain radio receiver has an overall (spot) noise factor $F = 120$ at 3.5 MHz.

(a) How much available input signal power S_i in dBm and E_s in microvolts are required for an output signal-to-noise ratio of unity if $B = 2000$ Hz.

(b) Repeat part (a) if $B = 500$ Hz.

Solution Equation (7-12) indicates that noise factor is a measure of how much the input signal-to-noise ratio is degraded (i.e., reduced) as it passes through a noisy linear system.

(a) For an output signal-to-noise ratio of unity, $S_o = N_o$ and $\text{SNR}_o = 1$, so that, from (7-12) $\text{SNR}_i = F$ and from (7-15), $S_i = \text{SNR}_i kT_0B$. Therefore, the available input-signal power (in watts) is

$$S_i = FkT_0B = 960 \times 10^{-18}\ \text{W}$$

Since dBm is a measure of power level relative to 1 mW,

$$S_{i,\text{dBm}} = 10 \log \frac{FkT_0B}{10^{-3}} = -174 + \text{NF} + 10 \log B$$

$$= -174 + 20.8 + 33 = -120.2\ \text{dBm}$$

If $R_s = R_L = 300\ \Omega$, $E_s = \sqrt{4R_sS_i} = 1.07\ \mu\text{V}$. This value of E_s or the value -120.2 dBm is the minimum discernible input signal below which the output signal becomes progressively more buried in the output noise.

(b) If B is reduced from 2000 Hz to 500 Hz, S_i is reduced by 4 times also, hence $S_i = 240 \times 10^{-18}$ W or -126.2 dBm. Further, E_s is reduced by $\sqrt{4}$ times, or $E_s = 0.54\ \mu\text{V}$. Decreasing the bandwidth B improves the sensitivity of the receiver by decreasing the noise delivered by the receiver.

7-3.1 Optimum Signal-to-Noise Ratio

For the circuit and numerical values of Fig. 3-9c, the output signal power is found to be

$$S_o = \left(\frac{g_m r_\pi E_s}{r_b + r_\pi + R_s}\right)^2 R_L = \frac{10^8 E_s^2}{(2600 + R_s)^2} \tag{7-16}$$

The total output noise power is found from Eq. (3-47) to be, after rearranging,

$$N_o = 3.23 \times 10^{-16} B \left[\frac{(2600 + R_s)^2 - 6.19 \times 10^6}{(2600 + R_s)^2}\right] \tag{7-17}$$

The output signal-to-noise ratio is

$$\text{SNR}_o = \frac{S_o}{N_o} = \frac{0.31 \times 10^{24}}{B}\left[\frac{E_s^2}{(2600 + R_s)^2 - 6.19 \times 10^6}\right] \tag{7-18}$$

When E_s is constant, it is seen that S_o is maximized and N_o is minimized, thus SNR_o is maximized, when $R_s = 0$. Usually, however, the actual source resistance R_g is constant and a matching network, such as an ideal transformer having a turns ratio n, is inserted between the source and the transistor. Thus, $E_s = nE_g$ and $R_s = n^2R_g$, and (7-18) becomes

$$\mathrm{SNR}_o = \frac{0.31 \times 10^{24} E_g^2}{B}\left[\frac{n^2}{(2600 + n^2R_g)^2 - 6.19 \times 10^6}\right] \tag{7-19}$$

This is maximized with the value n_{opt} for which $d(\mathrm{SNR}_o)/dn^2 = 0$. Hence,

$$n_{\mathrm{opt}} = \sqrt[4]{\left(\frac{2600^2 - 6.19 \times 10^6}{R_g^2}\right)} \tag{7-20}$$

If $R_g = 50\ \Omega$, $n_{\mathrm{opt}} = 3.89$, for which $R_s = 755\ \Omega$. If $R_g = 300\ \Omega$, $n_{\mathrm{opt}} = 1.59$ and $R_s = 755\ \Omega$. Further, SNR_i given by (7-15) becomes

$$\mathrm{SNR}_i = \frac{0.625 \times 10^{20} E_s^2}{R_s B} = \frac{0.625 \times 10^{20} E_g^2}{R_g B} \tag{7-21}$$

Therefore, the ideal lossless matching network does not change the input signal-to-noise ratio.

The noise factor of (7-12) becomes

$$F = \frac{\mathrm{SNR}_i}{\mathrm{SNR}_o} = \frac{2 \times 10^{-4}}{R_g}\left[\frac{(2600 + n^2R_g)^2 - 6.19 \times 10^6}{n^2}\right] \tag{7-22}$$

Note that the E_g and B terms cancel because of the ratio. Noise factor F given by (7-22) is minimized with the same n_{opt} given by (7-20) that maximized SNR_o. Thus, if $R_g = 50\ \Omega$, $n_{\mathrm{opt}} = 3.89$, and $F_{\min} = 1.36$.

7-4 OPERATING NOISE FACTOR

The standard noise factor is given in (7-11) as

$$F = \frac{N_o}{N_i G}$$

Separate N_o into the sum of two components, one being the noise power N_iG engendered at the output by the input termination at $T_0 = 290$ K, and the other being the noise power N_s introduced into the output by the internal noise sources of the system. Hence, the standard noise factor becomes

$$F = \frac{N_iG + N_s}{N_iG} = 1 + \frac{N_s}{N_iG} \tag{7-23}$$

Since $N_i = kT_0B$ and F, B, and G are all measurable, solve (7-23) for N_s to obtain

$$N_s = (F - 1)kT_0BG \tag{7-24}$$

The quantity $(F - 1)$ is known as the *excess noise factor* because it is a measure of the total noise that all noise sources interior to the system introduce into the output in excess of the output noise engendered by the input termination. Some error occurs in (7-24) because the input noise may be correlated to some extent with one or more of the internal noise sources, but the typical spread of values from one active device to another of the same type may overshadow the effects of any correlation coefficients. Further, correlation coefficients are seldom known or specified on data sheets. It is usual to assume all correlation coefficients are zero because of the simplicity and reasonable accuracy thereby achieved. Therefore, whenever the conditions are duplicated under which the standard noise factor is measured, the noise power N_s that the system itself introduces into the output will always be given by (7-24). For example, suppose a radio receiver has input noise from its antenna that is γ times higher than the standard matched thermal noise $N_i = kT_0B$ but with all other conditions being the same. Using procedures similar to the above, the *operating noise factor* F_{op} of the radio receiver is

$$\begin{aligned} F_{\text{op}} &= \frac{N_o}{\gamma N_i G} = \frac{\gamma N_i G + N_s}{\gamma N_i G} = 1 + \frac{N_s}{\gamma N_i G} \\ &= 1 + \frac{(F-1)kT_0BG}{\gamma kT_0BG} = 1 + \frac{F-1}{\gamma} \end{aligned} \tag{7-25}$$

where (7-24) has been substituted for N_s. Equation (7-25) shows that the operating noise factor of the noisy radio receiver approaches unity (as for an ideal noiseless system) when the input noise increases, that is, when γ increases. The reason for this is that as the input noise increases it tends to mask and prevail over the noise from the internal noise sources and causes their effects to appear less significant.

The operating noise factor can also be expressed in terms of the signal-to-noise ratios that apply to these noisy conditions. Thus,

$$F_{\text{op}} = \frac{\text{SNR}_i}{\text{SNR}_o}$$

If the minimum allowable output signal-to-noise ratio is taken to be unity for satisfactory operation of the receiver, $\text{SNR}_o = 1$ so that

$$F_{\text{op}} = \text{SNR}_i = \frac{\text{MDS}}{\gamma N_i}$$

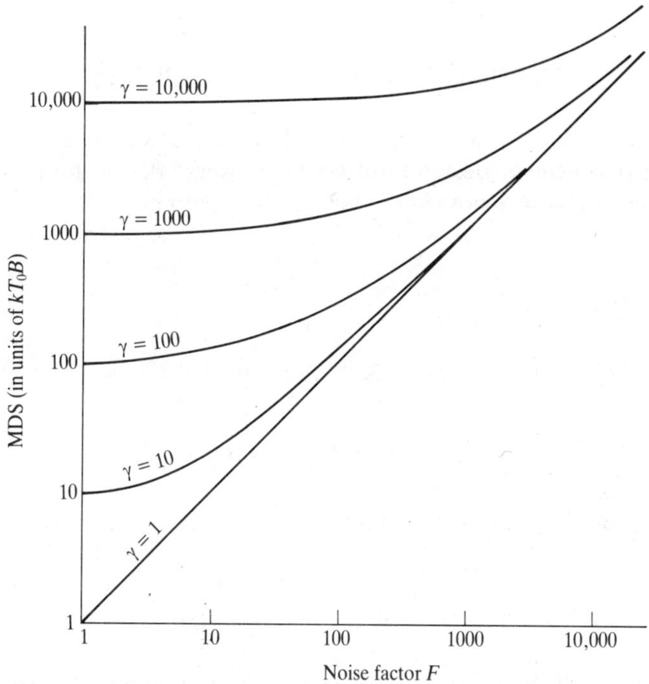

Fig. 7-2 Minimum discernible input signal power MDS vs. noise factor F for $\text{SNR}_o = 1$ with the input-noise multiplier γ as a parameter.

from which, using (7-25), the minimum discernible input signal power, MDS, is found to be

$$\text{MDS} = \gamma F_{\text{op}} N_i = (\gamma + F - 1)kT_0B \tag{7-26}$$

Equation (7-26) is plotted in Fig. 7-2 for several values of the parameter γ. It is seen that, as γ is increased, that is, as more noise is applied to the input of the receiver, the improvement in sensitivity caused by reducing F of the receiver becomes progressively less. This can also be seen quantitatively by rearranging (7-26) as

$$\text{MDS} = \gamma kT_0B + (F - 1)kT_0B \tag{7-27}$$

where the first term is the input noise and the second term represents N_s referred to the input by dividing N_s by G. In terms of dBm, (7-26) becomes

$$\text{MDS}_{\text{dBm}} = 10 \log (\gamma + F - 1) + 10 \log kT_0B + 30$$

Since the minimum value of S_i occurs when $F = 1$,

$$\text{MDS}_{\text{min,dBm}} = 10 \log \gamma + 10 \log kT_0B + 30$$

Subtracting,

$$\Delta \text{MDS}_{\text{dBm}} = 10 \log (\gamma + F - 1) - 10 \log \gamma \tag{7-28}$$

Suppose for a given γ we wish to find how much we must decrease F by improving the receiver design until we are within 1 dB of the most sensitive and noise-free receiver possible. Thus (7-28) becomes

$$\Delta \text{MDS}_{\text{dBm}} = 1 = 10 \log \frac{\gamma + F_{-1} - 1}{\gamma}$$

where F_{-1} represents the noise factor for which $\Delta\text{MDS}_{\text{dBm}} = 1$. Solving for F_{-1} yields

$$F_{-1} = \gamma(10^{1/10} - 1) + 1 = 0.259\gamma + 1 \tag{7-29}$$

with the following numerical results:

γ	1	10	100	1000	10,000
F_{-1}	1.26	3.6	26.9	260	2591

This shows that if the receiver is in a noisy location where the input noise is, for example, $1000kT_0B$, redesigning the receiver to make it more sensitive and less noisy by using lower-noise circuitry and devices to make $F < 260$ will improve sensitivity by no more than 1 dB.

7-5 EFFECT OF SOURCE TEMPERATURE ON NOISE FACTOR

The definition of noise factor F is based on the available thermal-noise power from a signal-source resistance at the standard temperature $T_0 = 290$ K. But in an operating system, the signal source is usually a receiving antenna and it may not be at temperature T_0, therefore the noise factor is affected.

An antenna has radiation resistance R_r and ohmic resistance R_a. Atmospheric and cosmic noise power at the antenna terminals can be represented by the thermal noise of the radiation resistance R_r at a temperature T_r called the noise temperature of the antenna's surroundings that radiate noise to it. The antenna may be highly directional and T_r is high when it is directed toward a high-noise region of the universe, or T_r is low when directed toward a low-noise region. Side lobes, sky waves, and ground waves may be significant and thereby influence the value of T_r. Also, the ohmic resistance R_a of the antenna structure is a source of thermal noise that depends on the antenna temperature T_a, which is also extremely variable.

The total mean-square noise voltage at the antenna terminals, given by

$$4kT_rR_rB + 4kT_aR_aB$$

may be represented by the thermal noise produced in a resistance $(R_r + R_a)$ at a temperature T_A. Thus,

$$4kT_A(R_r + R_a)B = 4kT_rR_rB + 4kT_aR_aB$$

from which is obtained the equivalent noise temperature of the antenna:

$$T_A = \frac{T_rR_r + T_aR_a}{R_r + R_a} \tag{7-30}$$

If the location or orientation of the antenna is changed, it is expected that T_A could be greater than, equal to, or even less than T_0, depending on the conditions. The available noise power from the antenna is obtained when its load resistance equals $(R_r + R_a)$ and that power is $\gamma kT_0B = k(\gamma T_0)B = kT_AB$.

Using steps similar to those taken above, the operating noise factor can be expressed as

$$F_{op} = \frac{N_o}{kT_ABG} = \frac{kT_ABG + N_s}{kT_ABG} = 1 + \frac{(F-1)kT_0BG}{kT_ABG} = 1 + (F-1)\frac{T_0}{T_A} \tag{7-31}$$

The operating noise factor is inversely proportional to the noise temperature T_A of the source, and the noise temperature T_A, in turn, is directly proportional to the amount of input noise. It is F_{op} that determines the degradation of signal-to-noise ratio as it passes through the system operating normally *in situ* rather than under test conditions. The temperature of the system itself is accounted for in its noise factor F.

7-6 THE FRIIS FORMULA

It is possible to express the overall noise factor of a multistage system in terms of the noise factors and power gains of the individual stages. Consider the cascade of three stages shown in Fig. 7-3. Let $N_i = kT_0B$ be the available thermal-noise power from the matched source resistance. It is assumed that conjugate matching exists throughout so that the power gains G_1, G_2, and G_3 are available power gains. The total output noise power of stage 1 is

$$N_{o1} = kT_0BG_1 + (F_1 - 1)kT_0BG_1 = F_1G_1kT_0B \tag{7-32}$$

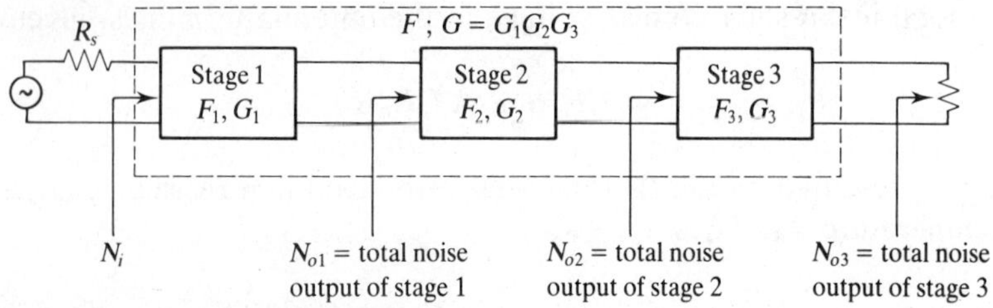

Fig. 7-3 Cascaded amplifiers used to determine the overall noise factor F. The temperature is $T_0 = 290$ K throughout.

The total output noise power of stage 2 is

$$N_{o2} = N_{o1}G_2 + (F_2 - 1)kT_0BG_2 = F_1G_1G_2kT_0B + (F_2 - 1)G_2kT_0B \tag{7-33}$$

The total output noise power of stage 3 is

$$\begin{aligned} N_{o3} = N_{o2}G_3 + (F_3 - 1)G_3kT_0B &= F_1G_1G_2G_3kT_0B \\ &+ (F_2 - 1)G_2G_3kT_0B + (F_3 - 1)G_3kT_0B \end{aligned} \tag{7-34}$$

The overall noise factor F is found, using (7-34) in (7-11), to be

$$F = \frac{F_1G_1G_2G_3kT_0B + (F_2 - 1)G_2G_3kT_0B + (F_3 - 1)G_3kT_0B}{G_1G_2G_3kT_0B}$$

from which is obtained

$$F = F_1 + \frac{(F_2 - 1)}{G_1} + \frac{(F_3 - 1)}{G_1G_2} \tag{7-35}$$

If these are n cascaded stages, the overall noise factor becomes

$$F = F_1 + \frac{(F_2 - 1)}{G_1} + \frac{(F_3 - 1)}{G_1G_2} + \cdots + \frac{(F_n - 1)}{G_1G_2 \cdots G_{n-1}} \tag{7-36}$$

This is known as the Friis formula. It is noticed that the relative importance of the excess noise factor $(F_n - 1)$ diminishes as the total power gain of the *preceding* stages increases. If there is sufficient power gain in the first one or two stages, the noise contribution of all the following stages is relatively minor. When the first stage performs some function, such as attenuation or mixing, with power gain that may be less than unity, the excess noise factor of the following stage or two becomes more prominent. Generally, the first stage contributes the most to the overall noise factor, and the last stage contributes the least.

As discussed earlier, the conditions under which F_n and G_n are individually measured can be duplicated simply by tuning everything for maximum output signal power. It has been found that there is an optimum source admittance $Y_{s,\text{opt}} = G_{s,\text{opt}} + jB_{s,\text{opt}}$ that minimizes a noise factor. Thus,

$$F = F_{\min} + \frac{R_{\text{ef}}}{G_s}[(G_s - G_{s,\text{opt}})^2 + (B_s - B_{s,\text{opt}})^2] \tag{7-37}$$

where $F_{\min}$ = minimum noise factor obtained by varying Y_s

$Y_s = G_s + jB_s$ = Thévenin equivalent source admittance, conductance, and susceptance,

$Y_{s,\text{opt}} + jB_{s,\text{opt}}$ = Thévenin equivalent source admittance, conductance, and susceptance for $F_{\min}$

R_{ef} = proportionality constant with dimension of ohms

The source admittance required for minimum noise factor differs from that required for available power gain. This mismatching of the input circuit generally decreases the output noise more than the output signal, but the improvement in noise factor may not always be significant.

7-6.1 Noise Factors of Lossless and Lossy Networks

Figure 7-4a represents a lossless passive network between R_L and R_s for which $R_s = R_{\text{in}}$ and $R_L = R_{\text{out}}$. Because of the input match, the thermal noise delivered to the network from R_s is the available noise power kT_0B.

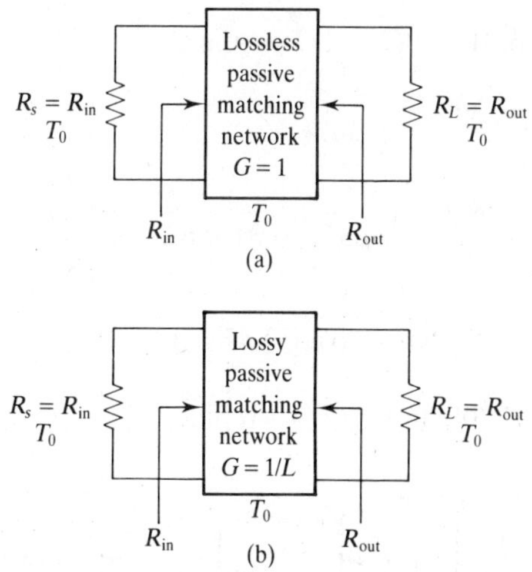

Fig. 7-4 (a) Lossless passive matching network has $F = 1$. (b) Lossy passive matching network has $F = L$ (L = loss).

Since the passive network is lossless and generates no noise of its own, it must deliver the same noise power kT_0B to R_L. The fact that there is a match at the output also means that the available noise power kT_0B is delivered to R_L. Thus, $F=1$ for the lossless passive network.

In Fig. 7-4b, the passive network has a numeric power loss $L>1$ or a numeric power gain $G<1$, where $G=1/L$. Because of the output match, the total thermal noise delivered to R_L is the available power kT_0B from the network. The output noise engendered by R_s is the available input noise power kT_0B times G. Therefore, the noise factor of the lossy passive network is

$$F=\frac{kT_0B}{kT_0BG}=\frac{1}{G}=L \tag{7-38}$$

Converting to decibels, it is seen that the noise figure NF equals the loss in decibels. Thus, a lossy passive matching network or a matched transmission line with a loss of 0.5 dB has NF = 0.5 dB because of (7-38), from which is obtained $F=1.122$.

Example 7-2

Determine the overall noise factor for the cascaded system shown in Fig. 7-5. Assume that appropriate matching exists throughout for the values specified.

Solution 1 Convert decibles to numeric values. Thus,

$$F_1=10^{0.3}=2 \qquad F_2=L_2=10^{0.2}=1.58 \qquad F_3=10^{0.4}=2.51$$

$$G_1=10^{0.8}=6.31 \qquad G_2=1/L_2=0.633$$

Therefore, from (7-35),

$$F=2+\frac{0.58}{6.31}+\frac{1.51}{6.31(0.633)}=2.47$$

and

$$\text{NF}=10\log 2.47=3.93\text{ dB}$$

Solution 2 Use (7-38) in (7-35) to replace F_2 by L_2 and replace G_2 by

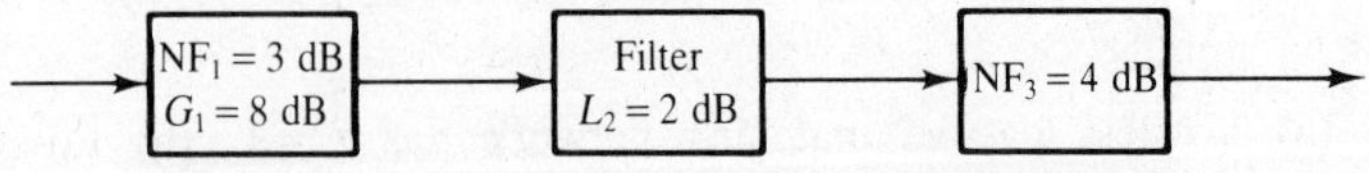

Fig. 7-5 Cascade of stages used in Example 7-2.

$1/L_2$ to obtain the combined form

$$F = F_1 + \frac{L_2 - 1 + L_2F_3 - L_2}{G_1} = F_1 + \frac{L_2F_3 - 1}{G_1} \tag{7-39}$$

Here, the numeric L_2F_3 acts like a noise factor F_2' in the Friis formula. Thus, L_2F_3 in decibels is the sum of the loss L_2 in decibels plus the noise figure NF_3. In other words, the filter loss can be accounted for by adding the filter loss in decibels to the noise figure of the following stage. This process combines these two into a single "second" stage having $NF_2' =$ 6 dB or $F_2' = 3.98$. The overall noise factor becomes

$$F = F_1 + \frac{F_2' - 1}{G_1} = 2 + \frac{3.98 - 1}{6.31} = 2.47$$

and the noise figure is

$$NF = 10 \log 2.47 = 3.93$$

as in Solution 1.

7-7 NOISE-FACTOR MEASUREMENT

A known resistance at a known temperature is an absolute standard noise source and is useful for measurement and calibration purposes. One such standard noise source employs two carefully matched resistive terminations, one immersed in a thermally insulated container of a cryogenic fluid such as liquid helium (4.2 K) or liquid nitrogen (77.3 K), and the other in a temperature-controlled oven set to the temperature of boiling water (373 K). In practice, two measurements of output noise powers from the system under test are necessary as represented in Fig. 7-6.

1. Measure the total output noise power N_1 when the noise source at "cold" temperature T_1 is connected to the system under test. Therefore

$$N_1 = kT_1BG + (F - 1)kT_0BG \tag{7-40}$$

assuming that the bandwidths associated with the two terms are the same.

2. Measure the total output noise power N_2 when the noise source at the "hot" temperature T_2 is connected to the system. Thus

$$N_2 = kT_2BG + (F - 1)kT_0BG \tag{7-41}$$

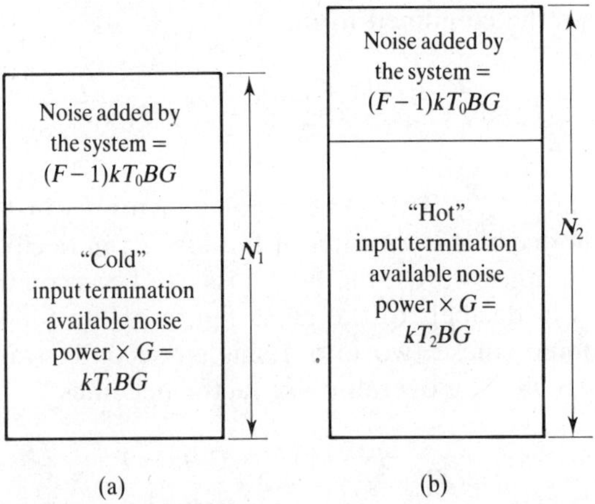

Fig. 7-6 Output noise powers N_1 and N_2 for a hot–cold standard noise source. (a) N_1 when the input termination is at the lower "cold" temperature T_1. (b) N_2 when the input termination is at the higher "hot" temperature T_2.

Notice that the BG product occurs in each term in (7-40) and (7-41): therefore if the ratio N_2/N_1 is taken, that product drops out provided that the bandwidth is the same for all terms. The ratio N_2/N_1 is known as the Y *factor.* Hence

$$Y = \frac{N_2}{N_1} = \frac{T_2 + (F-1)T_0}{T_1 + (F-1)T_0} \tag{7-42}$$

Solve (7-42) for the noise factor to obtain

$$F = 1 + \frac{(T_2/T_0) - Y(T_1/T_0)}{Y - 1} \tag{7-43}$$

Since the temperatures are known for the standard liquid-nitrogen noise source described above, $T_2/T_0 = 373/290 = 1.286$ and $T_1/T_0 = 77.3/290 = 0.267$, so that (7-42) becomes

$$Y = \frac{0.286 + F}{F - 0.733} \tag{7-44}$$

and (7-43) becomes

$$F = 1 + \frac{(1.286) - Y(0.267)}{Y - 1} \tag{7-45}$$

Equation (7-44) indicates that the values of Y, rounded off to two decimal places, vary from 4.83 when $F = 1$, to 1.11 when $F = 10$, then Y changes to only 1.03 when $F = 40$. This variation shows that using this hot–cold standard noise source for noise measurements is most effective when the noise factor is small, preferably less than about 10, the range for which the change in Y is the greatest.

The measurement of Y for a noisy system is most accurately made when the input noise source produces more available noise power than a resistive termination even at an elevated temperature. A noisy system tends to mask a small input noise. The above hot–cold standard noise source has $T_2/T_1 = 373/77.3 = 4.83$ and the responsibility for most of that ratio is due to the cold liquid-nitrogen temperature. If the cooling is eliminated to save the expense of the liquid nitrogen and the "cold" temperature is taken to be room temperature 290 K, $T_2/T_1 = T_2/T_0 = 373/290 = 1.286$, a very modest value.

7-7.1 Electronic Noise Sources

Very noisy sources can be fabricated from vacuum, gas-discharge, and solid-state processes and such devices are readily available. They can produce much more available noise power than resistive terminations even at elevated temperatures. When ON, the noise source available power is

$$P_{\mathrm{ON}} = kT_2B \tag{7-46}$$

where T_2 = equivalent ON noise temperature, kelvin. Again, two measurements of output noise powers are necessary as represented in Fig. 7-7. Thus,

$$N_1 = kT_0BG + (F - 1)kT_0BG = FkT_0BG \tag{7-47}$$

$$N_2 = kT_2BG + (F - 1)kT_0BG \tag{7-48}$$

from which is obtained

$$Y = 1 + \frac{(T_2 - T_0)/T_0}{F} \tag{7-49}$$

The ratio $(T_2 - T_0)/T_0$ is called the *excess-noise ratio*. Solve (7-49) for the noise factor to obtain

$$F = \frac{T_2 - T_0}{T_0}\frac{1}{Y - 1} \tag{7-50}$$

so that the noise figure is

$$\mathrm{NF} = 10 \log \left(\frac{T_2 - T_0}{T_0}\right) - 10 \log (Y - 1) \quad \mathrm{dB} \tag{7-51}$$

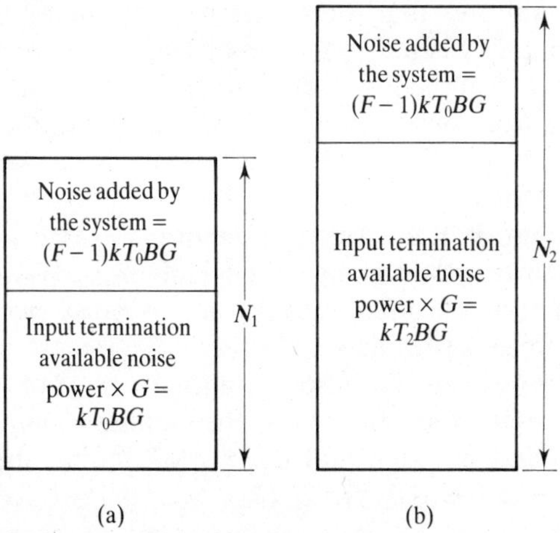

Fig. 7-7 Output noise powers N_1 and N_2 for an excess-noise source. (a) N_1 when the excess-noise source is OFF and at temperature T_0. (b) N_2 when the excess-noise source is ON and at the equivalent temperature T_2.

The excess-noise ratio in decibels is called ENR. Thus

$$\text{ENR} = 10 \log \left(\frac{T_2 - T_0}{T_0} \right) \tag{7-52}$$

Hence (7-51) becomes

$$\text{NF} = \text{ENR} - 10 \log (Y - 1) \quad \text{dB} \tag{7-53}$$

The excess noise often contains large-amplitude impulses that can destroy transistors. Hence, a loss pad of about 10 dB should be used between the excess-noise source and the system under test. The loss pad also helps reduce any indeterminate mismatch loss because the ENR is specified by the manufacturer for a 50-Ω termination. Consult the article by Swain and Cox (see Bibliography) for various errors in noise-figure measurements and how they can be accounted for in the design of a measuring system.

Example 7-3

A certain solid-state excess-noise source has ENR = 15.5 dB. Determine its equivalent ON noise temperature.

Solution From (7-52) solve for T_2 to obtain

$$T_2 = T_0(1 + 10^{\text{ENR}/10}) \quad \text{K}$$

Substitute ENR = 15.5 into this to find $T_2 = 10{,}579.6$ K. This noise source produces the same available noise power as a resistive termination at 10,579.6 K, provided that the resistance material does not burn up or melt away! Tungsten can be heated to 2700 K but most "hot loads" have been limited to about 1000 K because of practical difficulties with measuring the actual temperature accurately, assuring that the resistor material is homogeneous, maintaining a conjugate match over the frequency range of interest, accounting for temperature gradients that influence the average temperature of the resistor, and correcting for the attenuation and other effects of the connecting leads or transmission lines.

7-8 NOISE MEASURE

Suppose you have a radio or television receiver having a noise factor F_2 that appears to be too high. You wish to add a preamplifier having F_1 and $G_1 > 1$ ahead of the receiver in the hope that the overall noise factor of the combination will be less than the noise factor F_2 of the receiver alone. Is the desired improvement possible? To find out, use the Friis formula to express the noise-factor inequality as

$$F_1 + \frac{F_2 - 1}{G_1} < F_2$$

which can be rearranged as

$$F_1 < F_2 - \frac{F_2 - 1}{G_1}$$

Subtract unity from both sides of this and factor that result to obtain

$$F_1 - 1 < (F_2 - 1)\left(1 - \frac{1}{G_1}\right)$$

which can be expressed as

$$\frac{F_1 - 1}{1 - \dfrac{1}{G_1}} < F_2 - 1 \tag{7-54}$$

The quantity called *noise measure* M is defined to be

$$M = \frac{F - 1}{1 - \dfrac{1}{G}} \tag{7-55}$$

where $G > 1$ and $F > 1$. Thus, the overall noise factor of the system using the preamplifier will be less than the noise factor of the receiver alone if, from (7-54),

$$M_1 < F_2 - 1 \tag{7-56}$$

that is, the noise measure of the preamplifier must be less than the excess noise factor of the receiver alone, or, rearranging (7-56),

$$F_2 > M_1 + 1 \tag{7-57}$$

For example, if the preamplifier has $F_1 = 5$ and $G_1 = 6$, then (7-55) yields $M_1 = 4.8$. Using the preamplifier will improve the overall noise factor provided that the receiver has $F_2 > M_1 + 1 = 5.8$ from (7-57). The amount of improvement may be impressive if F_2 is large. Thus, if $F_2 = 40$, the overall noise factor in this case is found to be

$$F = F_1 + \frac{F_2 - 1}{G_1} = 5 + \frac{39}{6} = 11.5$$

The concept of noise measure is helpful when it is necessary to decide on the order in which several individual stages, each having known noise factors and power gains >1, are to be cascaded such that the overall noise factor is the smallest of the various combinations possible. For example, suppose two stages individually have F_1 and G_1, F_2 and G_2, and that they are to be connected in cascade in the proper order for the overall noise factor to be the smaller of the two possibilities. Represent this condition by

$$F_1 + \frac{F_2 - 1}{G_1} < F_2 + \frac{F_1 - 1}{G_2} \tag{7-58}$$

Rearrange (7-58) by placing terms containing F_1 on the left-hand side and terms containing F_2 on the right-hand side, then subtract unity from both sides, and factor that result to obtain

$$(F_1 - 1)\left(1 - \frac{1}{G_2}\right) < (F_2 - 1)\left(1 - \frac{1}{G_1}\right)$$

which can be written as

$$\frac{F_1 - 1}{1 - \frac{1}{G_1}} < \frac{F_2 - 1}{1 - \frac{1}{G_2}} \tag{7-59}$$

in which the noise measures of both stages are identified. Hence, the overall noise factor of the stages connected in the order 1–2 is smaller than when in the order 2–1 provided that $M_1 < M_2$. The first stage, where the noise performance is the most critical, should have the smaller noise measure and not the smaller noise factor. The smaller the value of M a stage has, the more suitable that stage is as a first stage.

When the same arguments are applied to a multistage system two stages at a time, it is quickly concluded that the individual stages should be connected in the order having $M_1 < M_2 < M_3 < M_4 < \cdots < M_n$ to cause the overall noise factor (and therefore also the overall noise measure) to be the smallest of any of the various sequences possible.

Example 7-4

Four stages A, B, C, and D individually have $F_A = 10$ and $G_A = 4$, $F_B = 6$ and $G_B = 2$, $F_C = 12$ and $G_C = 5$, and $F_D = 8$ and $G_D = 6$. In what order should the stages be connected to obtain the smallest possible noise factor of any of the sequences possible?

Solution Calculate the four noise measures to obtain $M_A = 12$, $M_B = 10$, $M_C = 13.75$, and $M_D = 8.4$. Thus, D should be the first stage, B the second, A the third, and C the fourth. Their overall noise factor is 9.81 and the overall noise measure is 8.85.

7-9 EQUIVALENT INPUT-NOISE TEMPERATURE

The concept of an equivalent (or effective) input-noise temperature, sometimes called simply the noise temperature, is a useful alternative to noise factor as a measure of noisiness of a system. It has been shown that the output noise power produced by the internal noise sources of a system is

$$N_s = (F - 1)kT_0BG \tag{7-60}$$

If (7-60) is divided by the power gain G, an equivalent input-noise power is obtained that is given by

$$N_e = \frac{N_s}{G} = (F - 1)kT_0B \tag{7-61}$$

and then the system is considered to be noiseless but still has power gain G. The equivalent input-noise temperature is defined to be the fictitious temperature T_e that the source resistance R_s must attain in order for its thermal-noise available power to equal that given by (7-61). Thus

$$kT_eB = (F - 1)kT_0B$$

from which is obtained

$$T_e = (F - 1)T_0 \tag{7-62}$$

For example, a typical low-noise amplifier suitable for the reception of satellite TV signals has an equivalent input-noise temperature of 120 K and, solving (7-62) for the noise factor, $F = 1.41$. The amplifier is said to be a 120° amplifier, meaning that it has an equivalent input-noise temperature of 120 K. An uncooled parametric amplifier can have $T_e = 290$ K, for which $F = 2$ and NF = 3 dB, and a cooled parametric amplifier can have $T_e = 17$ K, for which $F = 1.06$ and NF = 0.25 dB. A maser can have $T_e = 11$ K, for which $F = 1.04$ and NF = 0.16 dB.

The numbers for the equivalent input-noise temperature are not as awkward to use as the numbers for noise factor that differ from unity by only a few hundredths, or noise figures that are only a small fraction of a decibel. Improvements in radio-receiver performance from about 300,000 K to as low as 5 K have been prompted by the need for low-noise space communications.

Example 7-5

A parabolic dish antenna is tilted upward to receive a satellite TV signal. The equivalent antenna temperature is $T_A = 30$ K. Find the system operating noise temperature T_{op} when the receiver T_e is equal to (a) 17 K, (b) 290 K.

Solution The total input noise power is

$$kT_{\text{op}}B = kT_A B + kT_e B$$

from which is obtained the system operating noise temperature

$$T_{\text{op}} = T_A + T_e$$

(a) $T_{\text{op}} = 30 + 17 = 47$ K
(b) $T_{\text{op}} = 30 + 290 = 320$ K

The Friis formula can be expressed in terms of the equivalent input-noise temperature with the aid of

$$F - 1 = \frac{T_e}{T_0} \tag{7-63}$$

from (7-62). Thus, the Friis formula

$$F = F_1 + \frac{F_2 - 1}{G_1} + \frac{F_3 - 1}{G_1 G_2} + \cdots$$

after subtracting unity from both sides and using (7-63), becomes

$$T_e = T_{e1} + \frac{T_{e2}}{G_1} + \frac{T_{e3}}{G_1 G_2} + \cdots \tag{7-64}$$

where T_e is the overall equivalent input-noise temperature of the cascade.

PROBLEMS

Unless otherwise indicated, assume proper impedance matching throughout any cascade arrangement of components and that the temperature is 290 K.

7-1 Use Eqs. (7-3) and (7-4) to calculate the noise factor F for the amplifier of Fig. 3-9c when the signal source is matched to the amplifier. Use the previous numerical values.

7-2 The simplified circuit of an amplifier is shown in Fig. P7-2. The values are $r_b = 50\ \Omega$, $r_\pi = 750\ \Omega$, $g_m = 50\ \text{mA/V}$, $R_L = 5\ \text{k}\Omega$ and the bandwidth is 20 kHz. The transformer is ideal. Ignoring collector shot noise and thermal noise in R_L, calculate: (a) turns ratio $n = n_2/n_1$ for maximum power transfer from the signal source; (b) turns ratio for minimum noise factor; (c) turns ratio for minimum output signal-to-noise ratio.

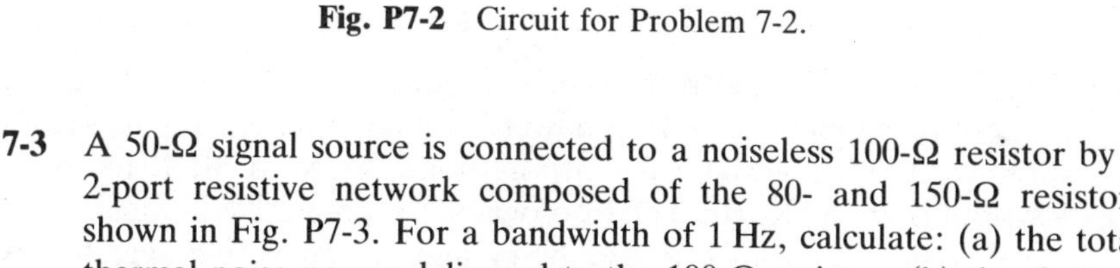

Fig. P7-2 Circuit for Problem 7-2.

7-3 A 50-Ω signal source is connected to a noiseless 100-Ω resistor by a 2-port resistive network composed of the 80- and 150-Ω resistors shown in Fig. P7-3. For a bandwidth of 1 Hz, calculate: (a) the total thermal noise power delivered to the 100-Ω resistor; (b) the thermal noise power delivered to the 100-Ω resistor by the 50-Ω source resistor; (c) the noise factor F of the 2-port network; (d) the noise figure NF of the 2-port network; (e) the ratio of the input signal power P_i to the output signal power P_o. (f) Compare the results of (c) and (e).

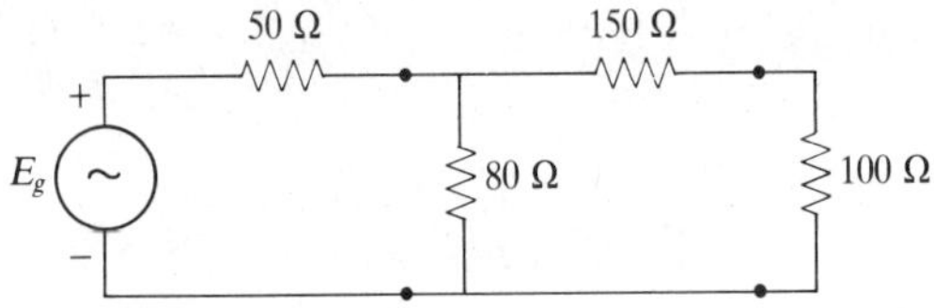

Fig. P7-3 Circuit for Problem 7-3.

7-4 A certain amplifier has a noise bandwidth of 4 MHz, a noise factor of 5, and a power gain of 10^6. Its input and output circuits are both matched to 500 Ω. Calculate the rms noise voltage across the 500-Ω load resistance.

7-5 A television receiver uses a semi-conductor diode in a difference mixer to produce the IF. The mixer circuit has a power loss of 7 dB and a noise figure of 9 dB. The IF amplifier following the mixer has a noise figure of 3.5 dB. Assume a bandwidth of 4.5 MHz. (a) Calculate the signal power in pW ($1\ \text{pW} = 10^{-12}$ W) and in dBm that must be applied to the mixer to obtain a signal-to-noise ratio of unity at the output of the IF amplifier under laboratory test conditions; (b) Explain why the noise figure of the lossy mixer does not equal its power loss.

7-6 A single-conversion superheterodyne TV receiver is connected to a 300-Ω signal generator. The receiver uses a semiconductor diode in a difference mixer to produce the IF. The mixer has a power loss of 8 dB and a noise figure of 10 dB. The IF amplifier that follows the mixer has a noise figure of 20 dB. Assume a bandwidth of 4.5 MHz. How much rms signal voltage in microvolts is required at the input terminals of the receiver to cause the signal-to-noise ratio at the output of the IF amplifier to be 20 for a good picture?

7-7 Three separate amplifier stages individually have $F_1 = 10$ and $G_1 = 4$, $F_2 = 6$ and $G_2 = 8$, $F_3 = 12$ and $G_3 = 2$. Calculate the overall noise factor when the stages are connected in the order: (a) 1–2–3; (b) 3–2–1; (c) 1–3–2; (d) 3–1–2; (e) for minimum overall noise factor.

7-8 You own a TV receiver having a noise factor of 12. It uses a transmission line with a 3-dB power loss between the antenna and the TV itself. (a) Calculate the overall noise figure in dB if the entire system is at 290 K: (b) You also own a TV preamplifier having $F = 8$ and $G = 11$. You hope to use the preamp to reduce the overall noise figure below that of part (a) by inserting the preamp either at the TV itself or at the antenna. Which is better and why?

7-9 A certain radio receiver has a measured noise figure of 25 dB. A preamplifier with a measured noise figure of 6 dB is to be added in front of the receiver to made the overall noise figure <8 dB.

Calculate the minimum permissible gain in dB that you would expect the preamp to have.

7-10 A radio receiver has a noise factor of 8 and a bandwidth of 2 kHz. For standard measuring conditions with 50-Ω matched input resistance: (a) calculate the minimum discernible input signal power in dBm for an output signal-to-noise ratio of unity; (b) how much input signal voltage produces an output signal-to-noise ratio of 10?

7-11 A certain amplifier having noise factor F_A and power gain G_A, and a certain attenuator having a power loss $L_2 > 1$ are to be connected in cascade. (a) Calculate the overall noise factor F_{2A} when the attenuator precedes the amplifier. (b) Calculate the overall noise factor F_{A2} when the amplifier precedes the attenuator. (c) What condition on G_A is required to always obtain $F_{A2} < F_{2A}$? Is this condition difficult to achieve?

7-12 Two amplifiers J and K individually have $F_J = 20$ and $G_J = 8$, $F_K = 8$ and $G_K = 20$. (a) Determine where an attenuator network N having a loss of 6 dB should be placed and the order of connecting the amplifiers in cascade with the attenuator such that the overall noise factor is the least possible. (b) Calculate F_{min}. Hint: Use the results of Problem 7-11.

7-13 Two amplifiers individually have $F_1 = 15$, $G_1 = 6$ and $F_2 = 8$, $G_2 = 14$. Determine where a 3-dB attenuator should be placed and the order of connecting the amplifiers in cascade with the attenuator such that the overall noise factor is the least possible.

7-14 Four separate amplifier stages individually have $F_1 = 4$ and $G_1 = 2$, $F_2 = 5$ and $G_2 = 4$, $F_3 = 7$ and $G_3 = 5$, $F_4 = 6$ and $G_4 = 7$. Using the concept of noise measure, determine the order in which the amplifiers should be connected to produce the least noise factor and calculate the least noise factor.

7-15 Three amplifiers, each characterized by their noise measures $M_1 = 5$, $M_2 = 10$, and $M_3 = 15$, and by their power gains $G_1 = 6$, $G_2 = 11$, and $G_3 = 16$, respectively, are to be connected in cascade to obtain the least noise factor F_{min}. Calculate the least noise measure M_{min}.

7-16 Calculate the noise factor F for each of the following equivalent input-noise temperatures in kelvin: (a) 2.5; (b) 25; (c) 250; (d) 2500; (e) 25,000; (f) 250,000.

7-17 (a) Calculate the noise figure (in dB) of an amplifier whose equivalent input-noise temperature is 120 K. (b) Calculate the minimum equivalent input-noise temperature of the cascade in Problem 7-12. (c) Calculate the equivalent input-noise temperature of an amplifier whose noise measure is $M = 25$ and whose power gain is $G = 15$.

7-18 The equivalent noise temperature of an antenna directed toward a certain region in space is 1500 K. The transmission line between the antenna and the receiver has a power loss of 6 dB. The receiver itself has an equivalent input-noise temperature of 200 K. For the entire system, calculate: (a) the operating noise figure (in dB); (b) the operating noise temperature.

7-19 The antenna and receiver of Problem 7-18 are used to receive a weak desired signal from the same direction as in Problem 7-18. The signal-to-noise ratio at the output of the receiver is to be unity. The original transmission line is to be replaced by a less lossy line to obtain a new sensitivity within 1 dB of the minimum possible for that noisy situation. (a) Calculate the new overall noise factor required of the transmission line and receiver. (b) Calculate the power loss in dB required of the new transmission line.

7-20 A typical TV receiving installation features an antenna connected to the receiver by a transmission line whose characteristic impedance is $Z_o = 300\ \Omega$. Assuming that the line is lossless and matched at both ends, the equivalent Thévenin circuit is shown in Fig. P7-20, where E_s represents the rms signal voltage induced in the antenna. For a temperature of 290 K and a noise bandwidth of 4.5 MHz, calculate: (a) signal power S delivered to R_{in} if $E_s = 100\ \mu$V; (b) noise power N delivered to R_{in} by thermal noise in Z_o alone; (c) signal-to-noise ratio S/N.

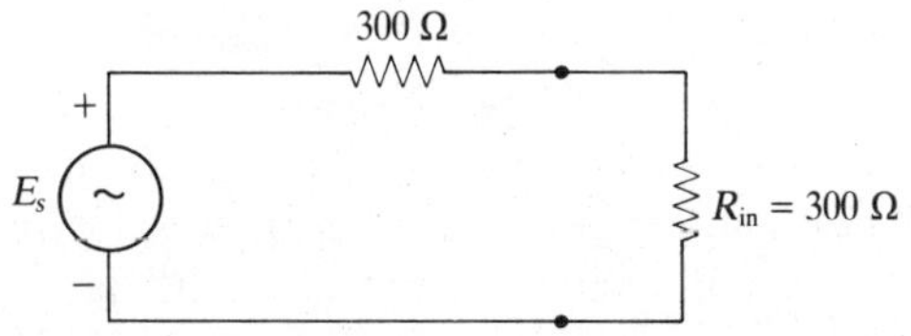

Fig. P7-20 Circuit for Problem 7-20.

7-21 When a resistor at 4.2 K is connected to the input terminals of a receiver, the output noise power is measured to be 0.1 mW. When another carefully matched resistor at 373 K is connected to the receiver, the output noise power is 0.35 mW. The receiver itself is at 290 K. Calculate the noise factor and the equivalent input-noise temperature of the receiver.

7-22 First, a 50-Ω resistor at 290 K is matched to the input of a receiver and the output noise power is measured to be 8 μW. Then, that resistor is removed and a semiconductor noise source having ENR = 15.5 dB is matched to the receiver through a 10-dB attenuator at 290 K. The output noise power is now measured to be 12 μW. Calculate the equivalent input-noise temperature.

8

THE SUPERHETERODYNE RADIO RECEIVER

The word "electron" was first used by Stoney in 1891 as the name for the enigmatic negative particle of electricity having unknown properties. The mysterious electron became a permanent fixture in the scientific scene when Thomson measured its charge-to-mass ratio in 1897. It was in 1910 that Millikan succeeded in determining the amount of electric charge on the electron.

When one ponders the facts that the electron is invisible, that it has an infinitesimal electric charge and an insufficient mass to be subject to significant gravitational force, and that experimental evidences of its particle and wave natures are conflicting, it is amazing that such an apparently intangible quantity could cause the tremendous impact that it has.

8-1 SOME HISTORY OF RADIO COMMUNICATIONS

Compared with other branches of engineering and science, electronic communications is not very old. Electromagnetic radiation was predicted theoretically by Maxwell in 1865 when he formulated his famous equations. Hertz verified that theory by producing electromagnetic waves experimentally in 1877.

Marconi began experimenting with a spark-gap transmitter and coherer receiver in 1895 and gradually increased the distance over which wireless telegraph messages could be sent by making a number of significant improvements in the system. It was Marconi who added the antenna and ground connections, designed a tuning system, and invented a magnetic detector to replace the coherer. In 1901 Marconi established the first transatlantic signal between England and Newfoundland.

During the early years of the 20th century, wireless apparatus consisted of spark-gap or electric-arc transmitters with coherer-type receivers. Spark-gap transmitters employed a capacitor connected across the secondary of an induction coil or a voltage-step-up power transformer having a high leakage reactance. Across the capacitor was a spark gap in series with the primary of a high-frequency transformer whose secondary was coupled to the antenna/ground system. When the capacitor voltage rose high enough, a spark jumped across the gap. This allowed the charged capacitor to discharge current into the inductance of the high-frequency transformer at a frequency determined by the capacitance and inductance. But the amplitude of the current decreased exponentially at a rate determined by the resistance, and this caused the bandwidth of the radiated output to be quite large. The system was suitable for Morse-code transmissions only.

Arc transmitters appeared in 1911. An electric arc has the property that if the arc current increases, the arc voltage decreases. Therefore, the arc acts like negative resistance, and negative resistance is a source of energy. When a capacitor and the primary of the high-frequency transformer are connected across the arc, energy is supplied continually to them by the arc, the ultimate source of that energy coming from the power source that operates the arc. Consequently, the high-frequency current has a fixed amplitude and it is known as a continuous wave, or CW. The bandwidth required is quite small.

A scientist by the name of Fessenden proposed to improve the wireless receiver of the day by a method that he called heterodyning. He coined the new word from the Greek words "heteros", meaning different, and "dynamis", meaning power. His scheme involved the nonlinear mixing of the incoming Morse-code signals with a locally generated, constant-amplitude sinusoidal signal, so that the newly created output signal was at the difference frequency between the two original input signals. The difference frequency was audible, and this allowed the receiver operator to recognize the dot-and-dash tones representing the Morse codes of the various characters. His proposal was doomed to failure in those days for the reason that it was far ahead of its time, there being no good way to generate the local signal except by placing a miniature wireless transmitter inside the receiver, and that made it a source of considerable interference.

However, Fessenden proposed to the General Electric Company that a high-speed alternator could be built to generate the continuous-wave, narrow-bandwidth signal that was desired. In fact, if the power level of the high-speed alternator could be made high enough, it would be a likely candidate to replace the highly damped, wide-bandwidth output signal from the wireless transmitters then in vogue. A young engineer by the name of Alexanderson was assigned to the project, and he supervised the construction of a model of the alternator that Fessenden had in mind. That model produced about 50 watts of continuous-wave power at a frequency of about 80 kHz. Successful demonstrations followed, and the output power was increased to 2000 watts and the frequency to 100 kHz. However, no significant market developed for the alternator, and Fessenden abandoned the project.

But the idea remained with Alexanderson, and with the backing of the General Electric Company, he proceeded to design and produce, in 1915, radio alternators capable of 200 kW at a frequency of about 150 kHz. Several of these alternators were installed around the world to support commercial, military, and governmental communications, and their future seemed assured. Alexanderson also designed and built some of the first vacuum-tube receivers of the day, known as the tuned-radio-frequency (TRF) type.

The new alternators challenged the well-established, world-wide wireless system, but they were not suitable for voice transmissions as were the new vacuum-tube circuits. An infant radio communications industry was developing. No one was in a good position to oversee the fledgling industry because the numerous patents were owned by several different companies and a few individuals. Consequently, detailed plans were made and carried out to pool the patent rights to almost everything pertaining to radio communications and put them under the control of a new company called the Radio Corporation of America, formed in 1915.

One of the first things RCA did was to propose a superpowered transmitter facility featuring ten of Alexanderson's alternators, the objective being to provide proprietary worldwide communications. The site was selected and two alternators were installed, along with construction of the antenna complex. But then the directors of RCA, taking cognizance of the developments in vacuum-tube technology and the growing interest in radio broadcasting, abandoned their alternator plans and concentrated their efforts on radio communications equipment built around the vacuum tube. The rapid improvements in the vacuum tube doomed the alternators to equally rapid extinction. Thus began the growth of radio broadcasting and the manufacturing of radio receivers for a mass market, commencing in the early 1920s. It is interesting to note, however, that a few of the alternators were still in service as late as the 1950s, and one or two may still be operational, occasionally, as museum curiosities.

A strong contributing factor to the development of radio technology was provided by the communications requirements of the military during the First World War, 1914–1918. According to his own written account Edwin Armstrong, a major in the U.S. Army Signal Corps stationed in France, was watching the ineffective gunnery of the Allied antiaircraft batteries one evening. He conceived of using a radio receiver to pick up the ignition noise of the incoming bombers and to use that system to control the antiaircraft guns.

Unfortunately, the vacuum tubes at that time were unable to amplify in the several megahertz frequency range where the ignition noise was most intense. They were only able to amplify fairly reliably, signals having frequencies up to about 100 kHz. He got around this difficulty by using a version of heterodyning originally conceived by Fessenden. Instead of mixing to produce an audible output frequency, a scheme now called direct conversion, he chose to produce a frequency considerably above the audio range, a supersonic intermediate frequency of about 50 kHz. He named his new creation the superheterodyne. Soon after the building of the first essentially complete superheterodyne, a name that now is often shortened to superhet, the war ended. And with that also ended the project to control the antiaircraft guns.

Armstrong was also responsible for developing and promoting the system of frequency modulation (FM). Armstrong also invented super-regeneration, in which the positive feedback around an amplifier kept the feedback system just on the verge of oscillation, so that the average gain of the amplifier was extremely large. He and deForest, the inventor of the triode vacuum tube, were engaged in a prolonged legal battle over who should be the recognized originator of the positive feedback or regeneration concept, the legal winner of that suit being deForest in 1934. However, Armstrong, deForest, Franklin, Meissner, and Round all discovered the vacuum-tube oscillator almost simultaneously in 1913–1914. Armstrong is acclaimed as being the chief promotor of the new superhet radio receiver. The concept has been universally accepted as the standard radio receiving system because its inherent advantages outweigh its disadvantages.

8-1.1 Early Radio Systems

Details about the coherer were initially described by Branley in 1880. It was he who first used the term "radio" in connection with his experiments. The coherer was a small glass tube containing metal filings loosely packed between electrodes at opposite ends of the tube. In this condition the resistance between the electrodes was high. When electrically connected in series with a battery and a relay winding, the current would be small and the relay contacts would not close. If an electromagnetic disturbance such as lightning or a strong wireless signal arrived at the device, the metal filings bound themselves more tightly, or cohered (hence the name), causing the resistance to decrease appreciably. This allowed the current in the relay winding to increase, and if the relay was sensitive enough, the relay contacts closed. This, in turn, operated an instrument to record the dots and dashes of Morse code on moving paper tape. When the electromagnetic field disappeared at the end of a dot or dash, the resistance of the coherer increased and the relay contacts opened. But the device had to be tapped regularly to prevent the filings from remaining cohered. This was done automatically with another relay arrangement.

The Fleming valve (a two-element vacuum tube or diode) was patented in 1904 as a detector of wireless signals. In 1906, Dunwoody and Pickard independently invented the crystal detector. It consisted of a sharp-pointed piece of wire, of such material as phosphor-bronze, called a cat's whisker, which pressed against a sensitive spot on the surface of a mineral such as carborundum, galena, or iron pyrites. Later, silicon and germanium were found to be useful as the crystal. The equivalent modern arrangement is known as an envelope detector.

In 1907 DeForest added a wire grating called a grid between the heated cathode and the cold anode of a diode vacuum tube. This three-element device, or triode, replaced the Fleming valve and the crystal detector because of the amplification it provided.

The early use of triodes was almost exclusively confined to radio broadcasting, telephone repeaters, and amateur radio services. The word "wireless" was superseded by the word "radio," and people who developed radio transmitters and receivers were called radio engineers. Many radio receivers in the early 1920s used several triode amplifier stages, each individually tuned with its own tuning capacitor and dial arrangement, that preceded the demodulator. These were called the tuned radio frequency (TRF) type of receivers.

8-1.2 Early Solid-state Electronics

The foundations for semiconductor electronics were laid by Plank's quantum theory in 1901, followed by Schrödinger's wave equation, Pauli's exclusion principle, and Heisenberg's uncertainty principle in the late 1920s. Maxwell–Boltzmann and Fermi–Dirac statistics were developed, the concept of the hole was introduced, and methods for producing intrinsic and extrinsic silicon were discovered.

In 1923 it was found that when a thin layer of cuprous oxide formed on a copper surface, the combination had a high resistance when the copper was positive and the oxide negative, but a low resistance for the opposite polarity. Stacks of several oxide disks were used in low-powered sources of DC for communications equipment and as low-frequency rectifiers for measuring instruments.

In the 1940s stacks of series–parallel connections of selenium rectifiers were widely adopted to supply DC power. A back electrode of aluminum was coated with a thin layer of the semiconductor selenium. A soft metal alloy front electrode with a low melting point was placed over this coating. By means of heat treatment and an electrical forming process, a barrier layer formed between the front electrode and the selenium. The selenium was considered to be p-type semiconductor and rectification occurred at the interface between the selenium and the front electrode.

After the crystal detector was discovered, high-frequency voltages could be rectified and detected for the first time by a solid-state device, although at a very low power level. A large amount of research was devoted to understanding, developing, and improving solid-state electronics. Out of a part of that effort came the recognition of a solid-state amplifying device called a point-contact *transistor,* announced in 1948 by Bardeen and Brattain. The name transistor was coined by Pierce from the words *trans*fer res*istor*. Then, in 1949, Shockley developed the theory of a junction-type transistor.

Although the integration of many electronic components into a compact semiconductor solid had been envisaged and hoped for earlier, it was in 1959 that Kilby fabricated for the first time a multivibrator circuit out of two pieces of silicon. It consisted of two transistors formed on top of a piece of silicon that contained the resistors. Then the capacitors were formed on a separate piece of silicon and the electrical connections were made with gold wires. Although it could not be called an integrated circuit by today's standards, it did introduce some concepts and processes that became the forerunners of modern microelectronics and demonstrated the feasibility of the procedure. Within a few years thereafter, integrated circuits began to be used in widespread applications where discrete transistors had once reigned supreme. Discrete transistors were being phased out just as they had displaced vacuum tubes.

Linear integrated circuits known as *op amps* appeared in 1960 and the first hand-held electronic calculator was marketed in 1963. The developments of integrated digital devices were accompanied by equally important advances in such things as printed-circuit techniques, light-emitting displays, lasers, floppy disks, personal computers, and optical-fiber communications.

8-2 THE SUPERHETERODYNE PRINCIPLE

The basic concept of the superheterodyne radio receiver is illustrated in Fig. 8-1. The incoming radiofrequency (RF) signal is combined with the sinusoidal output of a local oscillator in a mixer. This mixer is sometimes called the *first detector*. The output circuit of the mixer is usually tuned to the difference between the frequencies of the oscillator and the incoming signal. The sum frequency is also present, of course, and sometimes the output of the mixer is tuned to this sum frequency. The design of the receiver is such that when it is tuned to the frequency of another incoming signal, the frequency of the local oscillator is also automatically changed so as to maintain the same difference frequency as before, the process is called *tracking*. The difference frequency is called the *intermediate frequency* (*IF*) and the circuits that amplify these signals are called *IF amplifiers*.

Since the intermediate frequency is constant regardless of where the receiver is tuned, the tuned circuits in the IF amplifier are all fixed-tuned. Often, however, tuned circuits are replaced by crystal, ceramic, or mechanical filters. After processing by the IF amplifier, the signal is

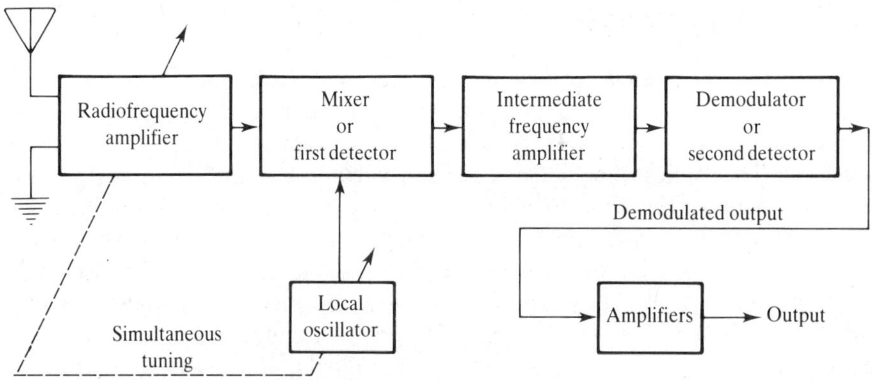

Fig. 8-1 Simplified block diagram of a superheterodyne radio receiver.

delivered to the demodulator (sometimes also called the *second detector*). Additional amplification follows the demodulator to provide the level of output desired. For example, if the output is voice or music, audiofrequency (AF) amplifiers will be needed to obtain the power to drive the loud speaker. If the output is the electrical equivalent of a television picture, video amplifiers are used.

A major disadvantage of the superhet is that any local oscillator within the receiver is also a miniature transmitter. If some of the oscillator output can find its way back to the antenna terminals of the receiver, then a weak signal will be radiated from the antenna at the frequency of the local oscillator. A small amount of radiation can occur directly from any of the RF and IF circuits, too. Other receivers, highly sensitive to weak signals and located nearby, can monitor the radiation, and if they have direction-finding apparatus, then the location of the subject receiver can be determined. Sometimes the local oscillator output will be modulated to some extent by the signal from the station that the receiver is tuned to. This means that the radiation from the local oscillator carries a version of the program from the original station, but at the local-oscillator frequency. These phenomena have applications in electronic countermeasures, surveillance, espionage, and counterespionage activities. Whether the consequences are considered good or bad depends on which side of the activities you are on. Such radiation can be minimized by shielding the local oscillator/mixer circuits especially, and even the entire receiver, by using double-balanced mixers; by keeping the oscillator power level as small as possible; by using buffer amplifiers between the antenna and the mixer/oscillator circuits; by improving the shape factor of the RF filters; by bypassing and filtering all incoming power leads, if any; by using a battery operated receiver inside a shielded room; and by "trapping" in the RF circuit at the oscillator and IF frequencies.

8-2.1 Oscillator Injection

Suppose the receiver is to pick up stations in the standard AM broadcast band. In the U.S., the carrier frequencies have been standardized to be every 10 kHz from 540 kHz to 1600 kHz. A typical value for the IF is 455 kHz. The tuning mechanisms for the various RF and oscillator circuits must be ganged together to produce the IF from the difference between their frequencies. Therefore, the oscillator could tune over the range 995 kHz to 2055 kHz, or from 85 kHz to 1145 kHz. When the oscillator frequency is set to be above the frequency of the incoming signals, it is said that *high-side injection* (*HSI*) is used. When the oscillator frequency is set to be below the frequency of the incoming signals, it is said that *low-side injection* (*LSI*) is used.

The frequency produced by the oscillator is determined approximately by the resonant frequency of its LC circuit components, or

$$f = \frac{1}{2\pi\sqrt{LC}}$$

Suppose that it is the capacitance that can be changed from C_{min} to C_{max} and that the inductance L remains constant. Then an LC circuit will be tuned from

$$f_{max} = \frac{1}{2\pi\sqrt{LC_{min}}}$$

to

$$f_{min} = \frac{1}{2\pi\sqrt{LC_{max}}}$$

Next, taking the ratio of the above, the capacitance ratio required is

$$\frac{C_{max}}{C_{min}} = \left(\frac{f_{max}}{f_{min}}\right)^2 \tag{8-1}$$

The RF circuits must tune from 540 kHz to 1600 kHz, so that $C_{max}/C_{min} = (1600/540)^2 = 8.78$. The minimum capacitance that is used to tune the inductor is the sum of the input capacitance of the amplifying device, stray wiring capacitance, the minimum capacitance of the tuning capacitor itself, and the capacitance of a "trimmer" capacitor used to make minor final adjustments in the resonant frequency of the RF circuit. If $C_{min} = 41.5$ pF, then $C_{max} = 8.78(41.5)$ pF $= 365$ pF, which has become a standard value for the RF circuit tuning capacitor.

If the oscillator uses high-side injection, then the tuning capacitor of the oscillator must have

$$\frac{C_{max}}{C_{min}} = \left(\frac{2055}{995}\right)^2 = 4.27$$

which means that C_{max} is about half that for the RF circuit.

If the oscillator uses low-side injection, then the tuning capacitor of the oscillator must have

$$\frac{C_{max}}{C_{min}} = \left(\frac{1145}{85}\right)^2 = 181.46$$

which means that C_{max} will be huge! The conclusion that can be drawn from this analysis is that it is better to use HSI rather than LSI when tuning the AM broadcast band because the tuning capacitance (or tuning inductance) range will be more reasonable.

On the other hand, if the receiver is to tune the FM broadcast band (88 MHz–108 MHz), and the IF = 10.7 MHz, it makes little difference whether HSI or LSI is used. The one to use must be chosen from other considerations than the range of the tuning variable, such as the facts that the oscillator frequency drifts more at higher frequencies and oscillators are usually more difficult to design at higher frequencies.

8-2.2 Frequency Conversion

It was shown in Section 2-9 that when the sum of two sinusoids such as $E_1 \cos 2\pi f_1 t + E_2 \cos 2\pi f_2 t$ is the input signal to a nonlinear mixer whose static characteristic is represented by a power series containing a squared term, then second-order intermodulation (IM) products at the sum and difference frequencies are present in the output, along with other possible components.

When the mixer is followed by a bandpass filter whose response is centered at either $f_2 + f_1$ or $f_2 - f_1$, then that particular component passes through the filter and the other is effectively eliminated.

In the superheterodyne radio receiver, $E_2 \cos 2\pi f_2 t$ can represent the constant-amplitude sinusoid supplied by the local oscillator. And $E_1 \cos 2\pi f_1 t$ can represent the signal received from an AM broadcast-band station in which E_1 is replaced by a term like $E_1(1 + m_a \cos 2\pi f_1 t)$. Consequently, each of the second-order IM products in the output of the mixer will be amplitude-modulated, but their carrier frequencies will have been changed to $f_2 + f_1$ or to $f_2 - f_1$. This action is known as *frequency conversion.*

When the bandpass filter response is centered at $f_2 + f_1$, *up conversion* is said to occur, and the arrangement is known as a *sum mixer.* When the bandpass filter response is centered at $f_2 - f_1$, *down conversion* is said to occur if $f_2 - f_1 < f_1$, but up conversion is possible if the numbers are such that $f_2 - f_1 > f_1$. This arrangement is known as a difference mixer. Down conversion is sometimes represented by DC and up conversion by UC.

The complete mixer may be referred to as a difference mixer using HSI/DC or perhaps as a sum mixer using LSI/UC. In either case, the desired output frequency of the mixer/filter combination at either $f_2 - f_1$ or $f_2 + f_1$ is called the *intermediate frequency,* usually represented simply by IF, as in Fig. 8-2a.

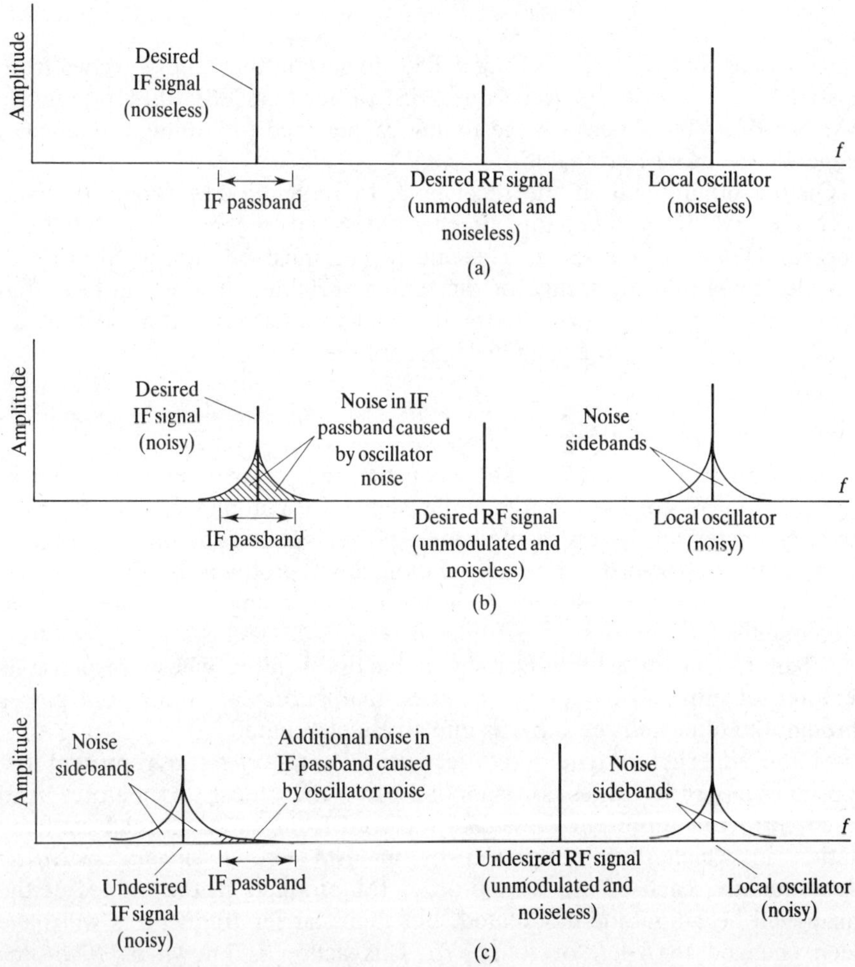

Fig. 8-2 (a) Difference mixing with high-side injection, down conversion. (b) Extension of (a) when oscillator noise is significant. (c) Reciprocal mixing occurs when an undesired signal (shown unmodulated and noiseless for simplicity) mixes with oscillator noise to produce additional noise in the IF passband.

8-2.3 Reciprocal Mixing

When the output voltage E_2 of the local oscillator varies because of noise produced by the oscillator components, then there are undesired noise sidebands accompanying the desired sinusoid in the oscillator output. A small fraction of the oscillator noise may spread into the RF and IF circuits to make the receiver more noisy. These noise sidebands also mix with the desired RF input signal to produce noise sidebands about the desired IF signal, as shown in Fig. 8-2b. Of course, the desired RF signal will have its own noise sidebands. When the oscillator noise sidebands also mix with undesired RF signals to produce additional noise in the IF passband, the process is called *reciprocal mixing*. The net effect of reciprocal mixing is to make the receiver more noisy and to make it appear to have a higher noise factor. This degrades the weak-signal performance of the receiver. The weak-signal performance of a low-noise receiver may be determined by reciprocal mixing caused by oscillator noise.

Even if the oscillator output is spectrally pure, it can mix with the inevitable noise sidebands of an undesired RF signal to produce additional noise in the IF passband. Mixing aside, a small fraction of the noise sidebands of both desired and undesired RF signals may spread into the RF and IF circuits.

8-3 SUPERHETERODYNE CHARACTERISTICS

The weakest link in a radio communications system is the receiver. When you consider that there are thousands upon thousands of transmitting stations around the world, operating in numerous modes such as TV, FM, AM, CW, teletype, and facsimile, including military, commerical, and radio telephone services, operating on so many different frequencies and with such a variety of power levels, the receiver has its job cut out for itself in selecting the one that is desired and rejecting all others.

When you tune to the frequency of a desired station, the receiver should have sufficient *sensitivity* to detect the presence of the desired signal at your location, sufficient *selectivity* to accept the station selected and to reject all stations operating on other frequencies, sufficient *fidelity* of reproduction of the demodulated output to be acceptable, and a sufficiently constant output level even though the signal from the station fades in and out.

More sophisticated design features will include such things as adjustable tone controls; freedom from annoying noise when tuning between stations or when the selected station goes off the air; no large-amplitude output noise spikes; the ability to adjust the receiver's bandwidth or tuning to decrease or eliminate interference; negligible drift of the tuning away from the selected station; a monitor of the input signal strength; a spectrum analyzer to continuously monitor the various input signals; a built-in noise-figure measuring system; ease and convenience of operation; and immunity to interference.

8-3.1 Sensitivity

Sensitivity refers to the ability of a receiver to respond to weak RF signals. It is often expressed as the strength of the RF signal in microvolts that results in a certain output signal-to-noise ratio. It is discussed in more detail in Chapter 10.

8-3.2 Selectivity

Selectivity is a measure of how well a receiver can select a desired station to the exclusion of all others. The early tuned-radiofrequency (TRF) receivers did not have this desirable characteristic to a great extent, but in those days there were not many radio stations to select from. The advent of the superheterodyne receiver improved on the selectivity problem, but it also introduced other undesirable actions known as *image responses,* discussed below, and *spurious responses,* discussed in Chapter 9.

Selectivity is often expressed by the shape of the IF filter response shown in Fig. 8-3. The filter may be composed of LC tuned circuits, crystal elements, ceramic elements, or mechanically or acoustically resonant elements, as determined by the application and the cost. Although not shown in Fig. 8-3, the filter will also have an attenuation or *insertion loss.* The *nose* bandwidth or *passband* is the difference between the frequencies for which the response is down by 3 dB from the maximum shown to be at 0 dB. The filter design determines the amount of ripple in the passband.

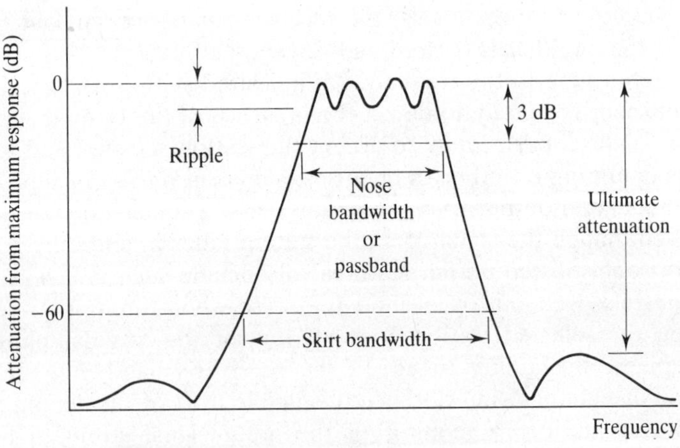

Fig. 8-3 Some selectivity specifications.

The response shape above and below the passband is called the *skirt*. The difference between the frequencies for which the response is down by 60 dB is called the *skirt bandwidth*. There are usually one or more frequencies on each side of the response curve at which the attenuation is theoretically infinite, and the response fluctuates somewhat beyond those frequencies. This gives rise to an ultimate attenuation that is often greater than 60 dB outside the skirt bandwidth.

Selectivity is often specified by a *shape factor* given by

$$\text{Shape factor} = \frac{\text{skirt bandwidth}}{\text{nose bandwidth}}$$

Selectivity is considered to be poor if the shape factor is greater than 10, reasonably good if the shape factor is between 3 and 6, and very sophisticated if the shape factor is less than 1.2. An ideal filter has infinitely steep skirts and a shape factor equal to 1. Sometimes the nose bandwidth is taken between the −6 dB rather than the −3 dB points when shape factor is specified, because this makes the shape-factor number smaller.

8-3.2.1 Image Response The earliest superhet receivers had independently adjustable tuning capacitors in the RF and the local oscillator circuits. If the incoming RF signal was at 1000 kHz and the IF was 100 kHz, for example, the local oscillator could be tuned to either 1100 kHz or 900 kHz to produce the desired IF from a difference mixer. Thus, for every received RF signal, there were found two settings of the oscillator tuning capacitor which worked equally well. One of these was called the *image* setting of the local oscillator. Of course, this phenomenon represents either LSI or HSI. The subsequent ganging together of the tuning capacitors eliminated the image setting.

As more stations began operating throughout the broadcast band, another annoying response appeared. If the incoming RF signal was from a weak distant station operating on 1000 kHz and the IF was 100 kHz, for example, then the local oscillator was automatically set to 1100 kHz to produce the desired signal that was delivered to the IF amplifiers. But suppose that an undesired signal from a strong local station on 1200 kHz also gets through the selectivity of the RF circuits and arrives at the mixer. It mixes with the oscillator frequency to produce an undesired signal at 100 kHz that will also pass through the IF amplifiers. Both of these IF signals are then demodulated, and you hear both stations simultaneously. The undesired RF signal on 1200 kHz that caused the interference is called the *image signal,* and 1200 kHz is called the *image frequency* of 1000 kHz, in this example. The output that is obtained from an input signal at the image frequency is called the *image response*.

In general, when HSI/DC is used, the image frequency is given by

$$f_{\text{image}} = f_r + 2\text{IF}$$

where f_r is the frequency to which the receiver is tuned.

In the above example, the receiver is tuned to a station on 1000 kHz. The desired signal is weak, perhaps because the station is far away. Because of the RF selectivity, a signal at 1200 kHz may be attenuated by, say, 10 dB. But if the signal at the image frequency 1200 kHz is 10 dB stronger than the other one, perhaps because of the proximity of that station to the receiver, then both signals produced by the mixer have the same amplitude. On the other hand, if the relative strengths of the two signals are reversed, it is possible that the image response will not be sufficient in amplitude to be noticeable. Because of this, interference that may be objectionable at one location and frequency caused by image response may not be noticeable at a location far removed from the other.

The image response of a receiver is said to be improved when the output produced by a signal at the image frequency is made to be less than it was before, (i.e., there is greater image rejection). Sometimes decreasing the RF bandwidth can place the image frequency far enough down on the selectivity curve to make the image response negligible. But the bandwidth can only be decreased to the point when further reduction will cause distortion of the desired signal, and the image rejection may still be too poor. Perhaps the best way to improve image response is to design the receiver initially to have as high an IF as possible. According to Eq. (8-1) when HSI/DC is used, the desired and image frequencies are separated by 2IF. If the IF is small, the image frequency falls higher on the RF selectivity curve than when the IF is large, so the image response is poor. Certainly, an IF of 455 kHz is much better in improving response than an IF of 100 kHz. But an IF of 40 MHz is even better, for two reasons. Not only does the image frequency fall far down on the RF selectivity curve, but the image frequency itself is so high that there are many high-powered stations operating at such high frequencies.

This brings us to one of the basic design principles for superhet receivers:

FOR GOOD IMAGE REJECTION, MAKE THE IF AS HIGH AS POSSIBLE.

8-3.2.2 Adjacent-Channel Selectivity The superhet is capable of very good adjacent-channel selectivity. That is, it has the ability to select a desired station and exclude undesired stations that may be operating in adjacent channels. The early tuned-radiofrequency (TRF) receiver was adequate until the broadcast band became crowded with so many stations that the receiver could no longer exclude undesired stations unless their operating frequencies were far removed from that of the desired station. This was because of the relatively broad passband of the RF tuned circuits.

There is no difference between the RF tuned circuits of the TRF and superhet receivers. Therefore the superior adjacent-channel selectivity of the superhet cannot be because of its superior RF circuits. Its superior selectivity is caused by its frequency-conversion process. This is explained in what follows.

Let us compare two superhets, A and B, each with HSI/DC, tuned to a desired station on 1000 kHz. Assume that an undesired station is in an adjacent channel on 1010 kHz. The desired and undesired frequencies are separated by 10 kHz out of a desired total of 1000 kHz, or 1 percent separation. Superhet A has an IF of 100 kHz (its local oscillator frequency is 1100 kHz) and superhet B has an IF of 500 kHz (its local oscillator frequency is 1500 kHz).

Superhet A produces a desired IF signal at 100 kHz and an undesired interfering signal at 90 kHz. Note that frequency conversion does not change the spacing between the desired and undesired signals. The desired and undesired IF signals are separated by 10 kHz out of the desired total of 100 kHz, or 10 percent separation.

Superhet B produces a desired IF signal at 500 kHz and an undesired interfering IF signal at 490 kHz. The desired and undesired IF signals in this receiver are still separated by 10 kHz as before but now that 10 kHz is out of a desired total of 500 kHz, or 2 percent separation.

Therein lies the fundamental difference in adjacent-channel selectivity of the two superhets. It is easier to distinguish between IF signals that are separated by 10 percent than by only 2 percent. Also, both receivers can provide better performance than a TRF receiver where the separation is only 1 percent. But superhet A can provide much better adjacent-channel selectivity than B. This leads to the next basic design rule for superhets.

FOR GOOD ADJACENT-CHANNEL SELECTIVITY,
MAKE THE IF AS LOW AS POSSIBLE.

8-3.2.3 Double Conversion The two basic design rules that have just been described are in direct conflict with each other, but both can be satisfied by designing the superhet to have two different IFs. The process, known as double conversion, is illustrated in Fig. 8-4. The first IF is made to be high for good image rejection and the second IF is made to be low for good adjacent-channel selectivity. If the receiver is tunable, then oscillator #1 must also be tunable to produce the constant first IF. Oscillator #2 then operates at a constant frequency.

For example, a certain superhet tunes from 2 MHz to 30 MHz. The first IF is 55 MHz, produced by a difference mixer with HSI. So oscillator #1 tunes from 57 MHz to 85 MHz. The image frequencies range from 112 MHz to 140 MHz. Not only are these image frequencies far removed from the desired RF signals, but there is little chance for image interference from high-powered VHF transmitters. If the second IF is 450 kHz, then the

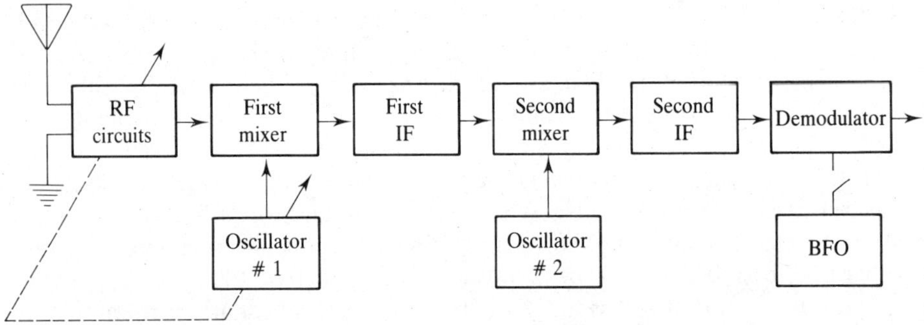

Fig. 8-4 Simplified block diagram of a double-conversion superhet. BFO = beat frequency oscillator.

frequency of oscillator #2 can be either 45.45 MHz or 44.55 MHz and a difference mixer is required.

The high first IF and low second IF should not be interchanged for a low first IF and a high second IF. If the interchange is made, good adjacent-channel selectivity is obtained from the low first IF, but then image rejection is poor. Further, the high second IF cannot undo the poor image response that has already occurred.

Figure 8-4 also includes a beat-frequency oscillator, or BFO, but this could be included in Fig. 8-1 as well. The BFO is used when receiving CW, or continuous wave, code signals. Each letter of the alphabet, each numeric digit, each punctuation mark, and each operating token, abbreviation, and shortcut, has its own unique Morse code composed of dots and dashes. For example, the letter A or a (no distinction is made between upper and lower case) is represented by a dot followed by a dash.

Suppose the speed of the CW operator at the transmitter is such that the telegraph key is pressed for 1/8 s and released to make a dot, then immediately the key is pressed for 3/8 s and released to make a dash. The transmitter output is on, then off, then on again, and then off, to represent the dot and dash of the letter. The output is controlled by the telegraph key. The envelope of the transmitted signal has a square-wave shape.

Without the BFO in the receiver, all that is heard by the receiver operator will be the thumps and clicks of the transmitter being keyed on and off. It is better if the receiver operator can hear a tone for each dot and dash. That is the purpose of the BFO—to provide for the tone. Suppose the receiver is tuned to the transmitter and that the second IF is 455 kHz. Then, whenever the key is pressed, a constant-amplitude 455-kHz sinusoid is delivered to the demodulator. If a BFO is present, its output at 456 kHz or 454 kHz mixes with the 455 kHz signals. Beating occurs as discussed in earlier chapters, the beat frequency being 1 kHz. If the amplitude of the BFO output is somewhat larger than that of the IF signal, the resulting envelope is almost sinusoidal. When this signal, with its varying envelope caused by beating, is demodulated, the 1-kHz tone is heard each time the key is pressed at the transmitter.

The frequency of the tone can be varied by changing the BFO frequency, or, if the BFO frequency is fixed, by retuning the receiver to change the frequency of the signal passing through the IF circuitry. Suppose an undesired tone is also present caused by an interfering station operating on a frequency that is only 100 Hz from that of the desired station. The interfering tone can be made to be inaudible by adjusting the BFO frequency or the receiver tuning to give zero beat for the undesired tone. The desired signal then has a tone of 100 Hz, but the interference of the undesired station cannot be heard.

8-3.3 Fidelity

In an audio sound system, true fidelity occurs when the amplified output waveform is a perfect replica of the input waveform. A good-quality high-fidelity (hi-fi) stereo music system represents a close approach to true fidelity.

In a radio receiver, true fidelity occurs when the demodulated and amplified output waveform is a perfect replica of what the modulated input signal actually represents. Fidelity of the overall radio system depends on how well the transmitter does its job, and whether propagation through the intervening space to the receiver influences the modulated wave unfavorably to affect the fidelity.

True fidelity is degraded by nonlinear components that introduce harmonic, intermodulation, and cross-modulation distortions; by insufficient dynamic range; by unequal amplification and nonconstant time delay of all frequencies contained in the waveform; by transient response that has long rise and decay times, or appreciable ringing; by assorted noises; and by various undesirable effects produced by the operation of mechanical devices such as earphones, loudspeakers, recording pens, and the human ear itself.

8-3.4 Automatic Gain Control (AGC)

The purpose of automatic gain control, or AGC, is to maintain a nearly constant output level even when the strength of the input signal changes. The same arrangement is sometimes referred to as automatic volume control (AVC). The strength of the input signal can change because of fading, or because the receiver is tuned from a weak station to a strong station or vice versa. Strong input signals do not need as much amplification in the receiver as do weak input signals, so they can afford to encounter less gain.

The AGC system is designed to control the overall amplification, or gain, of the receiver with a DC voltage that depends on the amplitude of the input signal. As the strength of the input signal increases, the DC control voltage also increases and then this, in turn, reduces the gain. As the strength of the input signal decreases, the DC control voltage also decreases and this, in turn, increases the gain. When the input signal strength falls below a certain value, such as that which produces a 10-dB output signal-to-noise ratio, the AGC action should cease. This is because such weak input signals need all the gain that is possible, but if there is no DC control voltage, there will be no compensation for fading or tuning effects. When the onset of AGC action is prevented until the input signal strength reaches a certain predetermined threshold, delayed AGC is said to occur.

Several diode circuits that can be used to produce the DC control voltage are shown in Fig. 8-5. Their operation is basically that of a half-wave rectifier having an RC filter. The input voltage for each circuit is taken near the end of the IF amplifiers but before the final narrow-band IF filter, if any. The DC voltage thus produced is fed back to preceeding IF and RF amplifiers to control their gains. The time constant of the RC filter must be short enough that the AGC voltage can follow intermittent changes in signal amplitude caused by fading or tuning effects, and long enough to prevent RF and AF voltages from being fed back to the controlled amplifiers. RF

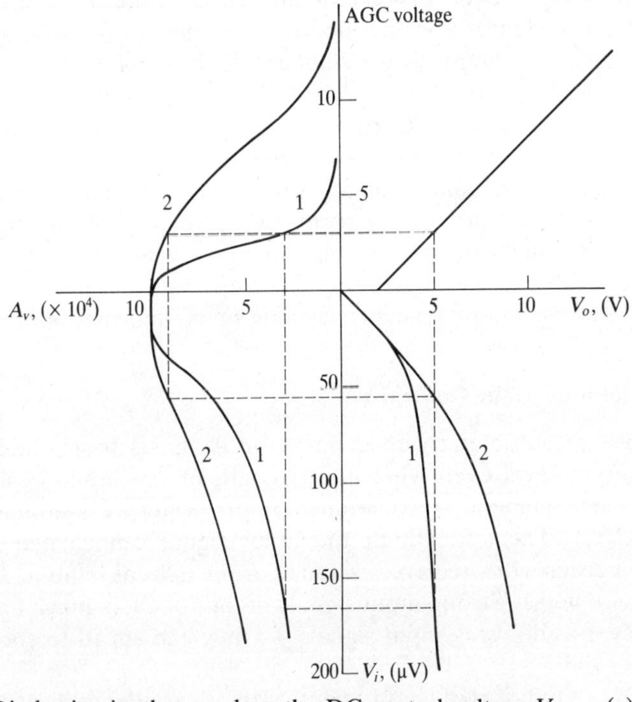

Fig. 8-5 Diode circuits that produce the DC control voltage V_{AGC}: (a) fast attack, slow decay; (b) slow attack, slow decay; (c) fast attack, fast decay; (d) slow attack, fast decay.

feedback can lead to possible instabilities in the forms of ringing and oscillation. AF feedback in an AM receiver can reduce the percentage modulation of the AM signal passing through the controlled amplifiers, primarily at the lower audio frequencies, and this antimodulation reduces the bass response. An RC decay time constant of about 0.25 s is satisfactory for an AM broadcast receiver, and of about 1 s for a short-wave CW receiver.

In Fig. 8-5a, capacitor C can charge rapidly through the low forward resistance of diode D1 to a voltage nearly equal to the peak value of the IF input voltage. The charging time constant is also known as the attack time constant, or simply as attack. This circuit is capable of fast attack. Then, when the input voltage decreases, D1 ceases to conduct, so that C must discharge through R (which includes any load resistance). If R is large, then C discharges slowly. The discharge time constant is also known as the decay time constant, or simply as decay. This circuit is also capable of slow decay.

In Fig. 8-5b, resistor R_a causes a slower attack than without it. The decay can still be slower than the attack if R is large.

The circuit of Fig. 8-5c is capable of both fast attack and fast decay, and resistor R has little effect on either. The capacitor can charge rapidly through the low forward resistance of diode D1 and then discharge rapidly through the low forward resistance of D2 when the input voltage decreases. Therefore the AGC voltage can follow rapid changes in the amplitude of the input signal. For that reason, this circuit is not suitable for AM receivers, but may be useful in FM receivers. By adding resistor R_a as in Fig. 8-5d, a slower attack can occur but a fast decay is still present.

Other kinds of automatic gain control are used in some receivers: noise limiters, noise blankers, squelch circuits, and hang AGC. Noise limiters are set to clip (i.e., limit) the output when its level exceeds a certain value because of the presence of an impulse of noise. Noise blankers act somewhat like an AGC system, but instead of simply reducing the gain for large input signals, they "gate off" the gain of a controlled stage when a noise impulse occurs. Squelch circuits "gate off" the gain of the audio output to prevent output noise when there is no input signal. Hang AGC employs three time constants for attack, hang (or hold), and decay. The hang voltage is audio-derived and, after a fairly long attack time and preset decay time (the hang time) of its own, controls a transistor gate that, in turn, makes the AGC decay time long when there is audio output and short when there is no audio output. Because of the hang time, short pauses in speech for a single-sideband transmission or between characters in a CW transmission, will not result in an increase in background noise. When there is no sustained input signal present, a burst of noise quickly increases the AGC voltage but then it quickly recovers.

8-3.4.1 An AGC System Analysis The steady-state operation of an AGC system in a fictitious receiver is depicted in Fig. 8-6. This is actually four

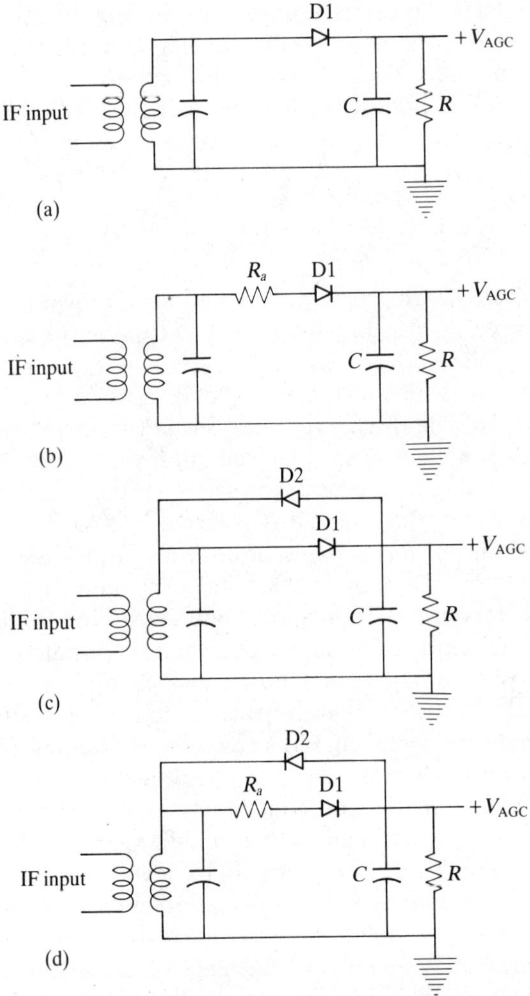

Fig. 8-6 Steady-state analysis of an AGC system. *1st quadrant*: AGC voltage vs. peak IF output volts. *2nd quadrant*: Overall system amplification vs. AGC voltage. *3rd quadrant*: peak input volts required vs. amplification. *4th quadrant*: peak output volts vs. peak input volts.

figures arranged around each of the four quadrants. The first quadrant shows the DC AGC voltage as ordinate vs. the peak value of the IF voltage applied to the AGC diode as abscissa. Delayed AGC is present because no AGC voltage is produced until the IF output voltage is greater than 2 V.

The figure in the second quadrant will look more reasonable if the page is rotated 90° clockwise. That graph is intended to show how the overall gain varies as the AGC voltage fed back to the controlled stages is changed. The overall gain A_v is shown as ordinate vs. the AGC voltage as abscissa. Two curves, marked 1 and 2, are shown for two different systems. The overall gain of the system represented by curve 1 is more sensitive to the AGC voltage than that for curve 2. Normally, the curves shown in the first two quadrants would be obtained by measurements on an actual operating system.

The curves shown in quadrant 4 are calculated from values obtained from the graphs in quadrants 1 and 2. For each value of overall gain and peak IF output voltage, there must be a corresponding peak input voltage applied to the antenna terminals of the receiver. For example, if $V_0 = 5$ V and $A_v = 3 \times 10^4$ for curve 1 in the second quadrant, then the peak input voltage is $V_i = 5/(3 \times 10^4) = 167\ \mu$V, shown as a point on curve 1 in the third quadrant. Also, using curve 2 in the second quadrant, when $V_o = 5$ V, $A_v = 9.2 \times 10^4$. Therefore, for this system, $V_i = 5/(9.2 \times 10^4) = 54\ \mu$V, shown as a point on curve 2 in the third quadrant. Neither of these curves is particularly meaningful by itself. They are used here primarily as an intermediate step to help in obtaining the final graphs in the fourth quadrant.

Finally, the graphs in the fourth quadrant will appear more sensible if the page is rotated from its original upright position 90° counterclockwise. Then the IF output voltage V_o (in volts) appears as ordinate vs. the input voltage V_i (in microvolts) as abscissa. No additional calculations are required for these curves. For example, if the output voltage is 5 V, the required input voltage from curve 1 in the third quadrant is 167 μV, and these two values locate a point on curve 1 in this quadrant. Or, if $V_0 = 5$ V and $V_i = 54\ \mu$V from curve 2, then the corresponding point on curve 2 in this quadrant is located as shown. When this process is repeated for several different values of V_o for both curves 1 and 2 of the second quadrant, the final input–output relationships that show the compression effect of AGC action will be obtained.

AGC voltage fed back to control the gain of preceding stages is said to be back-acting. This method cannot provide a perfectly constant output amplitude when the input amplitude varies because the effect on the preceding controlled stages follows after the change that occurs at the later stage where the AGC voltage is created.

AGC voltage fed forward to control the gain of later stages is said to be forward-acting. This method can provide a constant or even a drooping output characteristic when the input amplitude is increased.

The amount of AGC voltage is an indication of the strength of the input signal. Some receivers use a meter to measure the AGC voltage, but the meter is calibrated in arbitrary strength units S1 to S9. S9 represents an input signal strength of about 100 μV, each S unit below S9 represents a change of about 6 dB, and above S9 the meter is calibrated in dB. The calibration of an S meter is only approximately linear, and it is not an absolute calibration because the sensitivity of the receiver varies from frequency to frequency across a band and from band to band.

Some receivers also have a manual control of the AGC voltage. This allows the receiver operator to reduce the RF gain more than the automatic circuitry calls for when a strong input signal is present. This helps to prevent receiver overload caused by the strong signal.

8-3.4.2 Diode Attenuators and Switches The forward current of a semiconductor diode is given by the diode equation

$$i_D = I_o(e^{v_D/V_T - 1}) \tag{8-2}$$

where v_D = the forward diode voltage, $V_T = nkT/q$, and I_o = the reverse saturation current. The forward direction dynamic resistance r_d of the diode, at a certain Q point where $i_D = I_{DQ}$, is found from the reciprocal of the derivative of (8-2) with respect to v_D. The result is

$$r_d = \frac{V_T}{I_{DQ} + I_o} \tag{8-3}$$

If $V_T = 26\,\text{mV}$ at room temperature, with $I_o = 52\,\text{nA}$, and $I_{DQ} = 10\,\text{mA}$, then the resistance of the diode to small signals is $r_d \cong 26/10 = 2.6\,\Omega$. If the current is decreased to 0.01 mA, then $r_d \cong 2600\,\Omega$. The fact that the small-signal resistance of the forward-biased diode can be changed from only a few ohms to hundreds or even thousands of ohms simply by changing the forward-bias current suggests that the arrangement is suitable as a variable attenuator. However, when the amplitude of the signal becomes large, the diode is driven into nonlinear operation and into rectification.

If I_{DQ} is made small compared to I_o, then $r_d \cong 26/(52 \times 10^{-6}) = 500\,\text{k}\Omega$. This high value suggests that the diode resistance can approximate an open switch, especially in low-impedance circuits.

Unfortunately, the diode depletion capacitance complicates things. If this capacitance is 15.9 pF when the diode current is small, then, for a frequency of 5 MHz, there is a capacitive reactance of 2000 Ω in parallel with $r_d = 500\,\text{k}\Omega$. This is reason enough to rule out the use of that particular diode in an attenuator at high frequencies.

The diode capacitance can be reduced significantly by physically separating the p and n regions. This is accomplished in the PIN diode by fabricating a very lightly doped p region (i.e., almost an Intrinsic layer, hence the I in the name of the device), between the p and n regions. Not only does the lightly doped p layer decrease the diode capacitance, but, when there is no forward bias, it also causes the resistance between the p and n regions to be higher than in a pn diode. Then, increasing the forward bias on the PIN diode causes more current carriers to be injected into the middle layer, and this allows the resistance to decrease. PIN diodes are used primarily in high-frequency attenuators and switches at frequencies from about 10 MHz up into the gigahertz range.

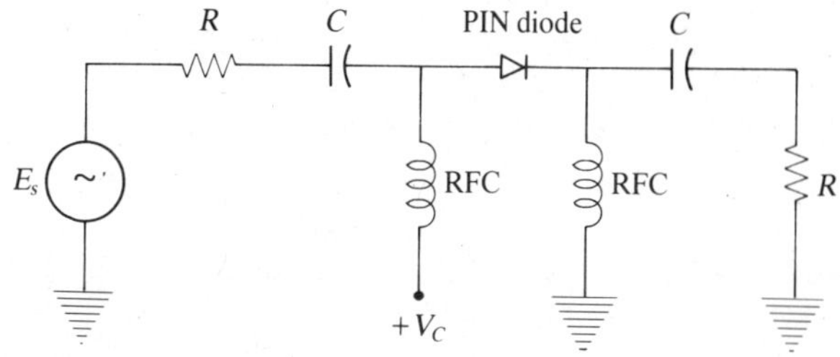

(a)

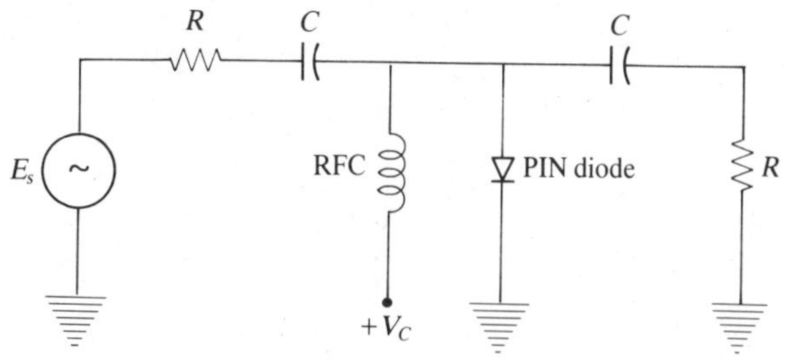

(b)

Fig. 8-7 PIN diode attenuators: (a) series; (b) parallel.

Series and parallel types of PIN attenuators are shown in Fig. 8-7. The dynamic resistance of either diode is controlled by the forward bias voltage V_c. The inductances marked RFC are called RF chokes and are used to pass the DC bias current for each diode and also offer a high impedance at the signal frequency. The capacitors are used for DC blocking, signal coupling, and bypassing. In an AGC system, the bias voltage V_c would come from a properly supplied AGC voltage.

When used as a switch, the bias voltage V_c would have to change the dynamic resistance r_d from a small value to a very high value. As the diode resistance is changed, in either system, the load resistance experienced by the signal source changes, and this may undesirably affect the voltage standing-wave ratio VSWR.

There are several different resistive circuits that can provide for variable attenuation and, at the same time, maintain the input and output resistances constant provided that the source and load resistances are constant. The bridged-T circuit of Fig. 8-8 is one such arrangement. The bridged-T circuit is symmetrical when the source and load resistances are the same.

Since attenuation occurs, $V_2 < V_1$. Define the attenuation ratio as $A = V_1/V_2$, so that $A > 1$. When expressed in dB, the attenuation is a positive number. After straightforward but tedious algebraic manipulations,

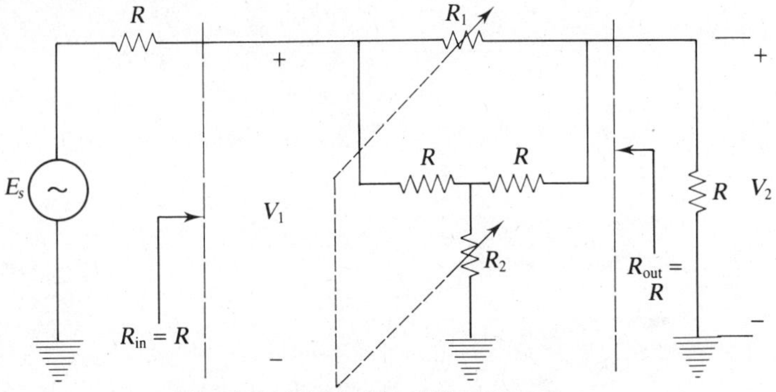

Fig. 8-8 The bridged-T attenuator.

it can be shown that $R_{in} = R_{out} = R$ provided that

$$R_1 = R(A - 1) \tag{8-4}$$

and

$$R_2 = \frac{R}{A - 1} \tag{8-5}$$

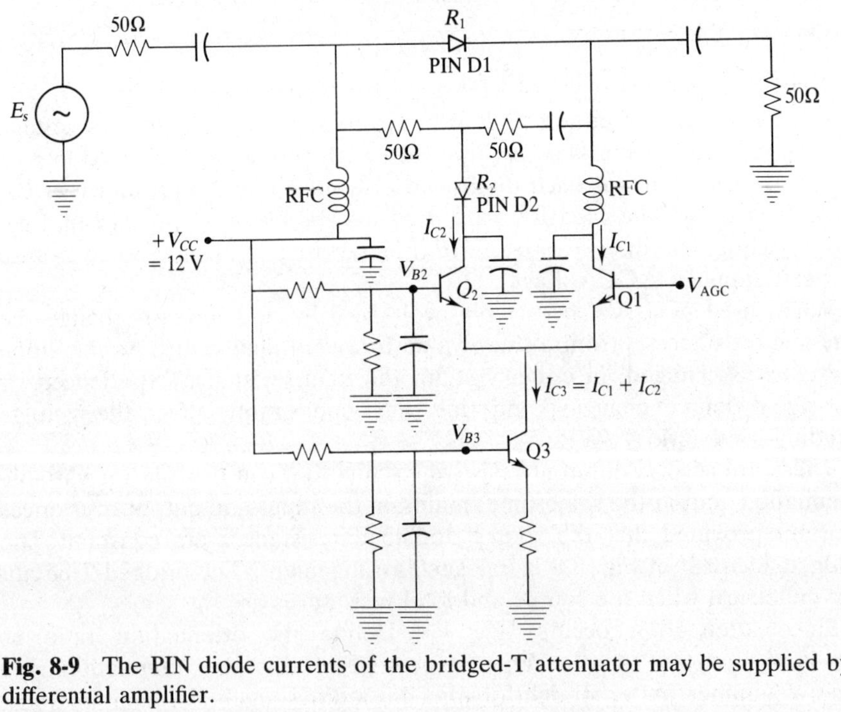

Fig. 8-9 The PIN diode currents of the bridged-T attenuator may be supplied by a differential amplifier.

from which is obtained

$$A = \frac{R_1}{R} + 1 = \frac{R}{R_2} + 1 \tag{8-6}$$

Also,

$$R_1 R_2 = R^2 = \text{a constant} \tag{8-7}$$

The amount of attenuation can be set by manually adjusting R_1 and R_2 via ganged rheostats having a common shaft or by switching in and out the proper resistances. The attenuation can be adjusted electrically by using PIN diodes for R_1 and R_2. One approach is shown in Fig. 8-9. There, the DC power supply of 12 V provides I_{C1} of transistor Q1 as bias current through diode D1, and I_{C2} of transistor Q2 as bias current through diode D2. The base voltage V_{B3} maintains the sum $I_{C1} + I_{C2}$ constant. When $V_{AGC} = V_{B2}$, $I_{C1} = I_{C2}$, and this makes $A = 2$, which is an attenuation of 6 dB. If $V_{AGC} > V_{B2}$, then $I_{C1} > I_{C2}$, R_1 decreases and R_2 increases, so the attenuation decreases. Similarly, if $V_{AGC} < V_{B2}$, the attenuation increases.

The manufacturer specifies how the dynamic resistance of a particular diode depends on the forward bias current at 25°C, usually as a graph like that in Fig. 8-10. It is a straight line over most of its useful range, but it tends to level off at very low and very high currents. The line can be represented by

$$\log r_d = -n \log I_{DQ} + \log r_1 \tag{8-8}$$

where the current is in milliamps, the slope is $-n$, and $r_1 =$ the resistance when $I_{DQ} = 1$. From Fig. 8-10, $r_1 = 27\ \Omega$ and $n = (\log 8900 - \log 27)/(\log 1 \times$

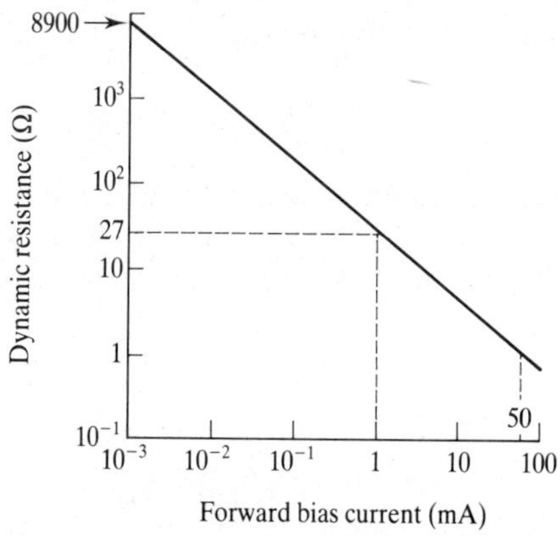

Fig. 8-10 Dynamic resistance vs. forward bias current of a typical PIN diode.

$10^{-3} - \log 0.001 \times 10^{-3}) = 0.84$. Thus, (8-8) becomes

$$r_d = \frac{27}{(I_{DQ})^{0.84}} \tag{8-9}$$

in which the current is in mA.

In Fig. 8-9, the resistances R_1 and R_2 of the bridged-T are the dynamic resistances of the two diodes, and the forward bias currents are I_{C1} and I_{C2}. Hence,

$$R_1 = \frac{27}{(I_{C1})^{0.84}} \tag{8-9a}$$

and

$$R_2 = \frac{27}{(I_{C2})^{0.84}} \tag{8-9b}$$

From (8-7), these combine into

$$R_1 R_2 = \frac{27^2}{(I_{C1} I_{C2})^{0.84}} = R^2$$

from which is obtained, for $R = 50\ \Omega$,

$$I_{C1} I_{C2} = \left(\frac{27}{50}\right)^{2/0.84} = 0.23 \tag{8-10}$$

This is a condition on the two forward bias currents that is required for this attenuator to remain matched.

But Q3 of Fig. 8-9 holds the sum of the two collector currents constant at a value I_{C3} (it is assumed here that the base currents of Q1 and Q2 are negligible). Consequently, for this circuit,

$$I_{C1} + I_{C2} = I_{C3} = \text{a constant} \tag{8-11}$$

where the constant I_{C3} is to be selected. The conditions represented by (8-10) and (8-11) cannot both be satisfied simultaneously over a control range, but they can be satisfied approximately. From (8-10),

$$I_{C2} = \frac{0.23}{I_{C1}}$$

which is the hyperbola in Fig. 8-11. From (8-11),

$$I_{C2} = -I_{C1} + I_{C3} \tag{8-12}$$

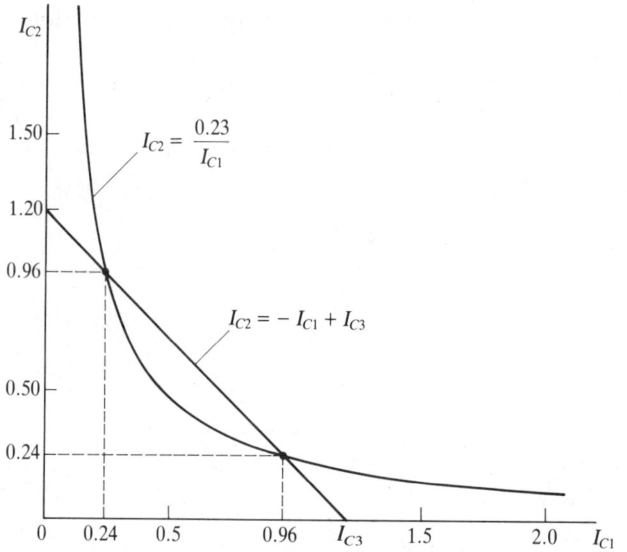

Fig. 8-11 Comparison of the conditions $I_{C1}I_{C2} = 0.23$ and $I_{C1} + I_{C2} = I_{C3}$.

which is the straight line in Fig. 8-11. If I_{C3} is selected such that the line is tangent to the hyperbola, then both of the conditions above are satisfied only at the one tangent point. If I_{C3} is selected such that the line crosses the hyperbola at two points, then both conditions are satisfied only at those two points. If I_{C3} is selected to be 1.2 mA, then

$$I_{C1} + I_{C2} = 1.2$$

When $V_{AQC} = V_{B2}$, the two collector currents are the same, so that $I_C = I_{C2} = 0.6$. From (8-9a) and (8-9b), $R_1 = R_2 = 27/(0.6)^{0.24} = 41.5\ \Omega$. Then $R_1R_2 = 1722$ instead of 2500.

The two points where the line crosses the hyperbola are located at $I_{C1} = 0.96$, $I_{C2} = 0.24$, and at $I_{C1} = 0.24$, $I_{C2} = 0.96$. At the first point, $R_1 = 27/(0.96)^{0.24} = 27.94\ \Omega$ and $R_2 = 27/(0.24)^{0.24} = 89.53\ \Omega$. Then $R_1R_2 =$ 2500 as expected. The attenuation can be calculated from (8-6). The result is $A = 1.56$ or 3.85 dB.

At the second point, $R_1 = 89.53\ \Omega$ and $R_2 = 27.94\ \Omega$, and $A = 2.79$ or 8.91 dB. The circuit of Fig. 8-10 can only approximate the condition $I_{C1}I_{C2} = 0.23$ at all other points, but it will provide attenuation there. The calculation of attenuation in the general case will not be pursued here.

Since this circuit does not maintain the matched condition over a control range, this arrangement for providing AGC action may be suitable when it is placed between an antenna and the receiver input terminals. This is because the antenna impedance varies greatly over a wide frequency range, and there is little hope of maintaining a matched condition over those frequencies anyway.

Instead of changing the dynamic resistances of PIN diodes by controlling the forward-bias currents, the forward-bias voltages can be controlled. The diode equation relates forward current to forward voltage. Thus, for two diodes,

$$I_1 = I_o e^{V_1/V_T} \tag{8-13a}$$

and

$$I_2 = I_o e^{V_2/V_T} \tag{8-13b}$$

Therefore,

$$I_1 I_2 = I_o e^{(V_1+V_2)/V_T} \tag{8-13c}$$

from which is obtained

$$V_1 + V_2 = V_{\text{SUM}} = V_T \ln \frac{I_1 I_2}{I_o} \tag{8-14}$$

If the temperature is constant, so that both V_T and I_o remain fixed, the condition $I_1 I_2 =$ a constant translates into the fact that the sum of the two forward voltages V_{SUM} must also remain fixed. If the diodes are the same type, then they have the same values for V_T and for I_o.

The PIN diode manufacturer usually specifies graphically how the forward current depends on the forward voltage. A typical graph is shown in Fig. 8-12 for 25°C. If this information is not available in the data sheets, laboratory measurements will have to be made so that V_T and I_o can be calculated and used in (8-14). Notice that the forward current in Fig. 8-12 is plotted on a log scale and that the forward voltage is plotted on a linear

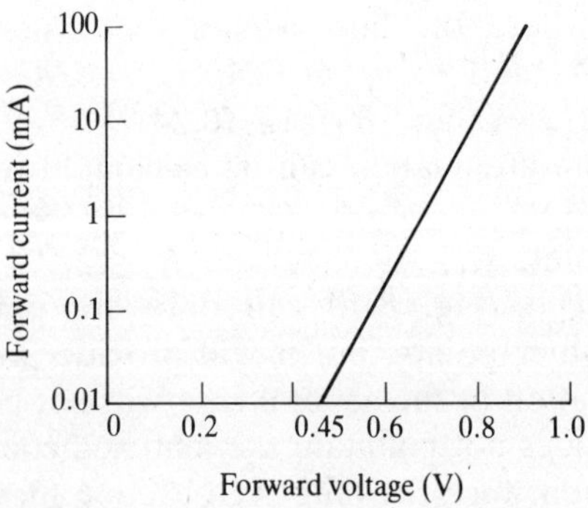

Fig. 8-12 Forward current vs. forward voltage characteristic for a typical PIN diode.

scale. These are the same axes that will be obtained after taking the logarithm of (8-13a) or (8-13b). Thus,

$$\log I_1 = \frac{V_1}{V_T} \log e + \log I_o$$

or

$$\log I_1 = \frac{0.434}{V_T} V_1 + \log I_o \tag{8-15}$$

This is a straight line when $\log I_1$ is ordinate and V_1 is abscissa. These coordinates agree with those of Fig. 8-12, therefore (8-15) represents the straight line in Fig. 8-12. Comparing (8-15) with $y = mx + b$, it is seen that the slope is $+0.434/V_T$. Taking numbers from Fig. 8-12, the slope is found to be

$$\text{Slope} = \frac{\log 10 - \log 0.01}{0.8 - 0.45} = 8.57$$

Equate this to the slope $0.434/V_T$ and solve to obtain $V_T = 50.64\text{ mV}$. At a forward voltage of 0.8 V the current is 10 mA. Thus,

$$10\text{ mA} = I_o e^{0.8/0.05064}$$

or

$$I_o = \frac{10 \times 10^{-3}}{e^{15.8}} = 1.37\text{ nA}$$

Consequently, from (8-14), for the bridged-T attenuator of the previous example with $I_1 I_2 = 0.23$, we find that

$$V_{\text{SUM}} = 50.64 \ln \frac{0.23}{1.37 \times 10^{-9}}\text{ mV} = 0.96\text{ V}$$

The proposed attenuator circuit is shown in Fig. 8-13a. The sum of the forward bias voltages of the diodes must equal 0.96 V. So, if one voltage increases, the other must decrease in such a way that $V_1 + V_2 = V_{\text{SUM}}$. The op amp circuit of Fig. 8-13b shows a method for using the control voltage V_1 and generating the other voltage V_2 so that their sum is constant. This situation can occur when only V_1 is initially available, and is equal, perhaps, to a DC AGC voltage V_{AGC} having positive polarity. If the polarity is negative, reverse both diodes.

In Fig. 8-13b, voltage V_1 is applied to the inverting input through R_1 and amplified by the inverting gain $-R_F/R_1$. The constant voltage V_{SUM} is

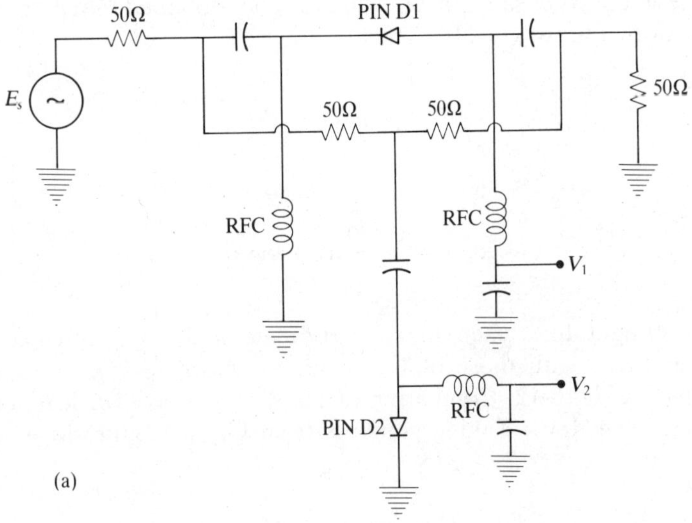

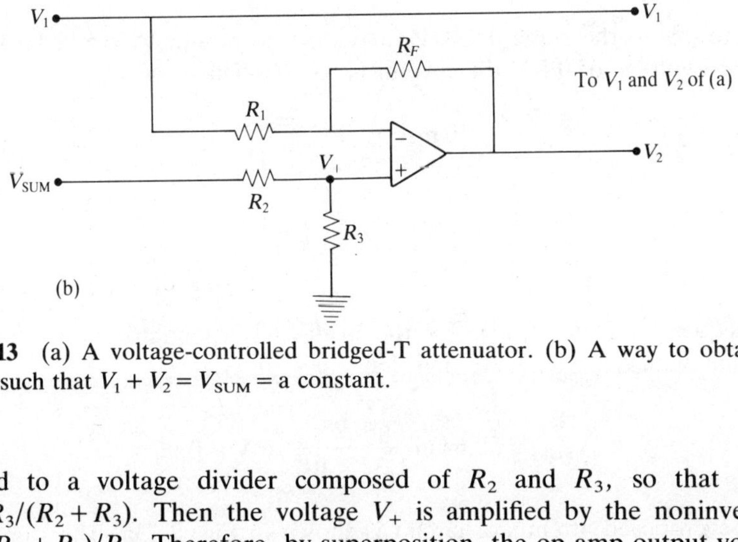

Fig. 8-13 (a) A voltage-controlled bridged-T attenuator. (b) A way to obtain V_1 and V_2 such that $V_1 + V_2 = V_{SUM} =$ a constant.

applied to a voltage divider composed of R_2 and R_3, so that $V_+ = V_{SUM}R_3/(R_2 + R_3)$. Then the voltage V_+ is amplified by the noninverting gain $(R_F + R_1)/R_1$. Therefore, by superposition, the op amp output voltage is

$$V_2 = -\frac{R_F}{R_1} V_1 + V_{SUM}\left(\frac{R_3}{R_2 + R_3}\right)\left(\frac{R_F + R_1}{R_1}\right) \tag{8-16}$$

Now, if we select $R_F = R_1$ and $R_2 = R_3$, (8-16) reduces to

$$V_2 = V_{SUM} - V_1$$

If V_1 and the V_2 thus produced are connected to the V_1 and V_2 terminals of Fig. 8-13a, then when $V_1 = V_{AGC}$ becomes more positive, V_2 becomes less positive. The dynamic resistance of D_1 decreases and that of D_2 increases, therefore the attenuation decreases. If this is the wrong direction, reverse the connections to V_1 and V_2. Instead of getting V_{SUM} from another voltage divider, the circuit is a little simpler if R_2 and R_3 form the voltage divider across a regulated DC source, and their values are adjusted to make $V_+ = V_{SUM}/2$.

8-4 ADVANCED DESIGN FEATURES

High-performance superheterodyne radio receivers have many other sophisticated design features, some of which are described in what follows.

8-4.1 Hang AGC

During a pause in speech or a short program break, the AGC system of an AM receiver still has the unmodulated carrier signal available to produce the AGC voltage. But in a SSB receiver, there is no carrier during a pause or break in the received program; and no carrier is transmitted between the dots and dashes of a received CW signal. However, it seems that this problem can be solved by using an AGC system having a long decay time of several seconds.

Unfortunately, if the decay time is lengthened this much, then, when tuning the receiver, a large-amplitude noise pulse, or the signal produced when tuning the receiver rapidly across a strong station, can cause a sudden increase of the AGC voltage and its attendant decrease of receiver gain. Therefore, because of the long decay time, the receiver is almost muted for several seconds so that weaker stations may not be noticed. This makes tuning more difficult and slow.

The problem can be eliminated in SSB and CW receivers by using what is called hang AGC. The simplified form of one such system is given in Fig. 8-14 to represent the action that takes place. The two input signals may be obtained from either the IF or audio sections, or from one of each. Assume that the receiver is not tuned to a station, that no AGC voltage is produced, and that the receiver gain is at its maximum. Now, an input noise impulse, whose duration is a few milliseconds, is received. Then, in the main circuit, C_1 charges through D1 and R_4 with an attack time of 5 ms. Also, in the hang circuit, C_2 charges through D2 and R_5 with an attack time of 50 ms. The gate voltage of the n-channel JFET Q2 does not have time to go negative enough to cut off that transistor, so Q2 is in its low-resistance state. That low resistance is assumed to be small compared to the value of R_3. Thus, R_1 and R_3 are in parallel. The noise impulse quickly disappears and C_1 discharges into 50 kΩ (R_1 in parallel with R_3) with a decay time of 50 ms instead of 5 s with R_3 removed.

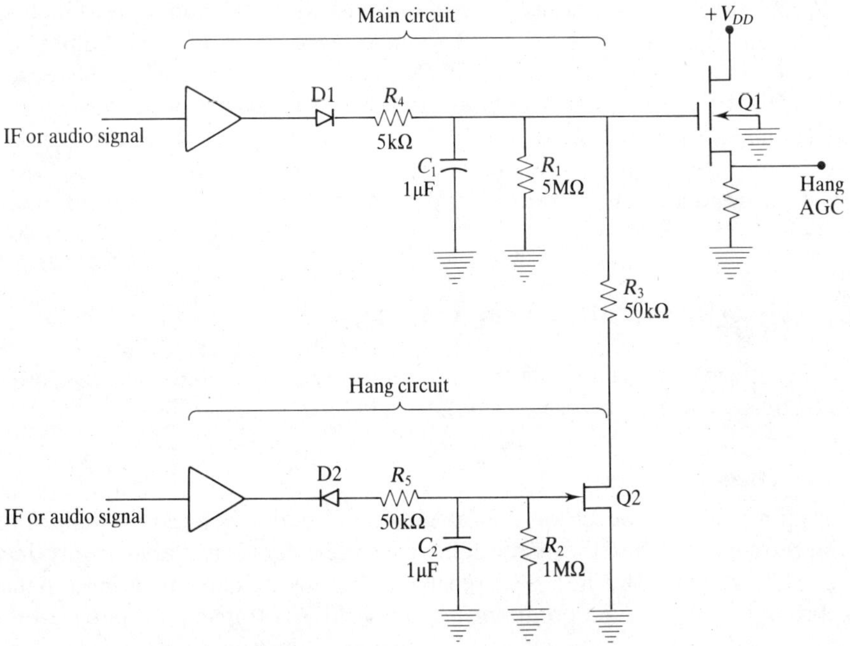

Fig. 8-14 Simplified hang AGC system.

With sustained input signals present, C_2 has more time to charge completely. With enough amplification before the hang circuit, and possibly a voltage-doubler rectifier too, the voltage across C_2 can drive Q2 well into cutoff. This effectively disconnects R_3 from the main circuit. The voltage across C_1 now has a decay time of 5 s to compensate for slow fading during a transmission. Noise impulses cause the same problem as before, so a noise limiter or blanker is called for.

When both input signals stop, during a pause or between a dot and dash, then C_2 discharges into R_2 with a time constant of 1 s and Q2 moves slowly towards its low-resistance state. This is the hang time, during which the voltage across C_1 does not change much because of its longer decay time. If there is no further transmission, then C_2 continues to discharge and this eventually drives Q2 into its low-resistance state. Then, C_1 discharges rapidly into 50 kΩ (R_1 in parallel with R_3) with a decay time of 50 ms. Thus, after the hang interval, the receiver gain increases rapidly.

8-4.2 Diplexers

Most electronic amplifier circuits are analyzed using the assumption that the loads are resistive and that straight lines called DC and AC load lines can be drawn. The DC load line is used to help locate the operating point of the amplifying device. Then, when a signal is applied, the operating path of the amplifying device is up and down along the AC load line.

This procedure given accurate results only when the loads are resistive. When a load contains any reactance at all, then the operating path changes from a straight line into an ellipse, as explained in Appendix D. It is shown there that if the load impedance is $Z = R + jX$, then: (a) the usual AC load line slope of $-1/R$ passes through the maximum positive and negative signal current points on the elliptical operating path; (b) a line whose slope is $-1/|Z|$ locates approximately the major axis of the ellipse; and (c) a line whose slope is $-R/|Z|^2$ passes through the maximum positive and negative signal voltage points on the ellipse. The shape and tilt of the ellipse depend on the phase angle $\theta = \tan^{-1}(X/R)$.

A disadvantage of the presence of an elliptical operating path is that, especially for large signal amplitudes, portions of the ellipse may enter either or both of the cutoff and saturation regions of the device output characteristics. This can occur even though the straight line representing the usual AC load line does not predict such action. Because of the nonlinearities thus introduced, the distortion components produced will be higher in amplitude than expected. This can occur whether the device is an amplifier or a mixer, and the distortion can increase as the load becomes more reactive. It can also cause an imbalance in an otherwise balanced modulator, mixer, or product detector. The existence of the phenomenon requires that attention be given to the nature of the terminations for balanced modulators, mixers, and product detectors at frequencies other than the one at the output desired.

Suppose a balanced mixer is used to convert a 5-MHz signal into one at 1.6 MHz using an oscillator output at 6.6 MHz. The desired output is at the difference frequency $6.6 - 5 = 1.6$ MHz, and the resistive termination can be provided at this frequency. But the mixer also produces an output at the sum frequency $5 + 6.6 = 11.6$ MHz, and a resistive termination should be provided at this frequency also.

There are several different circuits, called diplexers, that can perform this function. One such circuit is similar to the cross-over network of a hi-fi audio system that sends low frequencies to a woofer and the high frequencies to a tweeter. It uses a lowpass filter to send the desired 1.6 MHz signal to the output load resistor, and, in parallel, a highpass filter to send the 11.6 MHz signal to a dummy load resistor. Another circuit is shown in Fig. 8-15a. If $R = R_L = R_D = (L/C)^{1/2}$, then the input impedance $Z_{in} = R + j0$ at all frequencies removed from the self-resonant frequencies of the components themselves. The analysis of this circuit is reserved for Problem 8-15.

Another useful circuit is shown in Fig. 8-15b. It can be analyzed as follows, using $Y_{in} = 1/Z_{in}$.

$$Y_{in} = \frac{1}{R + j\omega L_1 + \dfrac{1}{j\omega C_1}} + \frac{1}{R + \dfrac{1}{j\omega C_2 + 1/j\omega L_2}}$$

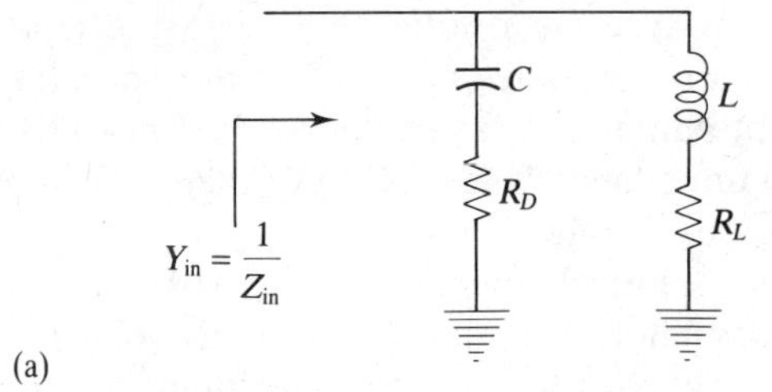

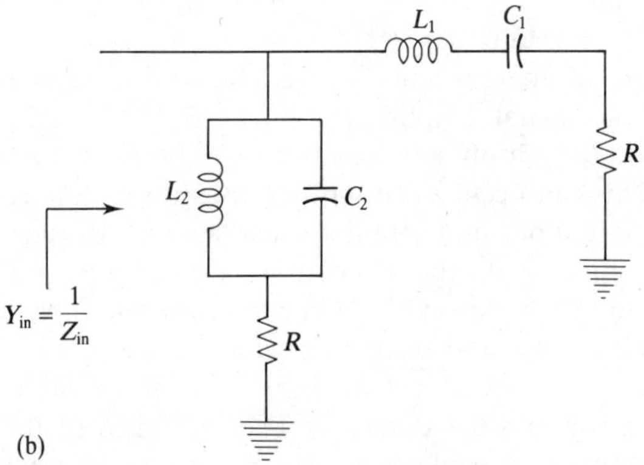

Fig. 8-15 Diplexers: when the component values are chosen properly, the input impedance will be resistive at all frequencies.

Next, rationalize the denominators separately to obtain,

$$Y_{in} = \frac{\omega^2 R C_1^2}{(1-\omega^2 L_1 C_1)^2 + \omega^2 R^2 C_1^2} + j\frac{\omega C_1(1-\omega^2 L_1 C_1)}{(1-\omega^2 L_1 C_1)^2 + \omega^2 R^2 C_1^2}$$
$$+ \frac{R(1-\omega^2 L_2 C_2)^2}{R^2(1-\omega^2 L_2 C_2)^2 + \omega^2 L_2^2} - j\frac{\omega L_2(1-\omega^2 L_2 C_2)}{R^2(1-\omega^2 L_2 C_2)^2 + \omega^2 L_2^2}$$

To force $Y_{in} = G + j0$, equate the sum of thc j terms to 0. Then

$$\frac{\omega C_1(1-\omega^2 L_1 C_1)}{(1-\omega^2 L_1 C_1)^2 + \omega^2 R^2 C_1^2} = \frac{\omega(L_2/R^2)(1-\omega^2 L_2 C_2)}{(1-\omega^2 L_2 C_2)^2 + \omega^2 (L_2^2/R^2)}$$

Next, equate coefficients of like powers of ω. This results in

$$L_1 C_1 = L_2 C_2$$

and

$$R^2 = \frac{L_2}{C_1} = \frac{L_1}{C_2}$$

Consequently, when these are substituted back into the expressions for Y_{in}, it is found that $Y_{in} = G + j0 = (1/R) + j0$. Therefore, the input impedance becomes $Z_{in} = R + j0$ at all frequencies removed from the self-resonant frequencies of the components. The desired resistive terminations are assured.

8-4.3 Bandpass Tuning

During the years following the Second World War, a lot of surplus military communications equipment became available. One of the receivers that became popular with radio amateurs had an IF of 915 kHz and a rather broad passband. As the amateur bands became more crowded and as single-sideband came into vogue, the broad passband was inadequate. Those were the days when selectivity was obtained by using singly and doubly tuned, synchronously and stagger-tuned LC circuits at low frequencies, such as 85 kHz. IF transformers for that frequency were readily available from other surplus equipment.

The scheme shown in Fig. 8-16 was devised to improve the selectivity of the surplus receiver. Assume, for example, that signal I and signal II are both present in the passband of the 915 kHz IF amplifier and that they overlap as shown. In the original receiver, both signals would be demodulated and the interference between them would be considerable. Either one could be the desired or the undesired signal. The new scheme interrupts the IF path and, instead, delivers the signals to a difference mixer. Then, with

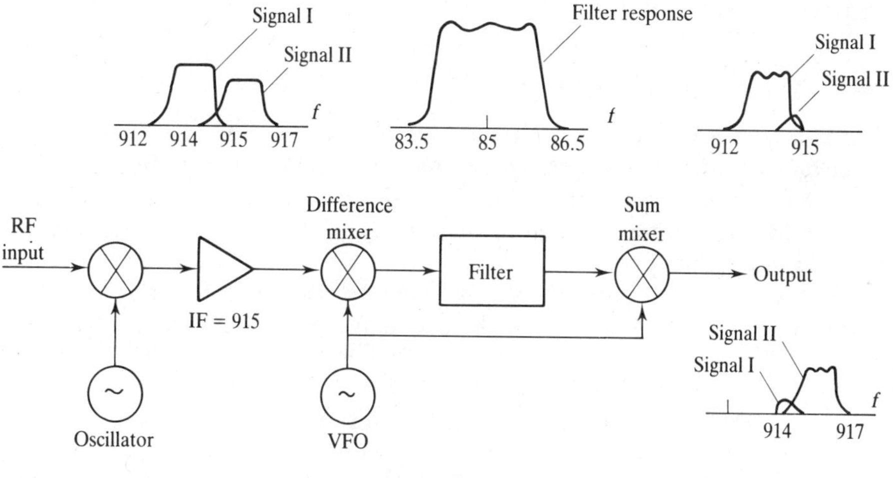

828.5 – 830.5 (all frequencies in kHz)

Fig. 8-16 The bandpass tuning concept. The upper output figure is obtained with the VFO at 828.5 kHz, and the lower output figure is obtained for the VFO at 830.5 kHz.

the proper frequency from the variable-frequency oscillator (VFO), the signals are down-converted into the 85-KHz range, where the filter selectivity comes into play.

Suppose signal I is the desired one. Then, if the VFO frequency is 828.5 kHz, signal I frequencies from 912 kHz to 915 kHz are converted into frequencies from 83.5 kHz to 86.5 kHz. The interfering signal II frequencies from 914 kHz to 917 kHz are converted into 85.5 kHz to 88.5 kHz, but no frequencies above 86.5 kHz pass through the filter.

To continue to use the remainder of the original receiver, the 85-kHz signals must be up-converted back to the 915-kHz range. So the output of the filter goes to a sum mixer along with the same VFO frequency used with the difference mixer. The outputs obtained for two VFO frequencies are indicated in Fig. 8-16. Interference has been reduced and could be eliminated, with accompanying distortion of the desired signal, if the original frequency of 914 kHz were converted into 86.5 kHz when signal I was the desired one, or 915 kHz converted to 83.5 kHz when signal II was the desired one.

By varying the VFO frequency, any 3-kHz portion of the original two signals can be made to sweep across and occupy the filter passband. This is the origin of the term bandpass tuning. The actual passband of the system does not change because it is set by the filter design.

The development of monolithic crystal filters having center frequencies from about 5 MHz to about 350 MHz has made the entire bandpass-tuning system of Fig. 8-16 outmoded. But the concept of converting the frequency of a desired signal, selected from a restricted band that contains many other signals, into a signal that can pass through an available filter, is still important; it is called *passband tuning*.

8-4.4 Variable Bandwidth

The ability of a receiver to reject unwanted signals is closely related to the receiver's bandwidth (or passband). As the bandwidth is reduced, more unwanted signals are rejected. But each situation determines how narrow the bandwidth can become before the quality of desired signal deteriorates to an unacceptable level.

The receiver operator should be able to adjust the bandwidth for optimum reception. One approach is to have several filters that can be switched in or out as needed. The AM filter has a bandwidth of about 10 kHz; the SSB filter bandwidth is about 2.3 kHz; the CW filter bandwidth is about 500 Hz; and the radioteletype filter bandwidth depends on the transmitter speed and frequency shift.

The bandwidth of a tuned circuit can be adjusted continuously if the circuit Q is varied by changing the resistance of the inductor with a potentiometer; more resistance means lower Q, which yields greater bandwidth. Less resistance means higher Q and narrower bandwidth. A circuit called a Q multiplier was developed that introduced negative resistance, from an amplifier with controlled positive feedback, into the tuned circuit to cancel some of the positive resistance, hence increasing the Q and decreasing the bandwidth still further.

An amplifier subject to negative feedback has a wider bandwidth than without the feedback, whereas positive feedback decreases the bandwidth. Not only does an amplifier on the verge of oscillation have a very high gain, it also has a very narrow bandwidth. The bandwidth, and the gain, can be varied by changing the feedback.

The bandwidth of a doubly-tuned transformer-coupled circuit depends on the amount of coupling; increasing the physical separation of the transformer windings narrows the bandwidth, and decreasing the separation increases the bandwidth.

A more modern method for producing a continuously variable bandwidth is illustrated in Fig. 8-17. Its operation is based on the facts that (1) the upper and lower sidebands of a signal are not inverted when passing through a difference mixer with low-side injection but they are inverted if the difference mixer has high-side injection, and (2) a lowpass filter can have a sharper cutoff characteristic if its cutoff frequency is low rather than high. The bandpass filters following the sum mixer and the last difference mixer are needed to eliminate other IM products. The effect of each operation on the sidebands of an input signal are shown at each stage when VFO1 is at 405 kHz and VFO2 is at 505 kHz. The final output spectrum has a bandwidth of 6 kHz. The steep cutoff characteristics are shaped by the lowpass filters. The bandwidth is changed by changing the frequencies of the two VFOs. If VFO1 is set to 406 kHz and VFO2 to 504 kHz, it is found that the final bandwidth becomes 8 kHz. Those frequencies must be ganged so that if one frequency increases by Δf Hz the other must decrease by the same amount to produce the desired effect. This system is complex and expensive to implement, and the receiver using it is subject to numerous undesired signals that are created by the four mixers and two oscillators.

8-4.5 The Wadley Loop

Frequency drift is one of the problems that arises when the frequency of an oscillator is continuously variable. The term frequency drift refers to the gradual change in the frequency of the oscillator after the tuning control has been set for a desired frequency. Frequency drift is caused by gradual changes in the values of components that determine the frequency of oscillation, produced by temperature variations as the components warm up or as the air currents change, and as they age. Frequency drift becomes greater as the oscillator frequency is made higher and the tuning range wider.

A superhet that has much frequency drift will have to be retuned to the desired station every now and then to keep the IF signal produced by the

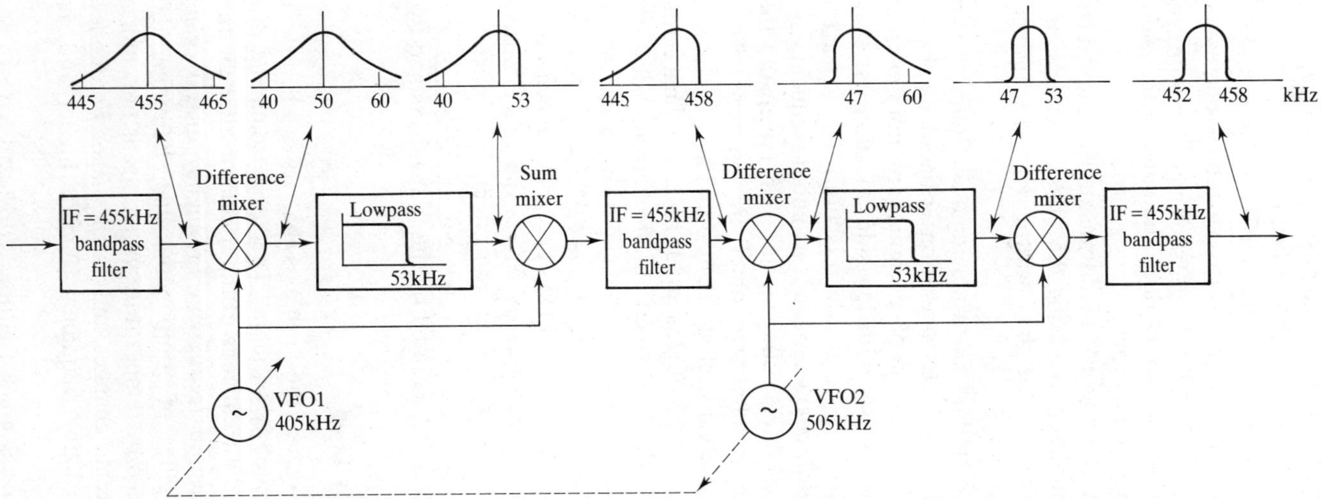

Fig. 8-17 A method for obtaining variable bandwidth. A spectrum at each location shows the effects.

drifting local oscillator within the IF passband, otherwise the station begins to sound distorted. The retuning is actually resetting the oscillator frequency back to the correct value determined by the station frequency and the IF. But this can mean that the RF circuit may now be out of tune, and, unless the RF passband is wide, the oscillator and RF circuits no longer track properly. Frequency drift can be minimized by maintaining the temperature of the oscillator environment constant, by using high-quality components, by good mechanical layout, by regulating the input DC voltage, by shielding the oscillator from undesired external influences, by using negative-temperature-coefficient components to compensate for the effects of those with positive temperature coefficients, by keeping the oscillator frequency low and the tuning range narrow, and by using the Wadley loop.

The Wadley loop is an ingenious method whereby the frequency of the signal delivered to the first IF in a multiple-conversion receiver is independent of the frequency drift of the tunable high-frequency oscillator. The basic scheme of the Wadley loop is represented in Fig. 8-18. The RF input is broadband, from 1 MHz to 30 MHz, and the receiver may also have a preselector. The MHz selector is actually an oscillator that can be tuned continuously from 41.5 MHz to 69.5 MHz. Its purpose is to provide the oscillator output so that difference mixer 1 can up-convert each 1-MHz segment of the broadband RF input into the IF_1 range of 40 ± 0.5 MHz. For example, if the MHz selector is set to 41.5 MHz, then an incoming RF signal at 1 MHz is converted to one at 40.5 MHz and a signal at 2 MHz is

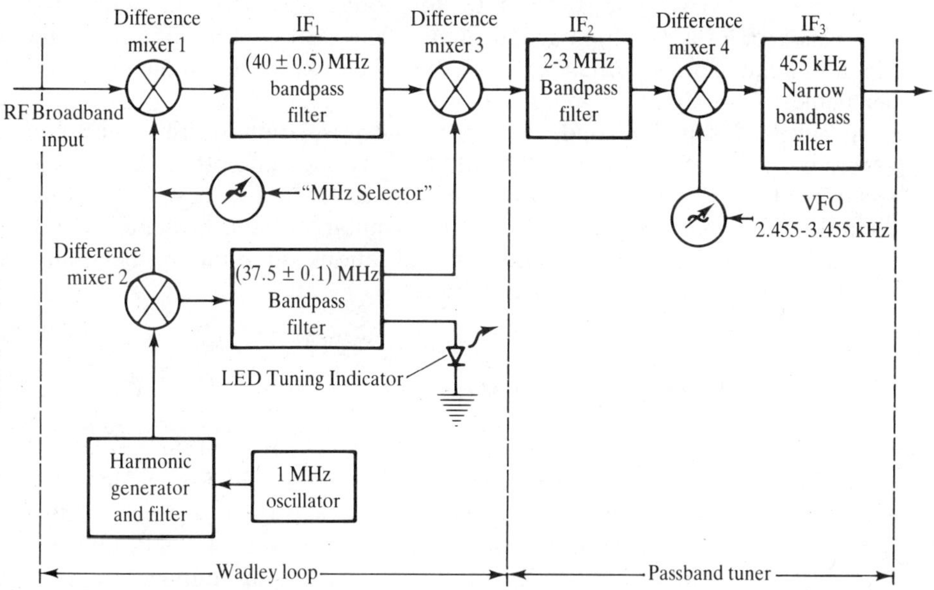

Fig. 8-18 Partial block diagram of a receiver using a Wadley loop and a passband tuner.

converted into one at 39.5 MHz, both of which pass through the IF_1, bandpass filter. Similarly, if the MHz selector is set to 69.5 MHz, incoming frequencies in the range from 29 MHz to 30 MHz are converted into 40.5 MHz to 39.5 MHz, both of which pass through the IF_1 filter.

The output of the MHz selector also mixes with the harmonics of a stable 1-MHz crystal oscillator in difference mixer 2. The harmonic generator clips the top and bottom of the oscillator output to produce a square wave. A square wave contains all even and odd harmonics, with their amplitudes varying inversely with the order of the harmonic. The harmonics enter a filter whose frequency response increases with frequency so that the output amplitudes of the harmonics are more nearly constant. When the output of difference mixer 2 is 37.5 ± 0.1 MHz, that signal passes through the bandpass filter, operates a tuning indicator light, and is applied to difference mixer 3.

The cancelation of the frequency drift of the MHz selector at the output of the Wadley loop can now be demonstrated. Suppose the MHz-selector frequency drifts by Δf. Then, the frequency of each of the resulting signals passing through the IF_1 filter and the 37.5 MHz filter will change by Δf. Thus, when they combine in difference mixer 3, the subtraction process causes the drift Δf of the MHz selector to cancel.

The output of difference mixer 3 contains all signals in a selected 1-MHz segment of the RF input that have been down-converted to the IF_2 range of 2–3 MHz. A desired signal in this range is chosen by mixing it in difference mixer 4 with the output from a variable-frequency oscillator, or VFO, that tunes from 2.455 MHz to 3.455 MHz, to produce the 455 kHz signals for IF_3. This process is called passband tuning. The restricted tuning range and the low-frequency operation of the VFO help keep its frequency drift negligible.

In spite of shielding and remote physical isolation of the harmonic generator, some of its output will still find a way back into other parts of the receiver. This will tend to mask the signals from any desired stations operating on those frequencies, but an advantage of the presence of the harmonics is that they can be used as calibration points for the beginning and end of the dial of the passband tuner.

8-4.6 The PLL in CB Transceivers

The Federal Communications Commission (FCC) established the original 23 AM channels of the 27-MHz Citizens Band (CB) Radio Service in 1958. Each channel was 10 kHz wide. At that time, the CB radio sets needed two quartz crystals for each channel—one for the transmitter frequency and one for the local oscillator frequency of the receiver. The combination of a TRANSmitter and a reCEIVER in the same enclosure is called a transceiver. The CBers purchased crystals for as many channels as they

TABLE 8-1 Citizens Band (CB) Channels

Channel	Frequency (MHz)	Channel	Frequency (MHz)	Channel	Frequency (MHz)
1	26.965	15	27.135	28	27.285
2	26.975	16	27.155	29	27.295
3	26.985	17	27.165	30	27.305
4	27.005	18	27.175	31	27.315
5	27.015	19	27.185	32	27.325
6	27.025	20	27.205	33	27.335
7	27.035	21	27.215	34	27.345
8	27.055	22	27.225	35	27.355
9	27.065	23	27.255	36	27.365
10	27.075	24	27.235	37	27.375
11	27.085	25	27.245	38	27.385
12	27.105	26	27.265	39	27.395
13	27.115	27	27.275	40	27.405
14	27.125				

needed or could afford to use, and plugged them into the appropriate crystal sockets.

In 1977, the FCC expanded the number of channels to 40 as listed in Table 8-1, which includes the original 23 channel allocations. The listed frequencies are for the transmitter frequencies and they are in the center of each channel. When the additional channels were approved, the frequencies of channels 24 and 25 were placed between those of channels 22 and 23. Channel 9 is reserved for emergency and motorist-aid communications only. A few channels for frequencies that are skipped are reserved for Radio Control (R/C) of model planes and boats. If the PLL had not appeared on the scene, up to 80 quartz crystals would be needed for each complete CB rig!

Consider the double-conversion receiver shown in Fig. 8-19. There, only one crystal oscillator is used with a PLL to provide all required local oscillator frequencies. The input circuit of the receiver is sufficiently broad to accept the carriers and sidebands of all 40 channels. When the lower frequency of the VCO subtracts from the higher incoming frequencies of a selected channel, that signal is converted to a carrier at 10,695 kHz plus sidebands. These combine with the crystal oscillator frequency of 10,240 kHz to produce the second IF of 455 kHz. This latter signal is then demodulated and amplified by circuits that are not shown.

Since the CB channels are 10 kHz wide, select 10 kHz as the reference frequency for the PLL. This requires a fixed counter of 1024 between the crystal oscillator and the phase detector. If channel 1 on 26,965 kHz is to be received, the VCO must be on 16,270 kHz (i.e., 26,965 − 16,270 = 10,695, which is the first IF). Therefore, the programmable counter must be set for

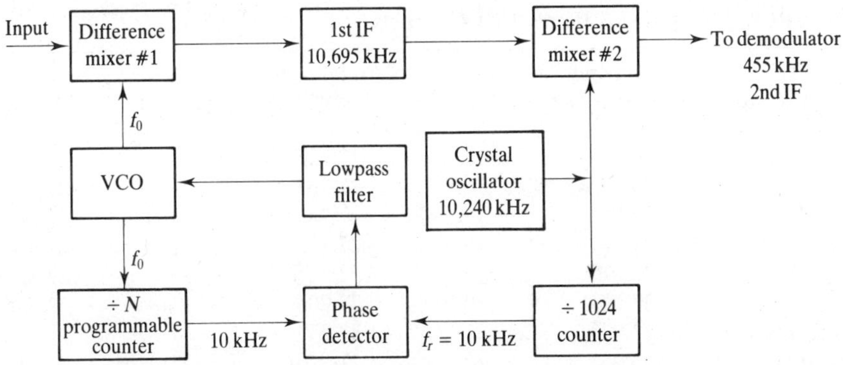

Fig. 8-19 Block diagram of a portion of a CB receiver.

a count of 1627. This causes the other input to the phase detector to equal $f_r = 10$ kHz for the loop to lock. When channel 1 is selected, by pushing a button perhaps, the circuit is designed so that this counter is set at 1627. It is left as an exercise for the reader to show that if channel 8 is selected, this counter is set at 1636.

8-4.7 An Automatic Tracking Receiver

Suppose a satellite in orbit about the Earth is monitoring some slowly changing physical phenomenon and transmitting the results back to a receiving location on Earth. Assume that the transmitter is using CW at a power level of 10 mW and a frequency f_s. The bandwidth required for the CW signal is only 10 Hz, say. As the satellite is aproaching the receiving location, Doppler shift can cause the received carrier frequency to be as much as 3 kHz higher. As the satellite recedes from the receiving location, the received carrier frequency can be as much as 3 kHz lower.

An ordinary receiver tuned to the original carrier frequency needs about 6 kHz of bandwidth because of the Doppler shift, much more bandwidth than the CW signal itself requires. This bandwidth allows more noise, too, and it could mask the low-powered desired signal. If the receiver had a bandwidth of only 10 Hz, its local oscillator would have to be tuned and retuned continually to convert the received signal into the narrow IF bandwidth. This is the action that the PLL performs automatically.

Consider the receiving system shown in Fig. 8-20. The signal from the satellite mixes with f_2 in a difference mixer to produce the IF. The converted signal passes through the narrow bandpass filter to provide input to the phase detector and to the demodulator. A crystal oscillator provides the reference for the phase detector that is equal to the desired IF as required for the loop to lock. The output frequency of the VCO is a subharmonic of the frequency f_2 needed by the mixer. The frequency multiplier can be a secondary PLL used for that purpose.

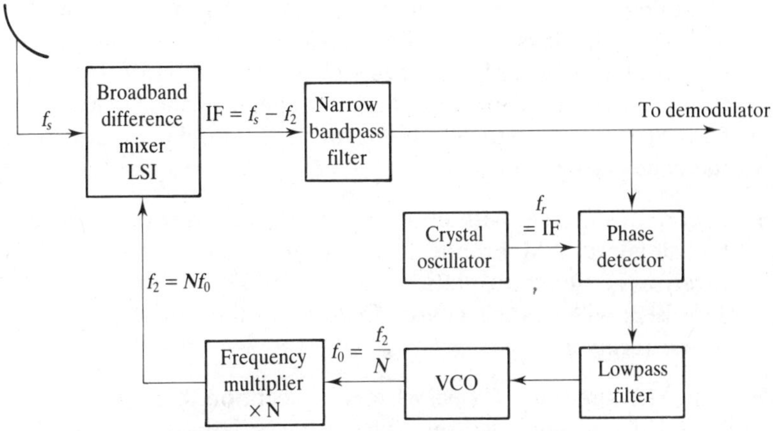

Fig. 8-20 An automatic tracking receiver.

As Doppler shift causes the received frequency to increase and decrease, the output of the phase detector causes the frequency f_2 to increase and decrease such that $f_s - f_2 = \text{IF} = \text{constant}$.

PROBLEMS

8-1 Draw the block diagram of a single conversion superheterodyne radio receiver suitable for demodulating the following signals: (a) lower-sideband; (b) CW (Morse code); (c) FM.

8-2 A superheterodyne FM receiver tunes 88–108 MHz. The IF is 10.7 MHz and high-side injection of the oscillator is used. Is difference mixing or sum mixing required? Why?

8-3 An AM superhet has an IF = 455 kHz, and it uses high-side injection. When the receiver is tuned to 540 kHz, a local AM station operating on 1450 kHz is heard. Explain how this image response can occur through no fault at the transmitter.

8-4 The receiver of Problem 8-3 is tuned near the high end of the AM broadcast band and a nonbroadcast-band station is heard. If a local transmitter is known to operate on 2.45 MHz, explain how that station is heard through no fault at the transmitter and determine the frequency to which the receiver is tuned.

8-5 The receiver of Problem 8-3 is tuned to a station on 1100 kHz. In addition to the desired signal, an interfering code signal is heard and identified as a radio amateur operating in the frequency range 3.5–4.0 MHz. It is known that spurious responses often occur in the receiver because harmonics of its own local oscillator are also present in the mixer. Explain how the code signal is also heard through no fault at the transmitter and determine the fundamental frequency of the code station.

8-6 A certain superhet with poor image rejection is used for reception on the standard AM broadcast band. It is found that each time the receiver is tuned to station A at 1000 kHz, another station B at 1350 kHz will also be heard. Determine the intermediate frequency of the receiver.

8-7 An AM superhet for short-wave reception has an IF = 1.6 MHz. High-side injection is used. The receiver is tuned to an AM station operating on 30 MHz. (a) Determine the local oscillator frequency and the image frequency. (b) Draw a block diagram of the receiver and label the frequencies in each part of the receiver, assuming that the station being received is amplitude modulated by a 5-kHz tone and that the bandwidth of the IF amplifier is 6 kHz.

8-8 Suppose you have a good quality AM radio receiver that tunes from 540 kHz to 1600 kHz. You wish to add some circuitry external to the receiver so that you can listen to the AM short-wave broadcast band 6000–6200 kHz by tuning the receiver only from 1000 kHz to 1200 kHz. No modifications are to be made internal to the receiver because it is still under warranty. Draw a block diagram showing how you could do this. Name all blocks and mark all required frequencies.

8-9 An amateur radiotelephone station operating on single sideband uses only the lower sideband, having a width of 3 kHz, for voice transmission. The (missing) carrier is at 7200 kHz. Your single-conversion receiver has an IF filter with a 3-kHz bandwidth extending from 453.5 kHz to 456.5 kHz. Draw a block diagram of the receiver, including one RF stage and a product detector where the missing carrier is inserted. Numerically label all frequencies throughout the block diagram such that proper demodulation can occur.

8-10 A radio amateur transmits a 2.3-kHz wide lower-sideband signal whose (missing) carrier is 3950 kHz. That signal is to be demodulated by a double-conversion receiver. The passband of the first IF extends from 1690 kHz to 1710 kHz, and the passband of the second IF extends from 100 kHz to 102.3 kHz. High-side injection is used in both mixers. (a) Draw a block diagram of the receiver showing blocks for the mixers, both IFs, the product detector, and all required local oscillators. (b) What should be the frequency of the reinserted carrier in the product detector for proper demodulation to occur?

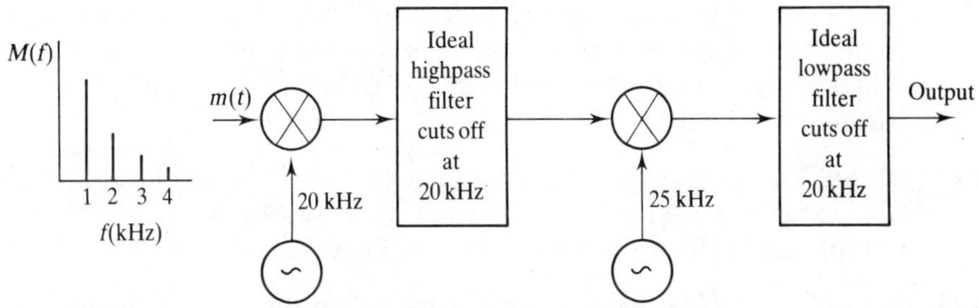

Fig. P8-11 System for Problem 8-11.

8-11 The system shown in Fig. P8-11 is a simplified speech scrambler. It can be used at the transmitter to provide some communications privacy from listeners with an ordinary AM receiver. It inverts normal speech frequencies—high frequencies become low frequencies, and low become high. The spectral components of a fictitious modulating signal $m(t)$ are given as $M(f)$. (a) By sketching the spectral components at each location, preserving relative amplitudes, demonstrate that the spectral components at the output are inverted from what they were at the input. (b) Briefly describe a system that a receiver can use to unscramble the scrambled signal.

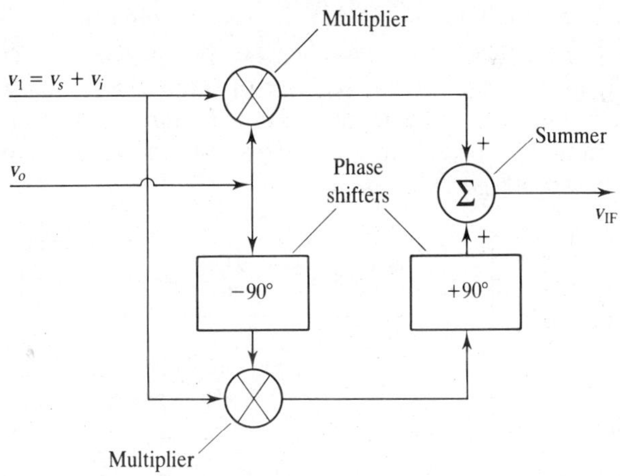

Fig. P8-12 Image-reject mixer of Problem 8-12.

8-12 A simplified block diagram of an image reject mixer is shown in Fig. P8-12. The RF input voltage v_1 is the sum of the desired signal $v_s = \cos \omega_s t$ and the undesired image signal $v_1 = \cos \omega_1 t$. The local oscillator delivers $v_o = \cos \omega_0 t$ and the oscillator frequency is above the desired signal frequency, that is, $f_{IF} = f_0 - f_s$. Determine what frequencies appear at the output v_{IF}. (The argument of a sine function can be negative if the image frequency is higher than the oscillator frequency, so that $\sin(-x) = -\sin(x)$.)

8-13 A receiver that uses the Wadley loop of Fig. 8-18 is tuned to a station operating on 3.4 MHz. Show that the signal passed to the IF_2 filter is free of drift in the frequency of the "MHz selector."

8-14 Perform a steady-state analysis of system 2 in the second quadrant of Fig. 8-6 assuming that a DC amplification of 3 follows the output of the AGC circuit represented in the first quadrant of Fig. 8-6.

8-15 Perform an analysis of the circuit in Fig. 8-15a to show that the input impedance $Z_{in} = R + j0$ provided that $R = R_L = R_D = (L/C)^{1/2}$.

8-16 Calculate the bandwidth produced by the system in Fig. 8-17 when VFO1 is at 404 kHz and VFO2 is at 506 kHz.

8-17 Repeat an analysis of the variable-bandwidth system in Fig. 8-17 when the lowpass filters cut off sharply at 57 kHz instead of 53 kHz.

8-18 A scanner is a type of receiver that automatically tunes to certain channels in sequence and stops when a signal is received, and begins to scan again when the signal ceases or scanning is initiated manually. Consider the fictitious system shown in Fig. P8-18, which uses a phase-locked loop to synthesize the local oscillator frequencies for a superhet. The desired frequencies are to vary from 159,985 kHz to 160,005 kHz in steps of 5 kHz. The programmable divider ratio N is to increment by 1 for each 5 kHz step. Determine the programmable divider ratio N for each frequency and insert the pairs in a table. (N should be an integer.)

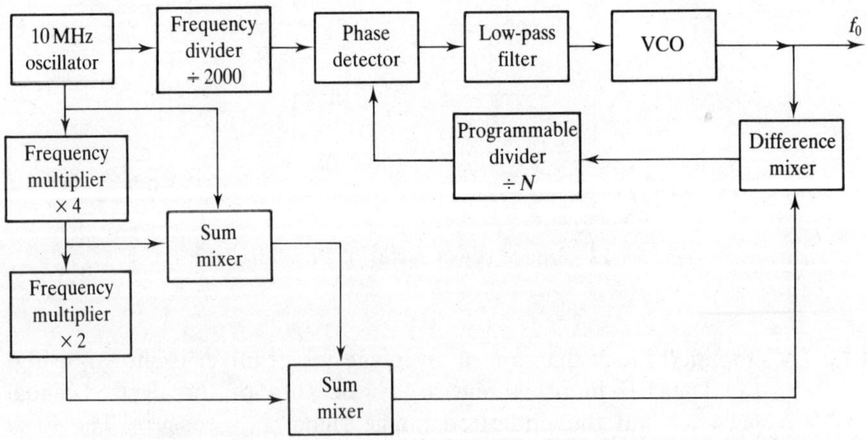

Fig. P8-18 Frequency synthesizer for Problem 8-18.

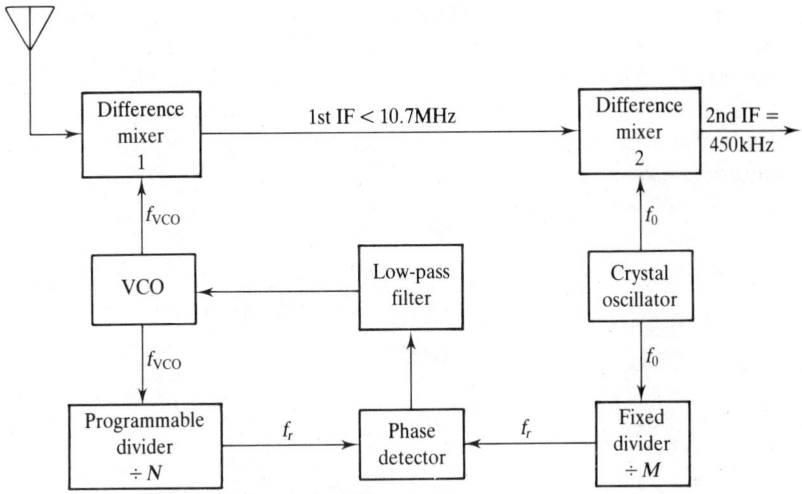

Fig. P8-19 Receiving system for Problem 8-19.

8-19 Consider the double-conversion receiving system shown in Fig. P8-19. In operation, it is supposed to automatically scan channels every 100 kHz throughout the 154–163 MHz FM band (police, taxi, weather, etc.) by adjusting the oscillator frequency of mixer 1 with a phase-locked loop featuring a programmable divider. Both difference mixers use low-side injection of their oscillator frequencies. (a) Choose f_r so that the programmable divider increments by 1 to scan adjacent channels starting at 154.05 MHz. (b) Choose an appropriate, constant value for the first IF < 10.7 MHz and as near 10.7 MHz as possible such that M (the fixed divider ratio and N) are each convenient integers. (c) Determine M. (d) Determine f_0. (e) Calculate a few values of N from the minimum to the maximum and enter the values in a table with the corresponding input-signal frequencies.

8-20 The simplified block diagram of an FM transmitter is shown in Fig. P8-20. The transmitter output operates on 96 MHz with a maximum

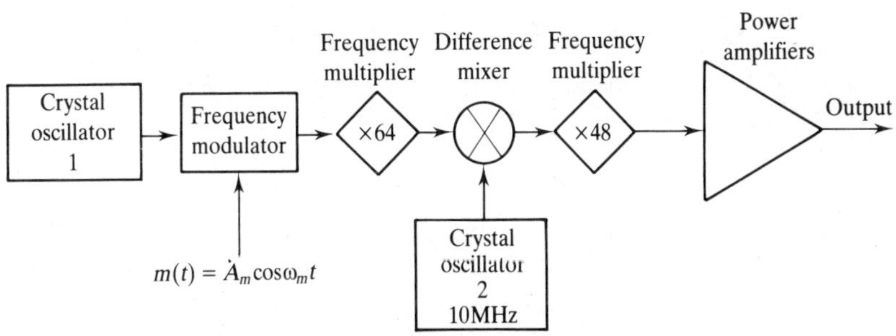

Fig. P8-20 Block diagram for Problem 8-20.

frequency deviation $\Delta f = 75\,\text{kHz}$. Calculate: (a) the possible frequencies required of oscillator 1; (b) the maximum frequency deviation Δf at the output of the frequency modulator before frequency multiplication occurs; and (c) the modulating frequency f_m such that the maximum phase deviation at the transmitter output is 360 *degrees*.

9

SPURIOUS RESPONSES

The word *spurious* is an adjective that means something is false, counterfeit, or not genuine. The word is usually used as the modifier in terms such as spurious response, spurious product, spurious output, spurious signal, or spurious emission. When applied to a radio receiver, a spurious response is any response, other than the desired one, caused by an internal deficiency in the receiver.

The word spurious is sometimes used as noun whose plural is taken to be *spurii* or *spuriae* (it appears both ways in the literature), and the noun is often abbreviated to the popular form *spur* (plural, *spurs*) because of the simplicity thereby afforded.

Spurious responses in radio receivers typically occur for reasons such as the following.

- Signals appear to be received on certain frequencies to which the receiver is tuned when in fact they are not actually being transmitted on those frequencies.
- Internally generated intermodulation components may heterodyne with desired signals to produce audible whistles during the tuning process. The resulting sounds are reminiscent of a tweeting bird, hence the effect is known as a *birdie*.
- Signals are sufficiently strong to "break through" into the IF stages, thus bypassing the RF circuits and their selectivity, and they cannot be eliminated by tuning the receiver.
- A strong interfering signal not only causes desensitization and possible blocking of a desired weaker signal but it can also transfer its

unwanted modulation to a weaker desired signal, the process being known as *cross modulation.* No special frequency relationship between the two signals is necessary, with the effect being observed for frequency separations up to many kilohertz.

- Signals appear to be received within the tuning range of the receiver, but they actually originate with the receiver's own oscillators instead of being externally received signals.

9-1 MIXER TERMINOLOGY

When the local oscillator frequency is higher than the signal frequency in a receiver mixer, *high-side injection* (HSI) of the oscillator output is employed. When the local oscillator frequency is lower than the signal frequency in a receiver mixer, *low-side injection* (LSI) of the oscillator output is used. The terminology high-side and low-side injections lose their meanings when two locally generated signals are combined in the mixer of a frequency synthesizer.

Down-conversion is said to occur when an input signal is converted to a lower frequency, and *up-conversion* occurs when an input signal is converted to a higher frequency. The mixer in which either conversion takes place is sometimes called a *frequency converter,* or, simply, a *converter.*

When the output circuit of a mixer is tuned to the difference frequency, which sometimes falls below the input signal frequency, down-conversion of the signal frequency occurs. The process is called *difference mixing,* and the mixer (including its associated circuitry) is referred to as a *difference mixer.*

When the output circuit of a mixer is tuned to the sum frequency, which

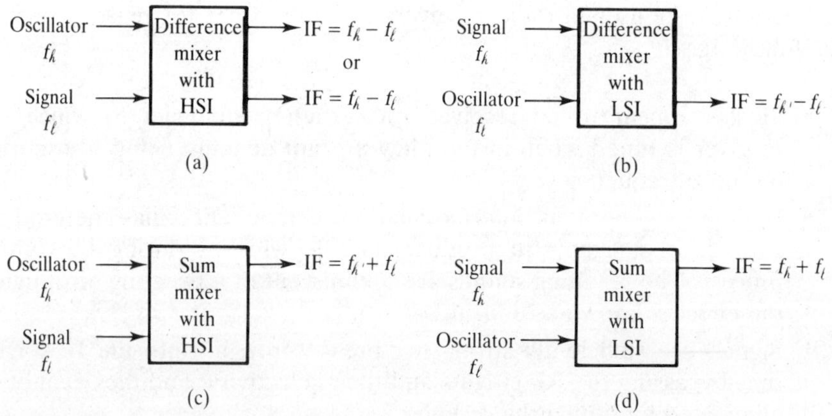

Fig. 9-1 Several mixing possibilities: (a) either up-conversion or down-conversion is possible depending on the relative values of f_h and f_ℓ; (b) down-conversion only; (c) and (d) up-conversions only.

must fall above the input signal frequency, up-conversion of the signal frequency occurs. The process is known as *sum mixing,* and the mixer (including its associated circuitry) if referred to as a *sum mixer.*

It is possible for the oscillator frequency of a difference mixer using high-side injection to be so high that the difference frequency is above the input signal frequency, so that up-conversion of the signal frequency occurs.

Several mixer relationships are shown in Fig. 9-1, where $f_{\hbar}$ represents the higher input frequency and f_{ℓ} represents the lower input frequency.

9-2 COINCIDENCE AND CONGRUENCE

When two sinusoidal voltages having frequencies f_{ℓ} (for the lower) and $f_{\hbar}$ (for the higher) are applied to a nonlinear mixer, harmonics of each sinusoid appear in the output of the mixer together with the intermodulation frequencies $\pm Lf_{\ell} + Hf_{\hbar}$ in a manner that depends on the nonlinearity of the mixer's static characteristic as discussed in Chapter 2. Here, we let L and H represent the positive-integer harmonic numers (including zero) that are associated with the frequencies f_{ℓ} and $f_{\hbar}$, respectively.

In a superheterodyne radio receiver, the desired output frequency from a difference mixer is the IF equal to $f_{\hbar} - f_{\ell}$, and the desired output frequency from a sum mixer is the IF equal to $f_{\hbar} + f_{\ell}$. The harmonics and intermodulation frequencies are the undesired spurious output frequencies, and they become troublesome when their frequencies are the same as or nearly the same as the frequency of the desired IF output signal.

Coincidence is said to occur when an undesired intermodulation frequency produced by the two signals at frequencies $f_{\hbar}$ and f_{ℓ} is the same as (i.e., coincides with) the desired IF output frequency produced by the same two input signals. The input signal frequency for which coincidence or near coincidence occurs becomes a spurious input frequency. The conditions for which coincidence and near coincidence occur are presented in the following sections.

Congruence is said to occur when an undesired intermodulation frequency produced by the two signals is the same as (i.e., is in congruence with) the design value of the IF for a particular receiver. The name *congruence* has been originated for use in this book to indicate the specific spurious responses described.

9-3 HARMONIC NUMBERS FOR COINCIDENCE

In what follows, the particular forms of the intermodulation frequencies are identified by type numbers for the two kinds of mixers.

DIFFERENCE MIXERS	SUM MIXERS
Type 1 refers to $L_1 f_{\ell} - H_1 f_{\hbar}$	Type 3 refers to $L_3 f_{\ell} - H_3 f_{\hbar}$
Type 2 refers to $H_2 f_{\hbar} - L_2 f_{\ell}$	Type 4 refers to $H_4 f_{\hbar} - L_4 f_{\ell}$

Each type can use either HSI or LSI. Also, since by definition $f_\ell < f_\hbar$, we have the additional requirement

$$0 < \frac{f_\ell}{f_\hbar} < 1 \tag{9-1}$$

to be applied to all four types. An exception can occur, however, in a sum mixer where $f_\ell = f_\hbar$ is possible.

9-3.1 Type 1 Coincidence

Type 1 coincidence in a difference mixer means that

$$L_1 f_\ell - H_1 f_\hbar = f_\hbar - f_\ell \tag{9-2}$$

where the subscript 1 represents type 1. Rearrange (2) as

$$\left(\frac{f_\ell}{f_\hbar}\right)_{\text{co}} = \frac{H_1 + 1}{L_1 + 1} \tag{9-3}$$

where the subscript co represents the coincidence condition. Using (9-3), (9-1) becomes

$$0 < \frac{H_1 + 1}{L_1 + 1} < 1 \tag{9-4}$$

Representative positive integers that satisfy (9-4) are given in the row for H and the column for L marked type 1 in Table 9-1. The fractions given in Table 9-1 are those values of $(f_\ell / f_\hbar)_{\text{co}}$ that satisfy (9-3). The table is readily extended to higher values of H_1 and L_1 as may be desired for type 1.

TABLE 9-1 Positive-integer Harmonic Numbers H and *L* with their Corresponding Fractional Values $f_\ell / f_\hbar$ for Coincidence in Four Types of Spurious Responses

L \ H		2	3	4	5	6	7	8	Types 2 and 4
		0	1	2	3	4	5	6	Types 1 and 3
3	1	1/2							
4	2	1/3	2/3						
5	3	1/4	2/4	3/4					
6	4	1/5	2/5	3/5	4/5				
7	5	1/6	2/6	3/6	4/6	5/6			
8	6	1/7	2/7	3/7	4/7	5/7	6/7		
9	7	1/8	2/8	3/8	4/8	5/8	6/8	7/8	
Types 2 and 3	Types 1 and 4								

Starting with any fractional entry in Table 9-1 (after removing any common factors it may have), other equal fractional values can be located by moving from that entry through a number of columns equal to the numerator and then through a number of rows equal to the denominator. The procedure is like using the slope concept on entries in the table rather than on a graph.

After locating equal fractional values of $(f_e/f_k)_{co}$ in Table 9-1, the corresponding pairs of values for L_1 and H_1 are obtained. Thus, for $(f_e/f_k)_{co} = 1/2$, Table 9-1 yields the individual pairs of values: $L_1 = 1$, $H_1 = 0$; $L_1 = 3$, $H_1 = 1$; $L_1 = 5$, $H_1 = 2$; $L_1 = 7$, $H_1 = 3$. When these pairs are plotted in Fig. 9-2 using scales for L_1 as abscissa and H_1 as ordinate suitable for type 1 from Table 9-1, a straight line is obtained whose slope equals $(f_e/f_k)_{co} = 1/2$.

Using similar procedures, other lines are drawn for several values of $(f_e/f_k)_{co}$ for type 1 from Table 9-1, including the limiting line for $(f_e/f_k)_{co} = 1$. It if found that all of the resulting straight lines for individually constant $(f_e/f_k)_{co}$ pass through the point $(-1, -1)$ and have a slope $(H_1 + 1)/(L_1 + 1)$. Each circle on a line for constant $(f_e/f_k)_{co}$ represents integer values of L_1 and H_1 for coincidence.

Since the order of an intermodulation product is $O = L + H$, other lines can be drawn as shown in Fig. 9-2 connecting those values of L_1 and H_1 for constant order O_1, where the subscript stands for type 1. This allows coincidence points for some specified order or range to be determined from Fig. 9-2.

9-3.2 Type 2 Coincidence

Type 2 coincidence in a different mixer means that

$$H_2 f_k - L_2 f_e = f_k - f_e \tag{9-5}$$

where the subscript 2 represents type 2. Rearrange (9-5) as

$$\left(\frac{f_e}{f_k}\right)_{co} = \frac{H_2 - 1}{L_2 - 1} \tag{9-6}$$

Using (9-6), (9-1) becomes

$$0 < \frac{H_2 - 1}{L_2 - 1} < 1 \tag{9-7}$$

Representative positive integers that satisfy (9-7) are given in the row for H and the column for L marked type 2 in Table 9-1. The fractions given in Table 9-1 are those values of $(f_e/f_k)_{co}$ that satisfy (9-6). The table is readily extended to higher values of L_2 and H_2 as may be desired for type 2.

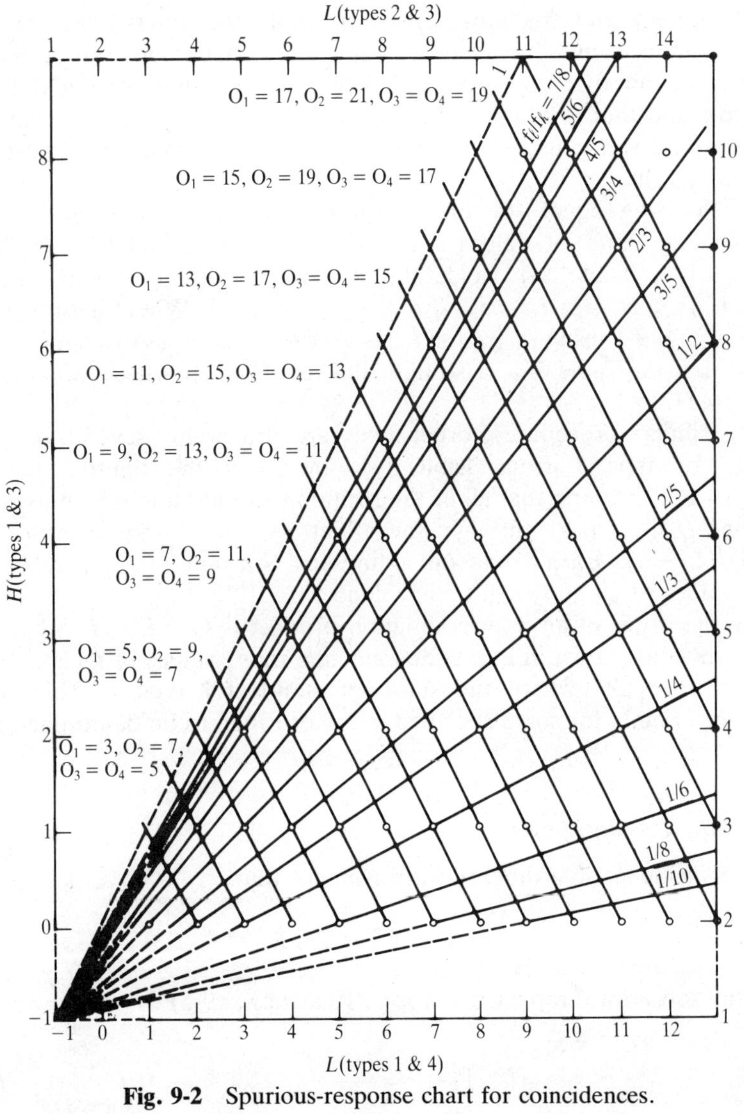

Fig. 9-2 Spurious-response chart for coincidences.

The method of locating equal relatively prime fractional values is the same as described for type 1. Thus, for type 2 with $(f_\ell/f_k)_{co} = 1/2$, Table 9-1 yields the individual pairs: $L_2 = 3$, $H_2 = 2$; $L_2 = 5$, $H_2 = 3$; $L_2 = 7$, $H_2 = 4$; $L_2 = 9$, $H_2 = 5$. When these values are plotted in Fig. 9-2 using scales for L_2 and H_2 suitable for type 2 from Table 9-1, a straight line is obtained whose slope equals $(f_\ell/f_k)_{co} = 1/2$.

Using similar procedures, other lines are drawn for several values of $(f_\ell/f_k)_{co}$ for type 2 from Table 9-1. It is found that all of the resulting lines for individually constant $(f_\ell/f_k)_{co}$ pass through the point (1, 1) on the scales for type 2 and have a slope $(H_2-1)/(L_2-1)$ equal to the particular values of $(f_\ell/f_k)_{co}$. Each circle on a line for constant $(f_\ell/f_k)_{co}$ represents integer values of L_2 and H_2 for coincidence.

Other lines are also drawn in Fig. 9-2 connecting those values of L_2 and H_2 for constant order O_2, where the subscript stands for type 2.

9-3.3 Type 3 Coincidence

Type 3 coincidence in a sum mixer means that

$$L_3 f_\ell - H_3 f_k = f_k + f_\ell \tag{9-8}$$

where the subscript 3 represents type 3. Rearrange (9-8) as

$$\left(\frac{f_\ell}{f_k}\right)_{co} = \frac{H_3+1}{L_3-1} \tag{9-9}$$

Using (9-9), (9-1) becomes

$$0 < \frac{H_3+1}{L_3-1} < 1 \tag{9-10}$$

Representative integers that satisfy (9-10) are given in the row for H and the column for L marked type 3 in Table 9-1. The fractions given in Table 9-1 are those values of $(f_\ell/f_k)_{co}$ that satisfy (9-9). The table is readily extended to higher values of L_3 and H_3 as may be desired for type 3.

The method of locating equal relatively prime fractional values is the same as described for type 1. Thus, for type 3 with $(f_\ell/f_k)_{co} = 1/2$, Table 9-1 yields the individual pairs: $L_3 = 3$, $H_3 = 0$; $L_3 = 5$, $H_3 = 1$; $L_3 = 7$, $H_3 = 2$; $L_3 = 9$, $H_3 = 3$. When these values are plotted in Fig. 9-2 using scales for L_3 and H_3 suitable for type 3 from Table 9-1, a straight line is obtained whose slope equals $(f_\ell/f_k)_{co} = 1/2$.

Using similar procedures, other lines are drawn for several values of $(f_\ell/f_k)_{co}$ for type 3 from Table 9-1. It is found that all of the resulting lines for individually constant $(f_\ell/f_k)_{co}$ pass through the point $(1, -1)$ on the scales for type 3 and have a slope $(H_3+1)/(L_3-1)$ equal to the particular constant value of $(f_\ell/f_k)_{co}$. Each circle on a line for constant $(f_\ell/f_k)_{co}$ represents integer values of L_3 and H_3 for coincidence.

Other lines are also drawn in Fig. 9-2 connecting those values of L_3 and H_3 for constant order O_3, where the subscript stands for type 3.

9-3.4 Type 4 Coincidence

Type 4 coincidence in a sum mixer means that

$$H_4 f_k - L_4 f_\ell = f_k + f_\ell \tag{9-11}$$

where the subscript 4 represents type 4. Rearrange (9-11) as

$$\left(\frac{f_\ell}{f_k}\right)_{\mathrm{co}} = \frac{H_4 - 1}{L_4 + 1} \tag{9-12}$$

Using (9-12), (9-1) becomes

$$0 < \frac{H_4 - 1}{L_4 + 1} < 1 \tag{9-13}$$

Representative positive integers that satisfy (9-13) are given in the row for H and the column for L marked type 4 in Table 9-1. The fractions given in Table 9-1 are those values of $(f_\ell/f_k)_{\mathrm{co}}$ that satisfy (9-12). The table is readily extended to higher values of L_4 and H_4 as may be desired for type 4.

The method of locating equal relatively prime fractional values is the same as described for type 1. Thus, for type 4 with $(f_\ell/f_k)_{\mathrm{co}} = 1/2$, Table 9-1 yields the individual pairs: $L_4 = 1$, $H_4 = 2$; $L_4 = 3$, $H_4 = 3$; $L_4 = 5$, $H_4 = 4$; $L_4 = 7$, $H_4 = 5$. When these values are plotted in Fig. 9-2 using scales for L_4 and H_4 suitable for type 4 from Table 9-1, a straight line is obtained whose slope equals $(f_\ell/f_k)_{\mathrm{co}} = 1/2$.

Using similar procedures, other lines are drawn for several values of $(f_\ell/f_k)_{\mathrm{co}}$ for type 4 from Table 9-1. It is found that all of the resulting lines for individually constant $(f_\ell/f_k)_{\mathrm{co}}$ pass through the point $(-1, 1)$ on the scales for type 4 and have a slope $(H_4 - 1)/(L_4 + 1)$ equal to the particular constant value of $(f_\ell/f_k)_{\mathrm{co}}$. Each circle on a line for constant $(f_\ell/f_k)_{\mathrm{co}}$ represents integer values of L_4 and H_4 for coincidence.

Other lines are also drawn in Fig. 9-2 connecting those values of L_4 and H_4 for constant order O_4, where the subscript stands for type 4.

Type 5 coincidence in a difference mixer means that $Hf_k + Lf_\ell = f_k - f_\ell$ and type 6 coincidence in a sum mixer means that $Hf_k + Lf_\ell = f_k + f_\ell$. Their analyses are reserved for Problem 9-2.

9-4 SIMPLIFIED SPURIOUS-RESPONSE CHART

Fortuntely, in use, it is not necessary to have the complete highly detailed illustration like Fig. 9-2 with many lines drawn for f_ℓ/f_k and O. All that is really needed is a set of rectangular coordinates with H as ordinate and L as abscissa for the types required, one line for the maximum order $\mathrm{O}_{\mathrm{max}}$ to be

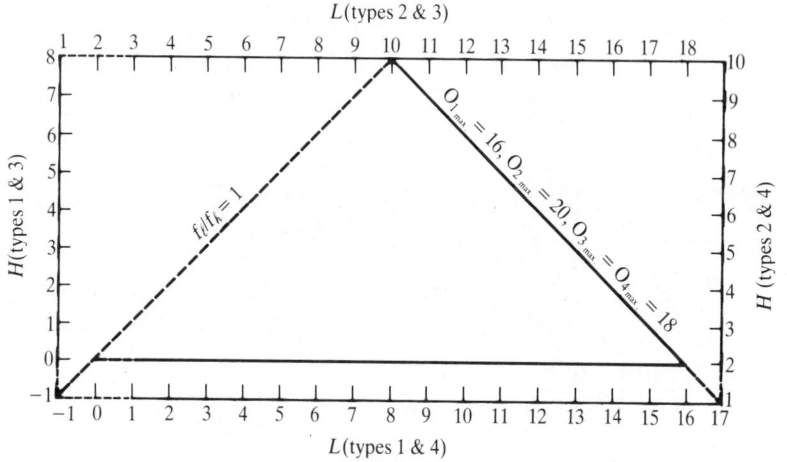

Fig. 9-3 A simplified spurious response chart. Coincidence points are at the intersections of integer values of the H and L coordinates.

considered, and the coincidence points below the lines for the $f_\ell / f_k = 1$ limiting value and O_{max}. Further, the coincidence points may be identified quite simply at the intersections of integer values of the L and H coordinates instead of by circles or heavy dots. The resulting figure, shown in Fig. 9-3 for $O_{max} = 16$ is quickly drawn for any maximum order. Notice that $O_2 = O_1 + 4$ and $O_3 = O_4 = O_1 + 2$. A line for any order can be quickly drawn by recognizing that it passes through the point $L_1 = O_1$, $H_1 = 0$.

Example 9-1

A certain superheterodyne receiver tunes the AM broadcast band 535 kHz to 1605 kHz. It uses HSI/DC with an IF = 455 kHz. (a) Calculate the ratios $(f_\ell / f_k)_{min}$ and $(f_\ell / f_k)_{max}$. (b) Determine what pairs of values for H and L yield coincidences for spurious IF signals of sixth order or less. (c) Calculate the frequencies of the input signals to which the receiver must be tuned for coincidences to occur.

Solution Since high-side injection is used, the incoming signal to which the receiver is tuned represents f_ℓ, therefore 535 kHz $\leq f_\ell \leq$ 1605 kHz. The oscillator frequency is always 455 kHz higher than f_ℓ, therefore 990 kHz $\leq f_k \leq$ 2060 kHz. Since difference mixing must be used for down conversion, only types 1 and 2 can be involved.

(a) $$\left(\frac{f_\ell}{f_k}\right)_{min} = \frac{535}{990} = 0.54, \qquad \left(\frac{f_\ell}{f_k}\right)_{max} = \frac{1605}{2060} = 0.78$$

(b) Prepared a simplified spurious response chart and draw the two lines for the ratios 0.54 and 0.78, noting that the slopes are the same as the

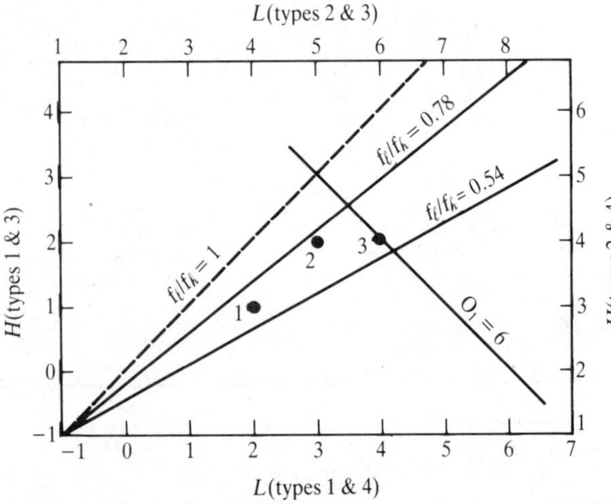

Fig. 9-4 The triangle formed by the lines for $f_e/f_k = 0.54$, $f_e/f_k = 0.78$, and $O_1 = 6$ encloses coincidence points that lead to in-band spurious responses in Example 9-1.

ratios. To aid in drawing the lines, it is convenient to use $535/990 = 4/7.4$ instead of 0.54, and $1605/2060 = 4/5.13$ instead of 0.78. Of course, if desired, other convenient numbers may be used to establish the slopes. Next, draw the line representing $O_1 = 6$. It is found that there are three coincidence points located inside or on the triangle formed by these three lines as shown in Fig. 9-4:

Point 1 for $O_1 = 3$, $(f_e/f_k)_{co} = 2/3$, $L_1 = 2$, and $H_1 = 1$

Point 2 for $O_1 = 5$, $(f_e/f_k)_{co} = 3/4$, $L_1 = 3$, and $H_1 = 2$

Point 3 for $O_1 = 6$, $(f_e/f_k)_{co} = 3/5$, $L_1 = 4$, and $H_1 = 2$.

These coincidence points will be encountered somewhere within the tuning range of the receiver, so that *in-band* spurious responses are said to result. Since $O_2 = O_1 + 4$, no type 2 coincidence points of sixth order or less are inside the triangle, therefore type 2 need not be considered unless higher order is allowed, in which case there may be more type 1 points. The most bothersome spurious response will likely be the one for the lowest order encountered.

(c) For point 1, $(f_e/f_k)_{co} = 2/3$ so that $f_k = (3/2)f_e$. Then, since IF $= f_k - f_e = 455$ kHz, we have $(3/2)f_e - f_e = 455$ kHz from which is obtained $f_e = 910$ kHz. If the receiver is tuned to a station or a signal generator operating on 910 kHz, the local oscillator is on $f_k = 1365$ kHz, and the desired IF $= (1365 - 910)$ kHz $= 455$ kHz as expected. Since $L_1 = 2$ and $H_1 = 1$ for point 1, a third-order type 1 spurious IF signal may be produced whose frequency is $L_1 f_e - H_1 f_k = [2(910) - (1)(1365)]$ kHz = 455 kHz. The frequency of the spurious IF signal coincides with the frequency of the desired IF signal, and this is the fundamental meaning of the term coincidence.

For point 2, $(f_e/f_k)_{co} = 3/4$, so that $f_k = (4/3)f_e$. Thus IF = 455 kHz = $f_k - f_e = (4/3)f_e - f_e = (1/3)f_e$, from which is obtained $f_e = 1365$ kHz. If the receiver is tuned to a signal generator operating on 1365 kHz, the local oscillator is on $f_k = 1820$ kHz and the desired IF = 455 kHz. Since $L_1 = 3$ and $H_1 = 2$ for point 2, a fifth-order type 1 spurious IF signal may be produced whose frequency is $L_1 f_e - H_1 f_k = [3(1365) - 2(1820)]$ kHz = 455 kHz, showing that coincidence occurs.

The same procedures applied to the sixth-order coincidence point 3, for which $(f_e/f_k)_{co} = 3/5$ with $L_1 = 4$ and $H_1 = 2$, shows that coincidence occurs when the receiver is tuned to a signal on 682.5 kHz (see Problem 9-3).

Whether the various IM products of different orders are actually created depends on the degree of curvature of the mixer static characteristic. Thus, third-order and above IM products do not occur if the mixer characteristic is strictly square-law shape.

Example 9-2

Use the third-order coincidence point of Example 9-1 to describe how a birdie is produced.

Solution Instead of tuning the receiver exactly to a station on 910 kHz, suppose that the receiver is detuned slightly so that the local oscillator is operating on $f_k = 1365.1$ kHz. Since the incoming signal frequency of the desired station is still 910 kHz, the desired IF = (1365.1 − 910) kHz = 455.1 kHz. The spurious IF = [2(910) − 1(1365.1)] kHz = 454.9 kHz, showing that coincidence does not occur. Both the desired and the spurious IF signals can pass through the IF passband if its bandwidth is adequate for normal AM broadcast-band signals. The desired IF signal is demodulated by the envelope detector as usual. But the 454.9 kHz and 455.1 kHz components produce an envelope variation at the 200-Hz beat frequency. This appears in the output of the envelope detector as a 200-Hz audible whistle. As the receiver is adjusted closer to the correct tuning, the frequency of the audible whistle decreases, and as the receiver is adjusted farther away from the correct tuning, the frequency of the audible whistle increases. When tuning rapidly through the station on 910 kHz, a receiver having a sufficient amount of third-order type 1 spurious response, in this case, produces the characteristic variable-pitch audible whistle known as a birdie.

9-4.1 Another Spurious-Response Chart

Other procedures have been developed to aid in the determination of interfering IM products. One of these procedures uses the chart shown in Fig. 9-5(a). The IM products for each of the chart lines are identified around the edges of the chart. The equations for the chart lines are easily derived as follows.

For type 1 IM products in a difference mixer whose $\mathrm{IF} = f_h - f_\ell$, we have $Lf_\ell - Hf_h = \mathrm{IF}$, from which is obtained

$$\left(\frac{\mathrm{IF}}{f_h}\right) = L\left(\frac{f_\ell}{f_h}\right) - H$$

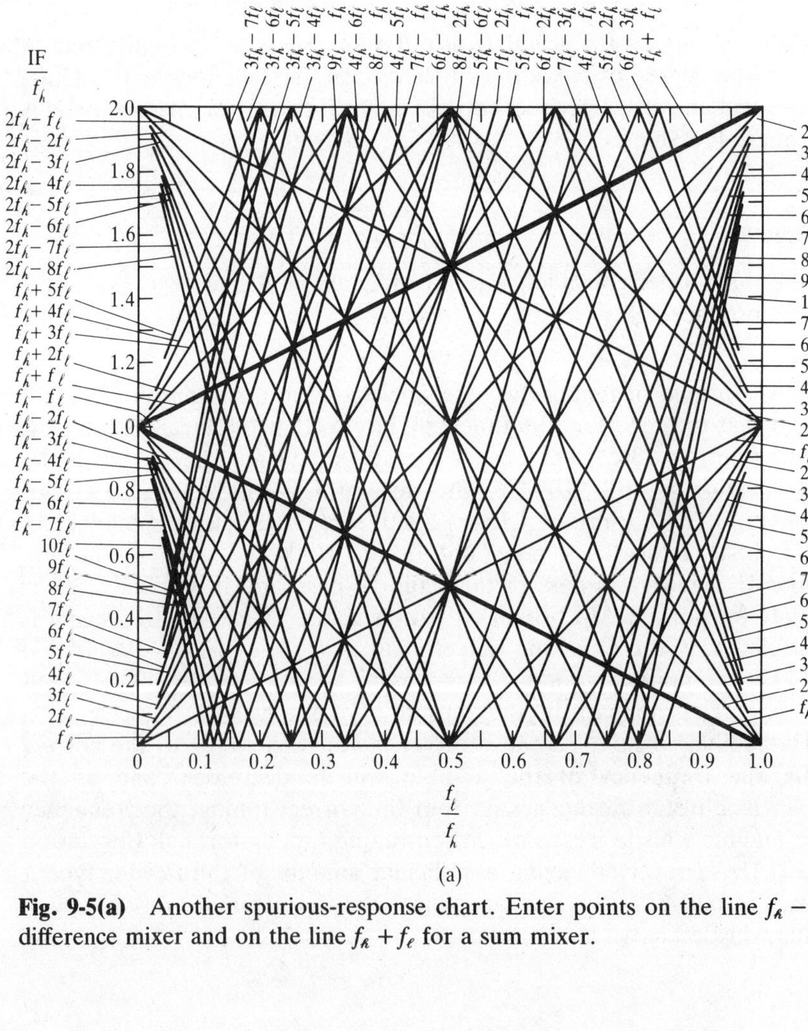

(a)

Fig. 9-5(a) Another spurious-response chart. Enter points on the line $f_h - f_\ell$ for a difference mixer and on the line $f_h + f_\ell$ for a sum mixer.

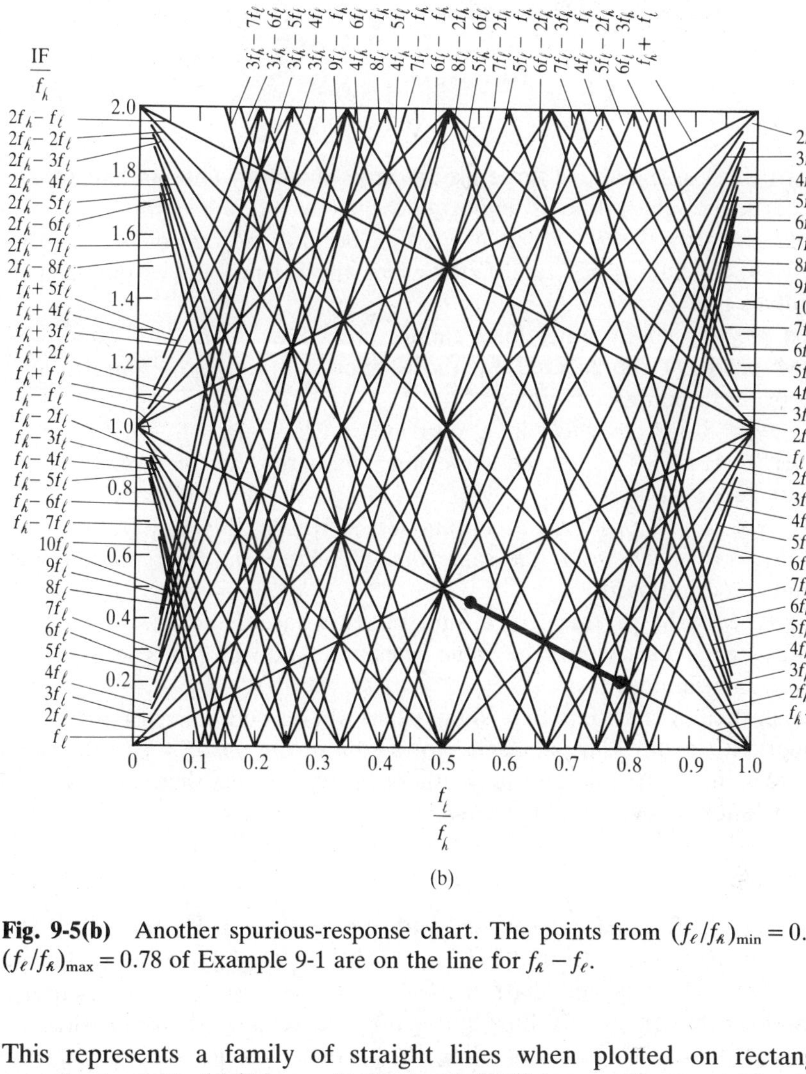

(b)

Fig. 9-5(b) Another spurious-response chart. The points from $(f_\ell/f_h)_{\min} = 0.54$ to $(f_\ell/f_h)_{\max} = 0.78$ of Example 9-1 are on the line for $f_h - f_\ell$.

This represents a family of straight lines when plotted on rectangular coordinates with (f_ℓ/f_h) as abscissa and (IF/f_h) as ordinate. A line has a positive slope $+L$ and a vertical-axis intercept $-H$.

For type 2 IM products in a difference mixer whose IF $= f_h - f_\ell$, we have $Hf_h - Lf_\ell = \text{IF}$, from which is obtained

$$\left(\frac{\text{IF}}{f_h}\right) = -L\left(\frac{f_\ell}{f_h}\right) + H$$

This also represents a family of straight lines on the same set of coordinates as above. The slope is $-L$ and the vertical-axis intercept is $+H$.

For type 5 IM products in a difference mixer whose IF $= f_h - f_\ell$, we have

$Hf_{\hbar} + Lf_{\ell} = \text{IF}$, from which is obtained

$$\left(\frac{\text{IF}}{f_{\hbar}}\right) = L\left(\frac{f_{\ell}}{f_{\hbar}}\right) + H$$

Many lines of the above three groups cross the chart line marked $f_{\hbar} - f_{\ell}$. This particular chart line is important when a difference mixer is used because the $\text{IF} = f_{\hbar} - f_{\ell}$. When a crossing occurs, the coordinates at that point yield the values of $f_{\hbar}$ and f_{ℓ} for which the IM product of the crossing line equals the IF.

Observe that many other lines cannot cross the $f_{\hbar} - f_{\ell}$ line. Therefore, the IM products represented by those lines will not cause interference problems in the difference mixer.

For type 3 IM products in a sum mixer whose $\text{IF} = f_{\hbar} + f_{\ell}$, we have $Lf_{\ell} - Hf_{\hbar} = \text{IF}$. This yields the same linear equation as for type 1, so the lines are superimposed.

For type 4 IM products in a sum mixer whose $\text{IF} = f_{\hbar} + f_{\ell}$, we have $Hf_{\hbar} - Lf_{\ell} = \text{IF}$. This yields the same linear equation as for type 2, so the lines are superimposed.

For type 6 IM products in a sum mixer whose $\text{IF} = f_{\hbar} + f_{\ell}$, we have $Hf_{\hbar} + Lf_{\ell} = \text{IF}$. This yields the same linear equation as for type 5, so the lines are superimposed.

The particular chart line that is important when a sum mixer is used is the one marked $f_{\hbar} + f_{\ell}$. This is because for a sum mixer the $\text{IF} = f_{\hbar} + f_{\ell}$. Many lines cross this chart line and may others cannot. Those that do cross give rise to possible interfering IM products.

Example 9-3

The use of the chart is illustrated with the receiver of Example 9-1. The receiver is tunable from $(f_{\ell}/f_{\hbar})_{\text{min}} = 0.54$ to $(f_{\ell}/f_{\hbar})_{\text{max}} = 0.78$. Locate those end points on the chart line for $f_{\hbar} - f_{\ell}$, because the receiver uses a difference mixer. If any lines cross the line segment between the two entered points, the corresponding IM products are identified along the chart edges. In this chart, there are 11 lines, representing IM products of orders up to 14, that cross the segment. If IM products of sixth-order or less are to be considered, only three lines cross the segment. These are found to be:

$$(4f_{\ell} - 2f_{\hbar}) \text{ at } f_{\ell}/f_{\hbar} = 0.6$$

$$(2f_{\ell} - f_{\hbar}) \text{ at } f_{\ell}f_{\hbar} = 0.667$$

$$(3f_{\ell} - 2f_{\hbar}) \text{ at } f_{\ell}/f_{\hbar} = 0.75$$

These agree with those found in Example 9-1.

9-5 IF SELECTIVITY

When the receiver is tuned so that the line for f_ℓ/f_h does not pass through, but is close to, a coincidence point located at or below O_{max}, it is possible that the spurious IF signal, produced from the L and H values for the nearby coincidence point, will fall inside the IF passband and cause interference.

Consider the third-order type 1 coincidence point 1 of Example 9-1, for which $(f_\ell/f_h)_{co} = 2/3$, $L_1 = 2$, $H_1 = 1$, and coincidence occurs when the receiver is tuned exactly to a station operating on 910 kHz. But if tuned instead to an adjacent-channel station on $f_\ell = 900$ kHz for which $f_h = 1355$ kHz, then $f_\ell/f_h = 0.6642$ and the spurious IF $= [2(900) - 1(1355)]$ kHz $= 445$ kHz, which is 10 kHz below the desired IF signal. If this spurious IF signal is not attenuated sufficiently by the IF selectivity characteristic, it will cause interference. In this case, tuning to a station in the adjacent channel led to the frequency of the spurious IF signal being separated from the desired IF signal by the same amount as the channel spacing. Such IF selectivity relationships are examined as follows.

Using values of L and H for a type 1 coincidence point,

$$L_1 f_\ell - H_1 f_h = \text{IF} = \text{a constant} \tag{9-14}$$

and for a type 2 coincidence point

$$H_2 f_h - L_2 f_\ell = \text{IF} = \text{a constant} \tag{9-15}$$

Regardless of the frequency to which the receiver is tuned, the IF is constant because of the receiver design. Thus, if the receiver is tuned to a signal at $(f_\ell + \Delta f)$ or at $(f_h + \Delta f)$ depending on whether HSI or LSI, respectively, is used, the spurious IF signal becomes, for type 1,

$$\begin{aligned}(\text{IF})_{sp} &= L_1(f_\ell + \Delta f) - H_1(f_h + \Delta f) \\ &= L_1 f_\ell - H_1 f_h + \Delta f(L_1 - H_1)\end{aligned} \tag{9-16}$$

and, for type 2,

$$\begin{aligned}(\text{IF})_{sp} &= H_2(f_h + \Delta f) - L_2(f_\ell + \Delta f) \\ &= H_2 f_h - L_2 f_\ell - \Delta f(L_2 - H_2)\end{aligned} \tag{9-17}$$

Then, using (9-14), (9-16) becomes

$$(\text{IF})_{sp} = \text{IF} + \Delta f(L_1 - H_1), \qquad \text{type 1} \tag{9-18}$$

and, using (9-15), (9-17) becomes

$$(\mathrm{IF})_{sp} = \mathrm{IF} - \Delta f(L_2 - H_2), \qquad \text{type 2} \tag{9-19}$$

Similar results are obtained for types 3 and 4.

The difference between the spurious IF and the desired IF is defined as

$$\delta f = (\mathrm{IF})_{sp} - \mathrm{IF} \tag{9-20}$$

Therefore, (9-18) becomes

$$\delta f = \Delta f(L - H), \qquad \text{types 1 and 3} \tag{9-21}$$

and (9-19) becomes

$$\delta f = -\Delta f(L - H), \qquad \text{types 2 and 4} \tag{9-22}$$

If $|\Delta f|$ represents the channel spacing over which a receiver tunes, then the quantity $|\delta f|$ represents how far away from the desired IF each spurious IF signal will be for adjacent channels. Except for some type 4 coincidence points, Table 9-1 reveals that $L > H$, so that $|\delta f| \geq |\Delta f|$. In order for the IF stages of the receiver to reject some of these spurious responses, it is necessary for the IF stages to have less bandwidth than $|\delta f|$. This, of course, is directly related to the IF selectivity. With SSB receivers, the spurious IF may fall in the unused sideband region and be attenuated in normal fashion.

For some type 4 coincidence points, $L = H$, that $\delta f = 0$, and for some other points, $H > L$, so that $\delta f = \Delta f(H - L)$.

Example 9-4

The fifth-order type 1 coincidence point 2 of Example 9-1 is obtained when the receiver is tuned to a signal on 1365 kHz. But instead of tuning to a signal at that frequency produced by a signal generator, suppose the receiver is tuned to a station operating on $f_\ell = 1360$ kHz. Calculate: (a) the frequency of the birdie; (b) the frequency to which the receiver should be off-tuned from the station on 1360 kHz to render the birdie zero-beat and therefore inaudible.

Solution (a) For $f_\ell = 1360$ kHz, the local oscillator is at $f_h = 1815$ kHz. The desired IF = (1815 − 1360) kHz = 455 kHz. The undesired spurious IF = $3f_\ell - 2f_h$ = [3(1360) − 2(1815)] kHz = 450 kHz. Hence, the spurious IF signal is 5 kHz below the desired IF signal and is near the edge of the IF passband for normal AM broadcast-band stations. Therefore, an annoying 5-kHz whistle or birdie is superimposed on the desired program.

Also, $\Delta f = (1360 - 1365)\ \text{kHz} = -5\ \text{kHz}$. Then, with $L_1 = 3$ and $H_1 = 2$ for the nearby coincidence point 2, we have, from (9-21),

$$\delta f = -5(3-2)\ \text{kHz} = -5\ \text{kHz}$$

as before. Thus, the undesired spurious IF = (455 − 5) kHz = 450 kHz. Trying to eliminate this spurious IF response by reducing the IF bandwidth also causes the desired IF signal to suffer from the reduced bandwidth.

(b) The birdie is at zero beat and therefore inaudible if

$$\text{Spurious IF frequency} = \text{desired IF frequency}$$

Detuning the receiver from the station on $f_\ell = 1360\ \text{kHz}$ by Δf kHz causes the local oscillator to be at $(1815 + \Delta f)$ kHz. Since the station frequency remains at 1360 kHz, the desired IF $= [(1815 + \Delta f) - 1360]\ \text{kHz} = (455 + \Delta f)\ \text{kHz}$. The spurious IF $= [3(1360) - 2(1815 + \Delta f)]\ \text{kHz} = (450 - 2\,\Delta f)\ \text{kHz}$. Consequently, for zero beat in this case,

$$(450 - 2\,\Delta f)\ \text{kHz} = (455 + \Delta f)\ \text{kHz}$$

or

$$\Delta f = -1.667\ \text{kHz}$$

That is, the receiver must be off-tuned to (1360 − 1.667) kHz = 1358.33 kHz. The desired IF signal is centered at (455 − 1.667) kHz = 453.33 kHz instead of 455 kHz. A critical listener may complain that the desired program is now somewhat distorted because the station is not tuned in correctly, but the birdie is inaudible. The distortion may be caused when some of the sideband frequencies are in the portion of the IF selectivity characteristic where the amplification is significantly less than that inside the IF passband.

9-6 RF SELECTIVITY

Another spurious phenomenon occurs when the receiver is tuned to any signal frequency f_r, at which a station may or may not be operating, and a strong undesired interfering station operating on *another* frequency $f_c = f_r + \Delta f$, mixes with the output of the local oscillator to produce an undesired IF signal that is in congruence with the receiver IF. Since it is assumed that the RF circuits (sometimes called the *preselector*) are tuned to the desired frequency f_r, the interfering spurious input signal, or, more simply, *spur*, will be at a frequency where the RF response (or gain) is less than the maximum possible at f_r. But this lack of RF response can be more compensated for if the spur has a large amplitude, thus rendering the interference discernible.

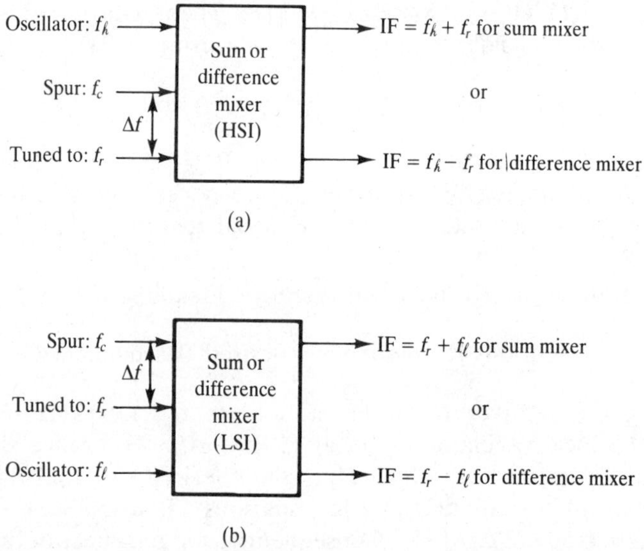

Fig. 9-6 Illustrating several possibilities whereby the spur frequency $f_c = f_r + \Delta f$ can produce the IF in a sum or difference mixer: (a) high-side injection; (b) low-side injection.

Figure 9-6 represents several possibilities that are to be considered. In all cases, the spur frequency f_c is taken to be Δf plus the frequency f_r to which the receiver is tuned. The values of Δf can sometimes be negative, as determined by the circumstances.

For type 1 with HSI, f_c replaces f_ℓ as the input signal frequency in the IM products, so that, using positive-integer values of L and H *that are not restricted just to coincidence points,* congruence occurs when

$$Lf_c - Hf_k = \text{IF} = \text{a constant} \tag{9-23}$$

In all cases, the desired IF does not change as the receiver is tuned to different frequencies because of the tracking of the local oscillator frequency. Since for type 1 with HSI we have $\text{IF} = f_k - f_r$, from which is obtained $f_k = \text{IF} + f_r$, (9-23) becomes

$$f_c = \frac{\text{IF}(H+1) + Hf_r}{L} \tag{9-24}$$

Thus, all combinations of $L \geq 1$ and $H \geq 0$ yield possible spur frequencies for any f_r.

Thus, with $H = 0$, the possible congruent spur frequencies are, from either (9-23) or (9-24).

$$f_c = \frac{\text{IF}}{L}, \qquad L = 1, 2, 3, \ldots \tag{9-25}$$

The condition called *IF breakthrough* occurs when $L = 1$ in (9-25), so that $f_c = \text{IF}$. If the skirt selectivity of the preselector is inadequate, a signal at $f_c = \text{IF}$ can be delivered to the IF amplifiers regardless of the frequency f_r to which the receiver is tuned. The likelihood of IF breakthrough is greatest when the receiver is tuned to a frequency nearest to the IF value than when it is tuned to other frequencies because then $\Delta f = f_c - f_r$ has the smallest magnitude. This occurs in Example 9-1 when the receiver is tuned to 535 kHz so that $\Delta f = (455 - 535)$ kHz $= -80$ kHz. This result is also found in Fig. 9-4 as an out-of-band coincidence point at $L_1 = 1$ and $H_1 = 0$. When the receiver is tuned to any other frequency in its tuning range, $|\Delta f| >$ 80 kHz.

Continuing with reference to Fig. 9-4 for Example 9-1, there is another out-of-band coincidence point at $L_1 = 3$ and $H_1 = 1$. Therefore, from (9-24),

$$f_c = \frac{455(1+1) + 535}{3} \text{ kHz} = 481.667 \text{ kHz}$$

so that $\Delta f = (481.667 - 535)$ kHz $= -53.33$ kHz. To check, when tuned to 535 kHz, the local oscillator frequency is 990 kHz. Then, [3(481.667) − 1(990)] kHz = 455 kHz, which is the IF. If the interfering station is actually operating on $f_c = 485$ kHz instead of 481.667 kHz, it may produce the IF when the receiver is tuned to another frequency found by solving (9-24) for f_r. Thus,

$$f_r = \frac{Lf_c - \text{IF}(H+1)}{H}$$

$$= \frac{3(485) - 455(1+1)}{1} \text{ kHz} = 545 \text{ kHz} \tag{9-26}$$

Therefore, $\Delta f = f_c - f_r = (485 - 545)$ kHz $= -60$ kHz. To check, when tuned to 545 kHz, the local oscillator frequency is 1000 kHz. Then, [3(485) − 1(1000)] kHz = 455 kHz, which is the IF.

The several possibilities for all four types with both HSI and LSI are analyzed in similar manner. The results are summarized in Table 9-2. In the IM products, f_c replaces the signal frequency f_ℓ when HSI is used, and f_c replaces the signal frequency f_k when LSI is used.

When desirable, it is possible quickly to rearrange the equations for f_c in Table 9-2 to express the RF selectivity as a percentage separation of f_c from f_r. The procedure involves subtracting f_r from both sides of each equation to obtain $\Delta f = f_c - f_r$ in terms of L, H, the IF, and f_r. Dividing those results by f_r, simplifying, and multiplying by 100 yields the equations for the percentage separation $(\Delta f / f_r) \times 100$ percent. The percentage separation is often a useful design criterion.

TABLE 9-2 Several Possibilities for Congruence Whereby the Receiver IF is Produced by a Signal at the Spur Frequency f_c When the Receiver is Tuned to f_r

	Difference Mixer	Sum Mixer
TYPES 1 AND 3 WITH HSI		
$Lf_c - Hf_h = \mathrm{IF}$	$\mathrm{IF} = f_h - f_r$	$\mathrm{IF} = f_h + f_r$
$f_c = \dfrac{Hf_h + \mathrm{IF}}{L}$	$f_h = f_r + \mathrm{IF}$	$f_h = \mathrm{IF} - f_r$
	$f_c = \dfrac{Hf_r + (H+1)\mathrm{IF}}{L}$	$f_c = \dfrac{(H+1)\mathrm{IF} - Hf_r}{L}$
TYPES 1 AND 3 WITH LSI		
$Lf_\ell - Hf_c = \mathrm{IF}$	$\mathrm{IF} = f_r - f_\ell$	$\mathrm{IF} = f_r + f_\ell$
$f_c = \dfrac{Lf_\ell - \mathrm{IF}}{H}$	$f_\ell = f_r - \mathrm{IF}$	$f_\ell = \mathrm{IF} - f_r$
	$f_c = \dfrac{Lf_r - (L+1)\mathrm{IF}}{H}$	$f_c = \dfrac{(L-1)\mathrm{IF} - Lf_r}{H}$
TYPES 2 AND 4 WITH HSI		
$Hf_h - Lf_c = \mathrm{IF}$	$\mathrm{IF} = f_h - f_r$	$\mathrm{IF} = f_h + f_r$
$f_c = \dfrac{Hf_h - \mathrm{IF}}{L}$	$f_h = f_r + \mathrm{IF}$	$f_h = \mathrm{IF} - f_r$
	$f_c = \dfrac{Hf_r + (H-1)\mathrm{IF}}{L}$	$f_c = \dfrac{(H-1)\mathrm{IF} - Hf_r}{L}$
TYPES 2 AND 4 WITH LSI		
$Hf_c - Lf_\ell = \mathrm{IF}$	$\mathrm{IF} = f_r - f_\ell$	$\mathrm{IF} = f_r + f_\ell$
$f_c = \dfrac{Lf_\ell + \mathrm{IF}}{H}$	$f_\ell = f_r - \mathrm{IF}$	$f_\ell = \mathrm{IF} - f_r$
	$f_c = \dfrac{Lf_r - (L-1)\mathrm{IF}}{H}$	$f_c = \dfrac{(L+1)\mathrm{IF} - Lf_r}{H}$

9-6.1 Type 5 Congruence

Type 5 congruence in a difference mixer with HSI and $\mathrm{IF} = f_h - f_r$ means that

$$Lf_c + Hf_h = \mathrm{IF} \tag{9-27}$$

from which is obtained, for $L \neq 0$,

$$f_c = \left(\frac{1-H}{L}\right)\mathrm{IF} - \left(\frac{H}{L}\right)f_r \tag{9-28}$$

But this is always negative except when $H = 0$, so that the only possible spur frequencies are given by

$$f_c = \frac{\mathrm{IF}}{L}, \qquad L = 1, 2, 3, \ldots \tag{9-29}$$

Type 5 congruence in a difference mixer with LSI and $\mathrm{IF} = f_r - f_\ell$ means that

$$Lf_\ell + Hf_c = \mathrm{IF} \tag{9-30}$$

from which is obtained, for $H \neq 0$,

$$f_c = \left(\frac{L+1}{H}\right)\mathrm{IF} - \left(\frac{L}{H}\right)f_r \tag{9-31}$$

This will always be positive if

$$\frac{L}{L+1} < \frac{\mathrm{IF}}{f_r} \tag{9-32}$$

which restricts the possible troublesome oscillator harmonics to the lower harmonic numbers.

9-6.2 Type 6 Congruence

Type 6 congruence in a sum mixer with HSI and $\mathrm{IF} = f_h + f_r$ means that

$$Lf_c + Hf_h = \mathrm{IF} \tag{9-33}$$

from which is obtained, for $L \neq 0$,

$$f_c = \left(\frac{1-H}{L}\right)\mathrm{IF} + \left(\frac{H}{L}\right)f_r \tag{9-34}$$

This will always be positive if

$$\frac{H}{H-1} > \frac{\mathrm{IF}}{f_r} \tag{9-35}$$

which restricts the possible troublesome oscillator harmonics to the lower harmonic numbers.

Type 6 congruence in a sum mixer with LSI and IF $=f_r + f_\ell$ means that

$$Lf_\ell + Hf_c = \text{IF} \tag{9-36}$$

from which is obtained, for $H \neq 0$,

$$f_c = \left(\frac{1-L}{H}\right)\text{IF} + \left(\frac{L}{H}\right)f_r \tag{9-37}$$

This will always be positive if

$$\frac{L}{L-1} > \frac{\text{IF}}{f_r} \tag{9-38}$$

which restricts the possible troublesome oscillator harmonics to the lower harmonic numbers.

9-7 COMBINED RF AND IF SELECTIVITIES

When the receiver is tuned to frequency f_r, it is possible that a strong interfering station operating on another frequency $f_x = f_r + \Delta f$ will produce an undesired IF signal at a frequency $f_y = \text{IF} + \delta f$ that will be within the IF passband or sufficiently far up on the IF response characteristic to render the interference discernible.

Figure 9-7 represents the several possibilities that are to be considered. In all cases, the spurious input frequency is $f_x = f_r + \Delta f$ and the spurious output frequency is $f_y = \text{IF} + \delta f$, where the value chosen for the IF is constant for a particular receiver.

For type 1 with HSI, f_x replaces f_ℓ as the input signal frequency in the IM products, so that, using values of L and H that are not restricted just to coincidence points,

$$Lf_x - Hf_h = f_y = \text{IF} + \delta f \tag{9-39}$$

Since for type 1 we have IF $= f_h - f_r$, in this case, from which is obtained $f_h = \text{IF} + f_r$, (9-39) becomes

$$f_x = \frac{Hf_r + (H+1)\text{IF} + \delta f}{L} \tag{9-40}$$

But part of this is exactly the same as the spur frequency f_c for type 1 with HSI given in Table 9-2. Therefore, (9-40) can be written as

$$f_x = f_c + \frac{\delta f}{L} \tag{9-41}$$

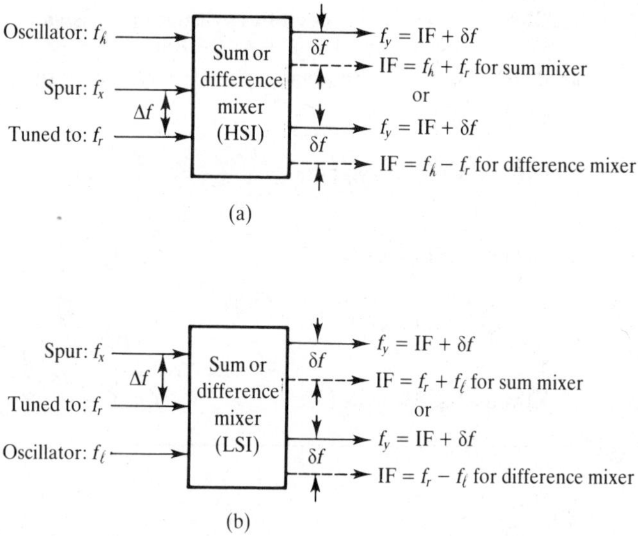

Fig. 9-7 Illustrating several possibilities whereby the spur frequency $f_x = f_r + \Delta f$ can produce a spurious frequency $f_y = \text{IF} + \delta f$ in a sum or difference mixer: (a) high-side injection; (b) low-side injection.

where

$$f_c = \frac{Hf_r + (H+1)\text{IF}}{L}$$

If the spurious IF frequency f_y is allowed to increase from the chosen IF value by the amount $+|\delta f|$, the spurious input frequency f_x that causes f_y increases by only $+|\delta f|/L$, a smaller amount that depends on the particular values of L.

For type 1 with LSI, f_x replaces f_h as the input signal frequency in the IM products, so that,

$$Lf_\ell - Hf_x = f_y = \text{IF} + \delta f \tag{9-42}$$

Since $\text{IF} = f_r - f_\ell$, in this case, then $f_\ell = f_r - \text{IF}$, and (9-42) becomes

$$f_x = \frac{Lf_r - (L+1)\text{IF} - \delta f}{H} \tag{9-43}$$

But part of this is exactly the same as the spur frequency f_c for type 1 with LSI in Table 9-2. Therefore, (9-43) can be written as

$$f_x = f_c - \frac{\delta f}{H} \tag{9-44}$$

TABLE 9-3 Several Possibilities Whereby a Signal at the Spur Frequency $f_x = f_r + \Delta f$ Produces an Undesired IF Signal at Frequency $f_y = \text{IF} + \delta f$ when the receiver is tuned to f_r

Types 1 and 3 with HSI	$f_x = f_c + \dfrac{\delta f}{L}$
Types 1 and 3 with LSI	$f_x = f_c - \dfrac{\delta f}{H}$
Types 2 and 4 with HSI	$f_x = f_c - \dfrac{\delta f}{L}$
Types 2 and 4 with LSI	$f_x = f_c + \dfrac{\delta f}{H}$

where

$$f_c = \frac{Lf_r - (L+1)\text{IF}}{H}$$

If the spurious IF frequency is allowed to increase from the chosen IF by the amount $+|\delta f|$, the spurious input frequency f_x that causes f_y *decreases* (because of the minus sign in (9-44) by only $|\delta f|/H$.

The several possibilities for all four types with both HSI and LSI are analyzed in similar manner. The results are summarized in Table 9-3, and they apply to both difference and sum mixers.

9-8 SPURIOUS-RESPONSE CHARTS FOR CONGRUENCE

To prepare the results given in Table 9-2 for graphical presentation, it is convenient to replace f_c with $f_r + \Delta f$ and to normalize each frequency to the IF. The results of that process are summarized in Table 9-4.

Specifically, for type 1 with HSI,

$$\frac{\Delta f}{\text{IF}} = \frac{H-L}{L}\left(\frac{f_r}{\text{IF}}\right) + \frac{H+1}{L} \tag{9-45}$$

The frequency ratios are linearly related in a rectangular coordinate system whose abscissa is the conversion ratio $x = (f_r/\text{IF})$ and whose ordinate is the spur frequency ratio $y = (\Delta f/\text{IF})$ required for congruency. Then, (9-45) is no more complicated than the familiar equation of a straight line

$$y = mx + b$$

TABLE 9-4 Frequencies of Table 9-2 Normalized to the IF and With the Spur Frequency f_c Replaced With $f_r + \Delta f$

	Difference Mixer	Sum Mixer
TYPES 1 AND 3 WITH HSI		
$Lf_c - Hf_{h} = \text{IF}$	$\text{IF} = f_{h} - f_r$	$\text{IF} = f_{h} + f_r$
$f_c = \dfrac{Hf_{h} + \text{IF}}{L}$	$\dfrac{\Delta f}{\text{IF}} = \dfrac{H-L}{L}\left(\dfrac{f_r}{\text{IF}}\right) + \dfrac{H+1}{L}$	$\dfrac{\Delta f}{\text{IF}} = -\dfrac{H+L}{L}\left(\dfrac{f_r}{\text{IF}}\right) + \dfrac{H+1}{L}$
TYPES 1 AND 3 WITH LSI		
$Lf_{\ell} - Hf_c = \text{IF}$	$\text{IF} = f_r - f_{\ell}$	$\text{IF} = f_r + f_{\ell}$
$f_c = \dfrac{Lf_{\ell} - \text{IF}}{H}$	$\dfrac{\Delta f}{\text{IF}} = \dfrac{L-H}{H}\left(\dfrac{f_r}{\text{IF}}\right) - \dfrac{L+1}{H}$	$\dfrac{\Delta f}{\text{IF}} = -\dfrac{L+H}{H}\left(\dfrac{f_r}{\text{IF}}\right) + \dfrac{L-1}{H}$
TYPES 2 AND 4 WITH HSI		
$Hf_{h} - Lf_c = \text{IF}$	$\text{IF} = f_{h} - f_r$	$\text{IF} = f_{h} + f_r$
$f_c = \dfrac{Hf_{h} - \text{IF}}{L}$	$\dfrac{\Delta f}{\text{IF}} = \dfrac{H-L}{L}\left(\dfrac{f_r}{\text{IF}}\right) + \dfrac{H-1}{L}$	$\dfrac{\Delta f}{\text{IF}} = -\dfrac{H+L}{L}\left(\dfrac{f_r}{\text{IF}}\right) + \dfrac{H-1}{L}$
TYPES 2 AND 4 WITH LSI		
$Hf_c - Lf_{\ell} = \text{IF}$	$\text{IF} = f_r - f_{\ell}$	$\text{IF} = f_r + f_{\ell}$
$f_c = \dfrac{Lf_{\ell} + \text{IF}}{H}$	$\dfrac{\Delta f}{\text{IF}} = \dfrac{L-H}{H}\left(\dfrac{f_r}{\text{IF}}\right) - \dfrac{L-1}{H}$	$\dfrac{\Delta f}{\text{IF}} = -\dfrac{L+H}{H}\left(\dfrac{f_r}{\text{IF}}\right) + \dfrac{L+1}{H}$

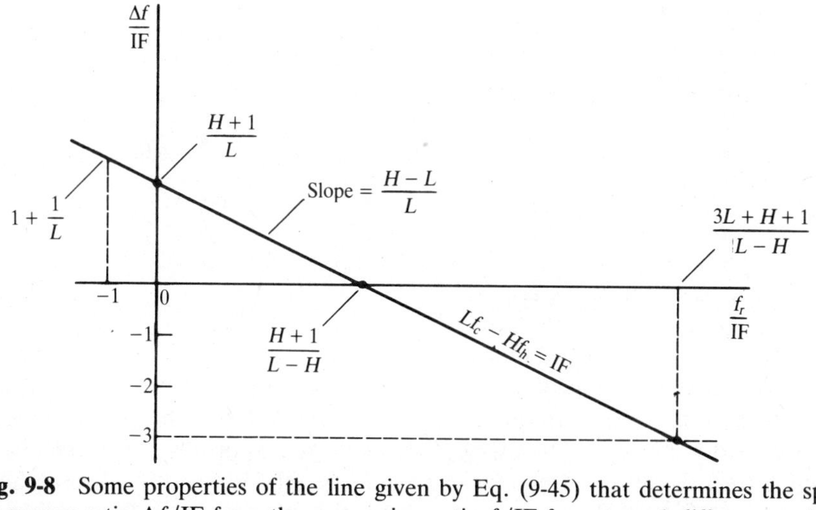

Fig. 9-8 Some properties of the line given by Eq. (9-45) that determines the spur frequency ratio $\Delta f/\text{IF}$ from the conversion ratio f_r/IF for a type 1 difference mixer with HSI.

where the slope $m = (H - L)/L$ and the y-axis intercept $b = (H + 1)/L$. Also, the x-axis intercept is found to be $-b/m = (H + 1)/(L - H)$. Slopes and intercepts from other entries in Table 9-4 are readily obtained in similar fashion. Only the portion of any line that extends into the right half-plane is usable, because the frequency ratio f_r/IF must be positive for physical realizability, but Δf may be either positive or negative. The straight-line representation of (9-45) is shown in Fig. 9-8. As H and L take on their various possible integer values, the locations and slopes of the corresponding lines change as shown in Fig. 9-9. Observe how the lines tend to converge for low values of the conversion ratio f_r/IF. This indicates that there should be fewer significant spurious responses if the values of f_r/IF are higher.

Example 9-5

A superhet with HSI has an IF = 455 kHz. It is tuned to f_r = 1365 kHz. Using Fig. 9-9, identify each spur frequency from IM products of sixth-order or less.

Solution Locate f_r/IF = 1365/455 = 3 in Fig. 9-9. The IM products of sixth-order or less that cross the vertical line at f_r/IF = 3 and the spurious frequencies are:

IM product	Δf/IF	$f_c = 1365 + \Delta f$ (kHz)
$3f_h - 3f_c$	0.667	1668.33
$2f_h - 2f_c$	0.5	1592.5
$f_h - f_c$	0	1365
$3f_c - 2f_h$	0	1365
$2f_c - f_h$	−0.5	1137.5
$4f_c - 2f_h$	−0.75	1023.75
$f_h - 2f_c$	−1.5	682.5
$4f_c - f_h$	−1.75	568.75
$f_h - 3f_c$	−2	455
$5f_c - f_h$	−2	455
f_c	−2	455
$2f_c$	−2.5	227.5
$3f_c$	−2.667	151.667

plus others farther removed.

Notice that many of the spur frequencies in Example 9-5 are far removed from the frequency 1365 kHz to which the receiver is tuned. Interference from these remote spur frequencies can be eliminated if the RF bandwidth is sufficiently narrow. One way to specify the spurious responses is to include only those spur frequencies within the RF bandwidth, or only those that fall within the RF tuning range of a tunable receiver. The additional relationships required for this specification are developed as follows.

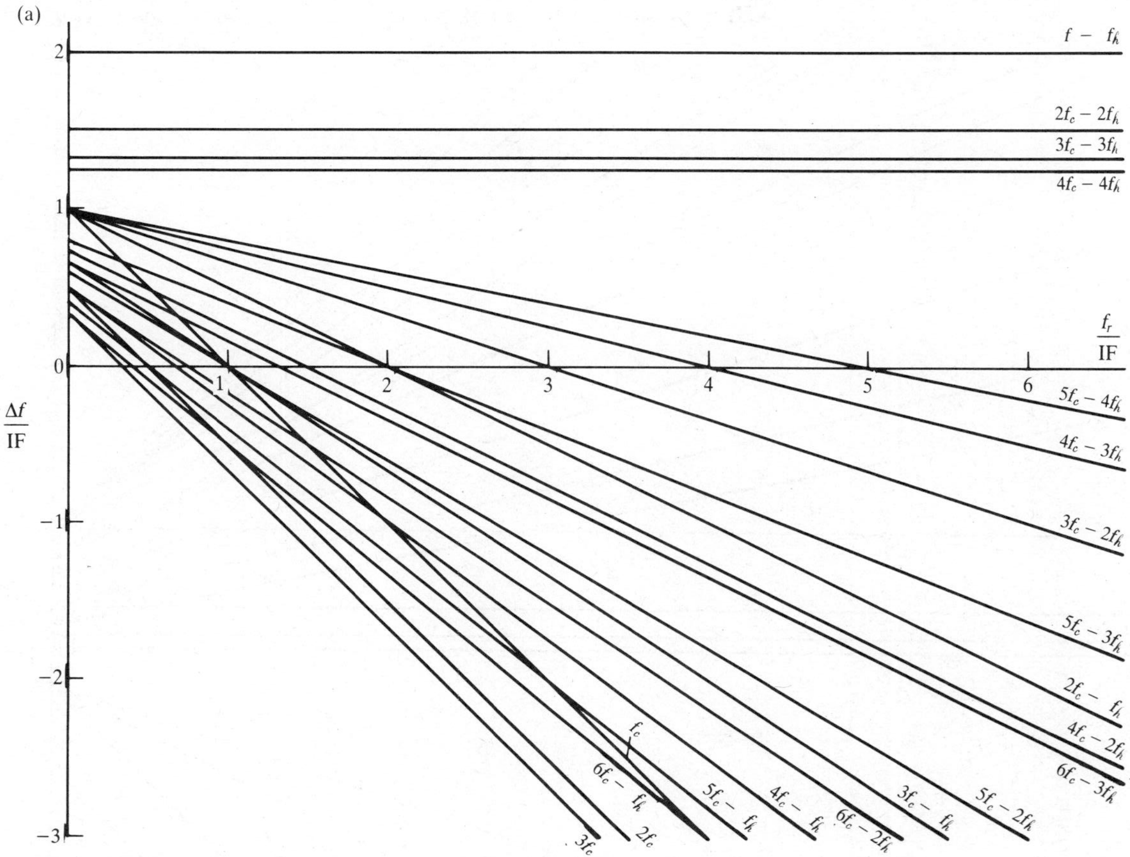

Fig. 9-9(a) Spurious-response chart for a difference mixer with HSI: type 1.

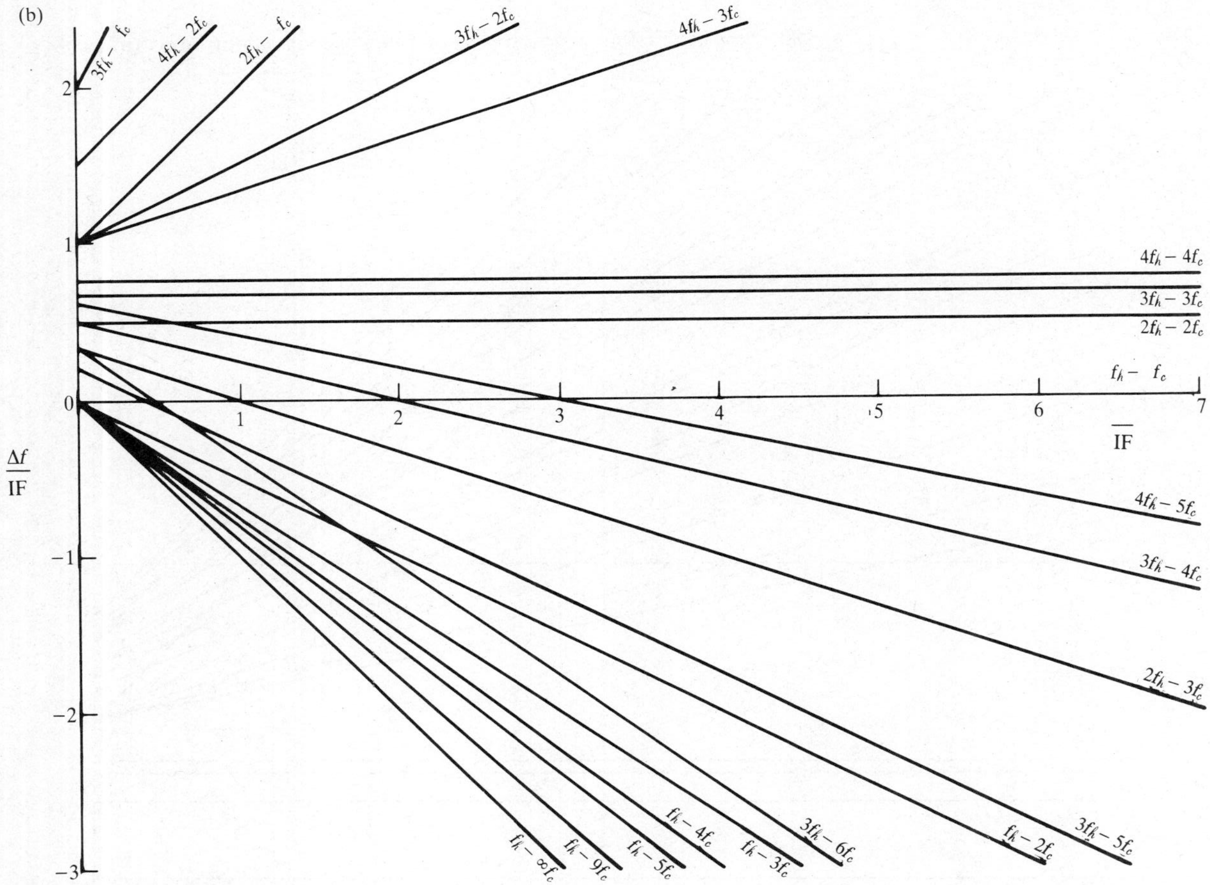

Fig. 9-9(b) Spurious-response chart for a difference mixer with HSI: type 2.

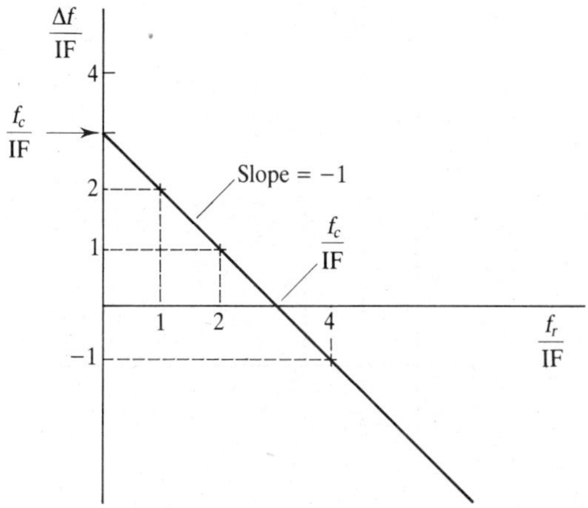

Fig. 9-10 Points along the line whose slope is -1 all have the same f_c as determined by the intercepts f_c/IF.

Let $\Delta f + f_r = f_c$, where f_c is to be constant. Divide this by the constant IF and rearrange to obtain

$$\frac{\Delta f}{\text{IF}} = (-1)\left(\frac{f_r}{\text{IF}}\right) + \frac{f_c}{\text{IF}} \tag{9-46}$$

When drawn on the coordinates of Fig. 9-9, this is a straight line whose slope is -1 and whose intercept on each axis is f_c/IF, as shown in Fig. 9-10. All points on this line have the same value of f_c. For example if $f_c = 1365$ kHz and IF $= 455$ kHz, then $f_c/\text{IF} = 3$. If the receiver is tuned to 910 kHz, then $f_r/\text{IF} = 2$ and $\Delta f/\text{IF} = 1$. Hence $f_c = 910 + 1(455)$ kHz $=$ 1365 kHz, as expected. Further, if the receiver is tuned to 540 kHz, then $f_r/\text{IF} = 1.187$ and $\Delta f/\text{IF} = 3 - 1.187 = 1.813$. Consequently, $f_c = 540 +$ 1.813(455) kHz $=$ 1365 kHz, as before.

To use this feature, let $f_{c,\max}$ and $f_{c,\min}$ represent the upper and lower limits, respectively, of the spur frequencies allowed to pass through either the RF bandwidth or the entire RF tuning range. Therefore, one line with slope -1 can be drawn through the intercepts at $f_{c,\max}/\text{IF}$ and another line with slope -1 through the intercepts at $f_{c,\min}/\text{IF}$ as in Fig. 9-11. This construction delineates a parallelogram whose corners and sides are determined by $f_{c,\max}/\text{IF}$ and $f_{c,\min}/\text{IF}$. When drawn on the spurious-response chart, only those IM products whose lines cross or touch any portion of the parallelogram will have spur frequencies between $f_{c,\max}$ and $f_{c,\min}$. The lines for all other IM products can be ignored.

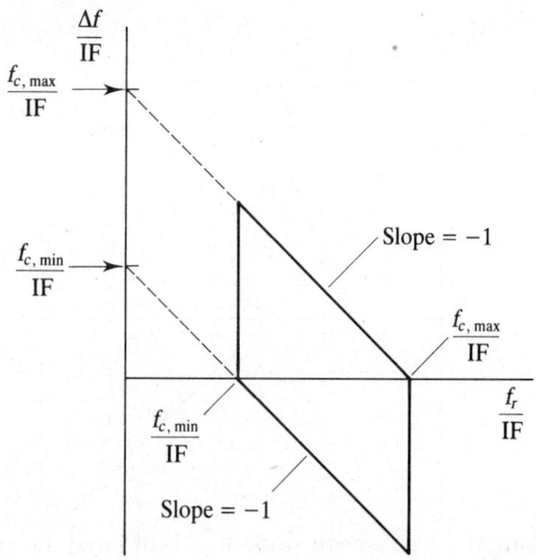

Fig. 9-11 Only those IM products whose lines intersect the parallelogram need to be considered.

Example 9-6

A superhet uses a difference mixer with HSI to obtain IF = 455 kHz. The receiver tunes from 540 kHz to 1600 kHz. Determine the spur frequencies of all IM products of sixth-order or less that occur within the entire tuning range.

Solution

$$\frac{f_{r,\min}}{\text{IF}} = \frac{f_{c,\min}}{\text{IF}} = \frac{540}{455} = 1.19$$

$$\frac{f_{r,\max}}{\text{IF}} = \frac{f_{c,\max}}{\text{IF}} = \frac{1600}{455} = 3.52$$

Although the parallelogram can be drawn on the charts in Fig. 9-9, that chart scale may cause the figure to be too small for accurate use. Also, separate charts are not readily available for general use. Therefore, draw the parallelogram to some convenient scale on a sheet of graph paper as in Fig. 9-12.

Each vertical side is limited to a length of $3.52 - 1.19 = 2.33$, and does not extend indefinitely as in Example 9-5. Now construct traces of the lines that touch or cross the parallelogram. This is the difficult part.

Consider IM products of the form $Lf_c - Hf_h$. From Table 9-4,

$$\frac{\Delta f}{\text{IF}} = \frac{H - L}{L}\left(\frac{f_r}{\text{IF}}\right) + \frac{H + 1}{L}$$

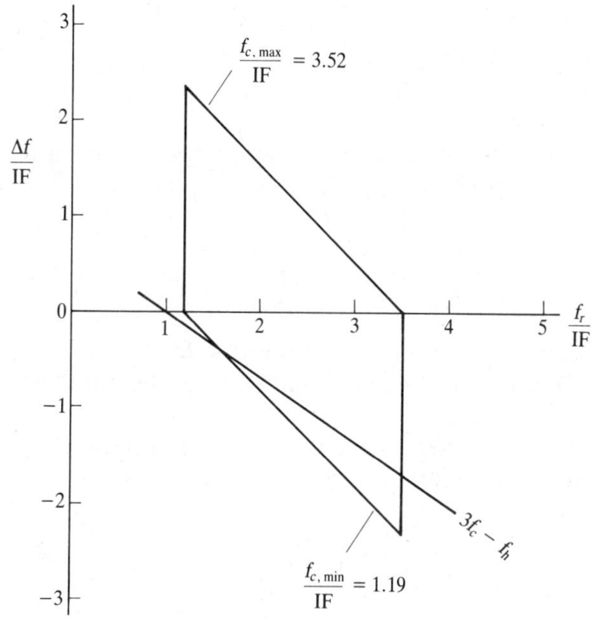

Fig. 9-12 Constructions for Example 9-6.

Which of these lines touch or cross the parallelogram? Try one that crosses the x axis at $f_r/\text{IF} = 1$. Then,

$$0 = \frac{H - L}{L}(1) + \frac{H + 1}{L}$$

from which is obtained

$$L = 2H + 1$$

The pairs of harmonic numbers are: $H = 0$, $L = 1$; $H = 1$, $L = 3$; $H = 2$, $L = 5$, etc. The line given by $H = 0$, $L = 1$ does not cross the parallelogram, so that pair can be discarded. The pair $H = 2$, $L = 5$ and the pairs for all higher values of H and L can also be discarded because they have higher orders than the imposed limit of 6. The line for the fourth-order pair $H = 1$, $L = 3$, passes through $f_r/\text{IF} = 1$ with a slope $(1 - 3)/3 = -2/3$. Locate the intercept and count squares to get the slope, and draw the trace marked $3f_c - f_h$ in Fig. 9-12. It intersects the parallelogram at $\Delta f/\text{IF} = -1.68$ and $\Delta f/\text{IF} = -0.38$. These correspond to $f_r = 1600$ kHz and $f_c = 837$ kHz at the high end of the tuning range, and $f_r = 712$ kHz and $f_c = 540$ kHz near the low end of the tuning range. Since many other traces need to be drawn too, we will discontinue this example until we determine a better way to draw only the traces that are needed.

9-8.1 Trace Construction

Consider the same IM products $Lf_c - Hf_h$ as in Example 9-6. To keep from being overwhelmed with subscripts and notation problems, let the vertical axis $\Delta f/\mathrm{IF} = y$, the horizontal axis $f_r/\mathrm{IF} = x$, $f_{c,\min} = x_1$, and $f_{c,\max} = x_2$ (see Fig. 9-13). These coordinates and the parallelogram are shown in Fig. 9-13. Then, the equation for the IM product lines becomes

$$y = \frac{H-L}{L}x + \frac{H+1}{L} \tag{9-47}$$

We wish to determine all possible values of H and L for which the lines given by (9-47) touch or cross the boundaries of the parallelogram. We will do this in steps.

First, IM product lines touch or cross the top of the parallelogram. Hence, at their intersection,

$$\frac{H-L}{L}x + \frac{H+1}{L} = -x + x_2$$

Solve this for x to obtain

$$x = \frac{L}{H}x_2 - \frac{H+1}{H}$$

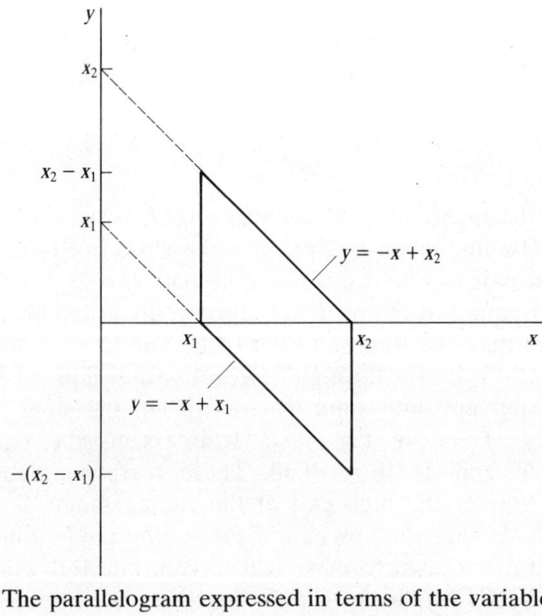

Fig. 9-13 The parallelogram expressed in terms of the variables x and y.

But we are interested only in the intersections that occur between x_1 and x_2, inclusive. Therefore,

$$x_1 \leqslant \frac{L}{H}x_2 - \frac{H+1}{H} \leqslant x_2 \tag{9-48}$$

Now examine both of the inequalities in (9-48) for L in terms of H. The results can be combined into

$$\frac{1+H(x_1+1)}{x_2} \leqslant L \leqslant \frac{1+H(x_2+1)}{x_2} \tag{9-49}$$

Now, going through the same process to find the IM product lines that intersect the bottom of the parallelogram results in

$$\frac{1+H(x_1+1)}{x_1} \leqslant L \leqslant \frac{1+H(x_2+1)}{x_1} \tag{9-50}$$

When the intersections of the IM product lines with the vertical side at $x = x_1$ are examined, it is found that L must be between the values given by left sides of the inequalities in (9-49) and (9-50). When the intersections with the vertical side at $x = x_2$ are examined, it is found that L must be between the values given by the right-hand sides of the inequalities in (9-49) and (9-50). Therefore, all intersections can be determined by using the smallest value given by the left-hand side of (9-49) and the largest value

TABLE 9-5 Acceptable Values of L in Terms of H, $x_1 = f_{c,\min}/\text{IF}$, and $x_2 = f_{c,\max}/\text{IF}$

	Difference Mixer	Sum Mixer
TYPES 1 AND 3 WITH HSI		
$Lf_c - Hf_\ell = \text{IF}$	$\frac{1+H(x_1+1)}{x_2} \leqslant L \leqslant \frac{1+H(x_2+1)}{x_1}$	$\frac{1+H(1-x_2)}{x_2} \leqslant L \leqslant \frac{1+H(1-x_1)}{x_1}$
TYPES 1 AND 3 WITH LSI		
$Lf_\ell - Hf_c = \text{IF}$	$\frac{1+Hx_1}{x_2-1} \leqslant L \leqslant \frac{1+Hx_2}{x_1-1}$	$\frac{1+Hx_1}{1-x_2} \leqslant L \leqslant \frac{1+Hx_2}{1-x_1}$
TYPES 2 AND 4 WITH HSI		
$Hf_\ell - Lf_c = \text{IF}$	$\frac{H(x_1+1)-1}{x_2} \leqslant L \leqslant \frac{H(x_2+1)-1}{x_1}$	$\frac{H(1-x_2)-1}{x_2} \leqslant L \leqslant \frac{H(1-x_1)-1}{x_1}$
TYPES 2 AND 4 WITH LSI		
$Hf_c - Lf_\ell = \text{IF}$	$\frac{Hx_2-1}{x_2-1} \leqslant L \leqslant \frac{Hx_2-1}{x_1-1}$	$\frac{Hx_2-1}{1-x_2} \leqslant L \leqslant \frac{Hx_2-1}{1-x_1}$

given by the right-hand side of (9-50). Hence, all possible values of H and L for which intersections only occur are related by

$$\frac{1 + H(x_1 + 1)}{x_2} \leqslant L \leqslant \frac{1 + H(x_2 + 1)}{x_1} \tag{9-51}$$

Similar procedures and interpretations lead to the results given in Table 9-5. Sometimes these inequalities result in extraneous traces that will not touch or cross the parallelogram. This is an expected consequence that is caused by replacing the four overlapping sets of inequalities (a set for each side of the parallelogram) by a single set for each type of mixer, as given in Table 9-5. Simply discard the extraneous traces when they occur. Now we can continue with Example 9-6.

Example 9-6 (continued)

Since $x_1 = 1.19$ and $x_2 = 3.52$, (9-51) becomes

$$\frac{1 + 2.19H}{3.52} \leqslant L \leqslant \frac{1 + 4.52H}{1.19} \tag{9-52}$$

Now, we simply select different values for H and use (9-52) to obtain appropriate values of L. For $H = 0$, no value for L can be found, so $H = 0$ is ruled out.

$H = 1$: $0.19 \leqslant L \leqslant 4.64$, which yields $L = 1, 2, 3, 4$
$H = 2$: $1.53 \leqslant L \leqslant 8.44$, which yields $L = 2, 3, 4$ (anything higher exceeds sixth-order)
$H = 3$: $2.15 \leqslant L \leqslant 12.24$, which yields $L = 3$ (anything higher exceeds sixth-order)

To draw the traces, use the slope $= (H - L)/L$, y-axis intercept $= (H + 1)/L$, and the x-axis intercept $= (H + 1)/(L - H)$ as found from Table 9-4. Organize these calculated values into a form like that of Table 9-6.

We must also consider the IM products $Hf_h - Lf_c$ for a type 2 difference mixer with HSI. From Table 9-5 we find that

$$\frac{2.19H - 1}{3.52} \leqslant L \leqslant \frac{4.52H - 1}{1.19}$$

$H = 0$: cannot be used because L is a negative fraction
$H = 1$: $0.34 \leqslant L \leqslant 2.96$, which yields $L = 1, 2$
$H = 2$: $0.96 \leqslant L \leqslant 6.76$, which yields $L = 1, 2, 3, 4$
$H = 3$: $1.58 \leqslant L \leqslant 10.55$, which yields $L = 2, 3$

TABLE 9-6 Calculated Values for Example 9-6

	For $Lf_c - Hf_h$							
H	1	1	1	1	2	2	2	3
L	1	2	3	4	2	3	4	3
x axis intercept	∞	2	1	2/3	∞	3	3/2	∞
y axis intercept	2	1	2/3	1/2	3/2	1	3/4	4/3
Slope	0	−1/2	−2/3	−3/4	0	−1/3	−1/2	0

	For $Hf_h - Lf_c$							
H	1	1	2	2	2	2	3	3
L	1	2	1	2	3	4	2	3
x axis intercept	0	0	−1	∞	1	1/2	−2	∞
y axis intercept	0	0	1	1/2	1/3	1/4	1	2/3
Slope	0	−1/2	1	0	−1/3	−1/2	1/2	0

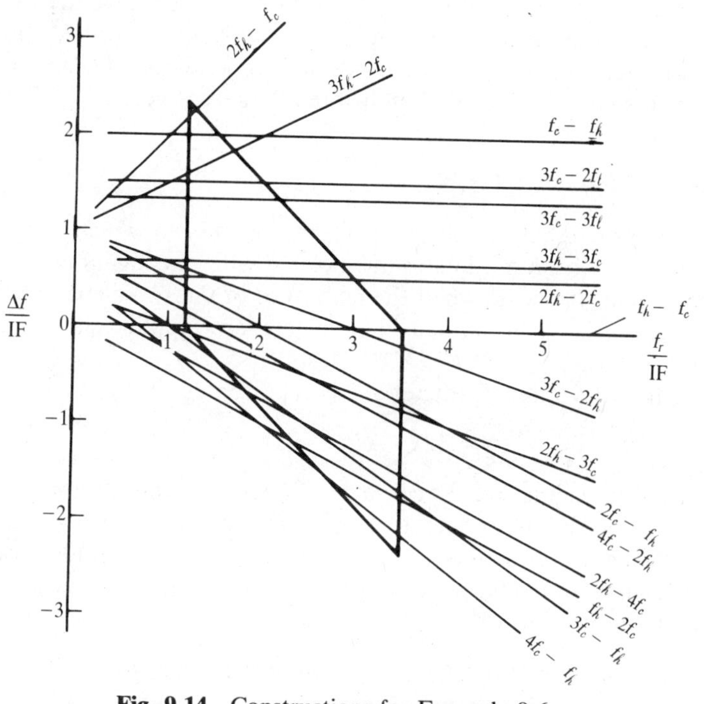

Fig. 9-14 Constructions for Example 9-6.

Other values of H yield IM products above sixth-order. To draw the traces, use the slope $= (H - L)/L$, the y-axis intercept $= (H - 1)/L$, and the x-axis intercept $= (H - 1)/(L - H)$. Now draw the traces and identify each one as shown in Fig. 9-14. Establish each slope by counting squares. No other traces intersect the parallelogram.

To examine the effect of the IM product $2f_h - f_c$ in Example 9-6, tune the receiver near the low end of the tuning range at $f_r = 546$ kHz, for which $f_r/\text{IF} = 1.2$. Locate where this intersects the trace for $2f_h - f_c$ in Fig. 9-14. That point corresponds to $\Delta f/\text{IF} = 2.2$, or $\Delta f = 1001$ kHz. Thus the spur frequency is $f_c = (546 + 1001)$ kHz $= 1547$ kHz, which is near the other end of the tuning range. Also, a line whose slope is -1 drawn through this point intersects the x axis at $f_c/\text{IF} = 3.4$, for which $f_c = 1547$ kHz. The oscillator frequency is $f_h = f_r + \text{IF} = 1001$ kHz. The frequency of the IM product is $2f_h - f_c = 1547$ kHz.

To examine the effect of the IM product $4f_c - f_h$, tune the receiver to 1600 kHz. Therefore, $f_r/\text{IF} = 3.52$ and the oscillator frequency is $f_h =$ 2055 kHz. The IM product corresponds to $\Delta f/\text{IF} = -2.137$ from which is obtained $\Delta f = -972.5$ kHz. (Of course, reading the graph's scale to 3-place accuracy is impossible.) Thus, $f_c = 1600 - 972.5 = 627.5$ kHz. Also, a line whose slope is -1 drawn through that point intersects the x axis at $f_c/\text{IF} = 1.38$, for which $f_c = 627.5$ kHz.

In both of the above situations, when the receiver is tuned near one end of the tuning range, interference from a spur frequency near the other end of the tuning range is possible for the allowed order of IM products. Whether interference actually occurs or not depends on the amplitude of the undesired signal and on the bandwidth of the receiver.

9-9 CROSS MODULATION

Cross modulation is the name given to the process in which a somewhat distorted version of the modulation associated with one signal is transferred to another signal. It occurs when the curvature of the static characteristic of the nonlinear system to which the signals are applied is at least of third-order.

Assume that the static characteristic is represented by

$$y = 12x - x^3, \qquad |x| \leq 2 \tag{9-53}$$

Let the desired signal be $E_1 \cos \omega_1 t = E_1 \cos \phi_1$. The desired signal is unmodulated for simplicity and to help demonstrate the cross-modulation phenomenon more effectively. Let the undesired amplitude-modulated interfering signal be $E_2(1 + m_a \cos \omega_m t) \cos \omega_2 t = E_2(1 + m_a \cos \phi_m) \cos \phi_2$. Then the output is

$$\begin{aligned} y = {} & 12[E_1 \cos \phi_1 + E_2(1 + m_a \cos \phi_m) \cos \phi_2] \\ & - [E_1 \cos \phi_1 + E_2(1 + m_a \cos \phi_m) \cos \phi_2]^3 \end{aligned} \tag{9-54}$$

This must be expanded so that all $\cos \phi_1$ terms can be collected to determine the output $y(\phi_1)$ at the frequency of the desired signal. Thus,

$$
\begin{aligned}
y(\phi_1) &= 12E_1 \cos \phi_1 - \frac{3E_1^3}{4} \cos \phi_1 - \frac{3}{2} E_1 E_2^2 \\
&\quad \times \left(1 + 2m_a \cos \phi_m + \frac{m_a^2}{2} + \frac{m_a^2}{2} \cos 2\phi_m\right) \cos \phi_1 \\
&= 12E_1\left[1 - \frac{E_1^2}{16} - \frac{E_2^2}{8}\left(1 + \frac{m_a^2}{2}\right)\right] \\
&\quad \times \left[1 - \frac{\dfrac{E_2^2 m_a}{4}\left(\cos \phi_m + \dfrac{m_a}{4} \cos 2\phi_m\right)}{1 - \dfrac{E_1^2}{16} - \dfrac{E_2^2}{8}\left(1 + \dfrac{m_a^2}{2}\right)}\right] \cos \phi_1
\end{aligned} \tag{9-55}
$$

This is similar to the standard form for an amplitude-modulated sinusoid given by

$$y = E_{1,\text{out}}[1 + m_{a,\text{out}} m_n(t)] \cos \phi_1 \tag{9-56}$$

except for the unimportant sign change preceding $m_{a,\text{out}}$ in (9-55) and the fact that the modulating signal in (9-55) contains some second harmonic amounting to $(m_a/4) \times 100$ percent. Hence, a somewhat distorted version of the modulation associated with one signal has been transferred to the other originally unmodulated signal.

The amplitude of the output voltage is, from (9-55),

$$E_{1,\text{out}} = 12E_1\left[1 - \frac{E_1^2}{16} - \frac{E_2^2}{8}\left(1 + \frac{m_a^2}{2}\right)\right] \tag{9-57}$$

which shows that compression and desensitization of the desired signal both occur.

If the modulating signal associated with the output at frequency ϕ_1 is taken to be $\cos \phi_m + (m_a/4) \cos 2\phi_m$, it must be normalized to its maximum value $1 + m_a/4$ to determine the product $m_{a,\text{out}} m_n(t)$. From this procedure is obtained

$$m_{a,\text{out}} = \frac{\dfrac{E_2^2 m_a}{4}\left(1 + \dfrac{m_a}{4}\right)}{1 - \dfrac{E_1^2}{16} - \dfrac{E_2^2}{8}\left(1 + \dfrac{m_a^2}{2}\right)} \tag{9-58}$$

If $E_1 = 0.1$ V, $E_2 = 0.9$ V, and $m_a = 1$ (i.e., values that assure $|x| \leq 2$ in this example), the amplitude of the output voltage is, from (9-57), $E_{1,\text{out}} = 1.017$ V instead of 1.2 V had the system been linear with no interfering signal. Also, the modulation index of the output voltage is, from (9-58) $m_{a,\text{out}} = 0.3$ or 30 percent.

The most effective way to combat cross modulation is to use devices that do not have third-order curvature in their static characteristics.

PROBLEMS

9-1 Two radio receivers A and B both tune the frequency range 150 kHz to 600 kHz. Each receiver has an IF = 1900 kHz. Receiver A uses difference mixing, and receiver B uses sum mixing. For each receiver: (a) calculate the frequency range of the local oscillator and determine whether high-side or low-side injection is required; (b) calculate $C_{\max}/C_{\min}$ for variable tuning capacitors in both the RF and oscillator circuits, and comment on any special considerations that apply to the tuning arrangement of receiver B; (c) calculate $(f_\ell/f_k)_{\max}$ and $(f_\ell/f_k)_{\min}$.

9-2 Define type 5 coincidence in a difference mixer as $Hf_k + Lf_\ell = f_k - f_\ell$ and type 6 coincidence in a sum mixer as $Hf_k + Lf_\ell = f_k + f_\ell$. (a) What conditions on H and L are required for both types? Tabulate the results. (b) Show how Table 9-1 can be modified so as to include types 5 and 6.

9-3 (a) Demonstrate that the sixth-order type 1 coincidence point 3 of Example 9-1 occurs when the receiver is tuned to a signal on 682.5 kHz. (b) Suppose the receiver is tuned near a desired station operating on 680 kHz. Show that the birdie produced by point 3 is at zero beat when the receiver is off-tuned to 678.33 kHz.

9-4 You wish to design a superhet to receive stations in the 3500 kHz to 4000 kHz range using high-side injection and down conversion. From the viewpoint of minimizing either in-band or out-of-band spurious responses of ninth-order or less, which should be better: an IF of 455 kHz or an IF of 1600 kHz? Why?

9-5 A certain superhet tunes the low-frequency range 180 kHz to 540 kHz. It uses high-side injection and down conversion to obtain an IF = 100 kHz. To what input frequencies should the receiver be tuned individually in order for the desired response and each spurious response of ninth-order or less to coincide.

9-6 The receiver of Problem 9-5 is tuned to a station on 180 kHz. (a) How far away (in kHz) from that frequency is a possible sixth-order spurious incoming signal? (b) How far away (in kHz) from the frequency of the desired IF signal is the spurious IF signal produced by the conditions for a third-order coincidence point?

9-7 A certain superhet tunes 520–1630 kHz to cover the AM broadcast band. The IF is 455 kHz and the local oscillator frequency is above the RF signal frequency. Of the various spurious responses possible, the one having the lowest order will be the most troublesome. Determine the frequency to which the receiver must be tuned to notice the worst spurious response.

9-8 A superhet has an IF of 455 kHz and the local oscillator frequency is above the RF signal frequency. It is tuned to a strong, local AM station on 1370 kHz and a "birdie" is heard. The "birdie" is determined to be a spurious response $(3f_\ell - 2f_k)$ mixing with the desired IF signal in the demodulator (or second detector or envelope detector). To what frequency should the receiver be tuned for the "birdie" to be zero beat?

9-9 A mixing system for a transceiver (i.e., a TRANSmitter–reCEIVER combination) that produces the oscillator frequency for a 144–148 MHz transmitter and the local oscillator frequency for the receiver with a 9-MHz IF is shown in Fig. P9-9. For each mixer, list the spurious responses of eighth-order or less that are possible.

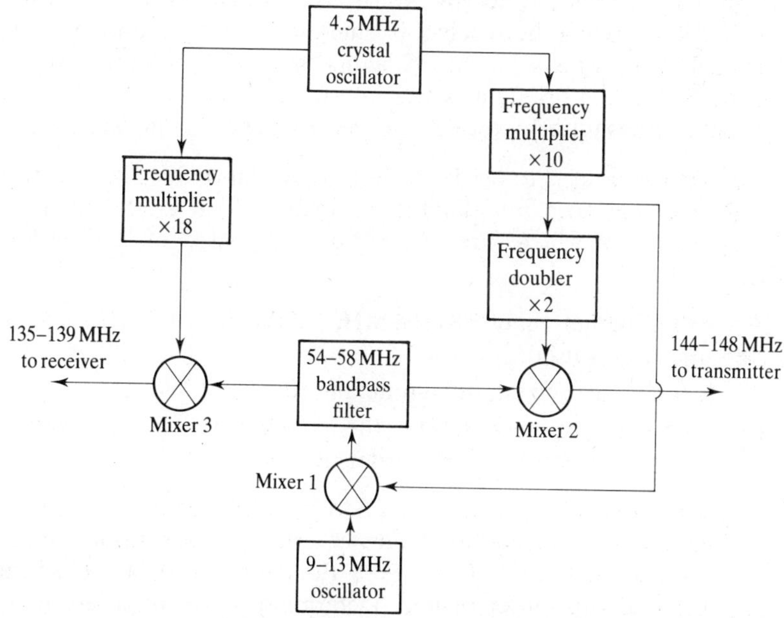

Fig. P9-9 Mixing system for Problem 9-9.

9-10 A certain down-conversion difference mixer has $f_\ell/f_k = 1/2$. (a) Calculate the IF in terms of f_ℓ only and in terms of f_k only. (b) Does this mixer require low-side or high-side injection of the oscillator output? (c) What possible practical operating problem can you identify with this situation other than various spurious responses from the chart?

9-11 A certain down-conversion mixer has an eighth-order type 1 coincidence point at $L = 6$ and $H = 2$ (see the spurious response chart). (a) Calculate the corresponding value of f_ℓ/f_k. (b) Determine whether this mixer requires low-side or high-side injection. (c) The RF signal circuitry is tuned to an input signal on 3500 kHz. Calculate the desired IF frequency and the type 1 undesired spurious frequency for coincidence.

9-12 An AM superhet using high-side injection has an IF = 1700 kHz. A third-order type 1 coincidence point is noticed just below the line for $(f_\ell/f_k)_{\min}$. Using the values of H and L for that point, calculate:

(a) The frequency of an out-of-band signal that can produce the IF when the receiver is tuned to (1) 3500 kHz; (2) 3510 kHz. There is no station on either frequency.

(b) The spurious IF frequency that is produced when the receiver is tuned to a station operating on (1) 3500 kHz; (2) 3510 kHz.

9-13 A single-conversion AM superhet has an IF = 455 kHz and uses high-side injection. A nearby AM transmitter is known to operate on 1830.5 kHz. When the receiver is tuned to an AM station on 920 kHz an annoying whistle is heard along with the programs of both stations. Explain briefly how this interference occurs through no fault at either transmitter and calculate the frequency of the whistle.

9-14 An AM superhet is to be designed to tune 500–1650 kHz. It is to use high-side injection and difference mixing. What is the highest IF frequency that will assure no spurious responses of sixth-order or less?

9-15 An AM superhet tunes 88–108 MHz. The IF = 10.7 MHz and low-side injection is used.

(a) Is difference mixing or sum mixing required? Why?

(b) To what frequency should the receiver be tuned to notice a spurious response of sixth-order or less?

9-16 A certain AM superhet tunes the AM broadcast band 535–1605 kHz. It uses down-conversion and high-side injection. A spurious response chart shows that if $(f_\ell/f_k)_{\min} \geqslant 0.8$ then there will be no spurious responses of sixth-order or less. Determine the IF required for such conditions.

9-17 All frequencies in the 3500–4000 kHz radio band are mixed with a fixed oscillator frequency of 8000 kHz and are up-converted to the 11,500–12,000 kHz band. List all of the spurious response frequencies, of sixth-order or less, produced within the 11,500–12,00 kHz band by a strong local station operating on 3998 kHz.

9-18 The input to a broadband difference mixer consists of all frequencies in the range 13.5 ± 1.5 MHz. The local oscillator is fixed at 10 MHz. The output of the mixer is to be selected by a tunable receiver. (a) Over what range of frequencies should the receiver tune? (b) Determine what spurious responses are produced by in-band signals and indicate their order up to and including the ninth.

9-19 In a certain system, it is desired to upconvert the 225–400 MHz UHF band to S-band in the general vicinity of 2200 ± 600 MHz. It is necessary to choose either low-side or high-side injection and either sum or difference mixing. The local oscillator is restricted to one of the following frequencies: 2000, 24000, or 2800 MHz. What choice of oscillator frequency and type of mixing minimizes spurious responses of sixth-order or less? Rank the possible choices as poor, good, or best.

9-20 Your first assignment as a new engineering employee is to design a two-meter incremental tuning transceiver with a 9-MHz receiver first IF. The Chief Engineer shows you the block diagram (see Fig. P9-20) of a mixing synthesizer being considered for frequency control. Of course, the transmitter and receiver are not both turned ON at the same time! Each frequency set is independently variable by increments, so the ratios f_ℓ/f_k have limits: minimum and maximum. (a) Does each mixer upconvert or downconvert? (b) What possible

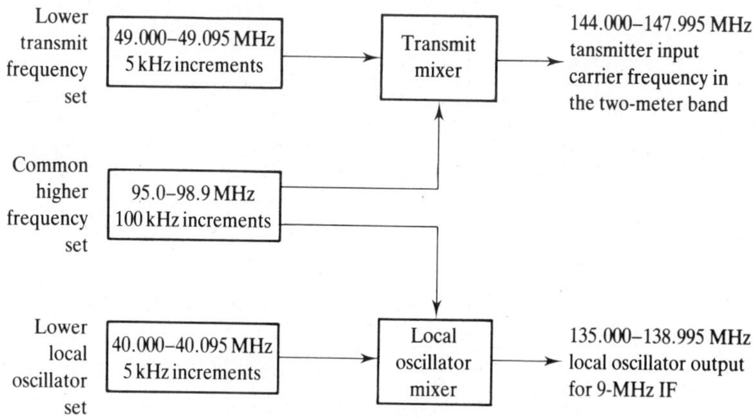

Fig. P9-20 System for Problem 9-20.

spurious response and harmonic problems can you identify for the Chief having sixth-order or less? (c) You propose a lower transmit set 32.000–32.095 MHz, common higher set 112.0–115.9 MHz, and lower oscillator set 23.000–23.095 MHz. Why?

9-21 A certain superhet uses a difference mixer with HSI to obtain an IF of 1700 kHz. One range of the receiver tunes from 3500 kHz to 4000 kHz. Draw the parallelogram for these conditions and the traces for all IM products of sixth-order or less that touch or cross the parallelogram.

10

INTERCEPT POINTS

10-1 IM PRODUCTS REVIEWED

When the sum of two sinusoids is applied to the input of a nonlinear device, the output contains harmonics of the original frequencies and various intermodulation (IM) products, as discussed earlier. The number of significant harmonics and IM products depends on the degree of curvature of the static characteristic for the particular nonlinear system involved. It is usual to let the amplitudes of the two input sinusoids be equal when using them together to test an amplifier or a radio receiver for the consequences of its nonlinearity. If the two input frequencies are sufficiently close and centered on the response characteristic, the outputs at these two frequencies will have the same amplitudes, the outputs at similar odd-order IM frequencies will also have the same amplitudes, and all of the frequencies will be in proximity. However, harmonics components and all even-order IM products will be far removed from the others.

To illustrate, let the two equal-amplitude input sinusoids have frequencies f_1 and f_2, and let $f_2 - f_1 = \Delta f$. Then, assuming that the input amplitudes are sufficient to drive the system under test into nonlinearity, some of the odd-order IM products are:

$$
\begin{array}{ll}
2f_2 - f_1 = f_2 + \Delta f & 2f_1 - f_2 = f_1 - \Delta f \\
3f_2 - 2f_1 = f_2 + 2\,\Delta f & 3f_1 - 2f_2 = f_1 - 2\,\Delta f \\
4f_2 - 3f_1 = f_2 + 3\,\Delta f & 4f_1 - 3f_2 = f_1 - 3\,\Delta f \\
\cdots \quad \cdots & \cdots \quad \cdots
\end{array}
$$

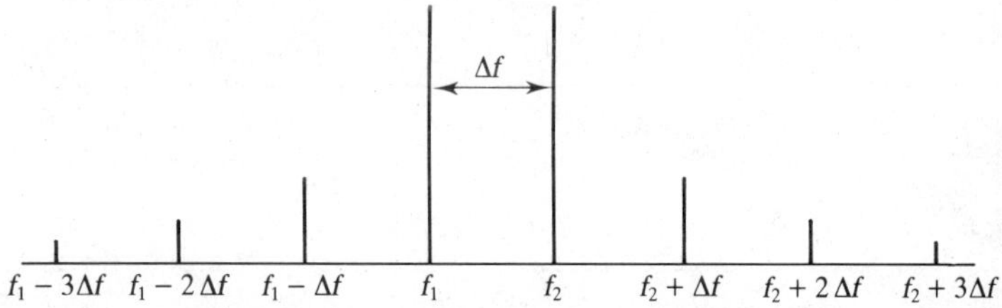

Fig. 10-1 Odd-order IM products above and below the desired output signals at f_1 and f_2.

and

$$
\begin{array}{ll}
2f_2 + f_1 = 3f_2 - \Delta f & 2f_1 + f_2 = 3f_1 + \Delta f \\
3f_2 + 2f_1 = 5f_2 - 2\,\Delta f & 3f_1 + 2f_2 = 5f_1 + 2\,\Delta f \\
4f_2 + 3f_1 = 7f_2 - 3\,\Delta f & 4f_1 + 3f_2 = 7f_1 + 3\,\Delta f \\
\cdots \quad \cdots\cdots & \cdots
\end{array}
$$

Notice that this latter set of IM products, involving the sums of the harmonic components, yields frequencies that are far removed from the original ones.

Some of the even-order IM products are:

$$
\begin{array}{ll}
f_2 - f_1 = \Delta f & f_2 + f_1 = 2f_2 - \Delta f \\
2f_2 - 2f_1 = 2\,\Delta f & 2f_2 + 2f_1 = 4f_2 - 2\,\Delta f \\
4f_2 - 4f_1 = 4\,\Delta f & 4f_2 + 4f_1 = 8f_2 - 4\,\Delta f \\
\cdots \quad \cdots & \cdots \quad \cdots
\end{array}
$$

Notice that all of these even-order IM products have frequencies that are far removed from the two original frequencies.

The conclusion that results from this analysis is that the odd-order IM products having frequencies $Hf_2 - Lf_1$ and $Lf_1 - Hf_2$ are the ones that are likely to be troublesome in a radio receiver because they are in the passband near the two original frequencies, as shown in Fig. 10-1, and therefore they cannot be eliminated by filtering. In both of these forms, the values of L and H must be such that their sum is an odd number.

10-1.1 dB and dBm

The decibel (dB) is a measure of how much larger or smaller the power P_2 is compared to the power P_1 and is defined as

$$N_{\mathrm{dB}} = 10 \log_{10} \frac{P_2}{P_1} \tag{10-1}$$

The powers P_1 and P_2 may be the input and output powers, respectively, of an electrical or electronic system. If $P_2 > P_1$ then the system has a power gain represented by a positive dB number N. If $P_2 < P_1$ then the system has a power gain represented by a negative dB number N. This is because the logarithm of a positive number less than unity is negative. It is usually more convenient to refer to a negative (in dB) power gain as being a positive (in dB) attenuation given by $\text{dB} = 10 \log P_1/P_2$.

If an amplifier has $P_1 = 0.1$ W and $P_2 = 10$ W, then the amplifier has a power gain of $10/0.1 = 100$ or 20 dB. The output power of 10 W is higher by a factor of 100 or by 20 dB than the input power of 0.1 W.

Equation (10-1) is also used to express the power P_2 in dB relative to a reference power P_1. The designation dBm is used for the decibel calculation when the reference power P_1 is taken to be 1 mW. Thus, $P_2 = 10\ \mu\text{W}$ becomes $10 \log (10 \times 10^{-6})/(1 \times 10^{-3}) = -20$ dBm. This means that 10 μW is 20 dB below 1 mW. Or, more simply, 10 μW corresponds to -20 dBm. A power equal to 1 mW is represented by 0 dBm.

Rearrange the above defining equation for N_{dB} as

$$\begin{aligned} N_{\text{dB}} &= 10 \log \frac{P_2/1\ \text{mW}}{P_1/1\ \text{mW}} \\ &= 10 \log \frac{P_2}{1\ \text{mW}} - 10 \log \frac{P_1}{1\ \text{mW}} \\ &= P_{2,\text{dbm}} - P_{1,\text{dBm}} \end{aligned} \tag{10-2}$$

This result means that when each of two powers is expressed in dBm or any common reference power, their difference is in decibels. Over the years there have been many such reference levels, including 1, 6, 10, 12.5, and 50 milliwatts, each used for a specific application.

One reason for using dB-related quantities instead of their actual numerical values is convenience in plotting. Suppose the powers P_2 and P_1 are related by

$$P_2 = a(P_1)^3$$

where a is a constant. The graph of P_2 vs. P_1 on rectangular coordinates is nonlinear. But if the logarithm of each side is taken and $\log P_2$ is plotted vs. $\log P_1$ on rectangular coordinates, or if P_2 is plotted vs. P_1 on log–log coordinates, the result is a straight line whose slope is equal to the exponent and whose vertical-axis intercept depends on the constant a. Another example of the use of dB is the Bode plot of the amplitude response of a system. An otherwise complicated graph becomes a series of straight-line segments that are easy to locate and construct. In addition, it is easy to keep track of a signal as it progresses through a multistage system by simply adding or subtracting the dB equivalent of each power gain or loss. Then finally, the last reason and certainly the least, is that using dB adds an additional layer of obscurity that can bewilder and confuse the uninitiated until they too can surmount the hurdle of becoming acquainted with dB and join the establishment.

10-2 THE THIRD-ORDER INTERCEPT POINT

As discussed earlier in Section 2.9, when two equal-amplitude sinusoidal signals are applied to a system, then owing to the linear term of the power series representation of the system the two output voltages are equal and are proportional to the input amplitudes $E_1 = E_2 = E$, provided that compression and desensitization are negligible. Further, the output voltages of each of the third-order IM products are equal and proportional to the cube of the input amplitudes $E_1 = E_2 = E$, owing to the cubed term of the power series, provided that the fifth- and higher-order terms of the power series are negligible.

Each input signal supplies an input power P_{in} (in watts) that varies as the square of E, which, when expressed in dBm (the reference being 1 mW), becomes

$$\begin{aligned} P_i &= 10 \log \frac{P_{\text{in}}}{10^{-3}} = 10 \log \frac{E^2}{2R_{\text{in}}\, 10^{-3}} \\ &= 10 \log E^2 - 10 \log R_{\text{in}} + 27 \end{aligned} \tag{10-3}$$

The output power of each fundamental component owing to the linear term is $P_{\text{out}} = A^2E^2/(2R_L)$. When expressed in dBm, this becomes

$$\begin{aligned} P_o &= 10 \log \frac{P_{\text{out}}}{10^{-3}} = 10 \log \frac{A^2E^2}{2R_L\, 10^{-3}} \\ &= 10 \log E^2 + 10 \log A^2 - 10 \log R_L + 27 \end{aligned} \tag{10-4}$$

The fundamental power gain is $G_P = P_{\text{out}}/P_{\text{in}}$. When expressed in dB, this becomes

$$\begin{aligned} G &= 10 \log G_P = 10 \log \left(A^2 \frac{R_{\text{in}}}{R_L} \right) \\ &= 10 \log A^2 + 10 \log R_{\text{in}} - 10 \log R_L \end{aligned} \tag{10-5}$$

Combine (10-3), (10-4), and (10-5) into

$$P_o = P_i + G \tag{10-6}$$

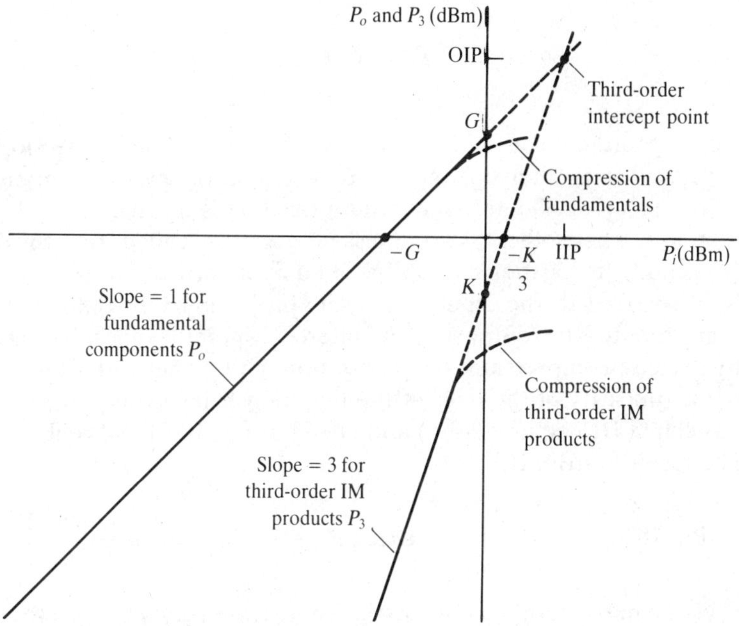

Fig. 10-2 The third-order intercept point is located where the lines for P_o and P_3 intersect.

When plotted on rectangular coordinates, with the scale for P_o (in dBm) as ordinate the same as the scale for P_i (in dBm), (10-6) is a straight line whose slope is unity and whose vertical axis intercept is G. The value of G determines where the line is located, as shown in Fig. 10-2.

The output voltages of the third-order IM products are both equal to $3CE^3/4$, owing to the cubed term of the power series, provided that the fifth and higher-orders in the power series are negligible. The output power at each third-order IM product is

$$P_{\text{out},3} = \frac{(\frac{3}{4}CE^3)^2}{2R_L}$$

When expressed in dBm, this becomes

$$P_3 = 10 \log \frac{(\frac{3}{4}CE^3)^2}{2R_L \times 10^{-3}} = 10 \log \frac{(\frac{3}{4}C)^2(E^2)^3}{2R_L \times 10^{-3}}$$

When combined with (10-3), this becomes

$$P_3 = 3P_1 - 56.5 + 20 \log C + 10 \log \frac{R_{\text{in}}^3}{R_L} \tag{10-7}$$

or

$$P_3 = 3P_i + K \tag{10-8}$$

where $K = -56.5 + 20 \log C + 10 \log (R_{in})^3)/R_L$. When plotted on the same rectangular coordinates as before, (10-8) is a straight line whose slope is 3, whose vertical axis intercept is K and whose horizontal axis intercept is $-K/3$. The value of K determines where this line is located.

The point where these two lines intersect is called the third-order intercept point. Its location can be specified by its projection IIP onto the P_i axis (i.e., referred to the input), or by its projection OIP onto the P_o axis (i.e., referred to the output). The intercept point cannot be measured directly because compression, desensitization, and higher-order IM products destroy the linearity of the lines at the high power levels required.

To calculate IIP, equate (10-6) and (10-7) using (10-5) and replace P_i with IIP. The result in dBm is

$$\text{IIP} = 10 \log A - 10 \log C - 10 \log R_{in} + 28.24 = 10 \log \frac{666.67A}{CR_{in}} \tag{10-9}$$

If a device whose static characteristic, represented by Eq. (2-44), is used with $R_{in} = 50$ ohms, the value of IIP is calculated to be 27 dBm. This corresponds to 5 V rms across R_{in}. From Fig. 2-4, it is seen that the device is driven well into saturation and cutoff.

10-3 MINIMUM DISCERNIBLE SIGNAL (MDS)

Noise factor F represents the ratio of the input signal-to-noise ratio S_i/N_i to the output signal-to-noise ratio S_o/N_o. The minimum acceptable value of S_o/N_o depends on the receiving system and on the experience of the receiving operator, if there is one. Some operators can make sense out of weak signals that are several dB below the output noise level N_o, while others have difficulties when the output signal S_o is even a little above the output noise level. This subjectivity can be eliminated from the mathematics by defining the minimum usable value of S_o/N_o to be unity. Hence, $F = S_i/N_i$, so that $S_i = FN_i$. This becomes the least input signal power that makes $S_o/N_o = 1$, or that makes $(S_o + N_o)/N_o = 2$, or 3 dB. When this S_i is expressed in dBm, it becomes, by definition, the minimum discernible signal, or MDS. Therefore,

$$\begin{aligned} \text{MDS} &= 30 + 10 \log FN_i = 10 \log FkT_0B \\ &= -144 + \text{NF} + 10 \log B \end{aligned} \tag{10-10}$$

where NF = 10 log F = noise figure of the system
B = bandwidth, Hz

If bandwidth is expressed in kHz, change -144 to -114 in (10-10).

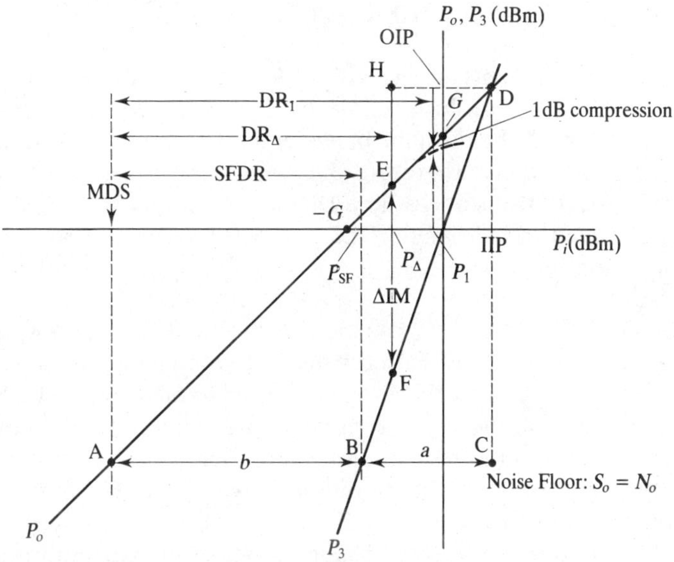

Fig. 10-3 Relationships between the minimum discernible input signal (MDS), noise floor at the output with $S_o = N_o$, and three dynamic ranges.

Consequently, the value of MDS becomes a measure of the sensitivity of the system. If the noise figure and the bandwidth are known, then MDS can be calculated and its value marked, as in Fig. 10-3. This establishes a noise floor with $S_o = N_o$ at the output, as shown in Fig. 10-3. Any output power below this level will be progressively buried in the noise and rendered increasingly difficult, if not impossible, to discern.

10-4 DYNAMIC RANGE (DR)

As the input signal strength is increased above MDS, the output signal power $S_o = P_o$ (in dBm) increases linearly above the noise floor. The third-order IM product P_3 (in dBm) increases at a rate three times faster than for P_o, but it is initially buried in the noise and its presence is not discernible. When the input signal strength rises to P_{SF}, the IM product P_3 rises to the noise floor. The difference (P_{SF} − MDS) is called the spurious-free dynamic range, SFDR, and it is measured in dB.

Sometimes dynamic range is defined to be $DR_1 = (P_1 - \text{MDS})$, where P_1 is an input signal strength at which the compression of the fundamental component is 1 dB (see Fig. 10-3).

Dynamic range is often also defined to be $DR_\Delta = (P_\Delta - \text{MDS})$, where P_Δ is an input signal strength at which the third-order IM product P_3 is below the power P_o by some specified amount $\Delta\text{IM} = P_o - P_3$, such as 60 dB (see Fig. 10-3).

10-5 CONSTRUCTIONS AND CALCULATIONS

Consider the constructions in Fig. 10-3. Along some output power level, such as shown at the noise floor, mark the distance between the lines for P_o and P_3 as b, and mark the distance between the line for P_3 and IIP as a. The slope of line AD is unity, therefore the length CD equals $a + b$, obtained from triangle ACD. The slope of line BD is 3, therefore the length CD also equals $3a$, obtained from triangle BCD. Hence, $a + b = 3a$, so that $b = 2a$, and this relationship holds true for any output power level below the intercept point.

The two lines for P_o and P_3 can be drawn readily if only one point on each can be located using measured values. Two equal-amplitude sinusoids (or "tones") are applied to the input of the system under test. Assuming that the attenuators on the signal generators are calibrated, the input power P_i in dBm of each sinusoid can be determined. The spectrum-analyzer display of the output will reveal the output power P_o in dBm of each desired component and also the amount ΔIM in dBm (taken to be a positive number) that the third-order IM product is below the desired output. This allows points E and F to be located. Also, the power gain G in dB equals $P_o - P_i$. This locates an intercept for the P_o line at G on the output power axis and at $-G$ on the input power axis. The lines may then be drawn with good accuracy and the intercept point can be determined graphically.

Drawing a horizontal line through point F shows that $b = \Delta\text{IM}$ and, since $a = b/2$, then $a = (\Delta\text{IM})/2 = \text{IIP} - P_\Delta$. This allows the input intercept point to be calculated from

$$\text{IIP} = (\Delta\text{IM})/2 + P_\Delta \quad \text{dbm} \qquad (10\text{-}11)$$

Referring to Fig. 10-3 at the noise floor, distance AC equals SFDR + SFDR/2 = 3(SFDR)/2 = IIP − MDS. Therefore, the spurious-free dynamic range is

$$\text{SFDR} = \frac{2(\text{IIP} - \text{MDS})}{3} \quad \text{dBm} \qquad (10\text{-}12)$$

The spurious-free dynamic range can be increased by using circuit designs that provide a larger IIP and/or a smaller MDS. Increasing IIP allows the system to respond better to higher input signal strengths. Decreasing MDS improves the sensitivity of the system by reducing the noise floor.

Again referring to Fig. 10-3 at the noise floor, distance BC = IIP − P_{SF} = (SFDR)/2. Substitute (10-12) into this and solve for the maximum input power for a spurious-free response. Thus,

$$P_{\text{SF}} = \frac{2\text{IIP} + \text{MDS}}{3} \quad \text{dBm} \qquad (10\text{-}13)$$

For input signal strengths larger than given by (10-13), the third-order IM products rise three times faster than the desired signals, and ultimately will cause objectionable interference if they become sufficiently strong.

Again referring to Fig. 10-3, let $|x|$ be the distance EH = HD. Then the distance FH is $|x| + |\Delta\text{IM}|$. But the slope of line FD is 3, thus $3\,|x| = |x| + |\Delta\text{IM}|$, from which is obtained $|x| = |\Delta\text{IM}|/2$. This gives the location of the output intercept point above the measured value of the desired output in dBm.

Consequently, the intercept point OIP, referred to the *output*, is

$$\text{OIP} = \frac{\Delta\text{IM}}{2} + P_\Delta + G \quad \text{dBm} \tag{10-14}$$

Then IIP, referred to the input, can be found by subtracting G (in dB) from the above. Thus,

$$\text{IIP} = \frac{\Delta\text{IM}}{2} + P_\Delta \quad \text{dBm}$$

which is the same as (10-11).

Example 10-1

A certain electronic system has a bandwidth of 10 kHz and input and output resistances of 50 Ω. Its input is matched to two 50-Ω signal generators through a resistive mixer whose attenuation is 6 dB. The two frequencies are $f_1 = 3800$ kHz and $f_2 = 3802$ kHz. When the calibrated attenuators on both signal generators are set to +4 dBm, the spectrum analyzer connected to the output of the system indicates that the level of both output signal components is 5 dBm, and that the third-order IM components in the passband are both 16 dBm below the signal components. No other IM components are discernible and compression has not occurred. When no input signals are present, the output noise level is determined to be −36 dBm. Calculate: (a) the input level P_Δ; (b) the power gain G in dB; (c) the intercept point IIP; (d) the minimum discernible input signal MDS; (e) the spurious-free dynamic range SFDR; (f) the noise figure NF; (g) the largest rms signal voltage at the input terminals of the system before the third-order IM products become discernible.

Solution

(a) The power supplied to the electronic system is equal to the power at the terminals of the signal generators minus the attenuation of the mixer. Consequently,

$$P_\Delta = 4 - 6 = -2 \text{ dBm}$$

The IM products are located at 3798 kHz and 3804 kHz.

(b) $G = P_o - P_\Delta = 5 - (-2) = 7$ dB.

(c) From (10-11), IIP = $(\Delta\text{IM})/2 + P_\Delta = 16/2 + (-2) = 6$ dBm.

(d) In Fig. 10-3, length CD = 36 + 7 + 6 = 49 dBm. Also, length AC = 49 dBm. Thus, MDS = −49 + IIP = −43 dBm. This corresponds to an rms input voltage of 1.58 mV.

(e) From (10-12), SFDR = 2(IIP-MDS)/3 = 2[6 − (−43)]/3 = 32.67 dBm.

(f) From (10-10), NF = MDS + 144 − 10 log B_{kHz} = −43 + 144 − 10 = 91.

(g) From (10-13), P_{SF} = (2IIP + MDS)/3 = (12 − 43)/3 = −10.33 dBm. This corresponds to an rms input voltage of 68.1 mV.

10-6 THE SECOND-ORDER INTERCEPT POINT

If the two equal-amplitude input signals that are delivered to the nonlinear device have frequencies that are quite widely separated, instead of relatively close together, second-order IM products may become bothersome. For example, consider a receiver in which an interfering signal is at $f_1 = 1$ MHz and the desired signal is at $f_2 = 140$ MHz. Then, the second-order IM products are at $f_2 + f_1 = 141$ MHz and $f_2 - f_1 = 139$ MHz, both of which are near the desired signal frequency. Also, the third-order IM products are at $f_2 + 2f_1 = 142$ MHz and $f_2 - 2f_1 = 138$ MHz.

The output voltages of the second-order IM products are both equal to BE^2, owing to the squared term of the power series, provided that the fourth and higher orders are negligible. The output power at each second-order IM product is

$$P_{\text{out},2} = \frac{(BE^2)^2}{2R_L}$$

When expressed in dBm this becomes

$$\begin{aligned} P_2 &= 10 \log \frac{(BE^2)^2}{2R_L\, 10^{-3}} \\ &= 20 \log E^2 + 10 \log B^2 - 10 \log R_L + 27 \end{aligned}$$

When combined with (10-13), this becomes

$$P_2 = 2P_i - 27 + 10 \log \frac{B^2 R_{\text{in}}^2}{R_L} \qquad (10\text{-}15)$$

and, therefore,

$$P_2 = 2P_i + K_2 \tag{10-16}$$

where $K_2 = -27 + 10 \log B^2(R_{in})^2/R_L$. When plotted on the same rectangular coordinates as before, (10-16) is a straight line whose slope is 2, whose vertical axis intercept is K_2, and whose horizontal axis intercept is $-K_2/2$. The value of K_2 determines where this line for P_2 is located.

The point where this line for P_2 intersects that for P_o is the second-order intercept point, as shown in Fig. 10-4. To calculate IIP_2, the intercept point referred to the input, equate (10-6) and (10-15) using (10-15) and replace P_i with IIP_2. The result in dBm is

$$IIP_2 = 10 \log \frac{A^2}{B^2 R_{in}} + 27 = 10 \log \frac{501A^2}{B^2 R_{in}} \tag{10-17}$$

Therefore, the second- and third-order intercept points differ by

$$IIP_2 - IIP = 10 \log \frac{501A^2}{B^2 R_{in}} - 10 \log \frac{666.67A}{CR_{in}}$$

or, combining terms,

$$IIP_2 - IIP = 10 \log \frac{AC}{1.33B^2} \tag{10-18}$$

The second- and third-order intercept points occur at the same place when

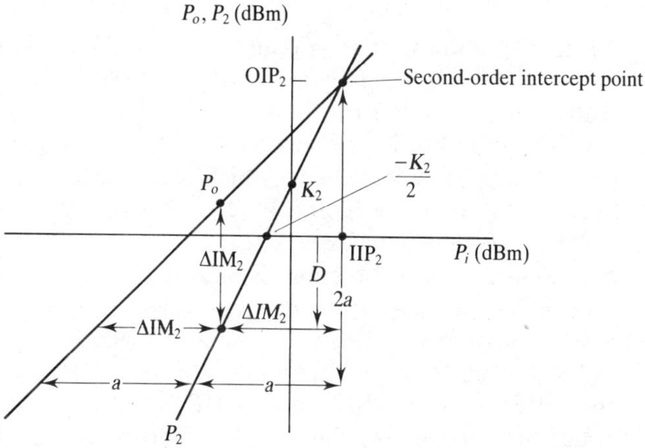

Fig. 10-4 Constructions for the second-order intercept point.

$1.33B^2 = AC$, from which is obtained,

$$B = \frac{\sqrt{A(3C)}}{2} \tag{10-19}$$

It has been assumed in this material that equal-amplitude input signals are delivered to the nonlinear device. If there is some selectivity between the input terminals of the system and the nonlinear device itself, then one of the input signals may be attenuated by some amount X dB more than the other signal. Consequently, the weaker signal would need to have its level increased by X dB to compensate for the attenuation so that both signals have the same amplitudes at the nonlinear device.

It is desirable that the system design be such that the line for P_2 passes through the noise-floor level at or to the right of point B in Fig. 10-3. This assures that the spurious-free dynamic range SFDR is indeed free of both second- and third-order IM products. If this occurs, it is easy to show (see Problem 10-3) that the required value of IIP_2 is

$$\text{IIP}_2 = \text{IIP} + \frac{\text{SFDR}}{2} \tag{10-20}$$

Example 10-2

Consider the design of an amplifier with input and output resistances of 50 Ω and a power gain of 20 dB. The spurious-free dynamic range of 80 dB is free of both second- and third-order IM products. The noise floor at the output is measured to be −105 dBm. From this information, the constants A, B, and C that are in the power series for the amplifier's input–output characteristic, can be calculated.

Solution Construct the lines for P_o, P_2, and P_3 approximately to scale, as shown in Fig. 10-5. The line for P_2 can be farther to the right of point B, but if it is to the left of point B the spurious-free dynamic range will be determined by the second-order IM products instead of the third-order IM products and it will become smaller, which is not desirable.

At the third-order intercept point F, $P_o = P_3$ with $P_i = \text{IIP}$. Thus, $\text{IIP} + 20 = 3\text{IIP} + K$, so that $\text{IIP} = 10 - K/2$.

At the second-order intercept point E, $P_o = P_2$ with $P_i = \text{IIP}_2$. Thus $\text{IIP}_2 + 20 = 2\text{IIP}_2 + K_2$, so that $\text{IIP}_2 = 20 - K_2$.

But $\text{IIP}_2 - \text{IIP} = 40$, and so substituting the above yields $K = 60 + 2K_2$.

Point E is located at $\text{IIP}_2 + 20$, therefore the noise floor is equal to $\text{IIP}_2 + 20 - 160 = \text{IIP}_2 - 140 = -105$, so that $\text{IIP}_2 = 35$ dBm. Then, working back through the above, we find $K_2 = -15$ dBm, $K = 30$ dBm, and $\text{IIP} = -5$ dBm. Also, MDS = −125 dBm.

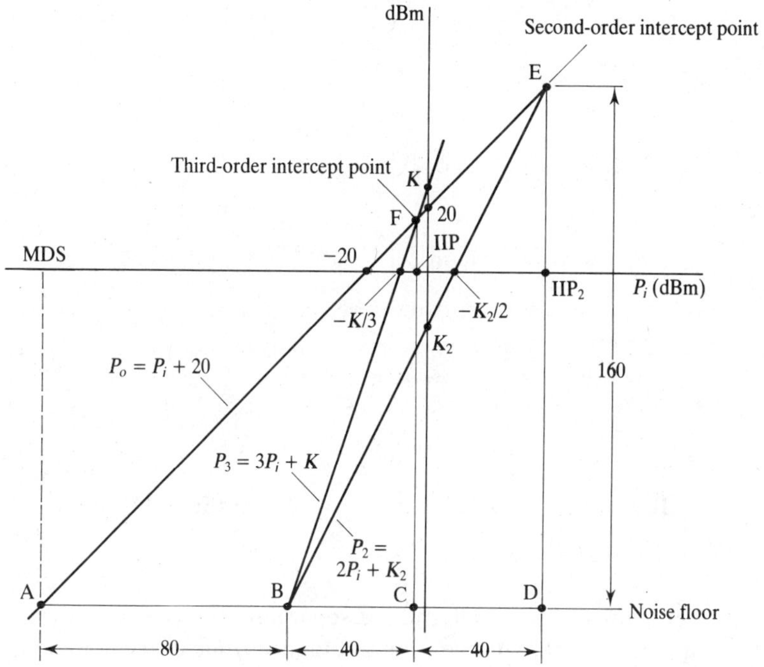

Fig. 10-5 Constructions involving separate second- and third-order intercepts IIP_2 and IIP, respectively.

Next, solve (10-5) for A to obtain

$$A = \sqrt{\frac{R_L}{R_{in}}}\,10^{G/20} \tag{10-21}$$

which then yields $A = 10$. Next, substitute this into (10-9) and solve that result for the coefficient C. Thus,

$$C = \frac{666.67}{R_{in}}\sqrt{\frac{R_L}{R_{in}}}\,10^{(G-2\mathrm{IP})/20} \tag{10-22}$$

which yields $C = 421.63$. Then substitute (10-21) into (10-17) and solve that result for B. Hence,

$$B = \frac{22.38\sqrt{R_L}}{R_{in}}\,10^{(G-\mathrm{IP}_2)/20} \tag{10-23}$$

which yields $B = 0.56$. Therefore, the static characteristic of the nonlinear amplifier is represented by

$$y = \pm 10x \pm 0.56x^2 \pm 421.63x^3$$

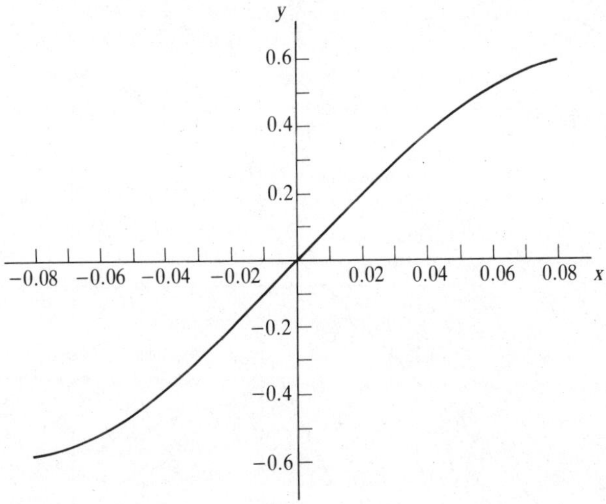

Fig. 10-6 The static characteristic $y = 10x + 0.56x^2 - 421.63x^3$.

The proper signs to use is determined separately by whether or not there is 180° of phase shift, whether the average output current increases or decreases as the input signal increases, and whether or not compression is observed. A graph of the function

$$y = 10x + 0.56x^2 - 421.63x^3$$

is shown in Fig. 10-6 to help visualize a possible nonlinear characteristic for this situation.

10-7 POWER, IM DISTORTION, AND INTERCEPT POINTS

Sometimes it is convenient to be able to express intercept-point relationships in forms other than those already used. For example, the third-order output intercept point can be expressed in terms of the third-order IM distortion level D (usually a negative dBm number) instead of the amount ΔIM that its level is below that of the output signal.

The third-order output intercept point was given in (10-14) as

$$\text{OIP} = \frac{\Delta\text{IM}}{2} + P_\Delta + G \quad \text{dBm}$$

Now substitute $P_o - D = \Delta\text{IM}$ and $P_o = P_\Delta + G$ into the above to obtain

$$\text{OIP} = \tfrac{3}{2}P_o - \tfrac{1}{2}D \tag{10-24}$$

where all three quantities are in dBm. But we know that

$$\mathrm{OIP}_{\mathrm{dBm}} = 10 \log (\mathrm{OIP})_{\mathrm{watts}} + 30$$
$$P_{o,\mathrm{dBm}} = 10 \log (P_{\mathrm{out}})_{\mathrm{watts}} + 30$$
$$D_{\mathrm{dBm}} = 10 \log (D_{\mathrm{out}})_{\mathrm{watts}} + 30$$

Substituting these into (10-24) and solving for the third-order intercept point in watts yields

$$\mathrm{OIP} = \frac{(P_{\mathrm{out}})^{3/2}}{(D_{\mathrm{out}})^{1/2}} \tag{10-25}$$

where all of these quantities are in watts. This shows how OIP in watts can be calculated from two-tone test measurements of the output signal power P_{out} in watts and third-order IM distortion level D_{out} in watts. Also, from (10-25), the distortion level produced when the OIP is known is

$$D_{\mathrm{out}} = \frac{(P_{\mathrm{out}})^2}{(\mathrm{OIP})^2} \tag{10-26}$$

where all quantities are in watts.

Now repeat the above steps for the second-order intercept point. From the constructions in Fig. 10-4, it is found that the second-order output intercept point is

$$\mathrm{OIP}_2 = \Delta\mathrm{IM} + P_o \tag{10-27}$$

where all quantities are in dBm. Also, since $\Delta\mathrm{IM} = P_o - D$, (10-27) becomes

$$\mathrm{OIP}_2 = 2P_o - D \tag{10-28}$$

where D is the second-order IM distortion level (usually a negative dBm number) and all quantities are in dBm. But

$$\mathrm{OIP}_{2,\mathrm{dBm}} = 10 \log (\mathrm{OIP}_2)_{\mathrm{watts}} + 30$$
$$P_{o,\mathrm{dBm}} = 10 \log (P_{\mathrm{out}})_{\mathrm{watts}} + 30$$
$$D_{\mathrm{dBm}} = 10 \log (D_{\mathrm{out}})_{\mathrm{watts}} + 30$$

Substituting these into (10-28) yields

$$\mathrm{OIP}_2 = \frac{(P_{\mathrm{out}})^2}{D_{\mathrm{out}}} \tag{10-29}$$

or

$$D_{out} = \frac{(P_{out})^2}{\text{OIP}_2} \tag{10-30}$$

where all quantities are in watts.

These relationships, in watts rather than in dBm, will be used in Section 10-7.2 when calculating cascaded intercept points.

10-7.1 IM Distortion and the Correlation Coefficient

Figure 10-7 depicts a nonlinear amplifier stage that produces IM distortion in its output with two tones applied to its input. The IM distortion will be either the second or third-order component, as determined by the test conditions.

Input power P_{in} (watts) is amplified to produce $P_{out} = GP_{in}$ (watts). If $D_{in} = 0$, then the total output distortion of the nonlinear amplifier is only what it creates, so $D_{out} = D_{amp}$ (watts). Then the OIP of this stage can be calculated from (10-25) or OIP_2 from (10-29), depending on the conditions.

If input distortion D_{in} is also present, then the output distortion is equal to the sum of the distortion generated by the amplifier, the amplified input distortion, and a term that depends on how the first two distortions are correlated. Thus,

$$D_{out} = \text{D}_{amp} + GD_{in} + 2C\sqrt{GD_{in}D_{amp}} \quad \text{watts} \tag{10-31}$$

where C is a correlation coefficient and $0 \leq C \leq 1$. If $C = 0$ (one extreme), then the power in an IM distortion component adds to the amplified input power of the same distortion component. Thus,

$$D_{out} = D_{amp} + GD_{in} \tag{10-32}$$

This action is akin to that when random noise power produced by a noisy amplifier is added to the amplified input noise power to obtain the total output noise power.

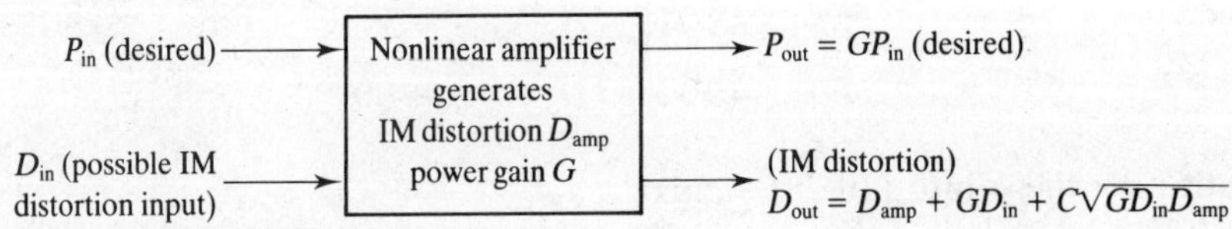

Fig. 10-7 Input and output powers (in watts) of a nonlinear amplifier subject to a two-tone test. The output IM distortion depends on the correlation coefficient C when D_{in} is present.

If $C = 1$ (the other extreme), then (10-31) becomes

$$D_{\text{out}} = D_{\text{amp}} + 2\sqrt{D_{\text{amp}} G D_{\text{in}}} + G D_{\text{in}} = (\sqrt{D_{\text{amp}}} + \sqrt{G D_{\text{in}}})^2 \qquad (10\text{-}33)$$

Voltage or current varies as the square root of power. So, when $C = 1$, the IM distortion voltages or currents, rather than their powers, add in a coherent fashion.

10-7.2 Cascaded Intercept Points

Assume that N amplifier stages are connected in cascade and that the third- and second-order output intercept points $(\text{OIP})_n$ and $(\text{OIP}_2)_n$, respectively, have previously been determined at the desired frequency for $n = 1, 2, \ldots, N$, (i.e., for each individual stage by separate measurements). We wish to calculate the overall output intercept points $(\text{OIP})_{\text{overall}}$ and $(\text{OIP}_2)_{\text{overall}}$ for the entire cascade in terms of the intercept points of the individual-stages. Of course, the terminating impedances for the individual-stage measurements should be the same as the terminating impedances of the stages when connected in cascade.

The following analysis is carried out for two stages at a time, starting with the first two. They are then subjected to the two-tone test, as shown in Fig. 10-8. No IM distortion is applied to stage 1, only the signal power $P_{\text{in},1}$ from one of the sinusoidal signal generators. The output of stage 1 contains the desired signal power $P_{\text{out},1} = G_1 P_{\text{in},1} = P_{\text{in},2}$, and a third-order IM distortion component $D_{\text{amp},1}$ generated by stage 1. Both of these are then amplified by stage 2. Therefore, $P_{\text{out},2} = G_2 P_{\text{in},2} = G_1 G_2 P_{\text{in},1}$, and the output IM distortion $D_{\text{out},2}$ depends on whether $C = 1$ or $C = 0$, the two extreme values of the correlation coefficient, or is somewhere in between.

If $C = 1$, then from (10-33),

$$D_{\text{out},2} = (\sqrt{G_2 D_{\text{amp},1}} + \sqrt{D_{\text{amp},2}})^2 \quad \text{W}$$

Therefore, from (10-25), the third-order output intercept point for these first two stages is

$$(\text{OIP})^{C=1}_{\text{first 2}} = \frac{(G_1 G_2 P_{\text{in},1})^{3/2}}{\sqrt{G_2 D_{\text{amp},1}} + \sqrt{D_{\text{amp},2}}} \quad \text{W} \qquad (10\text{-}34)$$

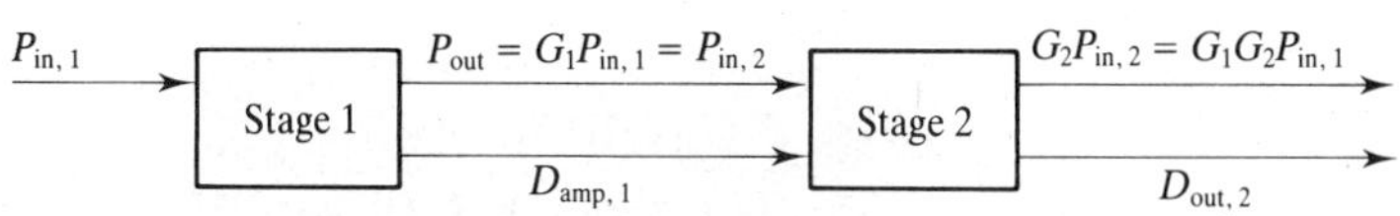

Fig. 10-8 A cascade of nonlinear amplifiers subject to a two-tone test.

But according to (10-26), $D_{\text{amp},1} = (G_1 P_{\text{in},1})^3/[(\text{OIP})_1]^2$ and, for the second stage, $D_{\text{amp},2} = (G_1 G_2 P_{\text{in},1})^3/[(\text{OIP})_2]^2$. Substituting these into (10-34) and simplifying yields

$$(\text{OIP})_{\text{first 2}}^{C=1} = \frac{G_2(\text{OIP})_1(\text{OIP})_2}{G_2(\text{OIP})_1 + (\text{OIP})_2} \tag{10-35}$$

The quantity $G_2(\text{OIP})_1$ is the third-order output intercept of stage 1 referred to the output of stage 2. There it combines with $(\text{OIP})_2$ as the product of the two divided by the sum of the two. This is similar to the way resistances in parallel combine to obtain the total. This is the way intercept points combine when they are referred to the same output location and the correlation coefficient $C = 1$.

If $C = 0$, then from (10-32), $D_{\text{out},2} = D_{\text{amp},2} + G_2 D_{\text{amp},1}$. Therefore, from (10-25), the third-order output intercept point for these first two stages is

$$(\text{OIP})_{\text{first 2}}^{C=0} = \frac{(G_1 G_2 P_{\text{in},1})^{3/2}}{(D_{\text{amp},2} + G_2 D_{\text{amp},1})^{1/2}} \tag{10-36}$$

As before, substituting for $D_{\text{amp},1}$ and $D_{\text{amp},2}$, then multiplying and dividing the $G_2 D_{\text{amp},1}$ term by $(G_2)^2$, and simplifying, (10-36) becomes

$$(\text{OIP})_{\text{first 2}}^{C=0} = \frac{G_2(\text{OIP})_1(\text{OIP})_2}{\sqrt{[G_2(\text{OIP})_1]^2 + [(\text{OIP})_2]^2}} \tag{10-37}$$

The denominator of (10-37) is less than that of (10-35). Therefore, the intercept point given by (10-37) for $C = 0$ is greater than the intercept point given by (10-35) for $C = 1$, as expected. The actual intercept point will be somewhere between these extremes.

The first two stages have now been combined into an equivalent stage whose power gain is $G_1 G_2$ and whose third-order intercept points are bounded by the value of C. This combined stage will now become the first stage, replacing the original one, and stage 3 of the cascade will become the second stage for the next calculations. Then (10-35) and (10-37) yield the overall intercept-point range for the first three stages whose power gain is $G_1 G_2 G_3$. These become the values for the next new first stage, and the original stage 4 becomes the next new second stage. This process is repeated until all N stages have been accommodated.

A similar procedure, using (10-29) and (10-30), allows cascaded second-order intercept points to be calculated. The results for two cascaded stages are

$$(\text{OIP}_2)_{\text{first 2}}^{C=0} = \frac{G_2(\text{OIP}_2)_1(\text{OIP}_2)_2}{G_2(\text{OIP}_2)_1 + (\text{OIP}_2)_2} \tag{10-38}$$

and

$$(\mathrm{OIP}_2)_{\text{first }2}^{C=1} = \left[\frac{\sqrt{G_2(\mathrm{OIP}_2)_1}\sqrt{(\mathrm{OIP}_2)_2}}{\sqrt{G_2(\mathrm{OIP}_2)_1} + \sqrt{(\mathrm{OIP}_2)_2}}\right]^2 \tag{10-39}$$

Example 10-3

Three amplifier stages, identified simply as 1, 2, and 3, have power gains or 10 dB, 6 dB, and 20 dB, respectively, and third-order output intercept points of −2 dBm, 5 dBm, and −10 dBm, respectively. They are connected in the order 1–2–3 to form a cascade. Calculate the overall third-order output intercept point.

Solution First, convert the dB-related values into watts-related values using

$$\frac{P_2}{P_1} = 10^{G_{\mathrm{dB}}/10}$$

and

$$(\mathrm{OIP})_{\mathrm{watts}} = 10^{[(\mathrm{OIP})_{\mathrm{dBm}}-30]/10}$$

Thus, $G_1 = 10$, $G_2 = 3.98$, $G_3 = 100$, $(\mathrm{OIP})_1 = 6.31 \times 10^{-4}$, $(\mathrm{OIP})_2 = 31.6 \times 10^{-4}$, and $(\mathrm{OIP})_3 = 10^{-4}$. Therefore, from (10-35),

$$(\mathrm{OIP})_{\text{first }2}^{C=1} = \frac{3.98(6.31 \times 10^{-4})(31.6 \times 10^{-4})}{3.98(6.31 \times 10^{-4}) + (31.6 \times 10^{-4})} = 0.0014$$

which corresponds to 1.46 dBm. Also, $G_1 G_2 = 39.8$. This combination becomes the new first stage, and the original stage 3 becomes the new second stage. Thus, using (10-35) again,

$$(\mathrm{OIP})_{\text{overall}}^{C=1} = \frac{100(0.0014)(10^{-4})}{100(0.0014) + (10^{-4})} \simeq 10^{-4}\ \mathrm{W}$$

which corresponds to −10 dBm. The overall power gain is 3980 or 36 dB. When calculations are continued using (10-37), then

$$(\mathrm{OIP})_{\text{overall}}^{C=0} = \frac{100(0.00197)(10^{-4})}{\sqrt{[100(0.00197)]^2 + (10^{-4})^2}} = 10^{-4}\ \mathrm{W}$$

or −10 dBm. The high power gain of stage 3, together with the output intercept point of the first pair, serve to make the overall third-order output intercept point essentially the same as for stage 3, regardless of the correlation coefficient.

10-7.2.1 Intercept-Point Measure Suppose the power gains and third-order output intercept points for two stages 1 and 2 are known. In what sequence should the stages be connected to produce the greater overall third-order output intercept point? Assume that the overall OIP of the sequence 1–2 is greater than for the sequence 2–1. Then, from (10–35), for $C = 1$,

$$\frac{G_2(\mathrm{OIP})_1(\mathrm{OIP})_2}{G_2(\mathrm{OIP})_1 + (\mathrm{OIP})_2} > \frac{G_1(\mathrm{OIP})_2(\mathrm{OIP})_1}{G_1(\mathrm{OIP})_2 + (\mathrm{OIP})_1} \tag{10-40}$$

This simplifies into

$$\frac{(\mathrm{OIP})_2}{G_2}(G_2 - 1) > \frac{(\mathrm{OIP})_1}{G_1}(G_1 - 1) \tag{10-41}$$

Each side of (10-41) contains values for only one stage and the forms on both sides are the same. Each side represents a third-order intercept-point measure for $C = 1$, so we will refer to this form as a 3point1 measure. When the 3point1 measure of stage 2 is greater than the 3point1 measure of stage 1, then the sequence 1–2 yields the greater third-order overall OIP.

Also, from (10-37), with $C = 0$,

$$\frac{G_2(\mathrm{OIP})_1(\mathrm{OIP})_2}{\sqrt{\{[G_2(\mathrm{OIP})_1]^2 + [(\mathrm{OIP})_2]^2\}}} > \frac{G_1(\mathrm{OIP})_2(\mathrm{OIP})_1}{\sqrt{\{[G_1(\mathrm{OIP})_2]^2 + [(\mathrm{OIP})_1]^2\}}} \tag{10-42}$$

which simplifies into

$$(\mathrm{OIP})_2\sqrt{\left(\frac{G_2^2 - 1}{G_2^2}\right)} > (\mathrm{OIP})_1\sqrt{\left(\frac{G_1^2 - 1}{G_1^2}\right)} \tag{10-43}$$

We will refer to the form on each side of (10-43) as a 3point0 measure. When the conditions of (10-41) and (10-43) are satisfied, then the sequence 1–2 yields the greater overall third-order intercept point for both extremes of the correlation coefficient.

Similar procedures, using (10-38) and (10-39), allow the conditions to be determined for making the overall second-order intercept point of two stages in the sequence 1–2 greater than for 2–1. Thus, for (10-38), with $C = 0$,

$$\frac{G_2(\mathrm{OIP}_2)_1(\mathrm{OIP}_2)_2}{G_2(\mathrm{OIP}_2)_1 + (\mathrm{OIP}_2)_2} > \frac{G_1(\mathrm{OIP}_2)_2(\mathrm{OIP}_2)_1}{G_1(\mathrm{OIP}_2)_2 + (\mathrm{OIP}_2)_1} \tag{10-44}$$

which simplifies to

$$(\mathrm{OIP}_2)_2\left(\frac{G_2 - 1}{G_2}\right) > (\mathrm{OIP}_2)_1\left(\frac{G_1 - 1}{G_1}\right) \tag{10-45}$$

The form on each side of (10-45) becomes a 2point0 measure. When (10-45) is satisfied and $C = 0$, then the overall second-order intercept point of the sequence 1–2 is greater than for the sequence 2–1.

From (10-39), with $C = 1$,

$$\left[\frac{\sqrt{G_2(\mathrm{OIP}_2)_1}\sqrt{(\mathrm{OIP}_2)_2}}{\sqrt{G_2(\mathrm{OIP}_2)_1}+\sqrt{(\mathrm{OIP}_2)_2}}\right]^2 > \left[\frac{\sqrt{G_1(\mathrm{OIP}_2)_2}\sqrt{(\mathrm{OIP}_2)_1}}{\sqrt{G_1(\mathrm{OIP}_2)_2}+\sqrt{(\mathrm{OIP}_2)_1}}\right]^2 \tag{10-46}$$

which simplifies into

$$\sqrt{(\mathrm{OIP}_2)_2}\left(\frac{\sqrt{G_2}-1}{\sqrt{G_2}}\right) > \sqrt{(\mathrm{OIP}_2)_1}\left(\frac{\sqrt{G_1}-1}{\sqrt{G_1}}\right) \tag{10-47}$$

The form on each side of (10-47) becomes a 2point1 measure. When the conditions of (10-45) and (10-47) are satisfied, then the sequence 1–2 yields a greater overall second-order intercept point than the sequence 2–1 for both extremes of the correlation coefficient.

10-7.3 Cascaded Intercept Points by an Alternate Method

A cascade of two nonlinear stages is shown in Fig. 10-9. The output signal voltage V_1 from stage 1 is also the input signal voltage to stage 2. The output signal voltage of stage 2 is V_2. Assume that the output signal voltage of each nonlinear stage can be represented by a third-order polynomial. Thus,

$$V_1 = A_1V_i + B_1V_i^2 + C_1V_i^3 \tag{10-48}$$

$$V_2 = A_2V_1 + B_2V_1^2 + C_2V_1^3 \tag{10-49}$$

The output signal voltage V_2 can be expressed in terms of the input signal voltage V_i by substituting the three terms of (10-48) into (10-49) wherever V_1 appears and then simplifying. The result is,

$$V_2 = A_1A_2V_i + (A_2B_1 + A_1^2B_2)V_i^2 + (A_2C_1 + 2A_1B_1B_2 + A_1^3C_2)V_i^3 + \cdots \tag{10-50}$$

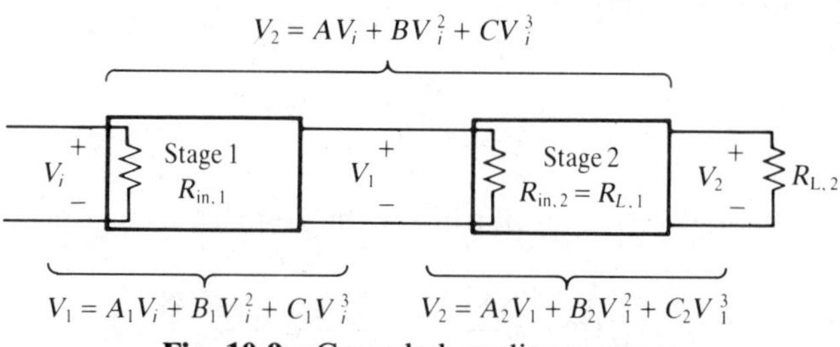

Fig. 10-9 Cascaded nonlinear stages.

The term $2A_1B_1B_2$ is assumed to be sufficiently small to be neglected (see Problem 10-9). Consequently, (10-50) becomes

$$V_2 = A_1A_2V_i + (A_2B_1 + A_1^2B_2)V_i^2 + (A_2C_1 + A_1^3C_2)V_i^3 + \cdots \quad (10\text{-}51)$$

$$= AV_i + BV_i^2 + CV_i^3 + \text{higher-order terms} \quad (10\text{-}52)$$

But (10-52) represents the overall input–output signal-voltage characteristic of the two-stage cascade. When terms of higher order than 3 are ignored, then $A = A_1A_2$, $B = A_2B_1 + A_1^2B_2$, and $C = A_2C_1 + A_1^3C_2$.

If the output intercept points of a single stage can be expressed in terms of the coefficients of its polynomial representation, then the overall intercept points of the two-stage cascade can be calculated. This will be done as follows.

The input power in watts of each sinusoid in a two-tone test is $P_{in} = E^2/(2R_{in})$. The desired fundamental output power is $P_{out} = A^2E^2/2R_L$. The power gain is $G = P_{out}/P_{in} = A^2R_{in}/R_L$. The output power of a third-order IM product is $P_{out,3} = (\frac{3}{4}CE^3)^2/2R_L$. The output power of a second-order IM product is $P_{out,2} = (BE^2)^2/2R_L$. These can be combined to obtain $P_{out,3} = (9C^2R_{in}^2/4A^2)GP_{in}^3$ and $P_{out,2} = (B^2 2R_{in}/A^2)GP_{in}^2$.

Let IIP be the value of P_{in} in watts at the third-order intercept point where $P_{out} = GP_{in} = G(\text{IIP}) = P_{out,3}$. Use the above relationships for the third-order input intercept point to obtain

$$\text{IIP} = \frac{2A}{3CR_{in}} \quad \text{W} \quad (10\text{-}53)$$

Therefore, the third-order output intercept point of a stage is

$$\text{OIP} = G(\text{IIP}) = \frac{A^2R_{in}}{R_L}\frac{2A}{3CR_{in}} = \frac{2A^3}{3CR_L} \quad \text{W} \quad (10\text{-}54)$$

Thus, for stages 1 and 2 of a cascade, $(\text{OIP})_1 = 2A_1^3/3C_1R_{L,1}$ and $(\text{OIP})_2 = 2A_2^3/3C_2R_{L,2}$, respectively.

Let IIP_2 be the value of P_{in} in watts at the second-order intercept point where $P_{out} = GP_{in} = G(\text{IIP}_2) = P_{out,2}$. Use the above relationships for the second-order input intercept point. Thus,

$$\text{IIP}_2 = \frac{A^2}{2B^2R_{in}} \quad \text{W} \quad (10\text{-}55)$$

Therefore, the second-order output intercept point of a stage is

$$\text{OIP}_2 = G(\text{IIP}_2) = \frac{A^2R_{in}}{R_L}\frac{A^2}{B^2 2R_{in}} = \frac{A^4}{2B^2R_L} \quad \text{W} \quad (10\text{-}56)$$

Thus, for stages 1 and 2 of a cascade, $(\mathrm{OIP}_2)_1 = A_1^4/2B_1^2R_{L,1}$ and $(\mathrm{OIP}_2)_2 = A_2^4/2B_2^2R_{L,2}$, respectively.

To determine the overall third-order output intercept point of the cascade of two stages, substitute $A = A_1A_2$ and $C = A_2C_1 + A_1^3C_2$ into the reciprocal of (10-54). This yields

$$\frac{1}{\mathrm{OIP}_{\mathrm{overall}}} = \frac{3(A_2C_1 + A_1^3C_2)R_{L,2}}{2(A_1A_2)^3} \tag{10-57}$$

If $R_{L,1} = R_{L,2}$, (10-57) becomes

$$\frac{1}{\mathrm{OIP}_{\mathrm{overall}}} = \frac{3C_1R_{L,1}}{2A_1^3A_2^2} + \frac{3C_2R_{L,2}}{2A_2^3} \tag{10-58}$$

or

$$\frac{1}{\mathrm{OIP}_{\mathrm{overall}}} = \frac{1}{A_2^2(\mathrm{OIP})_1} + \frac{1}{(\mathrm{OIP})_2}$$

If $R_{\mathrm{in},2} = R_{L,2}$, then $G = A_2^2$. Therefore,

$$\frac{1}{\mathrm{OIP}_{\mathrm{overall}}} = \frac{1}{G_2(\mathrm{OIP})_1} + \frac{1}{(\mathrm{OIP})_2} \tag{10-59}$$

After combining terms and taking the reciprocal, (10-59) becomes the same as (10-35) for $C = 1$. This method does not yield an overall third-order intercept point for $C = 0$.

Finally, substituting for A and $B = A_2B_1 + A_1^2B_2$ into the reciprocal of (10-55), using $R_{\mathrm{in},1} = R_{L,1} = R_{L,2}$, it is found that

$$\frac{1}{(\mathrm{OIP}_2)_{\mathrm{overall}}} = \left[\frac{1}{\sqrt{G_2}\sqrt{(\mathrm{OIP}_2)_1}} + \frac{1}{\sqrt{(\mathrm{OIP}_2)_2}}\right]^2 \tag{10-60}$$

After combining terms and taking the reciprocal, (10-60) becomes the same as (10-39) for $C = 1$. This method does not yield an overall second-order intercept point for $C = 0$.

PROBLEMS

10-1 An electronic system has a noise factor of 20, bandwidth of 10 kHz, and a power gain of 10^5. The input is matched to 75 Ω. Calculate in dBm: (a) the minimum discernible signal for $S_o = N_o$; (b) the noise floor at the output; (c) the spurious-free dynamic range if IIP = 0 dBm; (d) the maximum input power for a spurious-free response.

(e) Calculate the minimum and maximum values of the rms signal voltage (in μV or mV at the input terminals) that correspond to the spurious-free dynamic range.

10-2 An electronic system has a bandwidth of 5 kHz and input and output matched resistances of 75 Ω. The input is matched to two 75-Ω signal generators through a resistive mixer whose attenuation is 6 dB. The two frequencies are 7100 kHz and 7101 kHz. When the calibrated attenuators on the signal generators are both set to 0 dBm, the spectrum analyzer connected to the output of the system indicates that the level of both output signal components is +1 dBm, and that the third-order IM components are both 12 dBm below the signal components. No other IM components are noticeable and compression has not occurred. When no input signals are present, the noise floor at the output is measured to be −30 dBm. Calculate: (a) the input power level in dBm; (b) power gain of the system in dB; (c) input intercept point IIP in dBm; (d) minimum discernible input signal in dBm; (e) spurious-free dynamic range in dBm; (f) noise figure in dB, (g) maximum rms signal voltage at the input terminals before the third-order IM products become noticeable.

10-3 Use the graphical relationships between the various lines in Figs. 10-3 and 10-4 to demonstrate that:

(a) The spurious-free dynamic range SFDR is indeed free of both second- and third-order IM products if $\mathrm{IIP}_2 \geqslant \mathrm{IIP} + \mathrm{SFDR}/2$.

(b) The second and third-order intercept points are the same if $K_2 = (K + G)/2$.

10-4 The second and third-order input intercept points of a certain amplifier are each equal to 15 dBm. The power gain is 10 dB, the spurious-free dynamic range (determined by the second-order IM products) is 30 dB, and the input and output resistances are each 75 Ω. Draw the lines for P_o, P_2, and P_3 approximately to scale. Then, using geometrical relationships, determine (in dBm): (a) the noise floor; (b) MDS; (c) the values of K and K_2; (d) the input level for which P_2 is at the noise floor; (e) the constants A, B, and C that are in the power series representation of the input–output voltage characteristic of the amplifier.

10-5 Verify the conversions and cascaded intercept points of Example 10-2.

10-6 Using the individual stages identified in Example 10-3, form a cascade in the sequence 2–1–3. Calculate the third-order overall intercept points for $C = 1$ and $C = 0$.

10-7 Three amplifier stages, identified simply as 1, 2, and 3, have power gains of 10 dB, 8 dB, and 6 dB, respectively, and third-order output intercept points of −15 dBm, −12 dBm, and −9 dBm, respectively.

(a) Calculate the 3point1 and 3point0 measures. (b) Determine what sequence provides the greatest overall third-order output intercept points, and calculate the greatest values in dBm.

10-8 Three amplifier stages have power gains as given in Problem 10-7 and second-order output intercept points of −25 dBm, −21 dBm, and −18 dBm, respectively. (a) Calculate the 2point1 and 2point0 measures; (b) Determine what sequence provides the greatest overall second-order output intercept points, and calculate the greatest values in dBm.

10-9 Each of two identical amplifiers has the input–output characteristic shown in Fig. 10-6. Demonstrate that $|2A_1B_1B_2| \ll |A_2C_1|$ and that $|2A_1B_1B_2| \ll |A_1^3C_2|$ in Eq. (10-50).

APPENDIX A

DERIVATIONS PERTAINING TO BEAT PERIODS AND TRUE PERIODS IN THE LINEAR MIXING OF TWO SINUSOIDS

Assume that the beat period T_b equals the true period T_0. Then,

$$T_b = \frac{1}{f_2 - f_1} = T_0 = \frac{m}{f_2} = \frac{n}{f_1} \tag{A-1}$$

from which are obtained the harmonic numbers

$$m = \frac{f_2}{f_2 - f_1} \tag{A-2}$$

and

$$n = \frac{f_1}{f_2 - f_1} \tag{A-3}$$

These show that $(f_2 - f_1)$ must be the highest common divisor of both f_2 and f_1 such that the harmonic numbers are obtained from the ratios when the beat period equals the true period.

From (A-2) and (A-3), we find that

$$(m-1)f_2 = mf_1 \tag{A-4}$$

and

$$nf_2 = (n+1)f_1 \tag{A-5}$$

Divide (A-4) by (A-5) to obtain

$$\frac{m-1}{n} = \frac{m}{n+1}$$

which simplifies to

$$m = n + 1 \tag{A-6}$$

Thus, the harmonic numbers differ by 1 when the beat period equals the true period and if one of them is even the other must be odd.

Since, from (A-2), $f_2 = m(f_2 - f_1)$, and from (A-3), $f_1 = n(f_2 - f_1)$, then

$$\frac{f_2 + f_1}{f_2 - f_1} = \frac{m(f_2 - f_1) + n(f_2 - f_1)}{f_2 - f_1} = m + n \tag{A-7}$$

which is the sum of the harmonic numbers. Also, using (A-6) in (A-7),

$$\frac{f_2 + f_1}{f_2 - f_1} = m + n = 2n + 1 \tag{A-8}$$

which is always an odd integer.

From (A-1),

$$\frac{f_2}{f_1} = \frac{m}{n} \tag{A-9}$$

Then, using (A-6), first eliminating m and then eliminating n, this becomes

$$\frac{f_2}{f_1} = \frac{n+1}{n} = \frac{m}{m-1} \tag{A-10}$$

Therefore, when the beat period equals the true period, f_2/f_1 can be expressed as a ratio of even-to-odd or odd-to-even integers that differ by 1. This same result is obtained by solving (A-8) for f_2/f_1.

Using $f_2 = (m/n)f_1$ from (A-9), the beat period becomes

$$T_b = \frac{1}{f_2 - f_1} = \frac{n}{f_1(m-n)} \tag{A-11}$$

From (A-1), the true period is

$$T_0 = \frac{n}{f_1} \tag{A-12}$$

Dividing (A-12) by (A-11) yields

$$\frac{T_0}{T_b} = m - n \tag{A-13}$$

which shows that a sequence of $(m-n)$ beat periods must occur in order for the waveform of that sequence to have a period equal to the true period. Of course, if $m-n=1$ (the equivalent of (A-6)) the beat period is equal to the true period.

Sections of the (cosine) sinusoids that are multiplied to determine the resultant waveform are shown in Fig. A-1. As shown there, the time for half a beat period is $\dfrac{T_b}{2}=\dfrac{1/2}{f_2-f_1}$. The nodal point occurs at some fraction k (which can be either positive or negative depending on the situation) of a period of the higher frequency as measured from an instant at p complete periods of the higher frequency. Therefore,

$$\frac{1}{2(f_2-f_1)}=\frac{2(p+k)}{f_2+f_1} \tag{A-14}$$

from which is obtained

$$p+k=\frac{\dfrac{f_2}{f_1}+1}{4\left(\dfrac{f_2}{f_1}-1\right)}=\frac{f_2+f_1}{4(f_2-f_1)} \tag{A-15}$$

where p is an even or odd integer, and $-1<k<1$. For pure beating with f_2/f_1 given by the ratio of an even integer to an odd integer that differ by 1, then, from (A-10), $f_2/f_1=m/(m-1)$, in which $f_2>f_1$ and m is even. Thus, (A-15) becomes

$$p+k=\frac{\dfrac{m}{m-1}+1}{4\left(\dfrac{m}{m-1}-1\right)}=\frac{2m-1}{4}=\frac{m}{2}-\frac{1}{4} \tag{A-16}$$

Therefore, the nodal point in this case occurs when $p=m/2$ and $k=-1/4$, but the resultant waveform near the nodal point is illustrated in Fig. A-1(b).

For pure beating with f_2/f_1 given by the ratio of an odd integer to an even integer that differ by 1, then, again from (A-10), we use $f_2/f_1=(n+1)/n$, in which n is even. Thus, (A-15) becomes

$$p+k=\frac{\dfrac{n+1}{n}+1}{4\left(\dfrac{n+1}{n}-1\right)}=\frac{2n+1}{4}=\frac{n}{2}+\frac{1}{4} \tag{A-17}$$

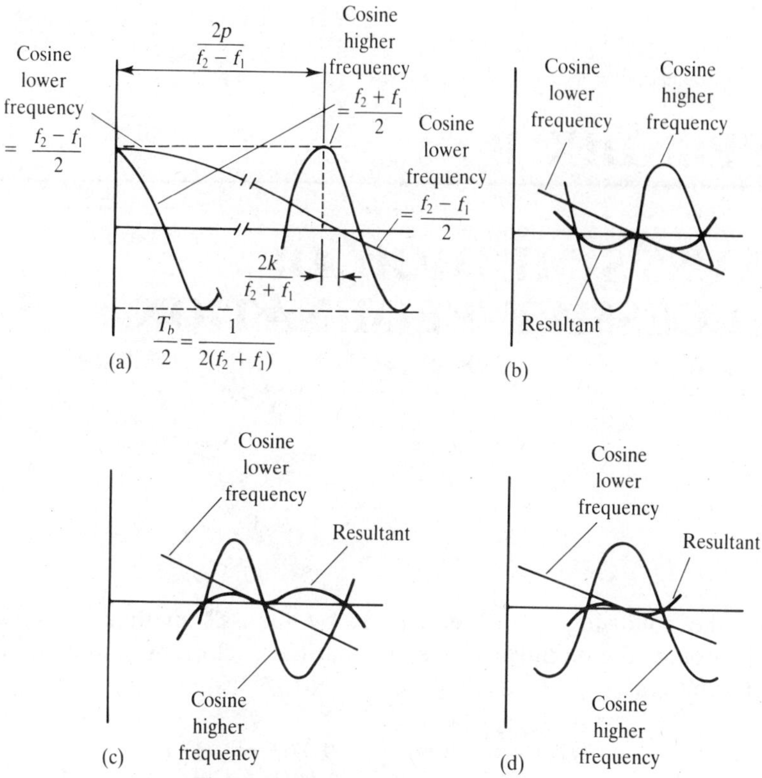

Fig. A-1 Sections of waveforms.

Therefore, in this case, the nodal point occurs when $p = n/2$ and $k = 1/4$, as shown in Fig. A-1(c).

If f_2 and f_1 are such that $(f_2 + f_1)/(f_2 - f_1)$ is divisible by 4, then (A-15) indicates that p is an even or odd integer and that $k = 0$. The resultant waveform near the nodal point is illustrated in Fig. A-1(d).

APPENDIX B

GAUSS'S METHOD OF SUCCESSIVE ELIMINATION

Solve the following set of equations by Gauss's method of successive elimination of the unknowns. Assume that the coefficients and constants are exact numbers:

$$60A + 708C + 9780E = 11.6 \tag{B-1}$$

$$708A + 9780C + 144708E = 126.4 \tag{B-2}$$

$$9780A + 144708C + 2217300E = 1690.282 \tag{B-3}$$

The coefficient of E in (B-3) is the largest of all coefficients, so solve that equaton for E to obtain

$$E = 7.7232 \times 10^{-4} - 652.6316 \times 10^{-4}C - 44.1077 \times 10^{-4}A \tag{B-4}$$

Substitute (B-4) into (B-1) and (B-2) and combine terms to eliminate E from those equations. Thus

$$16.8627A + 69.7263C = 4.1445 \tag{B-5}$$

$$69.7263A + 335.8986C = 16.0862 \tag{B-6}$$

Of these two equations, the coefficient of C in (B-6) is the largest, so solve that equation for C to obtain

$$C = 4.7890 \times 10^{-2} - 20.7581 \times 10^{-2}A \tag{B-7}$$

Then (B-4) becomes

$$E = -23.6313 \times 10^{-4} + 91.3662 \times 10^{-4} A \tag{B-8}$$

and (B-5) becomes

$$2.3888A = 0.8053$$

Therefore

$$A = 0.3371 \tag{B-9}$$

Finally, substituting this into (B-7) and (B-8) yields

$$C = -2.2086 \times 10^{-2} \tag{B-10}$$

and

$$E = 7.1682 \times 10^{-4} \tag{B-11}$$

Substituting these values back into (B-1)–(B-3) as a check of their validity results in near equalities. Of course, slight discrepancies are caused by rounding errors in the various computations.

APPENDIX C

TABLE OF INTEGRALS

$$I_n = \frac{1}{2\pi j} \int_{-j\infty}^{j\infty} \left| \frac{c_{n-1}s^{n-1} + \cdots + c_0}{d_n s^n + \cdots + d_0} \right|^2 ds$$

$$I_1 = \frac{c_0^2}{2d_0 d_1}; \qquad I_2 = \frac{c_1^2 d_0 + c_0^2 d_2}{2d_0 d_1 d_2};$$

$$I_3 = \frac{c_2^2 d_0 d_1 + (c_1^2 - 2c_0 c_2) d_0 d_3 + c_0^2 d_2 d_3}{2d_0 d_3(-d_0 d_3 + d_1 d_2)}$$

$$I_4 = \frac{\begin{array}{r} c_3^2(-d_0^2 d_3 + d_0 d_1 d_2) + (c_2^2 - 2c_1 c_3) d_0 d_1 d_4 \\ +(c_1^2 - 2c_0 c_2) d_0 d_3 d_4 + c_0^2(-d_1 d_4^2 + d_2 d_3 d_4) \end{array}}{2d_0 d_4(-d_0 d_3^2 - d_1^2 d_4 + d_1 d_2 d_3)}$$

$$I_5 = \frac{1}{2\Delta_5} [c_4^2 m_0 + (c_3^2 - 2c_2 c_4) m_1 + (c_2^2 - 2c_1 c_3 + 2c_0 c_4) m_2 + (c_1^2 - 2c_0 c_2) m_3 + c_0^2 m_4)]$$

where

$$m_0 = \frac{1}{d_5}(d_3 m_1 - d_1 m_2) \qquad m_3 = \frac{1}{d_0}(d_2 m_2 - d_4 m_1)$$

$$m_1 = -d_0 d_3 + d_1 d_2 \qquad m_4 = \frac{1}{d_0}(d_2 m_3 - d_4 m_2)$$

$$m_2 = -d_0 d_5 + d_1 d_4 \qquad \Delta_5 = d_0(d_1 m_4 - d_3 m_3 + d_5 m_2)$$

$$I_6 = \frac{1}{2\Delta_6} [c_5^2 m_0 + (c_4^2 - 2c_3 c_5) m_1 + (c_3^2 - 2c_2 c_4 + 2c_1 c_5) m_2 + (c_2^2 - 2c_1 c_3 + 2c_0 c_4) m_3 + (c_1^2 - 2c_0 c_2) m_4 + c_0^2 m_5]$$

where

$$m_0 = \frac{1}{d_6}(d_4 m_1 - d_2 m_2 + d_0 m_3) \qquad m_4 = \frac{1}{d_0}(d_2 m_3 - d_4 m_2 + d_6 m_1)$$

$$m_1 = -d_0 d_1 d_5 + d_0 d_3^2 + d_1^2 d_4 - d_1 d_2 d_3 \qquad m_5 = \frac{1}{d_0}(d_2 m_4 - d_4 m_3 + d_6 m_2)$$

$$m_2 = d_0 d_3 d_5 + d_1^2 d_6 - d_1 d_2 d_5 \qquad \Delta_6 = d_0(d_1 m_5 - d_3 m_4 + d_5 m_3)$$

$$m_3 = d_0 d_5^2 + d_1 d_3 d_6 - d_1 d_4 d_5$$

REFERENCE

H. M. James, N. B. Nichols, and R. S. Phillips, *Theory of Servomechanisms,* Vol. 25, M.I.T. Radiation Laboratory Series. McGraw-Hill, New York, 1947.

APPENDIX D

GENERALIZED OPERATING PATHS

Assume that two sinusoidal functions of time, x and y, having the same frequencies $\omega_0/2\pi$, act simultaneously at 90 space degrees, that is, in space quadrature. Let the y-directed space value be

$$y = Y \cos \omega_0 t \tag{D-1}$$

and let the x-directed space value be

$$x = X \cos (\omega_0 t + \theta) \tag{D-2}$$

Now, since $\cos \omega_0 t = y/Y$, then $\sin \omega_0 t = \sqrt{Y^2 - y^2}/Y$. Using the trigonometric identity $\cos(\omega_0 t + \theta) = \cos\theta \cos \omega_0 t - \sin\theta \sin \omega_0 t$, (D-2) becomes

$$x = X(y \cos \theta - \sqrt{Y^2 - y^2} \sin \theta)$$

which can be written as

$$Y^2x^2 - 2XY \cos \theta\, xy + X^2y^2 - X^2Y^2 \sin^2 \theta = 0 \tag{D-3}$$

From analytic geometry, an equation of the form

$$Ax^2 + Bxy + Cy^2 + Dx + Ey + F = 0$$

represents an ellipse if the discriminant $B^2 - 4AC < 0$. Thus, the path traced by (D-3) in the x–y plane is an ellipse if

$$(-2XY \cos \theta)^2 - 4X^2Y^2 < 0$$

or, finally, if

$$\cos^2 \theta < 1 \tag{D-4}$$

Inequality (D-4) is satisfied for all $|\theta| < 360°$ except for $\theta = 0$ or $\pm 180°$. There are three special cases to consider.

Case 1 When $\cos \theta = 0$, (D-3) becomes

$$Y^2x^2 - 2XYxy + X^2y^2 = 0$$

or

$$(Yx - Xy)^2 = 0$$

from which is obtained

$$y = \frac{Y}{X}x \tag{D-5}$$

This is a straight line whose slope is Y/X. It is inclined at an angle ϕ, relative to the positive x direction, given by

$$\phi = \tan^{-1}(Y/X)$$

This angle depends on the space values representing the amplitudes of the two sinusoids.

Case 2 When $\theta = \pm 180°$, (D-3) beomes

$$y = -\frac{Y}{X}x \tag{D-6}$$

This straight line has negative slope and is inclined at $-\phi$.

Case 3 When $\theta = 90°$, (3) becomes

$$\left(\frac{x}{X}\right)^2 + \left(\frac{y}{Y}\right)^2 = 1 \tag{D-7}$$

which is an ellipse when $X \neq Y$. When $X = Y$, the major and minor axes of the ellipse are equal space values, and the ellipse becomes a circle.

An application of this phenomenon is the well-known Lissajous figure. The two time functions are sinusoidal voltages that are applied to the horizontal and vertical deflecting systems of an oscilloscope. The path of the spot on the screen is a stright line, a circle, or an ellipse as determined by the phase angle θ between the two sinusoidal voltages.

FEATURES OF THE OPERATING PATH

The maximum possible y-directed space value is $y_0 = \pm Y$, obtained from (D-1). When this is substituted into (D-3), the corresponding x-directed space value is found to be $x_0 = \pm X \cos\theta$.

The maximum possible x-directed space value is $x_\infty = \pm X$, from (D-2). When this is substituted into (D-3), the corresponding y-directed space value is found to be $y_\infty = \pm Y \cos\theta$.

The y-directed intercepts y_i are found by setting $x = 0$ in (D-3) and solving for the corresponding values of y. Thus, $y_i = \pm Y \sin\theta$.

The x-directed intercepts x_i are found by setting $y = 0$ in (D-3) and solving for the corresponding values of x. Thus, $x_i = \pm X \sin\theta$.

Slopes of lines joining certain points are also identified in Fig. D-1. The line having slope m_0 joins points for which the slope of the ellipse is zero. The line having slope m_∞ joins points for which the slope of the ellipse is infinite.

Two other points on the ellipse can be found along a line whose slope m_G is the geometric mean of m_0 and m_∞. Thus,

$$m_G = \sqrt{m_0 m_\infty} = \sqrt{\left(\frac{Y}{X\cos\theta}\,\frac{Y\cos\theta}{X}\right)} = \frac{Y}{X} \tag{D-8}$$

If $y = y_G = (Y/X)x_G$ and $x = x_G$ are substituted into (D-3), it is found, after using some trigonometric identies, that at these two intersection points,

$$x_G = \pm X\cos(\theta/2)$$

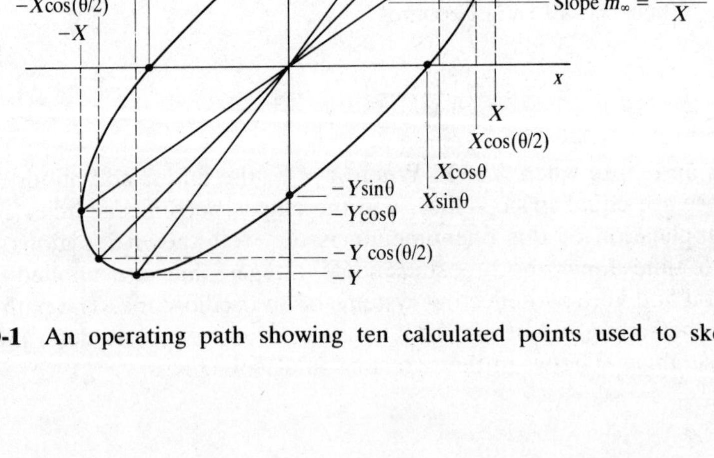

Fig. D-1 An operating path showing ten calculated points used to sketch the ellipse.

and

$$y_G = \pm Y \cos(\theta/1)$$

Therefore, if X, Y, and θ are known, then ten points on the operating path are easily calculated to quickly locate the ellipse and determine its extent and shape, as shown in Fig. D-1. Then the ellipse can be roughly drawn. At various instants during a cycle of operation, the elliptical path is followed clockwise when θ is negative and counterclockwise when θ is positive.

GENERALIZED LOAD LINES

Consider the two-port signal circuit shown in Fig. D-2. Assume that the output signal current is

$$i = I_m \cos \omega_0 t \tag{D-9}$$

Let $Z = R + jX$ represent the impedance that the passive load presents to the device at the frequency $\omega_0/2\pi$. Then,

$$v = -|Z|\, I_m \cos(\omega_0 t + \theta) \tag{D-10}$$

where

$$\theta = \tan^{-1}(X/R) \tag{D-11}$$

Thus, θ is positive when X is inductive and θ is negative when X is capacitive. These two sinusoids are measured relative to the Q point established by the DC operating conditions and by the DC load line.

It is often convenient to represent the output characteristics of a transistor graphically, with the current variable y-directed and the voltage variable x-directed. We wish to determine the operating path produced by the signal current and signal voltage sinusoids acting simultaneously in space quadrature on the i–v plane, that is, on the output characteristics of the device. This is the same situation as described above.

By comparing (D-9) and (D-1), it is seen that the original time-dependent variable y becomes i and amplitude Y becomes a space value representing I_m. By comparing (D-10) and (D-2), it is found that the

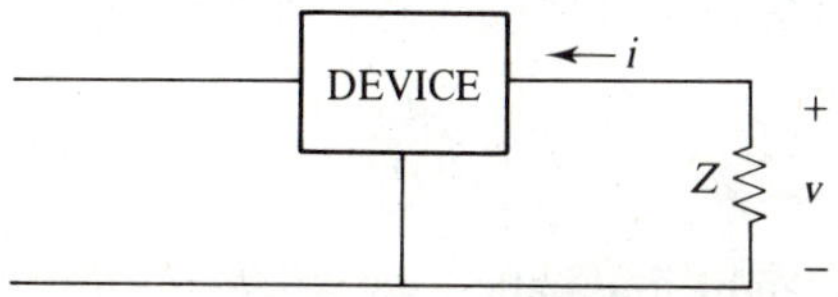

Fig. D-2 Output signal circuit of a simplified two-port device.

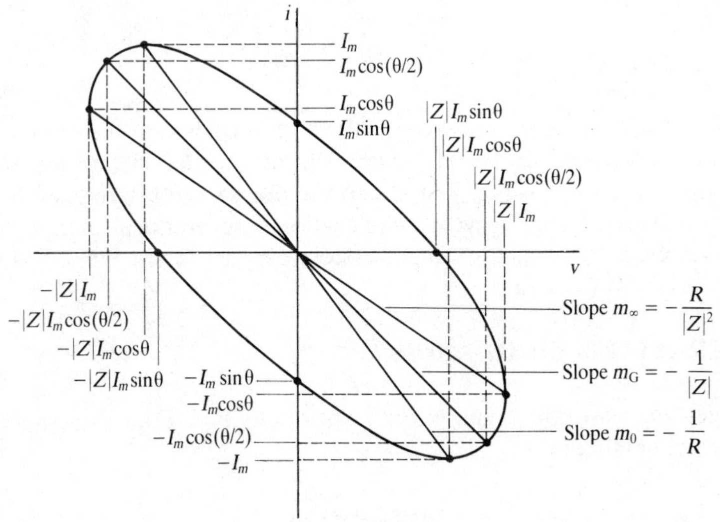

Fig. D-3 An operating path for a two-port circuit shown in Fig. D-2.

original time-dependent variable x becomes v and the amplitude X becomes a space value representing $-|Z|\, I_m$.

The operating path of the device is an ellipse if (D-4) is satisfied, or $\cos^2\theta < 1$. But, from (D-11), $\tan\theta = X/R$, so $\cos\theta = R/\sqrt{R^2 + X^2} = R/|Z|$. Therefore, an ellipse will always be obtained on the output characteristics whenever the load impedance contains any reactance at all, because then $R < |Z|$.

Since amplitude X is replaced with $-|Z|\, I_m$, the operating path of the device can be found by folding the ellipse of Fig. D-1 about the vertical axis, to obtain the ellipse of Fig. D-3. Then, the various slopes of the interior lines become:

$$\text{Slope } m_0 = \frac{Y}{X\cos\theta} = \frac{I_m}{-|Z|\, I_m(R/|Z|)} = -\frac{1}{R} \tag{D-12}$$

$$\text{Slope } m_\infty = \frac{Y\cos\theta}{X} = \frac{I_m R}{-|Z|^2 I_m} = -\frac{R}{|Z|^2} \tag{D-13}$$

$$\text{Slope } m_G = \frac{Y}{X} = \frac{I_m}{-|Z|\, I_m} = -\frac{1}{|Z|} \tag{D-14}$$

When the load does not contain any reactance, then $Z = R + j0$, so that $\theta = 0$, and the operating path given by (D-5), $y = (Y/X)x$, becomes

$$i = -\frac{1}{R}v \tag{D-15}$$

This represents the straight line called an AC load line. Its slope is $-(1/R)$ on the i–v plane whose origin is located at the Q point.

APPENDIX E

LISTINGS OF TWO COMPUTER PROGRAMS WRITTEN IN BASIC

1. Program to evaluate Bessel functions $J_n(x)$ for a specified value of x.
2. Program to determine the spectrum in angle modulation when the modulating signal is composed of two sinusoidal components.

```
10 'Program to evaluate Bessel functions Jn(x).
 Written in BASIC.
20 DIM J(27), Y(27)
30 CLS:INPUT''Enter the desired argument'';X
40 NMAX=INT(1.2*X+3.5) 'number of Bessel functions
 to calculate
50 LPRINT TAB(33);''*********''
60 PRINT TAB(33);''*********''
70 LPRINT TAB(33);'' n'';TAB(40);''Jn('';X;'')''
80 PRINT TAB(33);'' n'';TAB(40);''Jn('';X;'')''
90 LPRINT TAB(33);''---'';TAB(40);''---------''
100 PRINT TAB(33);''---'';TAB(40);''---------''
110 'Calculate the Bessel functions
120 Y(0)=1:Y(1)=0:C=0            'Initialize
130 N=INT (.65*X+17) 'Approx. min. number of
 recursions for each function
140 FOR R=NMAX TO 1 STEP -1
150 Y(R)=Y(R-1)
160 NEXT R
170 'Recursion
180 Y(0)=2*N*Y(1)/X-Y(2)
190 N=N-1
200 IF N=0 THEN 240
```

```
210 IF INT(N/2)< >N/2 THEN 140
220 C=C+2*Y(0)
230 GOTO 140
240 C=C+Y(0)
250 'Normalize
260 FOR R=0 TO NMAX:Y(R)=Y(R)/C:NEXT R
270 I=10000#          'Use 1000 for 3 decimal places,
 100000 for 5 decimal places
280 FOR R=0 TO NMAX:J(R)=INT(I*Y(R))/I
290 LPRINT TAB(33);R;TAB(40);J(R)
300 PRINT TAB(33);R;TAB(40);J(R)
310 NEXT R
320 END
10 ' SPECTRUM PROGRAM for angle modulation. Written in
 BASIC.
20 DIM Y(25),RJ(50),RK(50),JK(50,50),FREQ(50,50),
   L(50,2),P(50,2),Q(50,2)
30 KEY OFF:CLS
40 '
50 'ENTER initial data
60 '
70 PRINT''*** To correct an error type 99 at the NEXT
 ENTER***''
80 PRINT: INPUT''Enter the modulation index of the 1st
 function:'';B1
90 N=INT((3.24*B1+1.1)^.76) 'Number of Bessel
 functions >0.01-1st modulation index; recursion
 formula needs how many Bessel functions to calculate
100 PRINT''The number of Bessel values for the 1st
 function with the beta'';B1;''is'';N+1
110 INPUT''Enter the FREQUENCY (in kHz) of the 1st
 function:'';F1:IF F1=99 THEN 80
120 PRINT'' **Make the modulation index of the 2nd
 function different from that of the 1st function**''
130 INPUT''Enter the modulation index of the 2nd
 function:'';B2:IF B2=99 THEN 110
140 M=INT((3.24*B2+1.1)^.76) 'Number of Bessel
 functions >0.01-2nd modulation index; recursion
 formula needs how many Bessel functions to calculate
150 PRINT''The number of Bessel values for the 2nd
 function with beta'';B2;''is'';M+1
160 INPUT''Enter the FREQUENCY (in kHz) of the 2nd
 function:'';F2:IF F2=99 THEN 130
170 INPUT''Enter the number of decimal places . .
 (2,3,4,5)'';EX:IF EX=99 THEN 160
```

```
180 '
190 'Calculate the Bessel functions of positive
 orders
200 '
210 CLS:LOCATE 12,33:PRINT''PLEASE STANDBY''
220 FOR D=1 TO 2
230 B=((2-D)*B1)+((D-1)*B2):NM=INT((3.24*B+
 1.1)^.76)
240 'LPRINT TAB(33):''*******''
                    'Insert or remove the REMark marks
250 PRINT TAB(33);''*******''
                    '(') or REM to PRINT and/or LPRINT
260 'LPRINT TAB(33);'' n'';TAB(40);''Jn('';B;'')''
                                        'a function
270 PRINT TAB(33);'' n'';TAB(40);''Jn('';B;'')''
280 'LPRINT TAB(33);''--'';TAB(40);''------''
290 PRINT TAB(33);''--'';TAB(40);''------''
300 '
310 GOSUB 2000         'Subroutine to calculate the
                        Bessel functions
320 '
330 DP=10^EX:FOR X=0 TO NM
                'EX is the number of decimal places
340 L(X,D)=INT(DP*Y(X)+.5)/DP
                'Round off to EX decimal places
350 'LPRINT TAB(33);X;TAB(40);L(X,D)
                    'Insert or remove REM marks
360 PRINT TAB(33);X;TAB(40);L(X,D)
                    'to PRINT and/or LPRINT
370 NEXT:NEXT
                    'a function
380 LPRINT
390 PRINT
400 '
410 'Put END here to check the Bessel functions—
 include subroutine 1000
420 'Generate Bessel functions of negative order but
 with positive indexes
430 '
440 FOR D=1 TO 2:B=(2-D)*B1+(D-1)*B2:
   NM=INT((3.24*B+1.1)^.76):FOR X=1 TO NM
450 V=NM+1-X
460 P(X,D)=((-1)^V)*L(V,D)
470 NEXT
480 FOR X=NM+1 TO 2*NM+1
 'Shift indexes of the calculated Bessel functions
```

```
490 Q(X,D)=L(X-NM-1,D)
500 NEXT:NEXT
510 '
520 'Convert to double-sided Bessel functions of
  positive and negative orders
530 '
540 FOR X=1 TO 2*N+1:RJ(X)=P(X,1)+Q(X,1)
550 PRINT RJ(X);        'Insert or remove REM marks to
560 'LPRINT RJ(X);      'PRINT and/or LPRINT a function
570 NEXT
580 PRINT:PRINT
    'RJ(X) and RK(Y) are the double-sided spectra
                                             amplitudes
590 'LPRINT:LPRINT
600 FOR Y=1 TO 2*M+1:RK(Y)=P(Y,2)+Q(Y,2)
610 PRINT RK (Y);      'Insert or remove REM marks to
620 'LPRINT RK(Y);     'PRINT and/or LPRINT a function
630 NEXT
640 PRINT:PRINT
650 'LPRINT:LPRINT
660 '
670 'Put END here to check the positive-and-negative-
  order Bessel functions
680 'Include subroutine 1000. These values give the
  double-sided spectra.
690 '
700 FMAX=0;FOR X=1 TO (2*N)+1:FOR Y=1 TO (2*M)+1
710 JK(X,Y)=RJ(X)*RK(Y)
              'Calculate products of Bessel functions
720 FREQ(X,Y)=(X-N-1)*F1+(Y-M-1)*F2
              'Calculate side-frequency spacings from
               unmodulated center frequency
730 P$=''##      ##      ###.##      ##.###''
740 PRINT USING P$;X,Y,FREQ(X, Y),JK(X,Y)
750 'LPRINT USING P$;X,Y,FREQ(X,Y),JK(X,Y)
                                        'Display results
760 '
770 'Calculate the maximum side frequency with
  amplitude >=0.01
780 '
790 IF ABS(JK(X,Y))>= .01 THEN IF FMAX<FREQ(X,Y)
THEN FMAX=FREQ(X,Y)
  800 NEXT:NEXT
810 '
820 'Print out some data
```

```
830 '
840 'LPRINT:LPRINT''The highest side frequency with
  amplitude >=0.01 is'';FMAX;''kHz.''
850 'LPRINT:LPRINT''These are for a modulating
  frequency of'';F1;''kHz with a lower''
860 'LPRINT''modulation index of'';B1;''and'';N+
  1;''Bessel values, and a''
870 'LPRINT''modulation frequency of'';F2;''kHz with
  a higher modulation index of''; B2
880 'LPRINT''and'';M+1;''Bessel values.'':LPRINT
890 '(Put END here to check program this far — include
  subroutine 1000)
900 '
910 GOSUB 3000        'Calculate and print bandwidths
920 '
930 'Check for identical modulation-product
  frequencies
940 '
950 PRINT:PRINT''Check the following list to find the
  highest value of FREQUENCY>'';FMAX;''kHz''
960 PRINT''such that |SUM|>= 0.01. If any exist, then
  use the highest value to''
970 PRINT''increase the bandwidth of'';BW;''kHz to
  2(FREQUENCY_MAX) kHz.'':PRINT
980 'LPRINT:LPRINT''Check the following list to find
  the highest value of FREQUENCY>'';FMAX
990 'LPRINT''such that |SUM|>= 0.01. If any exist, then
  use the highest value to''
1000 'LPRINT''increase the bandwidth of'';BW;''kHz
  to 2(FREQUENCY_MAX) kHz.'':LPRINT
1010 '
1020 'GOSUB 4000
'Remove the REM mark to include the (slow) search
                                           routine.
1030 '
1040 LOCATE 22,55:PRINT''        '':BEEP:KEY ON:END
1050 '
2000 'Calculate the Bessel functions by recursion
2010 '
2020 Y(0)=1:Y(1)=0:C=0      'Initialize variables
2030 REC=INT(.65*B+17)
       'REC=approx. minimum number of recursions
2040 FOR I=NM TO 1 STEP -1
       'for each Bessel function calculation
```

```
2050 Y(I)=Y(I-1)          'This is the updating loop
2060 NEXT I
2070 'Recursion
2080 Y(0)=2*REC*Y(1)/B-Y(2)
        'This is the recursion formula
2090 REC=REC-1
2100 IF REC=0 THEN 2140
2110 IF INT (REC/2)<>REC/2 THEN 2040
2120 C=C+2*Y(0)
        'C=the normalizing variable
2130 GOTO 2040
2140 C=C+Y(0)
        'This is the final normalizing number
2150 'Normalize
2160 FOR I=0 TO NM:Y(I)=Y(I)/C
        'Array Y(I) contains the Bessel functions
2170 NEXT:RETURN
2180 '
3000 'Print bandwidths        'Insert or remove REM
                              marks to
3010 '                        'PRINT and/or LPRINT a
                              function
3020 IF F2>F1 THEN FHIGH=F2 ELSE FHIGH=F1
3030 BW=2*FMAX
3040 DELTAF=(B1*F1+B2*F2)
3050 CARSONBW=2*(FHIGH+DELTAF)
3060 PRINT''The bandwidth required to pass all side
  frequencies with amplitudes''
3070 PRINT''greater than 1% of the maximum possible
  value is'';BW;''kHz.''
3080 PRINT:PRINT''The maximum frequency deviation
  is''; DELTAF;''kHz.''
3090 PRINT:PRINT''The bandwidth by Carson's Rule
  is'';CARSONBW;''kHz.''
3100 'LPRINT''The bandwidth required to pass all side
  frequencies with amplitudes''
3110 'LPRINT''greater than 1% of the maximum possible
  value is'';BW;''kHz.''
3120 'LPRINT:LPRINT''The maximum frequency deviation
  is'';DELTAF;''kHz.''
3130 'LPRINT:LPRINT''The bandwidth by Carson's Rule
  is'';CARSONBW;''kHz.''
3140 RETURN
3150 '
```

```
4040 PRINT''and whose individual amplitudes add to
  obtain |SUM| >= 0.01.'':PRINT
4050 PRINT''X Y     I J     SUM         FREQUENCY kHz''
4060 PRINT''___     ___     ___         ______________''
4070 'LPRINT''The following pairs of harmonic numbers
  yield frequencies>'';FMAX;''kHz''
4080 'LPRINT''and whose individual amplitudes add to
  obtain |SUM| >= 0.01,'':LPRINT
4090 'LPRINT''X Y     I J     SUM         FREQUENCY kHz''
4100 'LPRINT''___     ___     ___         ______________''
4110 LOCATE 22,55:PRINT''** PLEASE STANDBY **''
4120 FOR X=1 TO N:FOR Y=1 TO 2*M+1
4130 MATCH=FREQ(X,Y)
                      'In Line 4170, LPRINT and PRINT can be
4140 TEMP=JK(X,Y) 'swapped or either one deleted. Use a
4150 SUM=0            'REM mark only on the last one listed.
4160 FOR I=N TO 2*N+1:FOR J=1 TO 2*M+1
4170 IF (I<>X) OR (J<>Y) THEN IF FREQ(I,J)=MATCH
  THEN SUM=SUM+TEMP+JK(I,J):IF MATCH > FMAX AND
  ABS(SUM) >= .01 THEN PRINT USING
  Q$;X,Y,I,J,SUM,MATCH:'LPRINT USING
  Q$;X,Y,I,J,SUM,MATCH
4180 NEXT:NEXT:NEXT:NEXT        'Insert or remove REM
                                marks to
4190 RETURN                     'PRINT or LPRINT a function
```

APPENDIX F

ANSWERS TO SELECTED PROBLEMS

1-2 (a) $T_0 = 5$ s; $f_0 = 0.2$ Hz; 50 & 51; (b) $T_0 = 10$ s; $f_0 = 0.1$ Hz; 81, 104, & 96; (f) $T_0 = 0.115$ s; $f_0 = 8.66$ Hz; 2 & 3

1-3 (b) (1) $T_b = 0.01$ s; $f_b = 100$ Hz; beat period = true period; (3) $T_b =$ 0.318 ms; $f_b = 3140$ Hz; 157 beat periods = true period

1-6 (a) not possible; (b) $f_2/f_1 = (3, 7, 11, \ldots)/(1, 5, 9, \ldots)$ for $x_{2,\max} = V_1 + V_2$ and $f_2/f_1 = (5, 9, 13, \ldots)/(3, 7, 11, \ldots)$ for $x_{2,\min} = -x_{2,\max}$. Examples: $f_2/f_1 = 3/1 = 9/3$ and $f_2/f_1 = 7/5 = 21/15$

1-7 (a) $T_0 = 1/3$ s; $x_{\max} = 9$ at $t = 0$ and every 1/3 s thereafter; $x_{\min} = -9$ at $t = 1/6$ s and every 1/3 s thereafter

1-8 (b) $T_0 = 1/8$ s; $f_0 = 8$ Hz

1-10 (b) $v_g = 0.36 \times 10^8$ m/s

1-13 (a) $R_1 = R_2 = 16.67\ \Omega$; $R_{in} = R_{out} = 50\ \Omega$; 6 dB; $v_o = (v_a + v_b)/2$

1-19 (a) $Q_{\max} = 0.1$; (b) $Q_{\max} = 0.4$

1-20 (a) 31.29 mW; (b) $m_a = 5\%$

1-21 $A = 0.65$ V

2-1

	Fixed	Time varying	Instantaneous	Dynamic	Linear	Nonlinear	Incrementally linear
(a)	×		×			×	×
(b)	×		×			×	
(c)		×	×			×	×
(d)	×			×	×		×
(e)	×			×		×	

2-2 $y = 20.48x + 3.84x^2 + 0.32x^3 + 0.01x^4$. Terminates

2-3 $y = 0.042x + 0.0013x^2 - 2.76x^3 + \cdots$

2-4 (b) $i = 0.025v + 3.13 \times 10^{-4}v^2 + 2.6 \times 10^{-6}v^3 + \cdots$ mA

2-6 $y = 0.3496x - 0.02206x^3 + 0.000697x^5$

2-9 (b) $\cos^9 \theta = (\cos 9\theta + 9 \cos 7\theta + 36 \cos 5\theta + 84 \cos 3\theta + 126 \cos \theta)/2^8$
(d) $\cos^{14} \theta = (\cos 14\theta + 14 \cos 12\theta + 91 \cos 10\theta + 364 \cos 8\theta + 1001 \cos 6\theta + 2002 \cos 4\theta + 3003 \cos 2\theta + 1716)/2^{13}$

2-10 (b) $\sin^9 \theta = (\sin 9\theta - 9 \sin 7\theta + 36 \sin 5\theta - 84 \sin 3\theta + 126 \sin \theta)/2^8$
(d) $\sin^{14} \theta = (-\cos 14\theta + 14 \cos 12\theta - 91 \cos 10\theta + 364 \cos 8\theta - 1001 \cos 6\theta + 2002 \cos 4\theta - 3003 \cos 2\theta + 1716)/2^{13}$

2-13 Add $10EX_0^5/16$ to the fundamental components
Add $5EX_0^5$ to the 3rd-harmonic components
Add $EX_0^5/16 \cos 5\theta$

2-15 (b) 2 dB of compression when $X_0 = 1.469$

2-17 $a_1 = 0.093mX_0$; $a_3 = -0.217mX_0$

2-24 $A = 10$ V

2-25 (a) $-1.78 \cos 2\pi f_1 t$; (b) $-0.84 \cos 2\pi(2f_2 - f_1)t$

3-3 (a) $kT/C(R_1 + R_2)$; (b) $kT/C(R_1 + R_2)$; (c) Each resistor delivers as much power as it receives, so the temperature of each is constant, as expected for thermal equilibrium

3-5 $v_0 = 3.22\ \mu$V

3-6 (a) 4×10^{-21} W; (b) -174 dBm

3-7 (a) 6.87×10^{-22} W/Hz; (b) 4×10^{-21} W/Hz; (c) 5.82; (d) 5.82

3-8 1. (a) $BW = 1$ MHz; (b) $B = 1$ MHz; 2. (a) $BW = 19$ kHz; (b) $B = 19$ kHz; 3. (a) $BW = 2$ MHz; (b) $B = 2$ MHz; 4. (a) $BW = 2.2$ MHz; (b) $B = 2.2$ MHz

3-9 (c) $BW = 4.68$ MHz; $B = 5.76$ MHz

3-10 $BW = 2.66$ kHz; $B = 2.89$ kHz

3-11 (a) 3 MHz; 3.5 MHz; 3.61 MHz; 5.5 MHz

4-1 (a) 1.54; (b) 3.33; (c) 1.43; (d) -4.33

4-3 (b) $J_7(2) = 1.75 \times 10^{-4}$; $J_8(2) = 2.29 \times 10^{-5}$

4-7 $D_6(p) = 10395 + 10395p + 4725p^2 + 1260p^3 + 210p^4 + 21p^5 + p^6$

4-12 (a) $n = 6$; (b) $n = 11$; (c) $n = 7$

4-13 (a) 1.78; (b) 1.33

4-16 $-20n$ dB/decade

4-16 (a) $n = 4$; (b) $n = 6$

4-17 $B_6(s) = (s^2 + 0.518s + 1)(s^2 + 1.414s + 1)(s^2 + 1.932s + 1)$
$B_7(s) = (s + 1)(s^2 + 0.446s + 1)(s^2 + 1.246s + 1)$
$\times (s^2 + 1.802s + 1)$

4-19 (b) $L = 11.26$ mH; $C = 0.011\ \mu$F; (c) 24.03

4-20 $SF = 8.33$; $L = 15.92$ mH; $C_1 = C_2 = 0.008\ \mu$F

4-22 (a) 148.36 kHz; (b) 1.13; 1.33; (c) 1.53

4-26

dB ripple	0.5	1.0	1.5	2.0	2.5	3.0	3.5	4.0
ϵ	0.35	0.51	0.64	0.77	0.88	1.00	1.11	1.23

4-27 (b) $n = 4$; (c) $n = 5$

5-3 9,750 kHz; 10,000 kHz; 11,000 kHz; 11,250 kHz

5-5 $f_c > 2$ W

5-8 3.73

512 (c) $m_{a-} = 54\%$

5-13 (a) 90%; (b) 72%

5-16 $A_u = 20 \log |2 + \Delta V/V|/|\Delta V/V|$

5-17 $R_1 = 5.56$ kΩ, $R_2 = 33.34$ kΩ, $R_3 = 16.7$ kΩ, $R_4 = 7.66$ kΩ, $R_5 = 1.27$ kΩ, $R_6 = 3.83$ kΩ, $C_1 = C_5 = 0.09\ \mu$F, $C_2 = 0.01\ \mu$F, $C_4 = 0.015\ \mu$F

5-18 $R_1 = R_5 = 1.67$ kΩ, $R_2 = R_4 = 10$ kΩ, $C_1 = 0.3\ \mu$F, $C_2 = 0.05\ \mu$F, $C_3 = 0.1\ \mu$F, $C_4 = 0.011\ \mu$F, $C_5 = 0.069\ \mu$F, $C_6 = 0.023\ \mu$F

6-1 (a) 0.1 radian or 5.73°; (b) $m(t)$ is periodic with $f_2/f_1 = 5/1$ and $m(t)_{\max} = -m(t)_{\min} = 3$ V; $\theta(t) = 17.19°$; (c) $f_1 = 3.18$ Hz; $f_2 = 79.58$ Hz; (d) $f = 82.76$ Hz

6-2 $\Delta\theta = 5.74°$; 0.5%

6-3 $m(t)_{\max} = 0.1$

6-4 (a) $f_1 = 60$ kHz; $f_2 = 20$ kHz; (b) $m(t)$ is periodic with $f_2/f_1 = 5/1$ and $m(t)_{\max} = -m(t)_{\min} = 4$ V; $f_{\max} = 80$ kHz; $\Delta\theta_1 = 753.98$ radians; $\Delta\theta_2 = 50.27$ radians; (d) $\Delta\theta = 804.25$ radians

6-7 (a) requires 8 pairs of side frequencies; $BW = 80$ kHz; (b) requires 2 pairs of side frequencies; $BW = 50$ kHz

6-8 (a) requires 1 pair of side frequencies; $BW = 10$ kHz; (b) requires 9 pairs of side frequencies; $BW = 54$ kHz

6-12 (a) $L_2 = 2.46\ \mu$H

7-1 1.62

7-2 (a) 4; (b) 2.48; (c) 2.48

7-3 3.67×10^{-21} W; (b) 0.384×10^{-21} W; (c) 9.56; (d) 9.8; (e) 10.31

7-4 6.3 mV

7-5 (a) 2.54 pW; −85.9 dBm; (b) the diode mixer introduces other noise in addition to thermal noise

7-6 261 μV

7-7 (a) 11.59; (b) 15.06; (c) 13.38; (d) 17.13; (e) 7.47 for the order 2-1-3

7-8 (a) 24; (b) better to put the pre-amp at the antenna because then the overall noise factor is smaller

7-9 10.3

7-10 (a) −127 dBm; (b) 0.63 μV rms

7-12 Order K-J-N; 8.97

7-14 Order 2-4-1-3; 6.46

7-16 (a) 1.0086; (b) 1.086; (c) 1.86; (d) 9.62; (e) 87.21; (f) 863.1

7-17 (a) 1.49 dB; (b) 2311 K; (c) 6767 K

7-18 (a) 3.24 dB; (b) 321.9 K

7-19 (a) 2.34; (b) 1.4 dB

7-20 (a) 8.33×10^{-12} W; (b) 18×10^{-15} W; (c) 463

7-21 $F = 1.49$; $T_e = 143.3$ K

7-22 1247 K

8-2 Difference mixing because sum mixing produces an IF higher than either of the other frequencies

8-3 When the signal at 1450 kHz arrives at the mixer along with the output from the local oscillator at 995 kHz (i.e., 540 + 455), their difference is 455 kHz which also passes through the IF circuitry

8-4 When the receiver is tuned to 1540 kHz, the local oscillator is at 1995 kHz. When the signal from the other station on 2450 kHz mixes with the oscillator on 1995 kHz, the difference frequency is 455 kHz, which also passes through the IF circuitry

8-5 When tuned to 1100 kHz, the oscillator frequency is 1555 kHz. Its 2nd harmonic, 3110 kHz, is also present in the mixer to combine with some other frequency of produce the IF. The interfering code signal is on 3565 kHz (i.e., 3110 + 455)

8-6 175 kHz

8-7 (a) oscillator: 31.6 MHz; image: 33.2 MHz; (b) zero output

8-8 An external oscillator and difference mixer is required

8-9 The 2nd oscillator must supply the missing IF carrier at 453.5 kHz

8-10 The 3rd oscillator must supply the missing IF carrier at 102.3 kHz

8-12 Terms involving the image that produce the IF, (i.e., $f_1 - f_o$), cancel, hence the name image-reject mixer

8-16 4 kHz

8-17 14 kHz

8-18

frequency	N
159,985	5997
159,990	5998
159,995	5999
160,000	6000
160,005	6001

8-19 (a) $f_r = 100$ kHz; (b) 1st IF = 10.65 MHz; (c) $M = 102$ to produce 100 kHz; (d) $f_o = 10.2$ MHz;
(e)

frequency	N
154.05	1434
154.15	1435
154.25	1436
⋮	⋮
162.05	1514
162.15	1515
162.25	1516

8-20 (a) 125 kHz; 187.5 kHz; (b) 24.414 Hz; (c) 11.94 kHz

9-1 (a) receiver A: 2050 kHz − 2500 kHz; receiver B: 1750 kHz − 1300 kHz; (b) receiver A: RF − 16; osc. − 1.49; receiver B: RF − 16; osc. − 1.81; the oscillator of receiver B must go to a lower frequency as the receiver is tuned to a higher frequency; (c) receiver A: $(f_e/f_k)_{min} = 0.073$; $(f_e/f_k)_{max} = 0.24$; receiver B: $f_e/f_k)_{min} = 0.086$; $(f_e/f_k)_{max} = 0.462$

9-4 455 kHz is better because the spurious responses are far out of the tuning range

9-5 200 kHz

9-6 (a) −15 kHz; (b) −20 kHz

9-7 910 kHz

9-8 1371.67 kHz

9-9 Mixer 1: $2f_e - 4f_k$; $6f_e$; $2f_k - 3f_e$; $5f_i$; Mixer 2: $4F_k - 4f_e$; $6f_e - 2f_k$; Mixer 3: $3f_k - 2f_e$; $4f_e - f_k$

9-10 (a) IF $= f_e$; IF $= f_k/2$: (b) low-side injection is required: (c) for low-side injection, the oscillator frequency can be the same as the IF $(= f_e)$ and it may be possible for the oscillator output to enter the IF circuitry

9-11 (a) 3/7; (b) low-side injection; (c) the spurious response is 6(1500) − 2(3500) kHz = 2 MHz, the same as the IF = 3500 − 1500 kHz = 2 MHz

9-12 (a) 3450 kHz; 3455 kHz; (b) 1800 kHz; 1810 kHz

9-13 The desired station on 920 kHz produces an IF = 455 kHz from an oscillator frequency of 1375 kHz. The undesired station on 1830.5 kHz mixes with the oscillator to produce an undesired IF = 455.5 kHz. The beat frequency is 500 Hz

9-14 166.67 kHz

9-15 (a) difference mixing is required; (b) there are no spurious responses of 6th order or less

9-15 133.75 kHz

9-17 $f_\ell - f_k = 11{,}990$ kHz; $3f_\ell = 11994$ kHz

9-18 (a) 2 – 5 MHz; (b) $2f_\ell - f_k$; $3f_\ell - 2f_k$; $4f_\ell - 3f_k$; $3f_k - 4f_\ell$; $5f_\ell - 3f_k$; $5f_\ell - 4f_k$; $4f_k - 5f_\ell$

9-21 O = 3, $H = 1$, $L = 2$: O = 5, $H = 2$, $L = 3$; the type 2 IM product with $H = 2$ and $L = 2$ is extraneous because it does not intersect the parallelogram

10-1 (a) −121 dBm; (b) −71 dBm; (c) 80.67 dBm; (d) −40.33 dBm; (e) 0.24 μV – 2.64 mV

10-2 (a) −6 dBm; (b) 7 dB; (c) 0 dBm; (d) −37 dBm; (e) 24.67 dBm; (f) 100; (g) 75.1 mV

10-4 (a) −35 dBm; (b) −45 dBm; (c) $K = -20$ dBm; $K_2 = -5$ dBm; (d) −15 dBm; (e) $A = 3.16$; $B = 1.45$; $C = 0.89$

10-6 $C = 1$: 0.998×10^{-4}; $C = 0$: 1×10^{-4}

10-7

	1	2	3
3point1	2.84×10^{-5}	5.31×10^{-5}	0.94×10^{-4}
3point0	3.14×10^{-5}	6.23×10^{-5}	1.22×10^{-4}

For the order 1-2-3 with $C = 1$: $\text{OIP}_{\text{overall}} = -11.2$ dBm and with $C = 0$: $\text{OIP}_{\text{overall}} = -9.53$ dBm

10-8

	1	2	3
2point0	2.84×10^{-6}	6.68×10^{-6}	1.19×10^{-5}
2point1	1.22×10^{-3}	1.56×10^{-3}	1.72×10^{-3}

For the order 1-2-3 with $C = 1$: $(\text{OIP}_2)_{\text{overall}} = -24.67$ dBm and with $C = 0$: $(\text{OIP}_2)_{\text{overall}} = -20.3$ dBm

BIBLIOGRAPHY

On Time-varying Components

D. G. Tucker, *Circuits With Periodically Varying Parameters.* D. Van Nostrand, London, 1964.

On Nonlinear Systems

George E. Anner, *Elementary Nonlinear Electronic Circuits,* Prentice-Hall, Englewood Cliffs, N. J., 1967.

R. Clay, *Nonlinear Networks and Systems,* Wiley Interscience, New York, 1971.

A. Gelb and W. E. Vander Velde, *Multiple-Input Describing Functions and Nonlinear System Design,* McGraw-Hill, New York, 1968.

Rome Air Development Center, *Nonlinear System Modeling and Analysis with Applications to Communications Receivers,* Signatron, Inc., Rome, 1973.

On Instantaneous and Dynamic Systems

Leon O. Chua, *Introduction to Nonlinear Network Theory,* McGraw-Hill, New York, 1969.

On Series Summations

L. B. W. Jolley, *Summation of Series,* 2nd edn., Dover, New York, 1961.

On Graphical Solutions of Nonlinear Systems

A. Preisman, *Graphical Constructions for Vacuum Tube Circuits,* McGraw-Hill, New York, 1943.

On the Method of Least Squares and Numerical Analysis

H. M. Nodelman and F. W. Smith, *Mathematics for Electronics,* McGraw-Hill, New York, 1956.

J. B. Scarborough, *Numerical Mathematical Analysis,* 4th edn., The Johns Hopkins Press, Baltimore, 1958.

On rms and Instantaneous Characteristics

S. Sabaroff, "Relationship Between RMS and Instantaneous Characteristics of Nonlinear Devices", *Proc. IEEE,* Vol. 55, Dec. 1967, p. 2167.

On Binomial Coefficients Identities

J. Riordan, *Combinatorial Identities,* Wiley, New York, 1968.

J. Riordan, *An Introduction to Combinatorial Analysis,* Wiley, New York, 1958.

On Amplitudes of Intermodulation Products

W. R. Bennett, "New Results in the Calculation of Modulation Products", *Bell System Technical Journal,* Vol. 12, pp. 228–243, 1933.

W. R. Bennett, "The Biased Ideal Rectifier", *Bell System Technical Journal,* Vol. 26, pp. 139–169, 1947.

R. G. Sea, "An Algebraic Formula for Amplitudes of Intermodulation Products Involving an Arbitrary Number of Frequencies", *Proc. IEEE,* Vol. 56, Aug. 1968, p. 1388.

R. G. Sea and A. G. Vacroux, "On the Computation of Intermodulation Products for a Power Series Nonlinearity", *Proc. IEEE,* Vol. 57, March 1969, p. 337.

On Noise

W. R. Bennett, *Electrical Noise,* McGraw-Hill, New York, 1960.

M. S. Gupta, *Electrical Noise: Fundamentals and Sources,* IEEE Press, New York, 1977.

Application Note 57-1, "Noise Figure Primer", Hewlett-Packard, Palo Alto, Calif. 1983.

B. K. P. Scaife, "Dispersion and Fluctuation in Dielectrics", Chapter in *Progress in Dielectrics,* Vol. 5, p. 158, Academic Press, New York, 1963.

H. L. Swain and R. M. Cox, "Noise Figure Meter Sets Records for Accuracy, Repeatability, and Convenience", *Hewlett-Packard Journal,* Vol. 34, no. 4, p. 23 April 1983.

On Mathematical Functions

Handbook of Mathematical Functions With Formulas, Graphs, and Mathematical Tables, edited by M. Abramowitz and I. Stegun, U.S. Government Printing Office, Washington, 1964.

H. L. Krall and O. Fink, "A New Class of Orthogonal Polynomials: the Bessel Polynomials", *Trans. Am. Math. Soc.,* Vol. 65, pp. 100–115, Jan. 1949.

On Filter Details

N. Balabanina, *Network Synthesis,* Prentice-Hall, Englewood Cliffs, N.J., 1958.

A. Budak, *Passive and Active Network Analysis and Synthesis,* Houghton-Mifflin, Boston, 1974.

L. Weinberg, *Network Analysis and Synthesis,* McGraw-Hill, New York, 1962.

On Radio Communications Receivers

J. B. Tsui, *Microwave Receivers with Electronic Warfare Applications,* Wiley, New York, 1986.

U. L. Rohde and T. T. N. Bucher, *Communications Receivers,* McGraw-Hill, New York, 1988.

On Modulation Processes

P. F. Panter, *Modulation, Noise, and Spectral Analysis,* McGraw-Hill, New York, 1965.

On Radio-Frequency Circuits

R. S. Carson, *High-Frequency Amplifiers,* Wiley Interscience, New York, 1982.

D. DeMaw, *Practical RF Design Manual,* Prentice-Hall, Englewood Cliffs, N.J., 1982.

J. Hardy, *High frequency Circuit Design,* Reston Publishing Co., Reston, Va. 1979.

W. H. Hayward, *Introduction to Radio Frequency Design,* Prentice-Hall, Englewood Cliffs, N.J., 1982.

On Phase Splitters

R. B. Dome, "Wideband Phase Shift Networks", *Electronics,* Vol. 19, p. 1121, Dec. 1946.

D. G. C. Luck, "Properties of Some Wide-Band Phase-Splitting Networks", *Proc. IRE,* Vol. 37, p. 147, Feb. 1949.

INDEX